The development of

Glass Bottle Making Machines

by

John Trevor Gledhill

The development of Glass Bottle Making Machines
by John Trevor Gledhill

Published by the Society of Glass Technology 2014

The objects of the Society are to encourage and advance the study of the history, art, science, design, manufacture, after treatment, distribution and use of glass of any and every kind. These aims are furthered by meetings, publications, the maintenance of a library and the promotion of association with other interested persons and organizations.

The Society of Glass Technology
9 Churchill Way
Chapeltown
Sheffield S35 2PY
UK

Registered Charity No. 237438

Web site: http://www.sgt.org

ISBN 978-0-900682-76-6

Introduction

At the end of the lecture I gave to the Yorkshire Section of the Society of Glass Technology on the 16th of November, 2006 at Allied Glass, Knottingley, Yorkshire, there were many comments made that this talk should be recorded somehow, either as the talk 'as was', or as a paper and put on to the website or published in the Society's Journal - Glass Technology. The problem here was that the presentation was a PowerPoint one with only a small amount of text being used as it was mainly pictorial so this approach was initially a non-starter, so I forgot about the request.

Much later after a great deal of thought I decided it could be done but it would have to be in the form of a book. That was my first mistake as the time that it has taken to compile this book has been mind blowing, but by doing it this way I would be able to incorporate additional material to that given in the talk. Unfortunately you find doing this type of exercise that the harder you look the more information you unearth so then it comes down to where do you draw the line, some people would say that it went too far while others would say that it did not go far enough. These latter types are the same type of people that request the recording of presentations; they just do not know the amount of time it takes.

Because of the time span covered much of the information comes from a trawl through previous publications. The main references in the bibliography are; The 'Journal of the Society of Glass Technology' - (1917 to 1959), and 'Glass Technology' - (1960 to date); 'Glass Machines. Construction and Operation of Machines for the Forming of Hot Glass'. Ed. W Giegerich and W. Trier; 'Die Glasfabrikation'. Ed. R. Dralle and G. Keppler Vol I 1926, Vol II 1931,; 'Revolution in Glassmaking' by W. Scoville and 'EMHARTGlass and The Story of Glass Packing' by John H. Edginton and Charles R. Hammond, and also from the various catalogues supplied by the machine suppliers and held in the Society library, and from the internet in particular information from Wikipedia, Trade Dictionarys, from 'Bottles and Extras' the web site of the Federation of Historical Bottle Collectors and that of the 'Society for Historical Archaeology'. Also here I would like to record my appreciation of the help that I received from the Society of Glass Technology's staff especially the Societies Librarian Mrs Mary Parker who is based at St. Georges Library and to Brian Lawrence and Ian Robertson who's information and guidance was a great help.

It must be stated here that the dating of certain events are not always 'cast in tablets of stone', this is due to discrepancies in what has already been published - authors always publish the facts as they understand them but sometimes they can be a little out in what is recorded history. This again comes down to their sources. The style writing is somewhat different to other historical tomes, I've written it as if I am sat down beside the reader and we are having a conversation or discussion about it.

Our journey takes us from the days of early hand making through to the modern day IS machine, i.e. a journey that travels for more than 300 years - from the late 17th century through to the early 21st century.

Obviously there will be machines that I have missed, some of these machines will be local and confined to a single company, or they maybe machines that were not used in this country, or due to the sparseness of recorded history they will be the ones that I do not know about. For any omissions I humbly apologise before hand. To add some interest to this dissertation I have included people's history where known, as asides, they will be about either their engineering skills or the company that they owned.

There are throughout this discourse inserts from patent applications, mainly British ones, not all of the patents that are registered come into fruition but I have included the extracts that are relevant to the text so that the reader can see the differences in approach that people made to the development of the machines. The inserts on patent applications come after 1918. The type of abstracts changes as the time goes by so we will stop at the end of 1955. The first five years of the abstracts

(1919 to 1923) are to be found in Appendix 4 itemised under the patent application. These abstracts and the ones after them can be obtained from the Society either at their address or via their website.

Where there is a lack of knowledge these inserts are a great help in showing peoples careers, who they worked for and when. This is especially true in the chapter on Feeders, Forehearths and Delivery Equipment. The illustrations when they are of 'good' to 'fair' quality have been made as large as the text will allow, this is so the reader (if there is one) can see as much detail as possible.

Glass making had taken place the same way for centuries until the development of the Siemen's furnace and the semi-automatic machine, the methods used were similar irrespective of the final article made, i.e. bottles, window glass and domestic and fancy tableware which were all made the same way.

This book will concentrate on the Glass Bottle Making Industry only, other more learned gentlemen than myself, can deal with the rest.

J. T. Gledhill.
Castleford,
United Kingdom

Contents

Chapter 1

The history of the development of Glass Bottle Making Machines

Before we start to discuss the main topic of this book let us look at some history regarding the melting of glass. No matter how large a bottle making machine is, if the furnace cannot serve it with sufficient glass continually then the machine will not be able to work to its full potential. This was not a problem when bottles were 'hand made' but it did become quite a problem when glass making changed to 'machine made' bottles, and the machines became larger.

The continuous automatic production by machines brought about the extinction of the pot furnace in the bottle and flat glass making industries, by its very method of 'intermittent' working the pot furnace was no longer capable of servicing these machines. This gave rise to the continuous 'tank furnace' which enabled the machines to operate twenty four hours a day, seven days a week, fifty two weeks a year.

History

After wine and beer started to be bottled in glass in the early to mid 17th century the growth of the container industry increased dramatically, and it soon became the major player in terms of tonnage melted in the glass industry. Compare these two facts, which take place nearly 300 years apart, they show that the amount of glass melted for containers as a percentage of the total made is, within loose rules, nearly a constant, firstly;
between 1681 and 1703 John Houghton an eminent member of the Royal Society wrote a series of letters to Parliament which were collected and published under the general title of "Husbandry and Trade Improvement". One of the subtitles in the book being

List of Glass Houses in England and Wales.
15th May 1696

An account of all the Glass Houses in England and Wales	The several Counties that they are in	The Number of houses	And the sorts of glass each house makes
In and about		9	For bottles
London and		2	Looking glass plates
Southwalk		4	Crown glass and plates
		9	Flint, green and ordinary
Woolwich	Kent	1	Crown glass and plates
		1	Flint, green and ordinary
Isle of Wight	Hampshire	1	Flint, green and ordinary
Topsham, Nr Exon	Devonshire	1	Bottles
Old Down, Nr Bath	Somersetshire	1	Bottles
Chellwood		1	Bottles and window glass
In and about		5	Bottles
Bristol		1	Bottles and window glass
		3	Flint glass and ordinary
Glocester	Gloucestershire	3	Bottles
Newham		2	Bottle Houses
Swansea in Wales	Glamorgan	1	Bottles
Oaken-Yate	Shropshire	1	Bottles and window glass
Worcester	Worcestershire	1	Flint green and ordinary
Coventry	Warwickshire	1	Flint green and ordinary
Stowerbridge	Worcestershire	7	Window glass
		5	Bottles
		5	Flint, green and ordinary
Nr.Liverpool	Lancashire	1	Flint, green and ordinary
Warrington		1	Window glass
Nottingham	Nottinghamshire	1	Bottles
Awsworth		1	Flint green and ordinary
Custom Moor, Nr Awsworth		2	Bottles
		1	Flint, green and ordinary
Nr. Silkstone	Yorkshire	1	Bottles
		1	Flint, green and ordinary
Nr. Ferrybridge		1	Bottles, flint, green and ordinary
King's Lynn	Norfolk	2	Bottles
Yarmouth		1	Flint, green and ordinary
Newcastle	Northumberland	6	Window glass
-upon-Tyne		4	Bottles
		1	Flint, green and ordinary

		90	

Table 1.1 Houghtons List of Glass Houses in 1696

"Full and Exact History of Trades - with the manner of their improvement". By 1696 it was the turn of the glass industry and in letter Number 198, dated 15th May 1696 Houghton listed all of the glass works in England and Wales which were working at that time, although quite exhaustive it is probably not the total.

The definitions of Houghton's letters for The Improvement of Commerce and Trade were as follows; in his terminology, flint meant lead glass which we assume means tableware, ordinary meant soda or potash-lime flint glass used for medicines (vials and other small bottles) and green again a soda or potash-lime glass which was used for bottles such as beers, wines and spirits. Beers and wines were also made in dark amber or dark green which was often referred to as 'black glass'.

Houghton, in Table 1,1 lists the glass houses by the product made of which there were 88 in total, designated as:

Flint/Ordinary/Green	27
Bottles	38
Window	14
Window/Bottles	3
Crown/Plate	5
Looking glass	2

If we assume equal working time for each glass for each of the works that made more than one type of glass then the number of bottle houses equates to 57.5, i.e. approximately 65% of the total. At this time the definition of a bottle house equated to one furnace, which would have had two, four or six pots so making an educated guess more difficult. Using the terminology that was used in the 19th century then if there were two furnaces on the site then there would have been two glass houses. The number of containers produced in 1696 was 2 880 000.

Whereas secondly in the early 1980s, nearly 300 years later, the total tonnage sold was approximately 2 300 000 of which 1 700 000 tons was for containers, giving a percentage of approximately 74%. As the first total gives only an indication of what the tonnage could have been, and it could have been lower depending upon the number of pots to a furnace, then it can be seen that there is an argument to say that 'very loosely' it is nearly a constant amount.

There are two differences between these two sets of figures, the first is that Houghtons were for England and Wales only while the 1980 figures are for the whole of the United Kingdom, Scotland in 1980 had a significant bottle making industry. The second difference is in the amount of 'flat glass' being made, in 1695 it was 100% of the usage, while in the 1980s approximately 50% was imported. If this total was increased to 100%, to make it comparable to 1695 then the percentage for containers would be more like 61%. Which now brings the totals much closer together.

In 1695 the furnaces were very much alike whether they followed the Northern or the Southern European design. The Northern design furnace was rectangular in shape that originally included 'wings' either two that would be on one side of the furnace or four one at each corner, Figures 1,1 and 1,2 for examples, while the Southern design furnace was circular. Sometimes these different furnace designs are differentiated as north of the Alps and south of the Alps. For protection the furnaces would have been housed in a wooden framed structure similar to a barn.

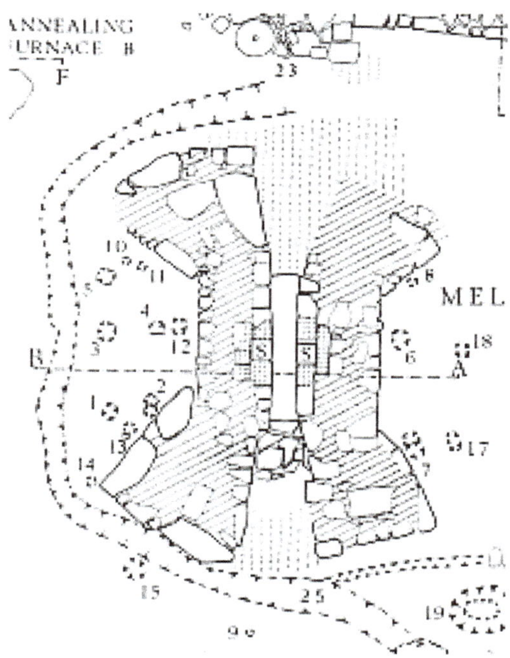

Figure 1,1. A plan of a four cornered, wood fired, furnace circa late 16th century. The furnace was at Rosedale, North Yorkshire

A two 'winged' furnace was used when

window glass was made by the 'broad glass' process. The open side enabled the blowers to swing the blown tubes of glass before they were cut and flattened.

Note this time the 'wings' are on opposite corners.

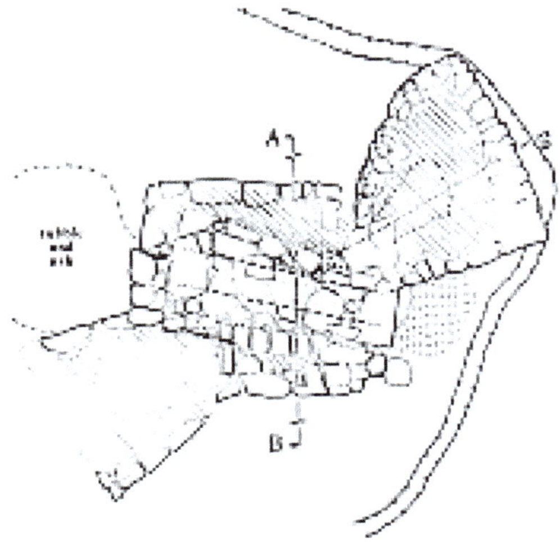

Figure 1,2. A plan of a two cornered, wood fired, furnace, circa late 16th century. This furnace was at Hutton-le-Hole, North Yorkshire

The next four pictures illustrate the changes to the glass making premises from about the mid 18th century, these plates are taken from Diderot's Encyclopédie of 1772 showing a 'Verrerie Angloise' glass works.

In Figure 1,3 we can see that the premises are now much more substantive with the furnace now enclosed by a brick built cone. Figure 1,4 shows a vertical section through the structure,

Figure 1,3. Outside view of the glass house

Figure 1,4. A vertical section though the centre of the glass house cone

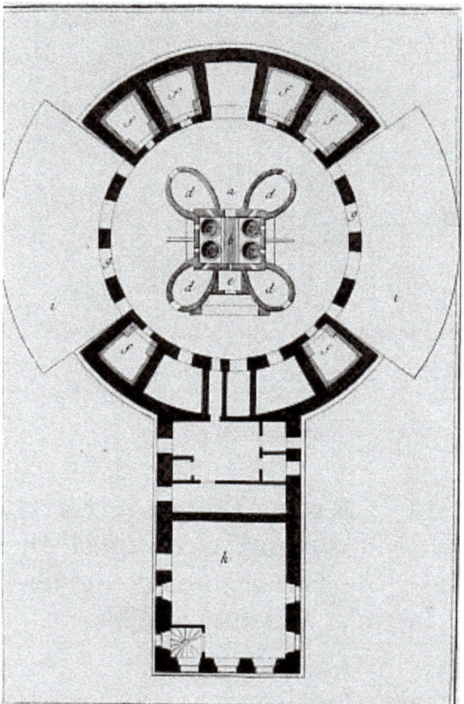

Figure 1,5 English cone furnace

Figure 1,5 shows a bottle-cum-window glass-cum-flint glass furnace while Figure 1,6 is of a plate glass furnace.

Figure 1.5 is a plan of an English cone furnace, a is the main oven, b is the grating for coal, c are pots in the furnace, d are the pot arches, e is the fritting chamber, and f are small annealing furnaces. In the bottle furnace, all of the wings were used as pot arches and the outside of the building is used as annealing chambers. This method of using separate annealing chambers is similar for the both the window glass and flint manufacture.

Figure 1,6. Is the plan of a plate glass furnace where HI is the siege for the pots. L is the siege for the curvettes in to which the glass was ladled. N are the curvettes, M the pots

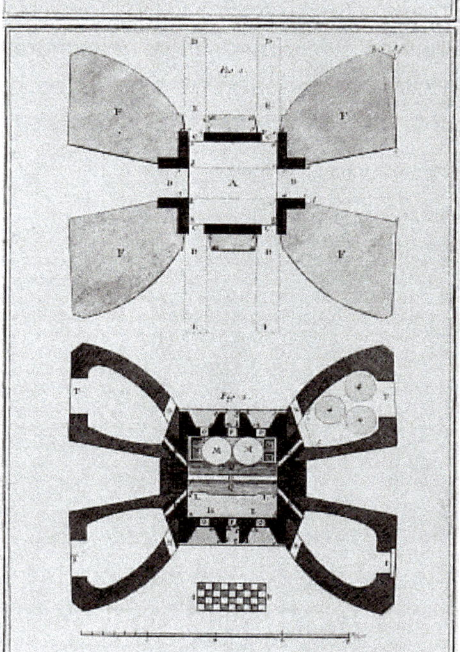

Figure 1.6 Plan of a plate glass furnace

in position, G space between sieges, O are the openings through which glass was transferred to the curvettes for pouring. T are pot arches, t is the side furnace where batch was treated before re-melting in the main chamber.

In the next three pictures that are from a later period, circa 1850, you can see that irrespective of the final product that these workmen are making, the methods are that being used are still the same. One difference from the earlier furnaces is that there are no 'wings' the annealing chambers are now all built into the outside walls of the building.

From about 1875 this glass melting method started to decline due to Siemens patents that gave rise to the regenerative tank furnace and the recuperative pot furnace. Over the next fifty years the glass industry gradually decided where it was going to go, the automated bottle and flat glass industries adopted the regenerative tank furnace while the 'hand made' industry such as the lead crystal manufacturers adopted the recuperative pot furnace.

Figure 1,7. Drawing of the interior of a typical Flint Glass house

Figure 1,8. Drawing of a typical Crown Window Glass furnace

Figure 1,9. Drawing of a typical Glass Bottle making shop

Chapter 2

Bottles made by the 'hand process' and the equipment that they used

Figure 2,1. A bottle making shop after 1855 using wooden moulds and the 'snap'

The method of producing glass bottles has not changed much in the last two millennia, it is a relatively simple but skilfull process using man's manipulative skills to shape the molten glass into the final article. The bottles were made by a group of five men working as a team, initially working at a pot but later on it would have been from the working end of a furnace. The team would consist of;

1, a gatherer,
2, a blower,
3, a glass maker,
4, a 'wetter off' and
5, a taker in.

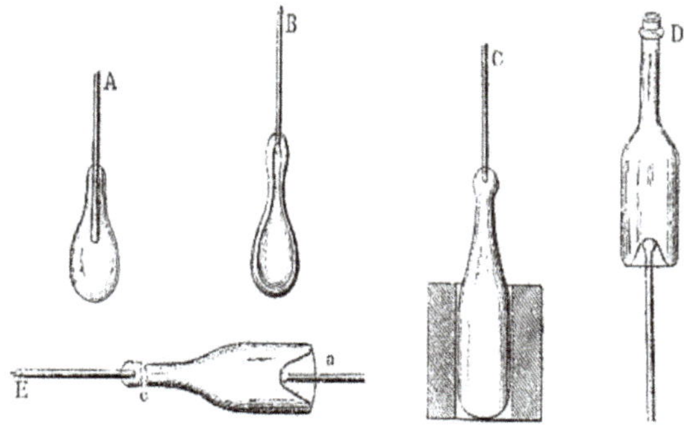

Figure 2,2. Pictorially the 'step by step' bottle making process

At Castleford, according to the local census returns, they used a different method, here the team consisted of a glass maker, two blowers, and two boys, one a 'wetter off' and the other a 'taker in'. This possibly could have been the normal method of manning for the bottle making industry. Going through the glass making process, using Figure 2,2 as a guide, which also uses a simplified mould. For the explanation we will use both the normal and the Castleford type of teams;

1. First a 'gather' of glass is gathered from the glass pot (or directly from the furnace) - by the gatherer or the glass blower. (A)
2. Then the gather is partially blown and swung until the neck of the bottle is formed, then it is rolled on a marver to bring the sides back to fit into the mould - by the gatherer or the glass blower. (B)
3. The partially blown glass is then placed in the mould and blown into shape - by the glass blower. (C)
4. The next stage is to form the base - using a punty iron the bottle bottom is pushed up inside the base, to form a 'pushup', and then the blowing iron is 'snapped off' just under where the glass and the blowing iron meet. The glass is now being held by the punty iron - done by the glass blower and a boy the wetter off. (D)
5. The glass is then passed to the glass maker, a small blob of glass is placed onto the neck of the bottle who then shapes

it into a finish, - done by the glass maker sat in his chair and a boy to add the blob of glass . (E)

6. And finally the bottle is placed in the annealing lehr - by the taker in.

For that system to work the following equipment was used:

The chair

The glass maker sits in what is known as a chair were he puts on the finish at the end of the bottle making process. Here you can also see a gathering or blowing iron and a punty iron.

Figure 2,3. A glass makers chair showing a blowing iron and a punty iron

Tools

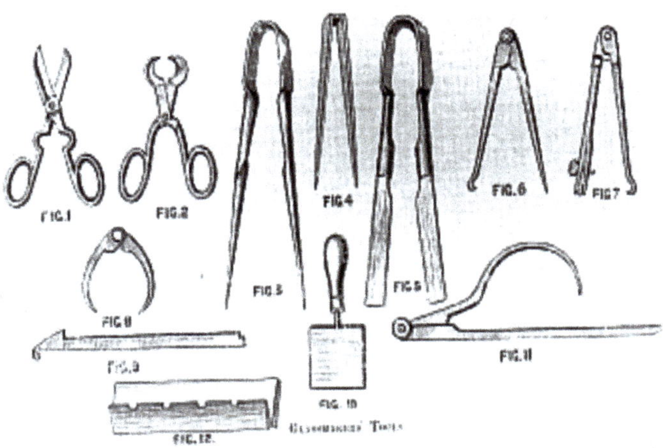

Figure 2,4. Tools for the manipulation of glass

To cut and manipulate the glass one or more of these tools would have been used, see Figure 2,4. Whereas for bottle making possibly only these four were used as shown in Figure 2,5.

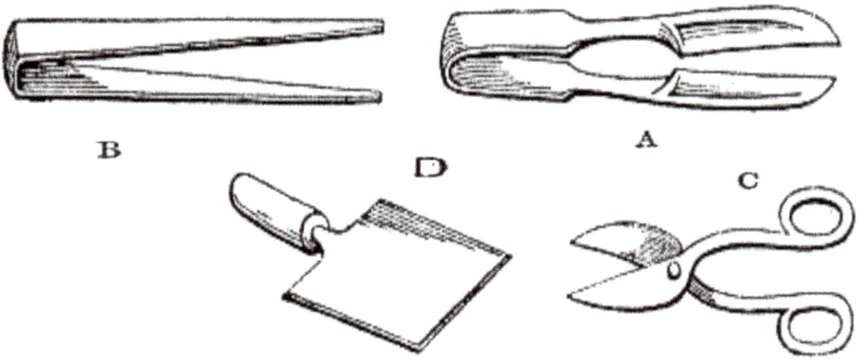

Figure 2,5. Possibly the bottle makers version of tools

Moulds
Moulds have been used for many years, originally they were made of wood such as these which were used at a glass house in the Black Forest in Germany.

The wood used was quite often from the pear tree, it was durable and long lasting. Others used were lime and apple. Eveson in his series 'Reflections' lists the results of tests with various kinds of wood for making the feet of wine glasses in Table 2,1.

Figure 2,6. Moulds from a Black Forest glass works circa 1600 to 1900. They are most likely to come from the latter half of the period

Wood	Comments
Agba (Malayan)	Good, but difficulty was experienced when the glass maker tried to re-scrape the foot-board.
Apple (small)	Satisfactory, but not quite as good as lime.
Beech	Unsatisfactory.
Cherry	Flaked slightly.
Elmus Ababa	Unsatisfactory: the feet of the wine glasses were ringy due to the pronounced grain of the wood.
Gurjun (Malayan)	Good, but difficulty was experienced when the glass maker tried to re-scrape the foot-board.
Holly	Unsatisfactory due to being too soft.
Hornbeam	Very unsatisfactory: flaked badly and burnt away very rapidly.
Horse chestnut	Unsatisfactory due to being too soft.
Iroko	Unsatisfactory.
Lignum vitae	Very unsatisfactory: warped badly when 'burnt'.
Lime	Satisfactory but not quite as good as pear.
Maple (English)	Unsatisfactory: too soft.
Maple (Canadian)	Usable but slightly soft: the footboard flaked slightly.
Oak	Unsatisfactory.
Pear	Excellent: the wood in stock was prevented from warping by placing weights on the cut lengths.
Sequoia sempervirens	Unsatisfactory: feet were ringy due to the grain.
Sycamore	Gave clear wineglass feet but wore away much too quickly.
Walnut (dark)	Gave clear wineglass feet but difficulty was experienced when the glassmaker tried to re-scrape the footboard.
Walnut (light)	Not as good as dark walnut.

Table 2,1. Results of different wood tests

Although not a perfect analogy they do give some indication of how the various woods would perform as moulds. Lime wood is wood from the lime tree, it is not to be confused with wood from the citrus fruit tree.

Quoting from his article;

"The carpenter was also closely involved in providing a vital service to some of the glass makers by making pearwood tools, called footboards, which were used when 'spinning' molten glass to form the feet of wineglasses. Pearwood is very suitable as it does not stain the glass and it also enables the glassmaker to produce clear crystal glass feet, free from toolmarks. Unfortunately, it warps badly and becomes twisted when kept in stock, and I thought it was about time a different and better wood was substituted. In later years, I carried out a series of experiments over a long period with full co-operation from a highly skilled glassmaker, only to discover after trying twenty different woods that pearwood was the best for the job! The results of the experiment may be useful to someone if only to provide information regarding substitution when pearwood is not available".

Figure 2,7. Painting of a Black Forest glass house

Later they would have used metal moulds - wooden moulds had been around from at least the early 1600s while metal ones only since the early 18th century, as witnessed by the following record from Wills' book:

> 'In 1752 a man was committed to Newgate, but later acquitted, for the crime of stealing 'one Brass Bottle mould' value 18 shillings, the property of Mr. Thomas Warren and Co., from the Glass-house in Temple Street'.

Warren's firm was established in Bristol and had a good reputation for its bottles and window glass.

In 1821 Rickett's patented his 3-part mould system. This system ensured that all of the bottles had identical dimensions on the outside the moulds used was made from either cast iron or brass.

The method of operation is as follows:

The actuating lever that controls the lifting and lowering of the bottle moulds 'bottom plate' was set to the down position. Then with the actuating mechanism for the 'shoulder and neck mould'

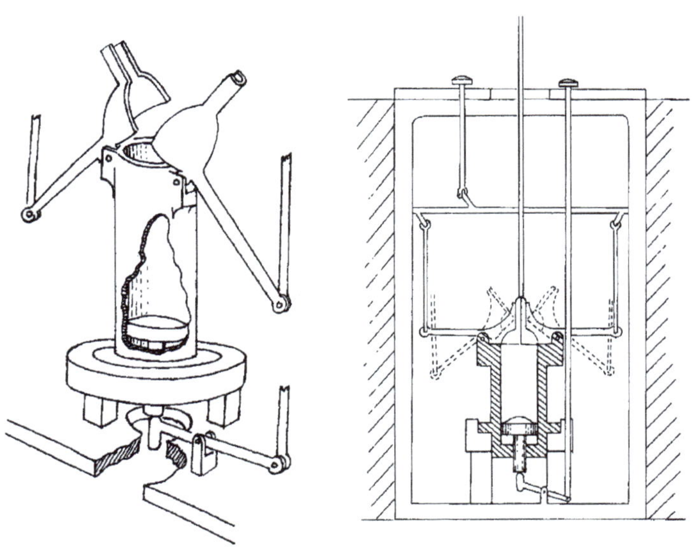

Figure 2,8. Rickett's three-part moulding system

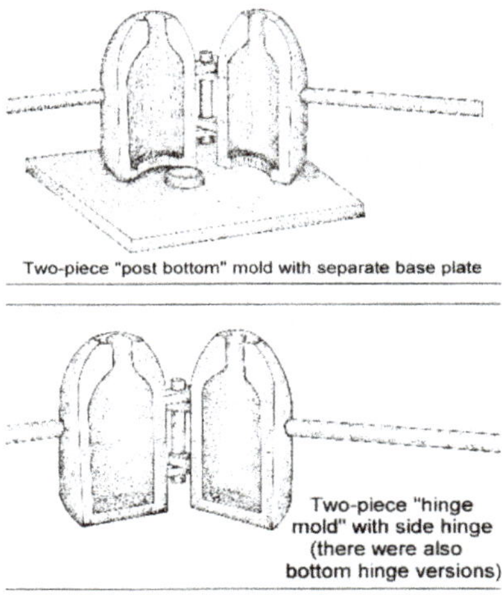

Figure 2,9 Side hinged moulds

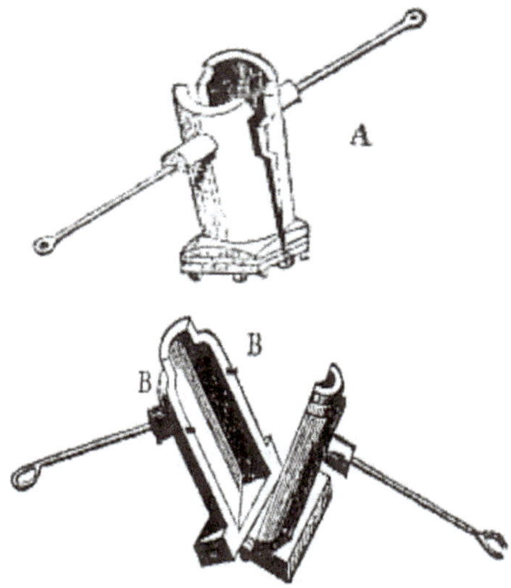

Figure 2,10 A bottom hinged mould showing it in the open and closed positions

halves open the gather of glass was inserted inside. The neck and shoulder moulds were closed and the bottle was blown. The upper two moulds were then opened and the bottle was raised up using the bottom plate actuator, and finally the bottle was removed from the mould.

From Rickett's patent the advances in mould making and the manner in which they were hung advanced with an ever-quickening pace. Figure 2,9 shows two side hinged moulds while Figure 2,10 shows a bottom hinged mould.

A later improvement to the hinging of the moulds was to facilitate some method of opening and closing them. This improvement was foot actuated as is shown in Figure 2,11.

Method of making moulded bottles

A gather of glass is made onto a hollow iron tube, its consistence is somewhere between putty and treacle, i.e. it will just stay on the tube. The glass is then rolled on a flat iron plate (marvered)

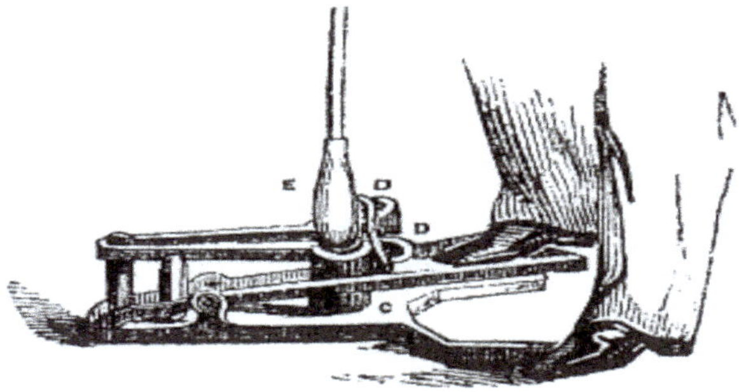

Figure 2,11. A foot operated mould opening and closing mechanism for a three piece mould

to obtain a cylindrical form. The neck is pinched by a small instrument and then placed into the mould which is at the feet of the blower, the mould is shut and the glass is blown up to the shape of the mould. The mould is opened and the glass - now in the form of a bottle - is withdrawn. The last operation is to put on the finish - carried out by the glass maker at the chair.

Mould 'A' as sketched, Figure 2,10, is termed an 'open and shut' mould, and is generally constructed in two exact halves, connected at the bottom by a hinge. For a round bottle the seams (joints) can be seen on the body, these lines are caused by the mould not closing exactly. The seam is not as unsightly when in the square or rectangular form (sketch 'B') as they are situated on opposite corners and are not as off- putting.

In the early part of the 19th century many manufacturers preferred no body seams so they used Ricketts or some other such mechanism which uses a three piece mould see Figure 2,8. Here the expanded ball of glass (E) is placed in to the mould (C), the two pieces of the neck are placed in position by pressing down on the pedal and the glass is blown to the shape of the mould. The finish is then put on as described earlier.

It is important for this method that the body is slightly tapered, the shoulder having a slightly larger diameter than the base. This is to ensure that the blown bottle can be removed from the mould.

All bottle moulds are required to be kept just below red heat when working to stop the glass surface appearing ruffled or wavy. This is done by a boy holding inside the mould a piece of red hot glass on a punty iron. Care needs to be taken to keep the metal moulds, whether they are brass or iron, a little below red heat, because if the moulds become overheated the glass will adhere (stick) to them.

If the opening and closing of the moulds was not done by the glass blower then he would have had an assistant to open and close the moulds.

The 'Snap'

The snap, Figure 2,12, was invented in 1857 and it can be seen being used in Figure 2,1. The invention of the snap did away with the glass that adhered to the base of the bottle. If the bottle was made without a push-up base or a very slight push-up the rough face of the punty would damage say the dining room furniture or it would make it very unstable.

This rough surface had to be smoothed to stop it doing any damage to furniture, which was a costly expense. The base of the bottle still had a scar but it was only where the iron had adhered to the surface it only was a slight surface mark.

The snap operates as follows, the snap is held vertical and a hot bottle placed into the cup, the handle, a tube into which the snap is fitted, is then pulled or pushed up to hold the bottle. Now it is possible to hold the bottle horizontally, without any damage, to apply the finish.

Figure 2,12 The 'Snapping device'

Other tools were used such as this 'lipping tool'. These were used to apply the finishes and their design would change

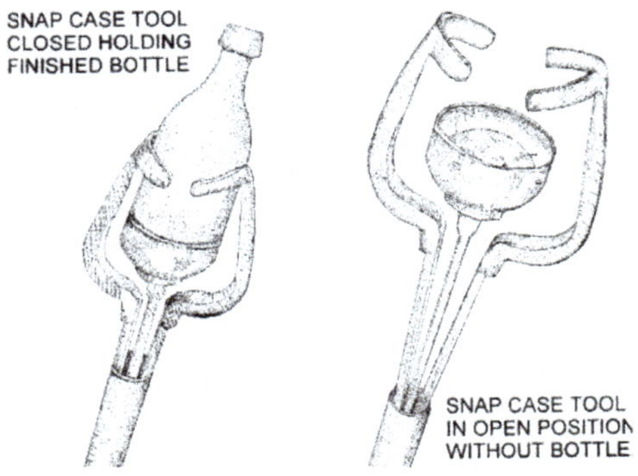

Figure 2,13. A more open 'snap'

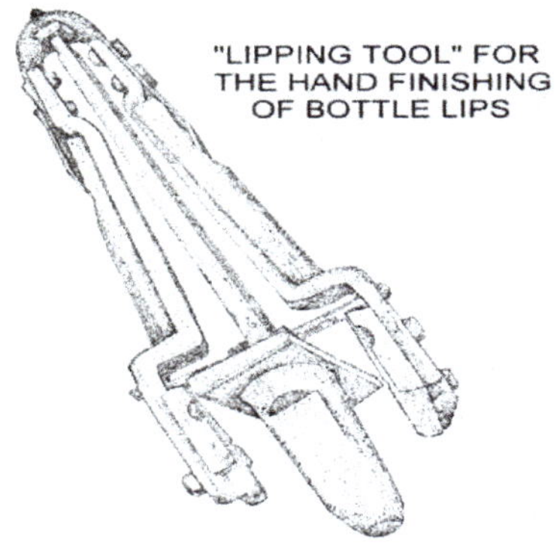

Figure 2,14. A hand finishing lipping tool

according to the finish being applied.

Sometimes the tools were to shape the 'bore' of the bottle such as the one in Hiram Codd's patent application for his Codd bottle.

In Codd's patent claim he describes how the groove for the rubber ring is to be formed by the tool he has invented, Figure 2,15. This lipping tool he describes as being similar to the ordinary tongs used to form the mouths of bottles except that when inserted into the mouth of the still molten bottle, the projections on the tongs form a groove around the interior of the mouth.

Figure 2,15. Hiram Codd's patented 'lipping tool'

He goes on to describe how the bottle was to be made:

'A bubble of glass is first blown and is roughly reduced to the desired shape by rolling and pressing it upon a stone. The roughly shaped bubble is then enclosed in a mould of the form desired, and the bottle is blown therein in the ordinary manner. When the bottle has been removed from the mould a glass marble previously heated is dropped into the bottle through the neck; the ring or head is formed at the top of the neck in the ordinary manner by means of the tool above described. After the bottle has been allowed to cool a ring of cork or other elastic material as for example, of vulcanised India rubber - is inserted into the groove from around the interior of the head, and the bottle is ready for filling with an effervescing or aerated liquid'.

Another Patent application for a similar lipping tool by William Wilson Macvay entitled 'Apparatus for Forming Groves in the Necks of Bottles' is shown in Figures 2,16 and 2,17.

A.D. 1880, 28th April. N° 1730.

Apparatus for Forming Grooves in the Necks of Bottles.

LETTERS PATENT to William Wilson Macvay, of the Firm of Sykes, Macvay, and Company, of Castleford, in the County of York, for an Invention of "IMPROVEMENTS IN TOOLS OR APPARATUS EMPLOYED FOR FORMING OR MAKING GROOVES IN THE NECKS OF BOTTLES."

PROVISIONAL SPECIFICATION left by the said William Wilson Macvay at the Office of the Commissioners of Patents on the 28th April 1880.

WILLIAM WILSON MACVAY, of the Firm of Sykes, Macvay, and Company, of Castleford, in the County of York. "IMPROVEMENTS IN TOOLS OR APPARATUS EMPLOYED FOR FORMING OR MAKING GROOVES IN THE NECKS OF BOTTLES."

The object of my Invention is to provide tools or apparatus for forming or making, preferably vertical, grooves in the necks of bottles, either separately or simultaneously with horizontal or annular grooves.

According to my improvements I use the tongs employed either in forming the necks of ordinary bottles, or those used in forming the annular grooves in the necks, similar to those for which Letters Patent were granted to William Brooke, dated 16th September 1871, No. 2447, and to Hiram Codd and Frederick Foster, dated 29th December 1873, No. 4268, and apply to that part of the tongs known as the plug or core a ring having any required number of projections on its periphery; these depend upon the number of the vertical grooves required.

The ring is passed into the bottle along with the plug or core, and the pressure required for forming the bottle forces the projections on the ring into the neck, and thus forms the vertical grooves. By the ring being loose on the plug or core it remains stationary with its projections in the grooves whilst the tongs are worked round to form the other parts of the necks.

Figure 2,16 Macvay's Patent application

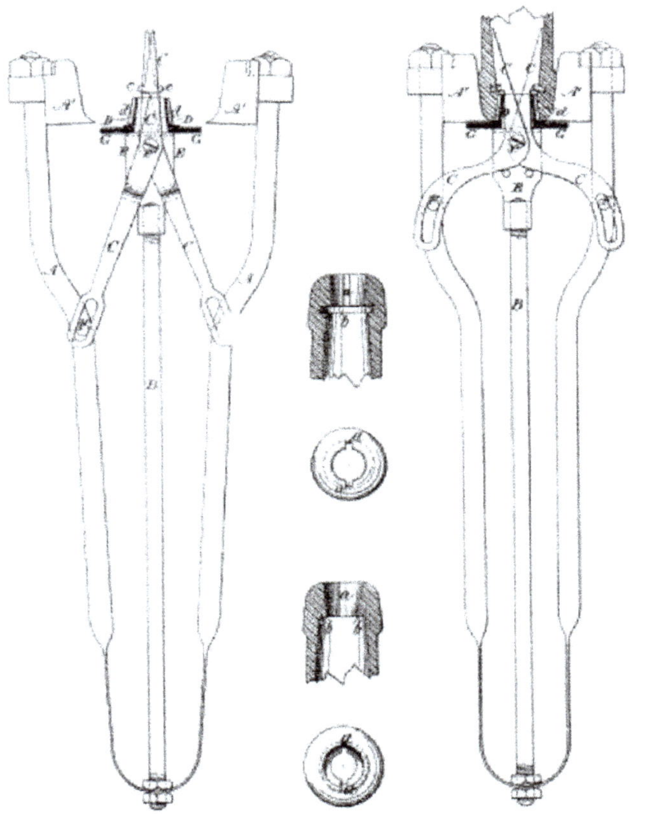

Figure 2,17. Drawing that accompanied W.W. Macvay's application

Chapter 3

Machine types and their equipment

Before we go on to the development of glass making machines, it may be worth while looking at where we are going, i.e. to today's modern machines. Although the modern machine has been around for more than eighty years todays machines look much different to the original ones. The original ones were pneumatically driven where as todays drives are electrical. The adjustments are 'push button' and the timing mechanism is electronic giving much better control as compared to the timing drum or some other similar device. Vast strides have been made to the cooling of the moulds and the ware handling is much improved.

Today most bottles are made on one type of machine that can make both the 'wide mouth' and the 'narrow mouth' types of containers. It is very rare that any of todays 'wide mouth' containers are made on a press machine but originally some were. Possibly the only place where some such machine is working is in a 'third world' country where the monies available for 'capital expenditure' and the size of the orders to be made do not justify the expense of purchasing a more modern machine.

The definitions or descriptions of the four processes are;

Press

A single step process, the article is pressed to the desired shape in a mould mechanically. Nothing else is used, i.e. compressed air for blowing, although compressed air would be used to operate the cylinder that applies the pressure for the press to press. The machines are of the single table type and for bottle making are probably now obsolete.

Press and blow

This is a two stage process, the first step is to manufacture a parison, which is done by pressing. Then the parison is transferred to the blow mould to be blown up to the finished size. Nowadays the machines are usually of the 'inline' type but some older types of single or twin table machines may still be in use.

Blow and blow

Again this is a two stage process, the first step is to make the parison which is made by blowing. Then the parison is transferred to the blow mould to be blown up to the finished size. Once again the machine may be of the 'inline' type although some single or twin table machines may still be in use.

Some types of feeder fed machines use vacuum for blowing, with the air in the atmosphere supplying the pressure for moving the glass. These types of machines are classified here as 'blow and blow' machines because they are making narrow neck bottles.

There is a certain type of machine in which the two parts of the process rotate around a single column and it uses vacuum as its blowing medium. Here referred to as blow and blow because they are making narrow neck bottles, but most times they are referred to as suction - suction machines.

Suction and blow

Here again is a two step process. The glass is sucked up out of a bath of glass to make the parison. Then the blank is transferred to the blow mould for completion of the bottle making cycle. This type of machine is now obsolete.

The machines could be classed as a single table machine as both parts of the process, i.e. as in the Owens machine, where the formation of the parison and then its subsequent blowing, rotate around a single columb. Or as a twin table machine, i.e. as in the Roirant A6 which other than how the of the blank is filled it operates as a feeder fed twin table machine.

The different types of mould equipment used

The following schematics show the different types of mould equipment used.

The press process

Figure 3,1 shows the basic mould equipment for a press machine. The sides of the vessel fits around the mould bottom plate, after the gob is placed inside then the plunger and the ring comes down. When the ring is resting on the top of the mould sides then downward pressure is exerted on the plunger to make the jar.

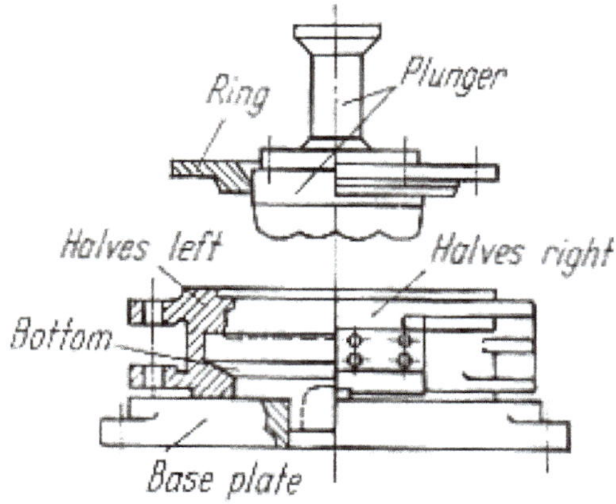

Figure 3,1. Basic press mould equipment

Press and blow process

Here the gob is loaded into the blank and the plunger comes down and presses the glass to the blank shape. The blank is transferred to the mould, the blowhead comes down and blows the glass up to the finished mould size.

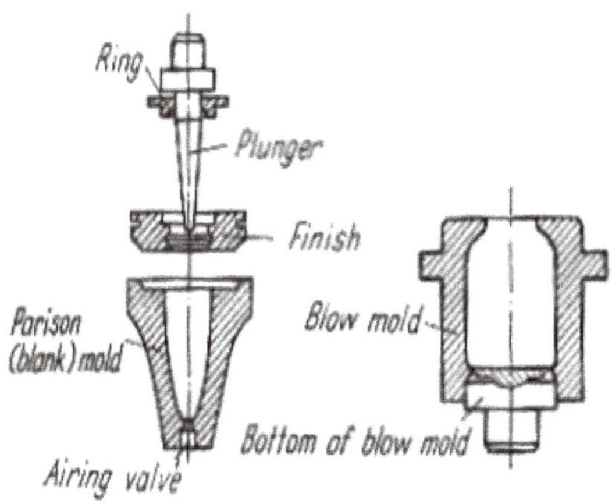

Figure 3,2. Basic press and blow equipment

Blow and blow process

Here the gob is loaded into the blank, the funnel is removed and is replaced by the bottom of the blank mould. Then a puff of air enters through the blank bottom to fill up the ring mould followed by a puff of air blown through the neck ring to form the blank. This is transferred to the blow mould and then the glass is blown to the finished size.

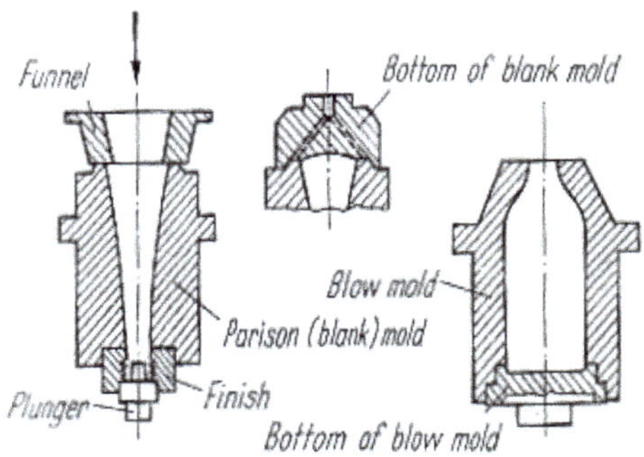

Figure 3,3. Basic blow and blow equipment

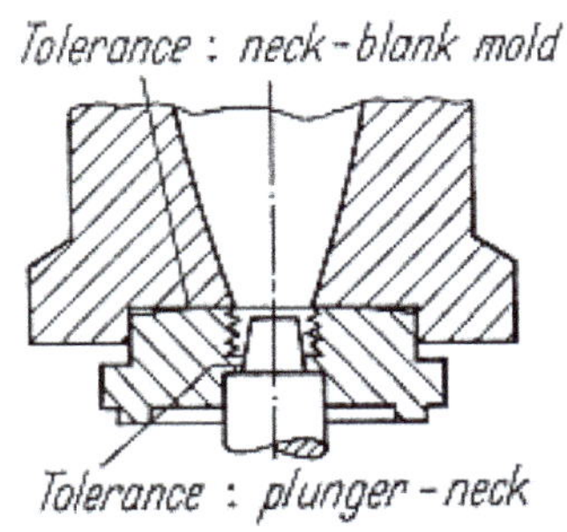

Figure 3,4. Detail view of neck ring

Suction and blow process

Figure 3,5 shows the basic suction and blow equipment, the parison passes over a bath of molten glass and fills itself by sucking up glass. Then the glass is sheared by a knife and the blank is passed to the blow mould where it is blown to the finished size.

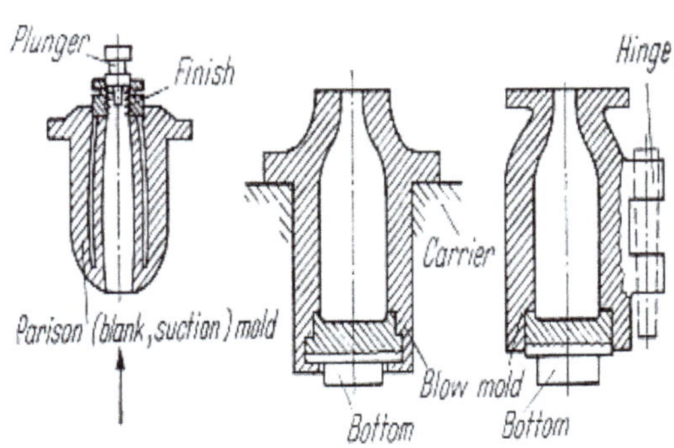

Figure 3,5. Basic suction and blow equipment

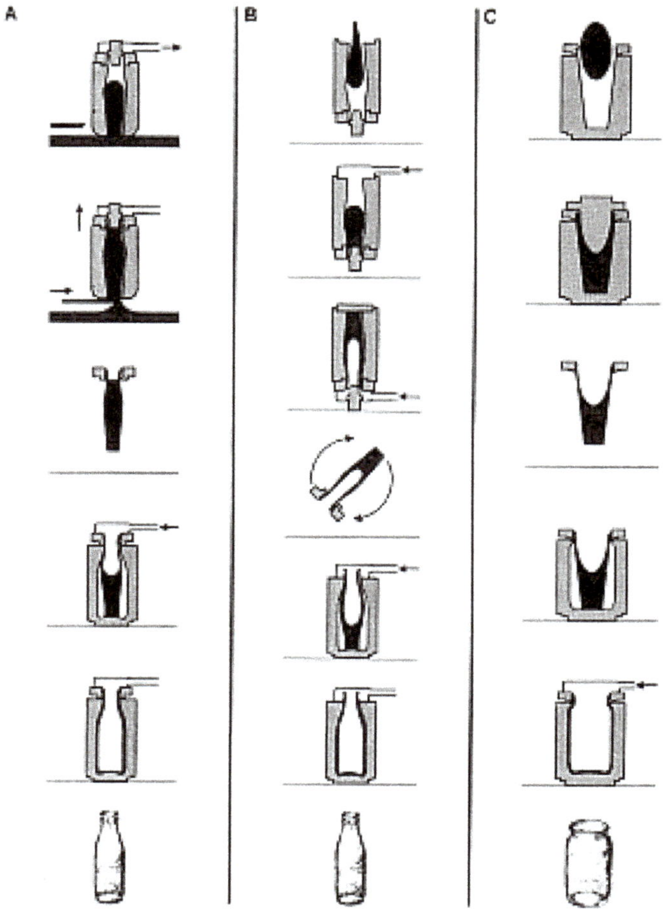

Figure 3,6. Historically the three main production processes for glass bottle production.
A = suction, B = blow & blow, and C = press & blow

Moulds

The design of mould equipment is a specialist skill, you have to make an article in a hot state, which is not at a constant temperature, so when it has cooled down it will conform to the exacting dimensional tolerances of the end product. As the mould equipment is often made from different metals then the differences in expansion also has to be taken into account.

Mould materials

In America the first patent for cast iron moulds was granted to Joseph Magoun in 1847 but it took until 1866 when chilled iron moulds were introduced before good results were achieved. Prior to this mainly wood and brass where used for moulds.

As the technology improved then more exotic materials became the standard, and during the last fifty years mould equipment of mixed materials has become the norm, i.e. iron moulds with a different mould bottom plate material, or the neck ring halves being different to the plug or plunger. Which makes for exacting tolerances and clearances between whatever mould materials is chosen.

The chemical composition of cast iron usually falls within these limits along with any alloying additives.

Usually contained in iron		Alloying additives	
C	2.7-3.8% (of which graphite 2.2-3.5%)	Cr	0.06-0.75%
		Ni	0.13-2.00%
Si	1.20-2.80%	Cu	0.10-0.54%
Mn	0.40-0.90%	Mo	0.06-0.60%
P	0.14-0.73%	Ti	0.10-0.25%
S	0.05-1.61%	V	0.15%

Table 3,1. Chemical composition of cast iron

In bottle making cast iron is the normal material used in the making of moulds, other mould materials used are basically the various alloys of iron, where the additions of the various alloying metals improve the life of the equipment. The cast iron has to be able to withstand the cyclic temperatures during its operation, for working temperatures between 500 and 700°C ordinary unalloyed cast iron is normally used, above this temperature up to about 1000°C, nickel and chromium alloys or pure nickel and chromium are used.

In every instance heat resistant grey cast iron is required and it is characterised by a relatively high carbon content, mostly in the graphite form, so graphitising elements are important. The additions to the iron play different roles the main ones are;

Carbon

This primarily determines the properties of the cast iron, if the eutectoid content becomes more than 0.86% then the irons strength decreases.

Graphite

A decrease in the strength of cast iron is caused by the notch effects of the graphite inclusions. This can be improved by the development of graphite that has special or discontinuous lamellar shape. The moulds tendency to scaling decreases with the addition of graphite.

Silicon

This is the second element after graphite to exert the greatest influence on the irons properties. It has a graphitising effect up to 3.5%, it also forms ferrites and in doing so decreases the hardness.

Manganese

This forms and stabilises the carbide above 0.5%, it increases strength and hardness and bonds sulphur in to the material. It increases the effect to the heat resistance of the iron and to its resistance to scaling.

Phosphorus

This improves the wear resistance but it also increases the brittleness of the iron and therefore its tendency to crack. Under certain circumstances it can effect the sticking of the glass by improper release.

Sulphur

This is innocuous up to 0.15%, it favours carbide formation, but in so doing it increases brittleness and the occurrence of hard spots.

Material	Applications
Cast iron without or with small additions of Ni, Cr, Cu, Mo, Ti, Al.	Press and hollow ware moulds, bottle glass moulds
Heat resistant steels; Ni-Cr-Fe, V5M, Ni-Cr-Wo-Cu-Fe, Cr-Co-Mn-Mo-Si-Fe.	Hollow and press ware - plungers, rings, and inserts.
Vanadium steels. Cr-Fe, Cr-Ni-Fe, Cr-Ni-Ti-Fe, Cr-Al-Ti.	Table rollers.
Bronze with or without addition of Al	Suction moulds, plungers, neck rings.
Nickel - chromium	Moulds for flacons
Monel (Ni-Cu)	Small moulds

Table 3,2. Commonly used mould materials with some of their applications

The effectiveness of the alloying additives are in the following directions; Al to 4.5%, Ti, Ni to 4.0%, Cu to 3.0% are graphitising. Cr is an outstanding carbide former. V, and also Ni, Mo, and Ti maintain carbide or pearlite and refine them. Cr, Ni, Mo, V, and Cu improves the strength of the iron, its hardness, scaling resistance and its constancy of volume.

There are a number of individuals who deserve mention, for their efforts in forwarding on the glass industry, they all started their own bottle machine making companies and they are Howard M. Ashley, Ben. Rylands, James Windmill, Michael

J. Owens, Frank O'Neill, Edward Miller, James W. Lynch, J. W. Miller, Emile Roirant and William B. Mitchell, and Carl. H. Rankine. Others always worked for a company such as W. A. Lorenz, Roy La France, L. D. Soubier, Karl Peiler. Where appropriate a brief synopsis on each of them is included; the positioning of which is in the most appropriate place.

Chapter 4

The rise and fall of the semi-automatic machine

Now we move on to semi-automatic machines, which we will divide into two parts, although they do overlap somewhat in the middle. The division taken is 1900, machines from about 1860 to 1900 were, in the majority, single head machines while from 1900 to 1925 they were all rotary multi-head machines.

Historically the main difference between the semi-automatic and the automatic machines is how the glass is fed to the machine. The semi-automatic variety contains both manually operated machines, i.e. where all of the operations are done by hand, and semi-automatic ones where most or all of the operations are done automatically once the machine has been set in motion.

Automatic flow machines did not come into being until Peiler patented the forehearth, as we know it today, in 1922. Some worked with earlier type forehearths but they were somewhat erratic in their operation.

Table 4,1 shows the main patents registered between 1859 and 1899, it must be stressed that the patentee was registering a patent, a working machine is a different thing. Sometimes patents are registered but a working model fails to materialise, for example the first three patents registered on our list, are those of;

A. Mein of St. Rollux, Glasgow in 1859, Patent Number 680.
C., G., W. and J. Kilner of Thornhill Lees in 1860, Patent Number 2079.
J. Bowron of Stockton-on-Tees in 1861, Patent Number 354.
All were for single mould 'blow' processes and all failed to become a working machine.

It was around this time, circa 1866, that Josiah Arnall the postmaster in Ferrybridge tried to interest Edgar Breffit, the lessor of the Ferrybridge Glassworks, in his new glass making

Development of semi-automatic machines in Europe 1859 to 1899

England Inventors	Type of Machine	Date of Patent	Date of Commercial Use	**Europe** Inventors	Type of Machine	Country of Origin	Date of Patent	Date of Commercial Use
Mein A.	Single Md	1859	--------	Bauer M.	Bl & Bl	Germany	1889	--------
Kilner G.G. W & J	Single Md	1860	--------	Hilde H.	Bl & Bl	Germany	1893 (d)	1896
Bowron J	Single Md	1861	--------	Maressier	--------	France	-----	Tried in 1894
Ashley H.M.	Pr & Bl / Bl & Bl	1886 -	1887 1890	Verney	Bl & Bl	France	1893 (d)	Tried in 1894
Windmill J.R.	Pr & Bl	1886	?	Bouchier C.	Bl & Bl	France	1897	1897
Rylands D.A. / Emmett C.	Pr & Bl	1889	pre 1896	Grote L.	Bl & Bl	Germany	1897	?
Dobson C.B.	Bl & Bl (a)	1893	--------					
Beckett R.T.	Bl & Bl	1893	--------					
Haley J.	Pr & Bl (b)	1897	1897					
Blue C.E.	Pr & Bl (c)	1897	1900					

Notes

(a) Modification of Ashley machine
(b) (5) dates of British Patents
(c) Probably invented in the USA but intrduced and used in Yorkshire
(d) British Patent

Table 4,1. European glass making machines patents, from 1859 to 1900

process. He had studied the hand making process when he was delivering the post to the works and had come to the conclusion that the finish had to be made first and the body of the bottle made last, i.e. it was press and blow.

Breffit considered this change too revolutionary, possibly the engineering was not good enough, remembering that Arnall was the postmaster and not an engineer, or the thought of making the finish first was just too advanced, so he declined the offer. Nothing further transpired until about twenty years later when Ashley, the manager of the Ferrybridge Ironworks, came to lodge with Arnall.

Continuing with the patents listed the next one is that of Windmill. James Richard Windmill's patent application was made on June 29th 1886, when he was 24 years old, his application was submitted three days before Arnall and Ashley submitted their patent on the 2nd July 1886.

So his was the first application for a bottle making machine, except that, it was only for the 'press and blow' process. Ashley's first application was also for a 'press and blow' process but his second patent application made later in the same year was for the 'blow and blow' process. So they both had a claim to the first operational machine. In Windmill's application, British Patent Number 8526, Figure 4,2, starts very explicitly;

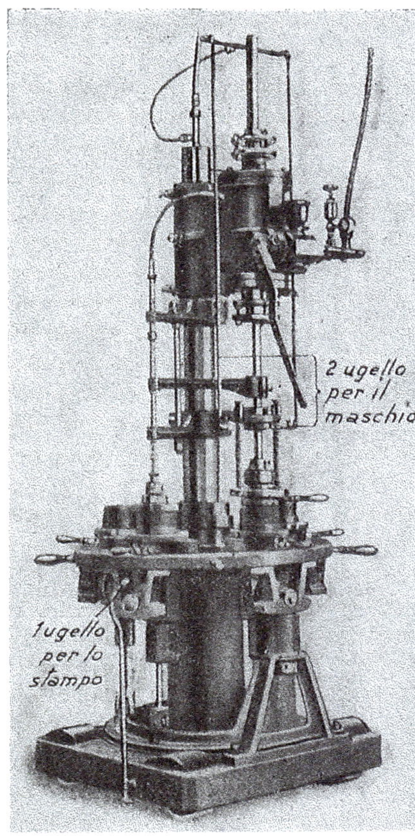

Figure 4,1. Windmill type press and blow machine

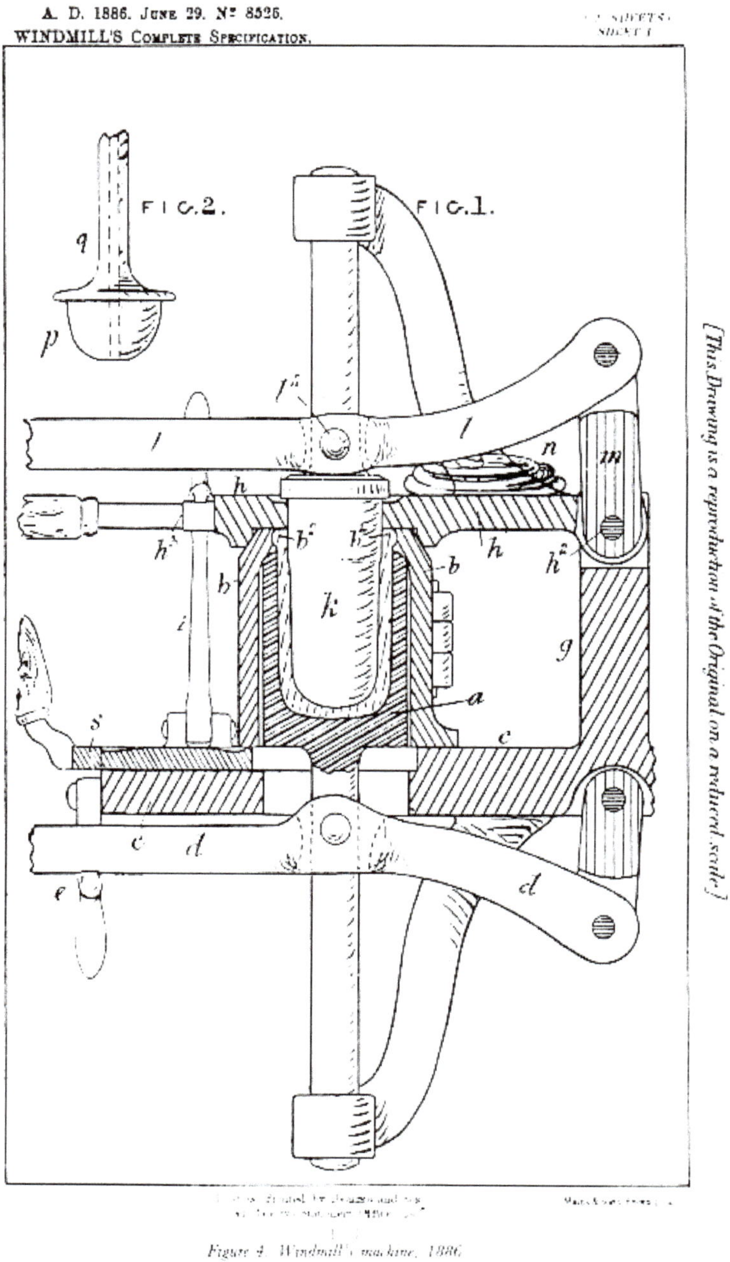

Figure 4,2. J. R. Windmill's patent application on 29th June 1886, Pat. No. 8526

"My invention consists in manufacturing bottles and jars and other like articles by the combination of pressing and blowing as hereinafter described ..."
this along with statements such as;
"The apparatus ... which I have found to answer well in practice ..."
indicated that the machine in the patent was already an operational machine.

The drawings were remarkable in their clarity, they possibly were made by James's brother Ernest who worked at nearby Stevens & Williams, and they infer that machine made jars had been made for a number of years at Brierley Hill.

The basic principle of this machine is the novel idea of fitting one mould inside the other, after the blank is pressed the blank mould slid out from inside of the blow mould and then the blowhead was put in position and the jar was then blown. The plunger that formed the parison inside the blank mould was operated by a hand lever.

James did not get any help from his firm's proprietor, Edward Woodward, who had several outside engineering interests, so he had to approach a friendly bank manager for a loan to finance the building of a prototype.

The Windmills were an old established family living in the Stourbridge area since at least 1622 when a daughter Mary was christened. They lived in Rowley Regis, which is about three miles from Brierley Hill church, they named the eldest son John, so successive John's children were christened in 1623, 1652, 1673 1709, and 1751.

Around this time 1750 to 1760 some of the Windmill's moved to the nearby coal mines at Oldbury or to the glassworks in Brierley Hill. Whether or not this was the family's introduction to the glass industry it is not known, but what is known is that glassmakers were living in Rowley Regis in 1647 so it is possible that some of the Windmill's family worked there.

It can hardly be a coincidence that in 1759 some Windmills appeared in Chelwood, south of Bristol where there was a glass house that had been started some one hundred years earlier by the Tyzacks of Stourbridge.

Parish records show that John Windmill (1772-1831) a bottle maker raised a family in Brierley Hill. One of John's sons was Edward (1808-1868); he followed his father's calling and worked at the Moor Lane glassworks and at least three of Edward's sons went into the bottle business. His eldest son was called William (1829-1877) who progressed through the ranks to become the manager of the bottle works, which by then had become Westwood & Moore.

Three of William's sons followed in their fathers' footsteps, and by 1881 his eldest, Edward (1853-1916) had followed him as manager of the bottle works. The younger brothers were William Charles (1859-1918) and James Richard (1862-1934). James worked first as a commercial clerk and then as a commercial traveller but was obviously an entrepreneurial character.

It is not known how efficient or for how long these machines worked as Westwood & Moore was sold around 1890 to James Wright and Co. who converted the works to a ceramic bottle manufactory and started a new glass bottle works, with a Sieman's gas fired furnace, at a neighbouring site in Delph. The life of the

Figure 4,3. James Windmill and his family in 1910

company was short, by 1894 the workforce was down to 36 men, and by 1899 the company was declared bankrupt.

As for James Windmill, sometime before the closure he started to work for Cannington & Shaw of St. Helens as a commercial traveller working out of their Birmingham office. One of his brothers, William, went to St. Helens as a bottle maker but the company that employed him is not known but it was a good chance that it was Cannington & Shaws.

Cannington & Shaws bought at auction some Ashley machines when the Ashley's assets were being sold off in 1899 and not long after converted some of them to the 'press and blow' process so it is not inconceivable that possibly the pair of them helped in the conversion.

The second application for a bottle making process was made by Josiah Arnall and Howard Ashley on July 2nd 1886 Patent Number 8677 which could be described as a 'press and blow' process.

They soon found that manufacture under this system had 'design' faults built into it, the glass was loaded into the mould and

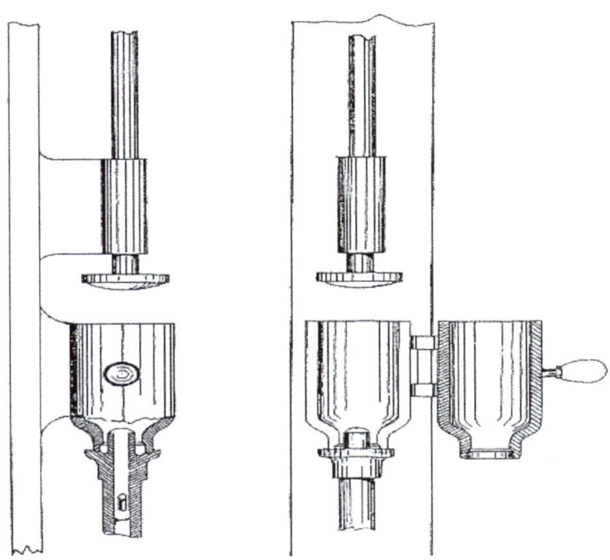

Figure 4,4. Arnall and Ashley's patent, Number 8677 applied for 2nd June 1886

pressed down manually by a loose fitting the ring, with the bottom plate still in place, air was passed through the neck ring plunger to inflate the glass up to the mould. After trial and error Ashley realised that mechanical production could only be possible with three moulds, ring mould, parison mould and blow mould and this change to the method enabled him to make narrow neck bottles.

Hand made process	Machine made process
Glass gathered	Glass gathered
Blank shape part blown	Neck ring formed
Body shape blown to final mould size	Body shape (glass blown up to correct size of blank)
Neck ring or finish put on	Blank shape blown up to final mould size

Table 4,2. Order of manufacturing hand vis semi-automatic process

The process devised has all the operations of the hand making process, see Table 4,2, it was very similar but the operations were carried out in a different order;

This led a second patent in November of the same year, but this time it was under Ashley's name alone. British Patent Number 14727.

The 'Plank' machine was Ashley's first "commercial" venture and its possible patent number was either 14727 or 3434 dated March 7th 1887, with the second the most likely.

The 'Plank machine' was difficult to operate and was very labour intensive. All of the mould equipment had to be lifted on to and off the machine and it took a team of seven people to operate it.

Figure 4,5. Ashley's 'Plank' machine

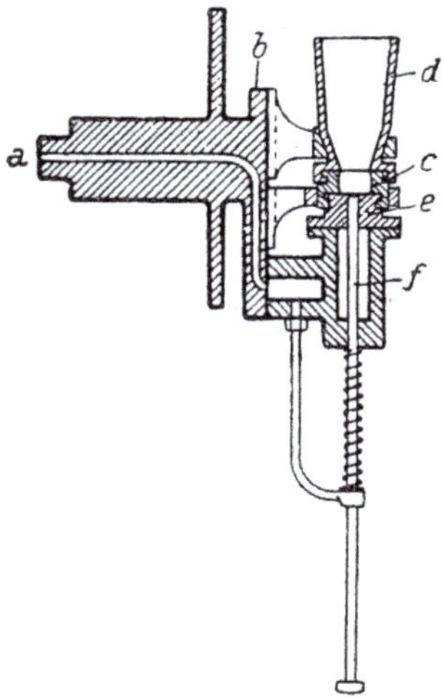

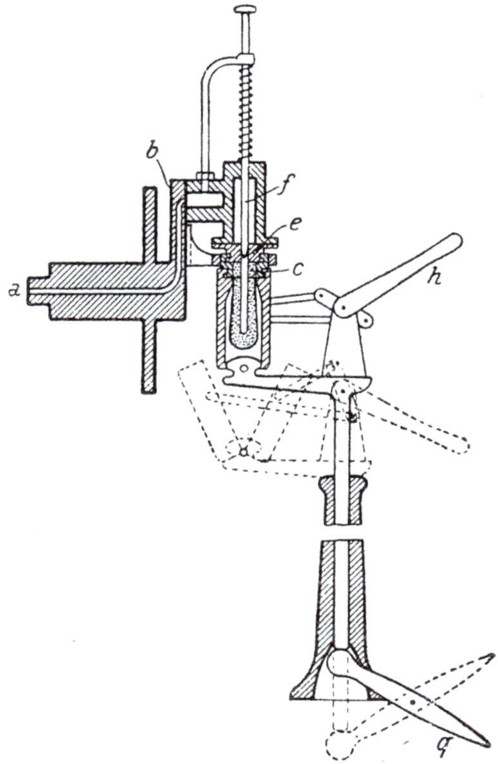

Figure 4,6. Preparatory mould (above)
Figure 4,7. Finishing mould (right)

In his March 1887 application Moody stated that Ashley expressed his thoughts as;

"I also find it necessary in some cases not to put the molten glass directly into the bottle mould, but to put it first into another receptacle to give it its first form, which receptacle I call a "Bloom" mould or "Parison" mould, and afterwards into the bottle mould.

I also find it necessary to arrange some of the parts of the machinery or of the moulds for forming the bottle on a revolving stand or table or on a sliding bench or gantry, in such a manner as shall admit of a number of parts being interchangeable".

Figure 4,8 is a later version of the same machine, its Patent Number was 3686, dated November 10th 1887. The machine was nick-named the 'Pillar' machine.

English described the action of the machines as follows:

"Both figures show that it consisted of a rotatable head, C, fitted to an upright plank or pillar, and that through the head passed a punch, D, which also acted as a compressed air pipe. The parison, G, and ring moulds, E, were mounted on the ends of pairs of tongs, F, and were placed in position over the air nozzle and operated by hand.

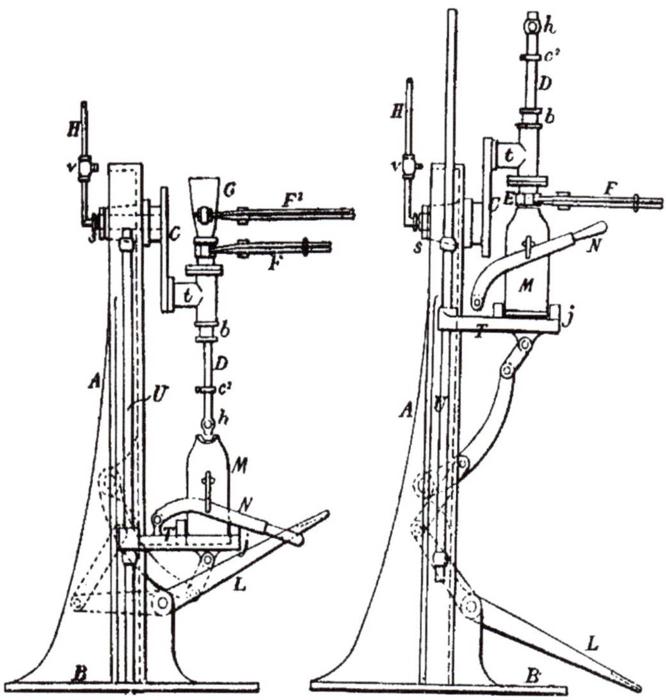

Figure 4,8. The 'Pillar' machine

Molten glass was poured into the parison mould, and the plunger, D, was pressed upwards into the glass to form the neck, using the cross piece on the stem as a handle. After completing the neck, the plug was removed, and a slight puff of air was given from the pipe, H, and controlled by the valve, V.

The parison mould halves were then opened and removed by hand, the head carrying the blank by the ring mould was rotated until the glass was hanging vertically, in which position it was paddled by hand or allowed to extend according to requirements. The blow mould halves, at first mounted on arms hinged to a long lever, fulcrumed at the back of the plank, and operated by a foot pedal, were brought round the blank and secured together by a catch pin. The compressed air valve, V, was opened to complete the blowing of the bottle."

Later, in the pillar machine the blow mould halves, M, were mounted on a rising table operated by a foot lever, L, and were closed by the lever, N.

With such a machine, one gatherer fed two units and could make 130 dozen soda-water bottles per day of ten and a quarter hours. It was not a significant increase in production but it proved that the machine made bottles were here to stay.

Ashley conducted trials with the 'Plank' machine, Figure 4,5, first at Ferrybridge and then at Armley in Leeds. The majority of the moneys to build the trial machines were supplied by John Hardcastle a charted accountant who practised in Leeds. The third set of trials was conducted at the Sykes, Macvay and the Codd Bottle Company site at Castleford with a public trial being held on the 16th December 1887.

It is not possible to ascertain when the Sykes, Macvay and the Codd Bottle Co. Ltd. acquired the patent rights of this machine, as the agreement only included Patents up to July 1887 so it is possible to assume that the trials started before the 10th November 1887 when his Patent Number 3686 was registered.

Towards the end of 1887 the Sykes, Macvay and the Codd Bottle Co. Ltd. and the company owned by Ashley and Hardcastle merged, with the new company named the Ashley (Machine Made) Bottle Co. Ltd. coming into existence on the 16th February 1888. The document is quite old and unfortunately it is on blue legal paper so it has been digitally enhanced to improve the clarity. The blue/black ink on dark blue paper is not good for easy reading.

You will notice that it took just over two months to complete and in the beginning the company was to be called the Sykes, Macvay (Machine Made) Bottle Co. Ltd. When it became the Ashley (Machine Made) Bottle Co. Ltd. is not known but the agreement shows the seal of the Ashleys Patent (Machine Made) Bottle Co. Ltd. a company that could have already existed.

Figure 4,9. is the front piece of the Agreement between H. M. Ashley and J. Hardcastle and the Sykes, Macvay and Codd Bottle Co. Ltd. Dated 7th December 1887. The next seven photographs complete the agreement. They are not numbered but they all belong to Figure 4,9, which can be read quite easily with a magnifying glass.

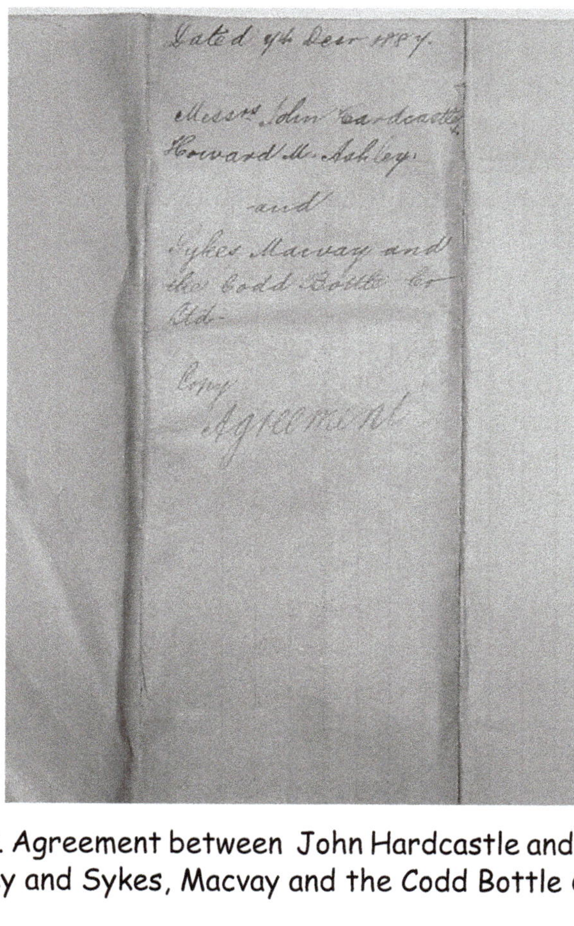
Dated 7th Decr 1887.

Messrs John Hardcastle
Howard M. Ashley

and

Sykes Macvay and
the Codd Bottle Co
Ltd.

Copy
Agreement

Figure 4,9. Agreement between John Hardcastle and Howard M. Ashley and Sykes, Macvay and the Codd Bottle Co Ltd

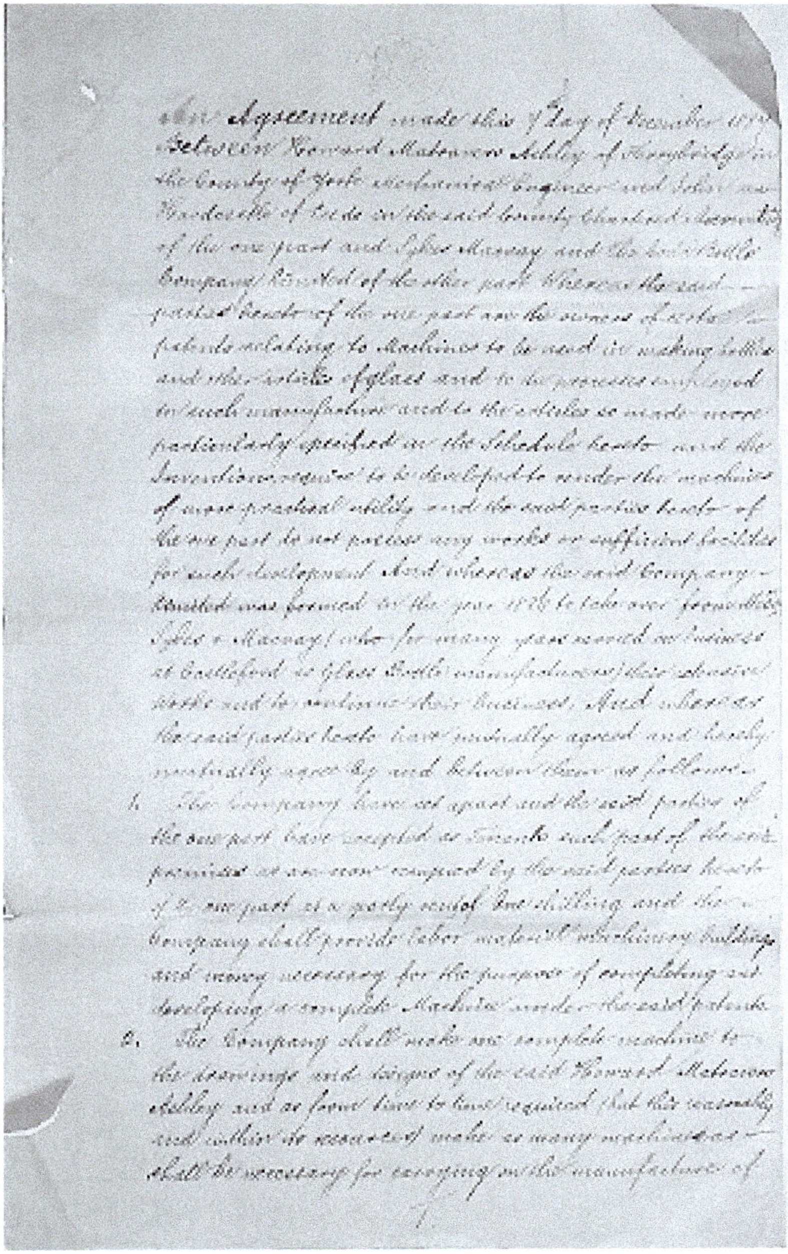

An Agreement made this 7 Day of December 18[illegible] Between Howard Matravers Ashley of Ferrybridge in the County of York Mechanical Engineer and John [illegible] Woodcock of Leeds in the said County Chartered Accountant of the one part and Sykes Macvay and the Castleford Bottle Company Limited of the other part Whereas the said parties hereto of the one part are the owners of certain patents relating to Machines to be used in making bottles and other articles of glass and to the processes employed in such manufacture and to the articles so made more particularly specified in the Schedule hereto and the Inventions require to be developed to render the machines of more practical utility and the said parties hereto of the one part do not possess any works or sufficient facilities for such development And whereas the said Company Limited was formed in the year 18[illegible] to take over from Messrs Sykes & Macvay (who for many years carried on business at Castleford as Glass Bottle manufacturers) their extensive works and to continue their business. And whereas the said parties hereto have mutually agreed and hereby mutually agree by and between them as follows—

1. The Company have set apart and the said parties of the one part have accepted as tenants such part of the said premises as are now occupied by the said parties hereto of the one part at a yearly rent of One shilling and the Company shall provide labor material machinery buildings and money necessary for the purpose of completing and developing a complete Machine under the said patents.

2. The Company shall make one complete machine to the drawings and designs of the said Howard Matravers Ashley and as from time to time required that this reasonably and within its resources [illegible] make as many machines as shall be necessary for carrying on the manufacture of

7

all bottles and other articles which can be most economically made by the machines than by the processes at present employed and shall thenceforth use such new machines instead of the old processes.

3. Subject to the provisions contained in clause 16 hereof the parties hereto of the one part agree to grant to the Company as after mentioned an absolute monopoly during the whole term of the Patents or any extension thereof of the right to make by themselves or employ other persons or firms to make the said machines for use by themselves or for sale to other persons or for the use of others to whom the parties hereto of the one part have sold.

4. And for the considerations herein appearing the parties hereto of the one part agree to grant to the Company the right to use during the whole of the term of the respective patents or any extension or extensions thereof all and singular and as many as the Company may at any time or times require the said machines for all purposes to which the same are applicable free from any royalty or payment other than as hereinafter provided.

5. The said parties hereto of the one part shall be entitled at any time during the existence of the patents to grant licences for the use of the machines to other manufacturers at a royalty of 4d per gross on all articles made or such other royalty as may be mutually agreed upon in writing from time to time one moiety of such royalty to belong to the Company and the other moiety thereof to the parties hereto of the one part and

6. Such licences to other Grantees shall contain a provision that the said licensees shall not sell any articles made by the machines under the patents (which shall be similar or competing articles to any from time to time made by the said Company) at net prices class for class below

2.

those charged by the Company.

7. The parties hereto of the one part shall be entitled at any time during the existence of the said patents to call upon the Company to supply machines to other manufacturers and the Company shall be entitled to charge for the machines so supplied such a price as will leave a profit of 100 per cent by way of royalty on the cost of each machine or such other royalty as may from time to time be determined by the parties hereto, and in the event of any difference arising between the Company and the parties hereto of the one part as to such cost the question in dispute shall be settled by arbitration in accordance with the provisions of the Common Law Procedure Act 1854 or any other Act relating thereto for the time being such profit to be divided between the Company and the parties hereto of the one part in equal moieties. In case of any default by the Company to supply any machines which the parties hereto of the one part have called upon the Company to supply to other Licensees within 3 months from the date of the order the parties hereto of the one part shall have liberty to supply such machines themselves.

8. In the event of any alleged infringement of the said patents and any action being taken by the said Company in respect of such infringement any law costs or other expenses incurred by the Company in protecting the right of making or using the said machines shall be born in equal moieties by the Company and the parties hereto of the one part

9. The Company shall issue to the parties hereto of the one part whenever called upon to do so fully paid up shares of the Company at the nominal or par value of £20,000 and also shall subject as after mentioned pay to the parties hereto of the one part a sum of £20,000 in cash. The Company propose to raise by a fresh issue of their

3.

shares and additional Capital of about £65,000 unless arrangements for the assignment of its business to a new Company to be called the Sykes Macvay (Machine made) Bottle Company Limited shall be completed and upon such sum being raised will immediately pay to the parties hereto of the one part the said sum of £20,000. But until the amount raised by the new issue of shares shall amount to the sum of £40,000 one moiety only of every sum received by way of additional share capital shall be payable until the said sum of £20,000 shall have been fully paid to the parties of the one part.

10. The Company shall be entitled so soon as the said £20,000 of shares shall have been issued to the parties hereto of the one part the said £20,000 in cash paid and the amount now due to the Companys bankers paid off and the Company's liabilities on the assets in respect they realized to call upon the parties hereto of the one part to assign absolutely the patents specified in the Schedule hereto and all amendments alterations or improvements therein subject nevertheless to the provisions hereinbefore contained in respect of one moiety of the Royalties to be received on machines and boxes made by licensees.

11. All the parties hereto agree that all amendments alterations or improvements of the said patents by any of the said parties hereto and all others who can be controlled by the Company or the parties of the one part shall be absolutely the property of the Company subject to the moiety of Royalties hereinbefore referred to.

12. Each of the parties hereto of the one part shall forthwith or as soon as practicable be appointed a Director of the Company subject to the provisions of the Articles of Association.

12a If the arrangements for assigning the business of the Company hereinbefore mentioned shall be carried through

4

the Company shall make such assignment subject to the rights of the parties hereto of the one part so as to place them as far as possible in the same position with regard to the proposed new Company as they will hold in relation to the Company under the Agreement and the parties hereto of the one part shall accept shares to the nominal value of £20,000 in such new Company in lieu of the obligation of the Company to allot them £20,000 of shares under the agreement

13. The said Howard Matravers Ashley shall be and act as the Menchanical Engineer of the Company and shall devote so much of his time as the Board may think necessary for superintending the making fixing and working of the machines whether on the Company's premises or on those of other Licensees for such a period and at such a salary as may be agreed upon between him and the Company

14. That if either of the said parties hereto requires any deeds executing for confirming or further assuring this agreement the same shall be executed by the said parties hereto and the expense thereof borne by the Company.

15. That the expression herein the said parties hereto of the one part shall be deemed to include ~~the~~ and be binding upon them and each of their executors and administrators and the expression herein the said Company shall be deemed to include its successors.

16. If at the expiration of 6 months from the date hereof the consideration contracted to be paid by the Company has not been paid then so long as such consideration remain unpaid the Companys monopoly of making and selling the machines and the licensing of other Bottle Manufacturers to use the same including the receipt of Royalties thereon shall be suspended until such consideration

5

has been paid.

Schedule hereinbefore referred to.

Patentee:	Number.
Arnall Ashley.	8677/86.
Ashley.	14727/86.
	3434/87.

Also

Ball stopper in name of Ashley. No. 7560/87 applied for in or about the month of July 1887. These two last are provisionally protected.

In witness whereof the said parties have hereunto set their hands and seals and the Company has hereunto affixed its Common seal the day and year first above written.

(Sd.) H. M. Ashley (L.S.) Sd. John Hardcastle (L.S.)

Seal of the Sykes
Macvay and the Codd
Bottle Co Ltd.

Signed sealed and delivered by the said Howard Matravers Ashley & John Hardcastle in the presence of

(Sd.) A. C. Dockerill.
Clerk.
160. Albion Road.
Newington Green N.

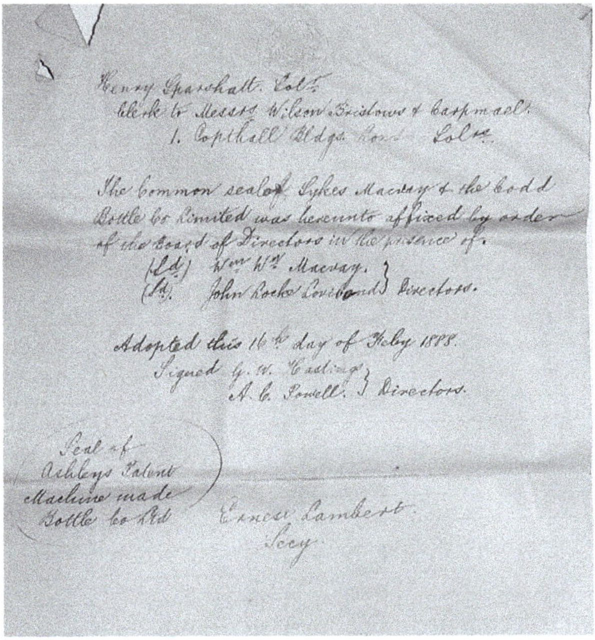
Henry Sparshatt. Solr.
Clerk to Messrs. Wilson Bristows & Carpmael.
1. Copthall Bldgs. [illegible] Solrs.

The Common seal of Sykes Macvay & the Codd Bottle Co Limited was hereunto affixed by order of the Board of Directors in the presence of.
(Sd.) Wm. Wr. Macvay. }
(Sd.) John Locke Lovibond } Directors.

Adopted this 16th day of Feby 1888.
Signed G. W. Casting }
A. C. Powell. } Directors.

Seal of
Ashleys Patent
Machine made
Bottle Co Ltd

Ernest Lambert.
Secy.

In the prospectus for the new company, the promoters asserted that the Ashley machine could make bottles at one tenth of the labour cost in comparison to the hand method of working, and the profits on the capacity of the Castleford works alone were estimated to be at £106,000 per annum or more than 80% of the capital of the company. It was further estimated that at least £50,000 per annum would be received as royalties for patent licences, as manufacturers who had already seen the machines were most anxious to obtain licences immediately. On issuing the prospectus, the capital of the company was quickly over-subscribed and more than £6500 had to be returned to the applicants for the shares.

During the time that negotiations for the formation of the new company was preceding a party of financiers and other interested people, including representatives of the Press travelled from London to Castleford to see the demonstration on the 16th December, with the following extract appearing in the Leeds Mercury the following day, the 17th.

THE FINANCIAL TIMES, MONDAY, FEBRUARY 13, 1888.

The LISTS will OPEN on MONDAY, the 13th inst., and CLOSE on or before THURSDAY, the 16th, for Town and Country.

Entire Foreign Patent Rights for Ashley's Invention for Bottle-Making by Automatic Machinery. Bottles produced at one-tenth the present cost of labour.

Estimated Profits from Royalties alone over £100,000 per Annum, equal to 65 per cent. on the entire Capital of the Company now issued.

The Vendors have so much confidence in the future success of the Company that they are prepared to receive the whole of the purchase-money in Shares, or Shares and Cash, at the option of the Directors of the Company.

EUROPEAN & AMERICAN MACHINE MADE BOTTLE COMPANY,

LIMITED.

Incorporated Under the Companies Acts, 1862 to 1887.

CAPITAL £600,000, IN 120,000 SHARES of £5 each.

Issue of £600,000, in 120,000 Shares of £5 each. Payable 5s. per Share on Application, 15s. on Allotment, £2 on April 5, and the balance on May 15, 1888, a large amount of which has already been applied for.

DIRECTORS

GEO. WOODYATT HASTINGS, Esq., M.P. [illegible]

COLONEL CHAS. TREVILLE [illegible]

F. R. [illegible]

[illegible]

BANKERS

[illegible]

SOLICITORS.

[illegible]

BROKERS

Figure 4,10 Advert for Ashley shares

"Glass Bottles"

At the Ashley (Machine Made) Bottle Co's new premises at the Sykes, Macvay and the Codd Bottle Co. at Castleford.

"Another familiar landmark is going. The glass bottle trade is in process of being melted down into new 'parisons' without blowpipes and blowers, and instead of five men being necessary to evolve an imperial receptacle for beer or aërated water, it almost looks as if five innocently occupied adults might discover pastime in watching the conjoined labours of a machine and a youth in placing bottles at the service of good liquor as fast as they can be counted --- Never since the days of the Pharaohs has anything so clever in glass-making been devised, nor anything so simple.

People may well ask, "Why did not some one think of this expedient before?" but, carried to the test of a patent, nobody ever has. It has remained for a Yorkshire man, Mr H. M. Ashley, of Ferrybridge, to revolutionise the trade, and as legal assistant in this enterprise, he has secured patents in Germany, Austria, France, Belgium, the United States, Canada, and other countries likely to value cheap bottles ---.

Time was not lost in escorting the guests to the shed where a humble beginning had been made in showing how a mould could be rendered obediently automatic with the assistance of a man and a boy, and as a measure of precaution, another man in addition. The conditions were not as perfect as they obviously will be when the mechanism is in full realisation of the inventor's ideas; but the merest novice standing could see that the invention is equal to the task, and is a grand simplification of a much sub-divided and imperfect operation as carried out in its present form.

It may be that the circular of invitation indulges in "high falutin" when it speaks of the idea as the "most marvellous invention which has been brought out since the Bessemer steel process was introduced," but it is a veritable stride in the ideas of mechanism.

It appears that England has not been holding her ground in bottle-making. This is attributable solely to the higher wages paid for labour in England, --- and the new apparatus which is to turn everything in our favour may be welcomed as a patriotic benefactor. The advantage now possible is certainly startling. It is stated that a gross of bottles, costing in labour by the present methods 3s. 10d. can now be manufactured at a labour cost of 3d. per gross ---

Of all the tasks in giving a bottle shape none seem to have changed so little as that of blowing. The Egyptians blew, as we know from their pictures, so did the Greeks and Romans, and everyone down to this day, and the merits of Ashley's invention is that is blows pretty skilfully with automatic mechanism ---

The new bottle-making machine steals a march on the blower - in fact, it gets several leagues ahead. It makes a bottle all in one continuous run. As shown yesterday, a little had to be left to the imagination, but only a little.

The molten glass-supply is brought on the end of a rod to the mould --- and dropped into the neck mould first of all. The neck is made primarily in the new process --- last in the old. Immediately upon falling into the neck mould a punch an inch or more in diameter rises and pushes up through the metal, the bottle appearing to rise in its formation. Then the punch sinks, and as it does so, a second interior punch rises and liberates a ring of air under gentle pressure. This has the effect of inflating the bottle. The position is then reversed, so that the air presses downwards, and the envelope is both forced downward and falls naturally with gravitation. As soon as the droop is sufficient, the bottle is enclosed in a mould, full pressure of air is turned on and the bottle is completed.

The advantage of this process lies in the perfection, uniformity, and rapidity of manufacture. There is no waste metal, and no inequality in the thickness of the envelope. By mouth blowing it is impossible to render two bottles exactly alike, any more than two soap bubbles. By the new process, it is not difficult to weigh the molten metal to a gram, and to inflate it to the full mould in quick succession.

As the machine will be constructed, six, ten, or more moulds will be charged successively from a crucible of metal fresh from the furnace, and the air blowing will be automatic. A man between the furnace and the moulds will be sufficient, and two other hands will do the rest, including storage in the annealing chamber. A turntable will revolve the moulds for their supply of metal. In the centre of the table there will be a supply of air laid on, with branches to each of the bottle moulds.

Average work affords this calculation. A furnace hole occupying five persons, produces seven gross per day, or thirty-five gross per week. The holes cost a little over 3s. 6p. per gross for labour. By the new system a hole will produce eighty gross per day at a cost of 3d. per gross for labour. In the works of Sykes & Co., sixty holes yield 420 gross per day, but the machine production will yield 4800 gross per day, and that will give a saving of £780 on the estimated output. Besides the saving in labour, which may offset the present advantages of Belgium and Germany, there is an enormous saving in waste material. The present system wastes about 33 per cent of the melting; and although the material is used afterwards, there is a deterioration in quality from the use of glass a second time."

An interesting development came about when English gave his paper entitled,

"The Ashley Bottle Machine. A Historical Note"

to the Society in Leeds on the 20th June 1923. The Science Museum saw the article or attended the lecture and deemed it of sufficient importance to warrant acquiring one for display in the Museum. The original one would be their best option but they were not able to find it or any other one. So they set about making a replica. They would not have been able to succeed with out the help of Fred Sweeting, the then manager of Breffits.

Fred's father Christopher had worked at Breffits before him, so he would have known a great deal about the machines origins. Fred was able to supply drawings and even a photograph, together with first hand advice from some of the workers who had built and used the first machine.

The replica is in the Science Museum and Figure 4,11 shows a

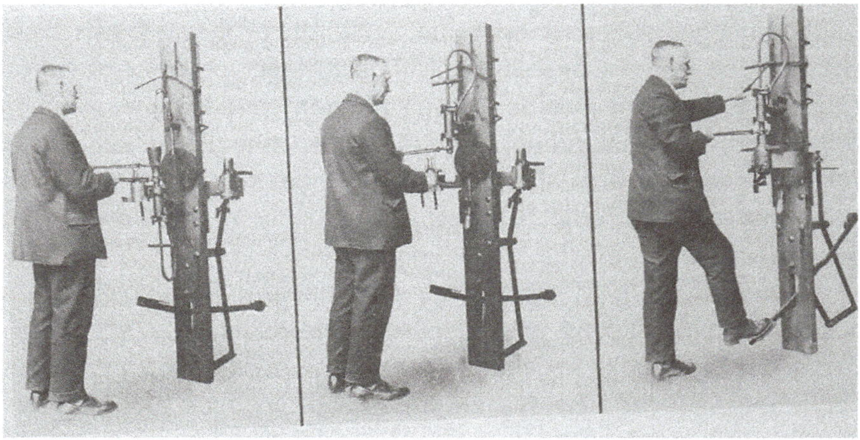

Figure 4,11. A member of the Science Museum demonstrating in 1929 how the 'Plank machine' worked'

member of the museum staff demonstrating it in 1929.

Ashley's next move was to try to remove the deficiencies of the single unit 'plank' and the 'pillar' machines. Here he tried to design a machine in which several bottles were being made at the same time, with the whole operation being worked by a gatherer and a blower.

Figure 4,12 shows Ashley's attempt at a 'rotary' machine as against the fixed vertically operated machines that were already in use. The machine was built in Sheffield by the Yorkshire Engine Company and its operation again described by English was as follows:

"The machine had four parison moulds, m_2, four corresponding ring moulds, and one blow mould, m_2. The blowing heads, R and D, were mounted on arms at the circumference of a table, E, and were capable of inversion about a shaft, a. The m ring moulds and parison moulds were hinged to the same arms that carried the blowing heads. Passing through each blow-head was a hollow plunger, P, which was normally held in the "out" position by a compressed spring. The mould table was rotated step by step at a pre-determined rate by

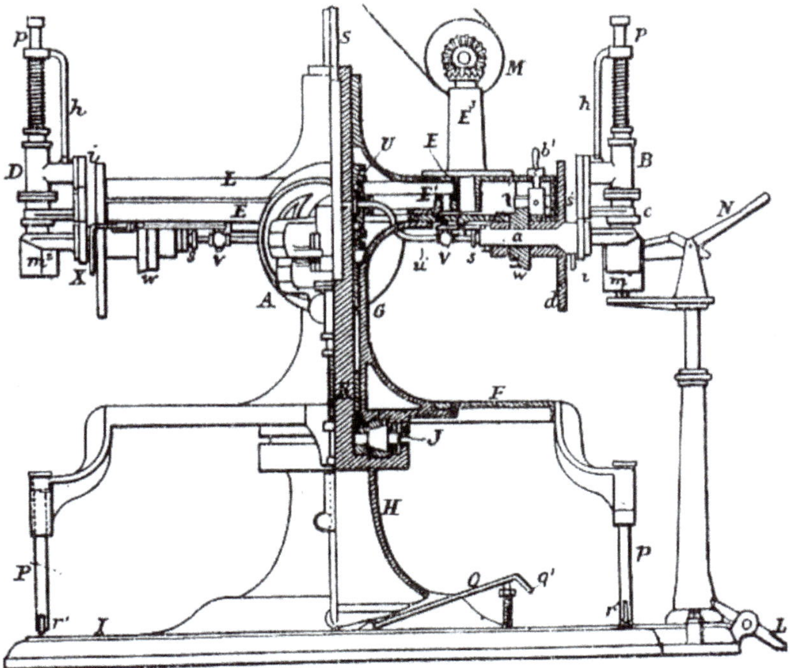

FIG. 3.

Figure 4,12. Ashley's last patent application for a rotary machine. Dated January 3rd 1890, Patent Number 129

means of a friction gearing, M.

In the first position, glass was filled into the inverted parison mould, and then the intermittent rotation carried the table round one step. In doing so, a subsidiary table or framework, F, rotating with the main table, caused a sliding push rod to run up the fixed inclined plate, Q, thus forcing the plunger, P, into the soft glass and forming the neck of the bottle.

At the beginning of the next step forward, the plunger was quickly withdrawn, the spring keeping the push rod roller on the sharp decline, q, and at the same time a puff of air was given to the blank. During the remainder of this step of the rotation, the parison mould and the blow-head complete were inverted mechanically by the co-operation of a fixed table, L, with a friction wheel, w, fixed on the axis, a, of the crank arm carrying the blowhead, etc. Towards the end of this inverting movement, the parison mould halves were opened wide by means of eccentric gears attached to the fixed disc, d, and the partly formed blank was left hanging freely from the ring mould. The blow mould, which was mounted on an entirely separate stand, was brought up and closed round the blank by means of levers, L', and N. Compressed air for blowing the bottle was brought from the main, S, to a central distribution box, and through the hollow shaft, a, to the blowhead and was controlled by the valve, V. After the completion of the blowing operation, the mould was opened by hand, and removed by releasing the foot pedal, leaving the finished bottle hanging from the ring mould.

The third step of the intermittent rotation caused slides to open the ring mould halves and allow the bottle to fall on to a suitable receptacle, after which the blowhead and its attached parts were again mechanically inverted and the parison mould closed ready to receive another charge of glass."

This machine was not a success, and unfortunately three other machines were ordered before the first one was tested, but it only required small design modifications to make a somewhat similar machine a success. Here Ashley was fortunate that a draughtsman from the Sheffield company was present at the works and he became more familiar with the process, so he along with Ashley designed the changes. The major modifications carried out were;

The four units were replaced by three.

The inversion and opening of the parison moulds was done by compressed air operating a piston as the speed of rotation meant the original method was too slow in its operation.

The mechanical intermittent rotation was also dispensed with and the operator had to rotate the machine by hand to suit his rate of working.

The single blow mould was retained it now was mounted on a horizontal shaft and rocked into position.

Later, in 1892, a sliding blow mould table was used. According to English, in 1892 Ashley had 10 of these three head machines at work, using two people to operate and capable of making 180 dozen bottles per day, but according to Turner, Dralle on his visit to Castleford to look at a Siemans furnace said that there where twenty two machines working, a fact which is also confirmed by Moody.

This new type of machine appeared about 1889 and it was called the 'Pillar and Table' machine at that time twenty two of the 'Plank' type machines, which had been made in the Ashley fitting shop, were already working. Is there some confusion here in the number of machines?

A possible explanation regarding the number of machines can easily be explained if Dralle visited the factory early in the year and further rotary machines were installed later on in the year after his visit. This is one of the problems that can arise when people only talk about the year and don't include the month when it is known.

The fitting shop also built the first 'Pillar' machine but about 1890 all of the machines were now being built by the Providence Iron Foundry in Castleford, the engineering works and iron foundry of Joshua Horne. He built for the Ashley (Machine Made) bottle company a total of thirty-one single mould machines, twenty-six in the completed state and five were delivered is parts, which were put together by the factorys workforce.

Ashley designed at least two more machines with the patent applications being made in 1890. One was for a rotary type machine, Figure 4,13 and the other Moody likened it to an IS machine, Figure 4,14. The rotary type Patent Number 129 may have been the redesigned one that succeeded the original unsuccessful rotary design.

According to Moody the actions of the machine were as follows;

> "The pneumatic cylinder, b, moved the whole blank frame sideways, so each set of three moulds in turn came above the blow mould, which is not shown here, The blank moulds were

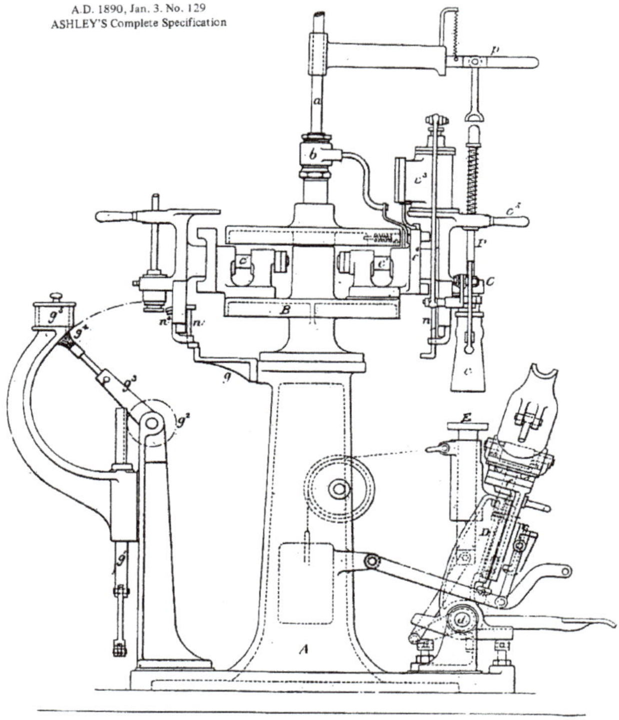

Figure 4,13. Ashleys first rotary machine.
March 1st, 1889 Patent Number 3673

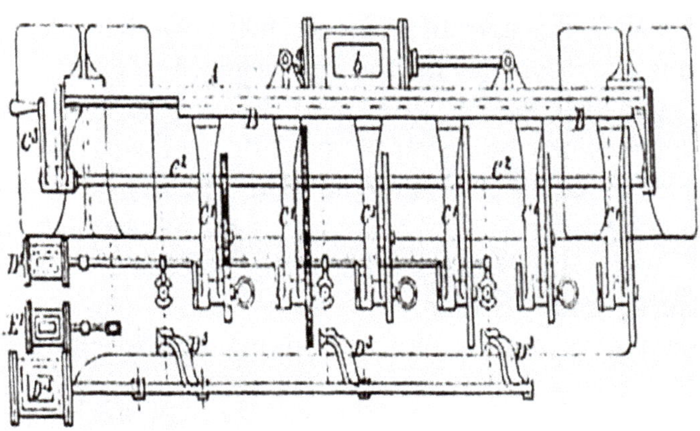

Figure 4,14. Ashley's 'in line' machine, showing the blanks only.
Patent application date 23rd July 1890 Patent Number 11 547

inverted by hand wheel, C1, geared to rotate the two sets in opposite directions. The blow moulds moved sideways on the rocking frame D3".

It could be said that Ashley was an exceptional engineer being that he filed nine patent applications in four years. A listing of them is shown in Table 4,3.

Patent Number	**Year**	**Date**	**What for**
8 677	1886	2nd July	First patent.
14 727	1886		Improvement to first patent. Possible patent of the Plank machine.
3 434	1887	1st March	Probable patent of the Plank machine.
7 566	1887		Ball stopper
18 661	1888		Improvements to machine.
3 686	1888 or 1889	10th November	Pillar machine.
3 673	1889	1st March	Rotary machine.
129	1890	3rd January	Rotary machine
11 547	1890	23rd July	Inline machine

N.B. Patent Number 8 677, 2nd July 1886 joint application with J. Arnall

Table 4,3. H. M. Ashley's Patents. 1886-1890

Despite the glowing prospects given in the prospectus when the Ashley (Machine Made) Bottle Co. Ltd. was formed the company soon hit hard times, to such an extent that in November of 1890 a committee was formed to report on its workings and its lack of success. Extracts from this report which was issued in January 1891 which clearly shows why the company did not make the anticipated profits as stated in the prospectus. For example, the committee states:

"Notwithstanding the sanguine statements in the prospectus, which were certainly calculated to lead subscribers to believe that the patent was then an ascertained success, it now appears that the at date of formation of the company it was really in a complete experimental stage, no complete machine having been made. ----

--- No licences for the use of the machine such as are referred to in the prospectus have been granted in connection with the patents, although several applications for licences have been received from other bottle makers."

In discussing the balance sheet presented in 1889, the report says:

"The cost of production has been excessive, and this fact your committee believes to be attributable to the lax management which has prevailed in carrying on the business of the company. The general laxity of management seems to have been increased by the division of the responsibility of the conduct of affairs between two managing directors and the mechanical engineer, who, your committee is led to understand, have not worked amicably together.

The looseness of management seems to have been particularly apparent in the machine department, which your committee may remark was only responsible for 13 per cent of the good bottles turned out during the year ending 31st July, 1890, as against 87 per cent, made by the hand department.

It would appear that early in 1888, soon after the formation of the company, the first four-mould machine was ordered from the Yorkshire Engine Company. Before this had been tested, three more were ordered from the same company. When these machines were delivered they were found to be unworkable, and now, the committee is informed, are worth perhaps £15 as old metal, although they cost nearly £1100.

Later on, a considerable number of single-mould machines were ordered from Mr. Horne, twenty-six in a completed state and five in parts, which were put together by the companies own mechanics. Of these machines, it would appear that eleven were sold to a foreign company, and of the remainder only four are now working, the others having, your committee is informed, been cast aside, although it is believed they might have been worked with advantage.

Your committee considers that the Ashley patent machines have never had a proper chance under the present management of developing their capabilities, and is surprised at the small progress which has been made since the formation of the company in the direction of such development ----.

It appears to your committee that if other makers can obtain profits, your company should be able to do so, while, with the machines under the control of a competent engineer, the profits would be considerably increased."

As a result of this report, a new board of directors was appointed, with the view of improving the management of the company, but still the venture was not a financial success, and at a meeting of shareholders held in September, 1894, for a voluntary wind up and reconstruction of the company. There had been a loss of £2500 on the year's working; of this £2200 was said to have been lost through a prolonged coal strike, and the remainder due to foreign competition. In his speech at this meeting, the chairman (Mr. A. Blackhouse), said;

"The bottle machine never had fair play, and he was afraid it never would so long as the works were carried on at Castleford, where the union was in the very strongest position, and no bottle hand who was a member of the union would either work himself, or allow his lads to assist in the working of any of the company's machines.

They were therefore dependent on non-union labour, and very strong hostility was shown towards their men. When the men did their duty to the machine it worked exceedingly well -----. The machine was the most valuable asset of the company, but the fact was that they had not sufficient capital to carry on their business ----. It was a pity that this patent had ever been linked with such miserable works as those at Castleford."

After reconstruction of the company, matters did not improve sufficient, so as the result of a order given in the High Court of Justice by Mr Justice Kekewich, on November 17th, 1894, the works of the company were sold by auction, the date of the sale being March 22nd, 1899.

According to English the bottle making machines were bought by Messrs. Bagley & Co. of Knottingley and Messrs. Cannington & Shaw of St. Helens. But according to Hodkin they were all purchased by Bagley & Co. of Knottingley who later sold half-shares in the machines to a Lancashire company.

The forgoing text has been a quite comprehensive record about this first venture by a set of individuals to form a company to make bottles by machine, and it highlights the many pitfalls that can befall such an enterprise. With hindsight they never had a chance, they must have been quite articulate giving many promises but failing to come up with the results.

The following statement comes to mind when thinking about the performance of the people as managers and directors of the company;

"They were long on talk and short on action"

The problems they had to overcome were;

- At the start of the venture there was not a tried and tested working machine available. This smells as a little bit of sharp practice.
- There were no lines of demarcation between the two managing directors and the engineering manager, i.e. the fitting shop foreman. It would seem, possibly, that the works had been divided into 'machine' and 'hand' sections with a managing director for each but nobody told them that they could not interfere with each others operation. Also the mechanical engineer was being treated as a person of equal standing.
- The companies engineering staff, i.e. the mechanical engineering manager, was not up to the task of keeping the machines in working order. Possibly he was not qualified enough to carry out his duties.
- The lack of planning, in so much, the ordering of three

additional machines before the first one was tested out in the field. This smacks of blind faith or pure arrogance on those who sanctioned the order.

- There were no means of keeping control of the operating costs of the machines, possibly the engineering staff did not think that they were responsible for output.

Then there were the outside influences;

- Castleford was the centre of the Glass Makers United Trade Protection Society the original trade union, and its members were only bottle makers. The protection society (the union) started in Castleford and later it spread out its influence throughout the land. The Castleford and District, and the National offices for all of the time that they existed were located at 2, Wesley Street in Castleford.
  Its objective was to keep its members in employment with a good re-numeration for the work being done, so it would have been violently against its members losing their jobs. With hindsight a place where the union were not as strong as they where in Castleford may have led to earlier acceptance of the machines.
- The National and Yorkshire disputes, local or county wide, with the union.
- Other none related disputes which affected the glass industry i.e. those between the miners and the colliery owners.

As an aside J. Lumb & Co. also worked in the town and when they started to manufacture bottles by machine and by 1904 they stopped making by hand. This stopped any internal friction between the machine and the hand makers and also any industrial 'blackmail' from the union. There was not a machine making union when they started their 100% all machine made philosophy, maybe they remembered what had happened to the Ashley (Machine Made) Bottle Co. about 15 years earlier.

The pressing process

The pressing of glass was the first process to be mechanised. In

1814 crude pressing into moulds in England and the Netherlands was practised, in 1820 the first blowing with compressed air was recorded at Baccarat in France, and in America in 1827 the first side-lever press was used.

It can be assumed that pressing glass in the beginning was done without a mould ring and it was in the course of its development that due to the variations in weight the mould was covered with a ring to force the glass to fill the whole cavity. Once the automatic feeder was invented, along with centrally loaded gobs that were homogeneous in temperature then the ring mould was not required. Added to this if the mould and plunger were uniformly heated then very good rings were obtained.

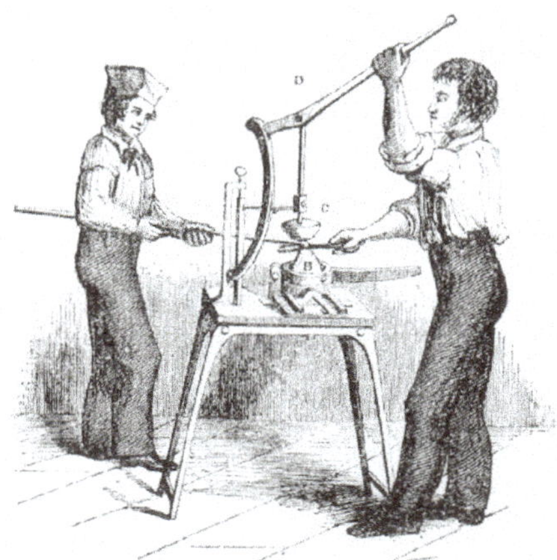

Figure 4,15. A pre-1850 press machine

Pressed glass became competition for the cut glass industry, and many of the items made were a 'poor mans' copy of what that industry produces. But with the advent of the 'pop up' mould bottom and eventually hinged moulds some of the limitations of the process were overcome and new markets developed.

For instance bottles, wide mouth or narrow neck, can be made on a press, two presses working side by side made are required, one to make the upper part of the bottle and the other to make the lower part. Then using the residual heat in the glass they were fused together, after annealing part of the solid neck is cut off and the hole then was opened up by grinding. In the light of the automation in the container industry which came about sixty years later this method would have been short lived.

The following, shown in Figure 4,15, is a simple press where the gather of glass is placed into the mould B and cut to the required weight. The plunger C is then lowered down onto the glass and pressure is exerted to the plunger pressing the glass into the required form.

As pressing of containers is non-existent nowadays except for the manufacture of lids for food bottling jars we will only touch on this process, and show only old single presses which in this case are of German origin.

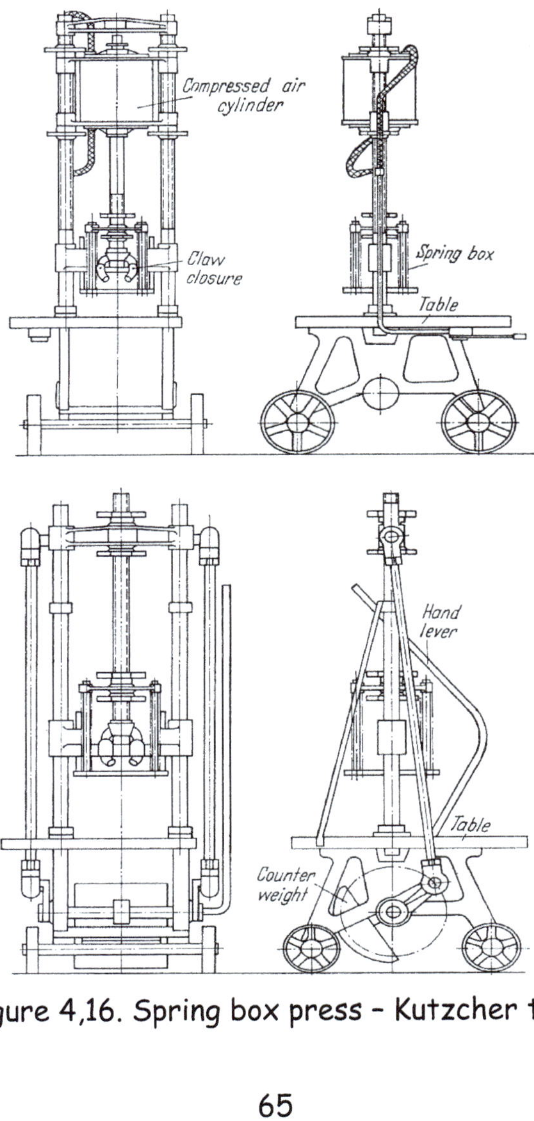

Figure 4,16. Spring box press - Kutzcher type

There were two types of press which originated very early on, they were the;

The spring press

Where the lower ring is fastened to the mould ring. The spring pressure can be regulated and causes the ring to sit tightly onto the mould. Muscle power has to overcome the pre-stress of the spring.

The eccentric press

If large pressures are required such as for large pressed ware the eccentric press is more suitable because the part of the press carrying the mould covering ring is locked immediately after contact with the mould.

In 1881 Fahdt of Dresden patented a spring press that had adjustable springs, claws for locking the plunger, and air cooling. A fully semi-automated press was manufactured in 1898 by W.J. Miller the United States of America. Please note that in America

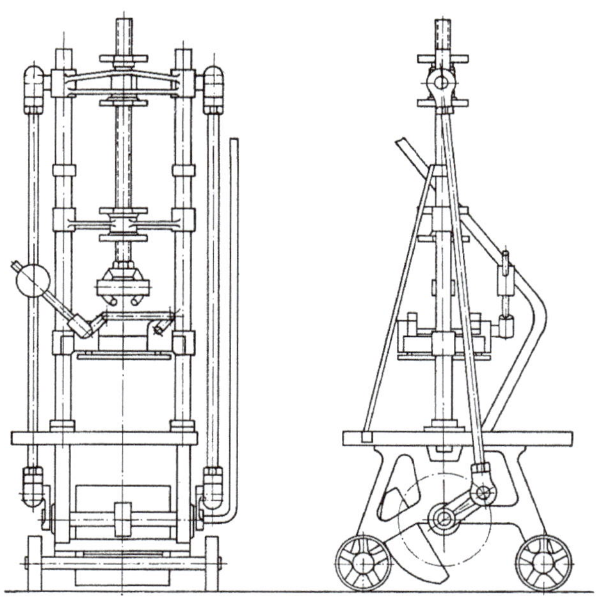

Figure 4,17. Eccentric press - Kutzscher type

they talk about the machine only when they make this statement. In Europe it has to include some automatic glass feeding device as well before it can be classed as an automatic machine.

In contemporary pressed glass manufacture manual and automatic presses have developed in to three groups;

1. The Spring cage press.
   Manual spring cage press,
   Air pressure spring cage press.

2. The Eccentric press.
   Manual Lever eccentric presses,
   Air pressure eccentric presses.

3. The Toggle press.

Meanwhile in America a patent was applied for in 1875 by James S. and Thomas B. Atterbury, of Pittsburg (Amer. Patent Number 139 993), whose objective was to make jugs with a press-on

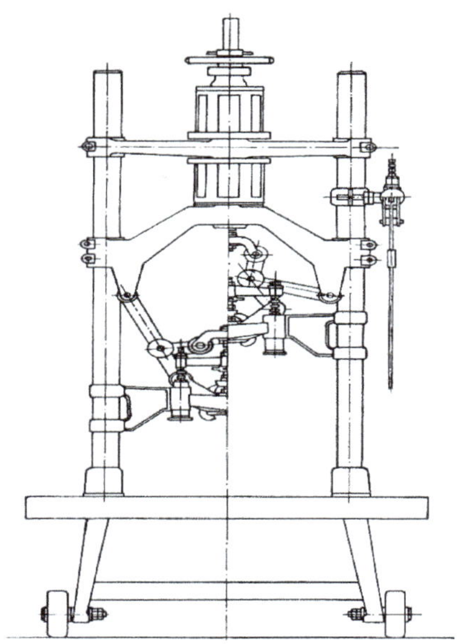

Figure 4,18. Toggle press - Putsch type

handle, there seems to be no positive outcome from this.

Where; a = the mould, c = the ring and d = the blowhead.

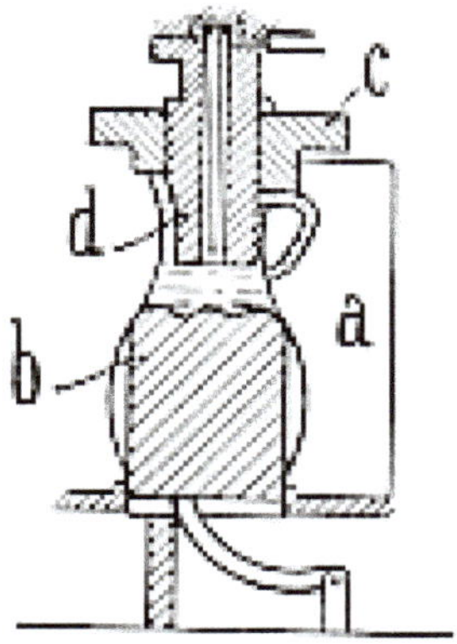

Figure 4,19.
Attenbury,s Patent

Atterburys idea was quite remarkable he tried to use part of the two part blow mould as a press mould. The bottom of the press mould which after prepressing is simply drawn down, remains the bottom in both stages; the prepressing and the blowing. The compressed air for the blowing was blown through the plunger after withdrawal of the bottom. It is assumed that the exit hole in the plunger was closed by a valve during pressing.

In 1876 A. R. Weber, also had a patent granted in Britain, it was similar to the machine of James Bowron but where the pressure on the glass was used to form a more perfect finish.

In Bowron's machine Bowron wanted to fill the glass into an upside down blow mould, next to reverse the mould and then blow. The counter weighted bottom was not used in the filling of the upside down mould, it was used to press the glass up into the neck.

Whereas Weber placed the upright mould into a one part lower mould and a two part upper mould the bottom of the lower mould was turnable. The glass was loaded into the lower mould and pressed by means of the bottom towards the neck, and finally blown while the bottom is withdrawn.

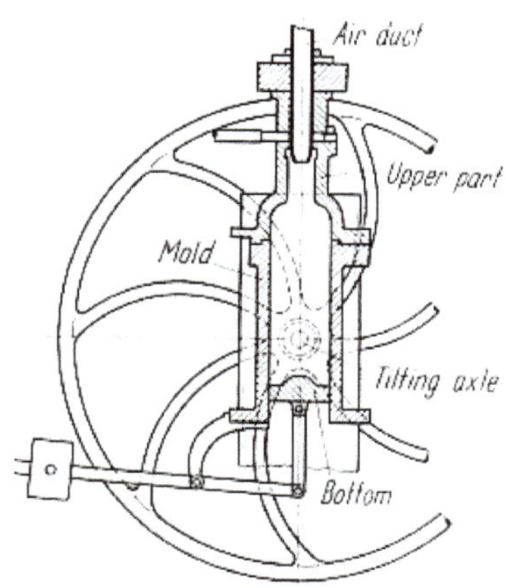

Figure 4,20. Bowron's Patent.
Brit, Pat, No. 354, 1861

In 1881 August, Philip Arbogast applied for a patent of his method of the press and blow process which was granted on the 11th June 1882 (Amer. Patent Number 260 819). The claim that it was the first machine ever built was never challenged because he himself had never built a machine that was successful.

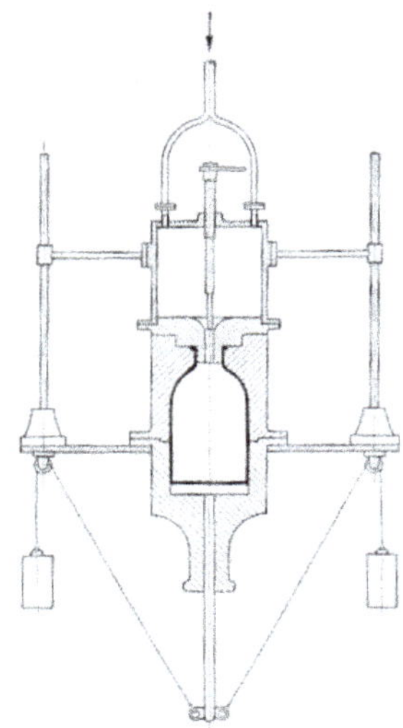

Figure 4,21. Weber's Patent

In his application he for his patent he claimed;

> "improvement in the manufacture of glassware, consisting in pressing the mouth or neck to finished form with a dependent mass of glass, then withdrawing the plunger, then removing the article from the press mould, and finally inserting it in a separate mould and blowing to form the body".

Here are disclosed for the first time the three essential elements, which would be incorporated in all future machines for making bottles mechanically;

The formation of the mouth of the neck - the so called 'finish' - is necessary the first process.

The formation of a parison is a requisite intermediate stage in the machine manufacture of a container.

The blowing operation completes the formation of the article.

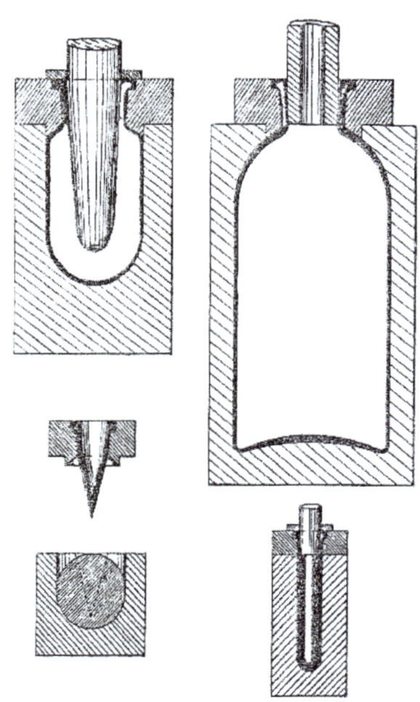

Figure 4,22. The Arbogast Process

Arbogast having failed to make a success of his patent sold it to the D. C. Ripley Company of Pittsburgh in 1885 for a pittance. Ripley built and used machines based on the Arbogast principle, but only for a short time, because his factory was union controlled and the union fixed small outputs and high rates of wages, thus making the use of the machines uneconomic.

In 1891 the Ripley Company was absorbed by the United States Glass Company, which acquired the Arbogast patent and proceeded to grant licences to non-union firms. In 1893 this led to the Enterprise Glass Company beginning to make vaseline jars on press and blow machines licensed under the Arbogast patent. The machines were also used by Messrs Taylor of Huntingdon, West Virginia, for the manufacture of jars.

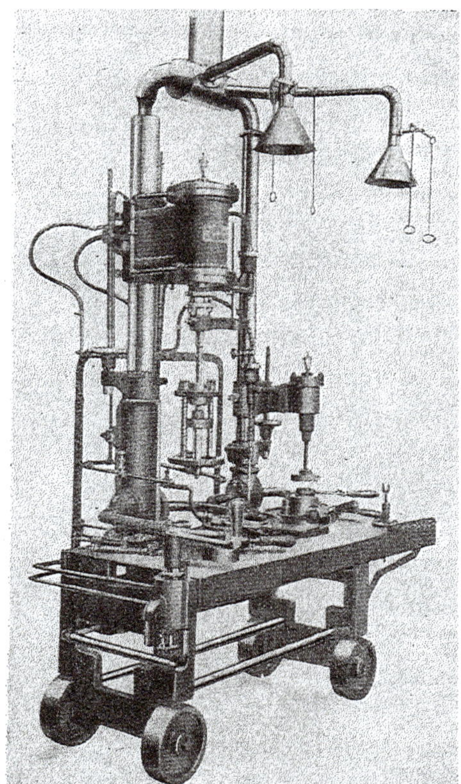

Figure 4,23. Arbogast press and blow machine

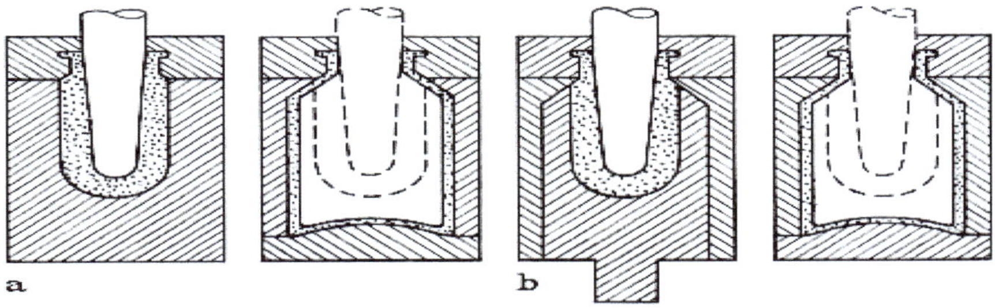

Figure 4,24. The classical press and blow methods Arbogast vs Windmill. a = Arbogast and b = Windmill. Notice Windmill's sliding blank mould

Figure 4,24 shows schematically the two press and blow processes side by side, the Arbogast process, on the left uses a conventional press blank/blow mould arrangement whereas in the Windmill process on the right the blank slides out of the mould and then a plate is put on the bottom of the blow mould.

Back in England Dan Rylands became the co-owner of the Hope Glass Works at Stairfoot in Barnsley, part of which he inherited from his father on his death in 1881. He sold his interests in his own glassworks, the Dove and Dearne in Wombwell and the Bleach Works at Stairfoot, Barnsley, to give him the finance to purchase the rest of his fathers half share. He soon severed his partnership with Hiram Codd which resulted in Codd forming an partnership with the Sykes and Macvay Bottle Co. of Castleford.

In March 1888 Dan Rylands and C. Emmett applied for a patent for an eight head press and blow rotary machine using Windmills sliding moulds (British Patent 3263, 3rd March). Rylands already had acquired a licence to use Windmill's patent, his idea was to make Codd bottles with it, but this was not successful until at least some forty years later by Redfearn Bros. of Barnsley.

Rylands was an innovative man having made some forty seven patents in twelve years, some of them had no practicable application, glass railway sleepers in 1887 (British Patent 4974, 4th April) and glass compressed air cylinders in 1890 (British Patent 17 117, 27th October). While others just did not work for example in 1887 (British Patent 17 543, 21st December) he and B. Stoner patented a variation to one of Ashley's unsuccessful applications, they had like Ashley no success.

He was also granted many for the hand making process of bottles, they were mainly in and around the neck of the Codd bottle which formed a major part of his business.

Rylands also was a man driven by the financial aspects of business and any way that he could make more money for himself he would do so, for example many of his patents that he was granted he had them licensed to his company for them to benefit in their operations, but also to himself financially.

His ventures finally brought him down and after a poor business decision regarding the old Oaks coal mine, which he had re-named the Rylands Main Colliery, he sank two new shafts but the output was not up to expectations. By 1893 his funds had run out and his creditors applied for his liquidation.

Outside of England there had been patents in Europe by Bauer, in Germany, in 1889, for a machine somewhat similar to that patented by Weber in America (Ger. Patent 45 912) and after 1891 a number of patents by Hilde of London who constructed a machine that was worked by Siemens in Dresden for quite some time.

His first patent was German, Patent Number 58 961, 1st January 1891, and it followed Ashley's principles, others were applied for up to 1906 amongst others was a British Patent in

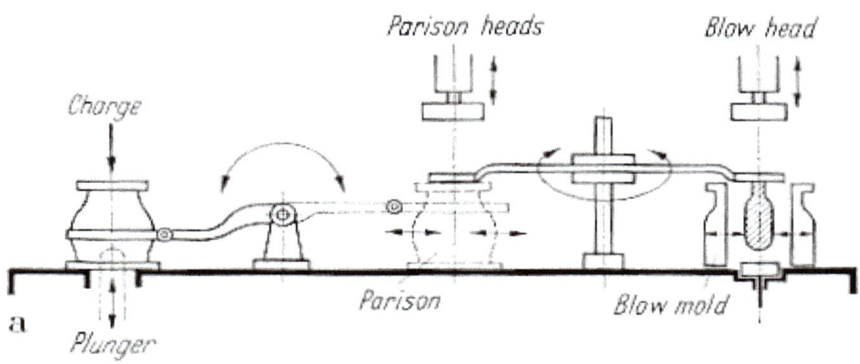

Figure 4,25. Schematic of Hilde's bottle making machine

1893. His machines were not very successful they wasted too much glass due to how the gob was sheared. He cut the glass coming out of the parison.

Other patents ascribed to him were an Austrian one Öst. Pat. No. 6954, and three German ones D.R. Pat. No's 136 530, 138 859, and 136 648. Later there were the Am. Pat. No. 784 744, and the Brit. Pat. No. 1 973 in 1902, and Öst. Pat. No. 26 327, D.R.Pat. No. 185 052 in 1906 and D.R. Pat. No. 256 816, 263 920 in 1911.

In 1893 C. B. Dobson had a British patent granted for modifications to an Ashley machine, there is no record of when or if it went into commercial production but it may have been used 'in-house'. The author assumes that this would have been at

Figure 4,26. Grotes machine patented in 1897. The machine was of limited commercial value

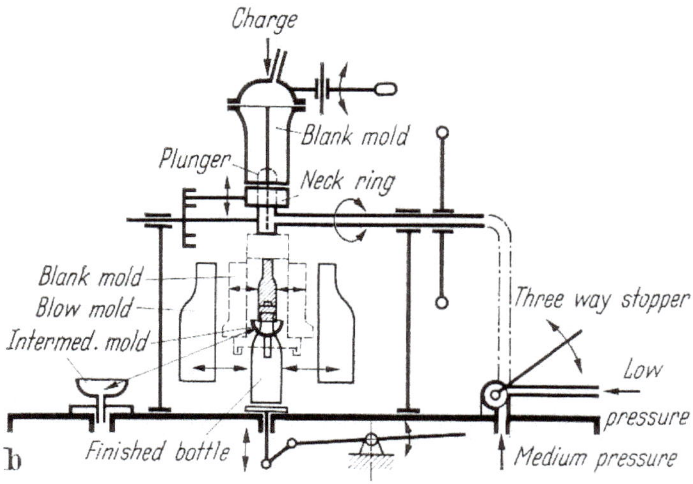

Figure 4,27. Schematic of Boucher's machine

the works of Dobson & Nall in Barnsley.

In 1897 Grote developed a machine at Wolfrathshausen in Bavaria and attempted to market it in London (Brit. Patent 12 392, 1897). Grote's constructions were all based on Ashley's differing only in how the application were activated and manipulated. The last machine to be built by him was noteworthy as the opening and closing of the moulds was done pneumatically.

In France Boucher in Cognac was granted a French Patent 262 149 on the 7th August 1896 and an addition to it on the 12th September 1897.

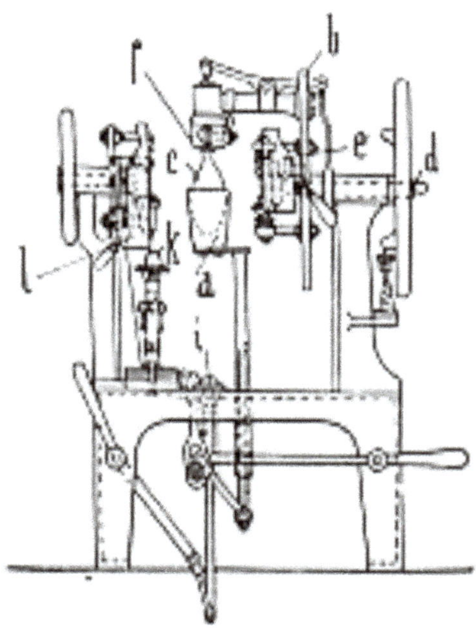

Figure 4,28. An early Boucher machine, possibly the original

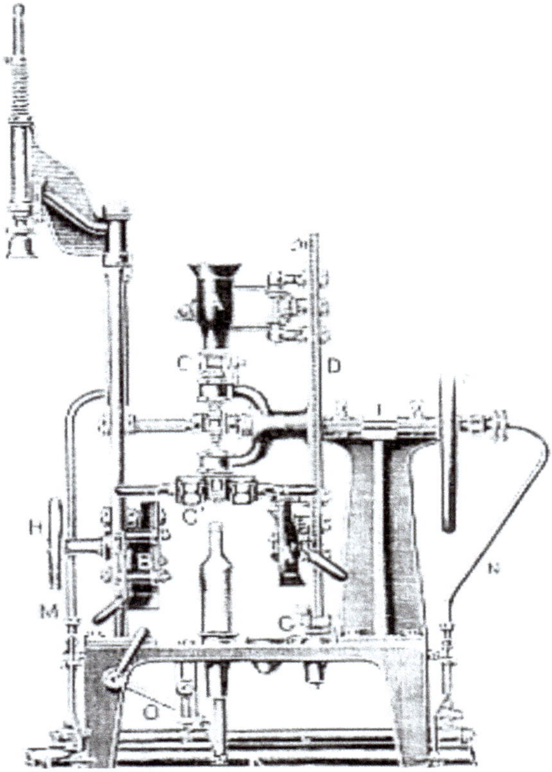

Figure 4,29. A later Boucher machine.
Patent applied for in 1915, Number 7969

Claude Boucher, Maitre-Verrier at Verriers de Cognac started his experiments in 1892, five or six years after Ashley and by 1897 he had a machine running successfully.

In 1898 when he was assured of success, he communicated the news to several of his engineering friends and acquaintances in Great Britain, France, Germany, Belgium etc. and invited them to see the machine in action. Later he developed a double headed mould machine. While these machines were a commercial success in France and other romance countries very little progress was made in Germany with them.

Figure 4,28 shows a slide of an early Boucher machine while Figure 4,29 shows a much later one.

Figure 4,30. A shop working Foster-Boucher Machines

The quoted output of one of these double-headed machines was 120 bottles of 700 gram weight per hour.

John Forster an engineer from St. Helens, in Lancashire attended the demonstration and was so impressed that he became determined to obtain the licence to make these machines. He purchased the patent rights in 1899 but found it difficult to persuade British manufacturers to buy it. In 1900 he decided to enter into glass bottle manufacture and built himself a small furnace melting green glass which he equipped with Boucher machines. He was then able to prove that the machines would actually make bottles which confounded his critics.

By 1901 Forster had made improvements to the machine, patented them, and started to sell the machines as Forster or Forster/Boucher machines. By 1907 he had seven of the improved machines working in St. Helens. One improvement was a patent that enabled the machine to become the first machine to make internal screw finishes.

He sold many of his machines and in 1908 George Younger & Sons Glass Works in Alloa, Scotland was completely equipped with Foster machines. Other companies in Great Britain to use them where Bagley & Co., of Knottingley, the Kinghorn Bottle Co., of Kinghorn, Fife, Brooks of Leeds, Walkers of Irvine and the United Kingdom Glass Co. of London.

They were also used abroad for example an order for twenty four machines was placed with then in 1915 to be delivered to South Africa. Forsters designed and built glass factories in China and South Africa and had received inquiries from Argentina, Australia, Canada, India, Italy and Turkey, so it would be safe to assume that any other orders that were placed with them would include their machines.

Figure 4,31. Charles Blue machine. Ca. 1900. This is a photograph of an improved 'five mould' Blue machine

Back in America in 1896 Charles E. Blue of the Wheeling Mould and Foundry Company, built a press and blow machine for the Atlas Glass Company - later to become Hazel-Atlas Glass Company (Amer. Patents 607 947, 607 948, and 607 949, Brit. Patent 1897, Ger. Pat. 90 216, and 102 845)

There are two tales regarding what happened next:

1. Blue had obtained a licence to use the Windmill process, but it is not clear who granted him the licence - James Windmill, Dan Rylands, or the Rylands Company.
2. Or following litigation between the United States Glass Company and the Atlas Glass Company, Blue was forced to take a licence under the Arbogast patent. In addition he found himself

to be in conflict with two English patentees Windmill and Rylands who had covered the telescopic blank mould.

The Blue machine unlike the Arbogast one did not require the transference of the parison into another blow mould. Blue came to a satisfactory arrangement with the two English patentees and his machines became widely used in the early part of the twentieth century, for fruit jars and other wide mouth containers. At least one machine was used in England at the Kilner factory installed in 1900.

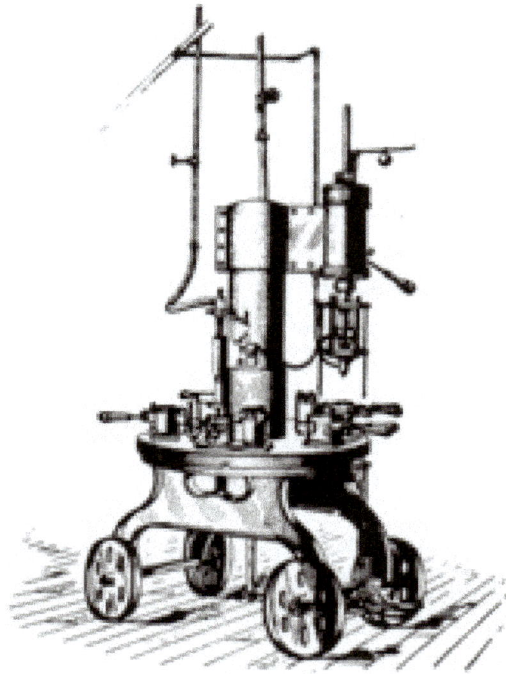

Figure 4,32. F.C. Ball's semi-automatic machine

It is most likely that the second tale is the most accurate one.

Also in America in 1897 (Patented in 1898) F.C. Ball invented a semi-automatic machine which was claimed to be the worlds first, see Figure 4,32. Another machine was invented by the same company in 1900 - the Ball-Bingham machine and as before this machine was claimed to be the first automatic machine, see Figure 4,33. These were for the manufacture of wide-mouth jars.

These are typical American claims as they had no idea as to what was happening in the rest of the world, but even in America in the year before (1896) Charles Blue had already built a machine for Atlas Glass Company and as far as the second machine goes how can it be automatic when it is obviously fed by hand gathered glass.

Figure 4,33. The Ball-Bingham machine

In 1897 William Haley an immigrant from America built Jacksons of Knottingley a similar press and blow machine. He later went to work for Turners of Saville Town, Dewsbury where in 1899 he built them a machine. Once again there are two schools of thought regarding which of these companys Haley worked for first.

Two machines where tried out in France, both without success, they were the Vernay (Brit. Patent 1893) and the Maussier machines. Both were tried in Lyons in 1894 and the Vernay was also tried out the following year at Folembray.

Table 4,4 shows a list of the main patents taken out in Europe

Development of Semi-Automatic machines in Europe 1900-1923

England				Europe				
Inventors	Type of Machine	Date of Patent	Date of Commercial Use	Inventors	Type of Machine	Country of Origin	Date of Patent	Date of Commercial Use
Horne, J.	Bl & Bl	1901	1901	Heerrt, H. W.	Bl & Bl	Austria	1900	?
Forster, J.	Bl & Bl	---	1901	Severin, H.	Bl & Bl	Germany	1900	---
Clegg, A. E.	Bl & Bl	1902	1902	Schiller, A.	Pr & Bl	Germany	1905	1906
Simpson/ Bradshaw	Bl & Bl	---	1905	Schiller, A.	Suc & Bl (b)	Germany	1906	1907
Tourres, A.	Bl & Bl	1908	1908	Barrez, C.	Bl & Bl (a)	France	1907	1908
Harisson	Bl & Bl	1909	1909	Wolf, J.	---	Germany	---	1911
Knowles/ Inman	Bl & Bl	1909	1909	Kutscher, F. W.		Germany	---	1913-1914
Sweeting/ Hardman	Bl & Bl	1923	1923					

Notes: (a) British Patent. (b) Suction for formation of the "finish"; basically a Blow and Blow machine.

Table 4,4. European patents 1900-1923

from 1900 to 1923, which can be assumed to be the end of the semi-automatic machines developments. It was not the end of the semi automatic machines however as they went on being used right up to the 1980s.

In England in 1901 two patents were taken out, one by Forster who we have already mentioned and the other by Joshua Horne of Castleford. Earlier we said that around 1890 Horne started to make Ashley machines for the Ashley (Machine Made) Bottle Company until they ceased to trade in 1894.

He continued to manufacture the machines for Breffit's at the Aire and Calder Glass Works in Castleford, and modified the Ashley machine sufficiently to call it the Breffit-Horne. According to Moody he took out a patent on the 26th October 1897 (British Patent 24 786) so this patent was probably was for that machine. In 1904 he sold a number of three station machines to Kilner Bros. of Conisborough.

In 1901 according to Turner he had made sufficient changes to the design that he took out another patent this time for his own machine and then went on to sell then world- wide. The machines were quite common in Yorkshire and Lancashire in England. They were operated at Castleford, Knottingley, Thornhill Lees, Barnsley, St. Helens, Manchester, and Newport (Monmouthshire), while machines were also supplied to France (Douai), Germany (Gerresheim), and the USA (Philadelphia). All told some 200 of these machines were absorbed by the glass industry between 1901 and about 1917 where difficulties arising from the war put an end to the firm's operations.

Joshua Horne and his partner Joseph Brewerton were first listed in the 1857 Trade Directory as co-owners of the Providence Foundry in Aire Street, Castleford. Their trades had been listed in the various Trade Directories as;

- 1857 iron founders and engineers,
- 1861 iron founders and millwrights,
- 1862 millwrights and engineers,
- 1866 engineers, millwrights, machinists and brass and iron founders,

- 1867 engineers and iron founders.
- 1875 engineers.

Somewhere between 1875 and 1877 the business became listed under the name of Joshua Horne. Notice the change of emphasis now to include glass.

- 1877 engineers and iron founders
- 1888 engineers, iron founder, maker of stopper presses and plates for patent stoppers for glass bottles, & manufacturer of air compressors for testing and filling bottles,
- 1892 glass engineer and iron founder,
- 1904 glass engineer and iron founder,
- 1908 engineer and iron founder.

There are no entries in the next directory of 1922 but as various other authors have indicated they stopped selling the machine around 1917, maybe Joshua Horne had either died or retired, he must have been at least eighty years old in 1917.

Again in 1901 another Castleford firm John Lumb & Co. Ltd. placed two patents Brit. Patent Numbers 21 158 and 21 521, for a glass bottle making machine and then the following year (1902) placed Patent Numbers 4 145, 26 262 and 26 333 which was an amendment. The machine in question was the Simpson and Bradshaw.

Colonel Thomas William Simpson returned from fighting in the South African War (1899-1902), 'Willie' as he was known in the factory worked with William Bradshaw to develop their machine which used some of Grote's ideas. The moulds which were attached to a large vertical mounting plate moved vertically as against the horizontal movement of the previous machines. Other ideas included the use of 'fined moulds' for cooling.

By 1904 Lumb's was in to bottle making with machines in such a way that the management declared that "in future all bottles would be machine made". Commencing with eight machines it had risen to 23 machines in 1907 and by 1914 the total had grown to 38 machines of different varieties. In 1910 they applied for another patent, Number 14 003 which the author surmises was for the "Flyer" which was probably the first machine to carry the finished bottle passed the gathering point.

In 1923 a patent was taken out regarding the Simpson & Bradshaw machine.

In 1902 A.E. Clegg of Leeds applied for a patent for a blow and blow machine it was operational in the same year but no further details are to hand. See the Abstracts for other Clegg patents.

Meanwhile back in Europe H.W. Heerdt of Austria patented an Austrian one, (Öst. Pat. 818), a German one (Ger. Pat. 105 842 on the 22nd of May 1898) and an English one (Brit. Pat. 2790 in 1900) there are no records as to its success.

In 1902 Severin obtained German patents D.R.Pat. numbers 127 298, 136 895, and 156 964 then in 1903 he was granted patents D.R. Pat. 161 557 on the 5th June 1903 and then Brit. Pat. 9 489.

The main difference in his machine was that he opened the parison mould as quickly as possible, left it in place and then simply reversed the blank. This reduced the gobs large temperature inhomogeniety, and some of the cooling devices used by his predecessors could be eliminated. The glass could be worked cooler and the output was higher. A typical yield of 900-1150 bottles of 450 grams weight could be produced in 8 hours.

Figure 4,34 shows that the Severin machine had two moulds

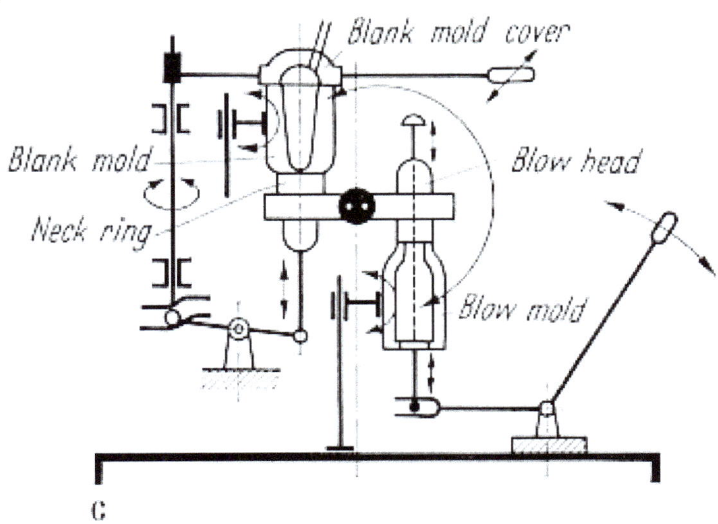

Figure 4,34. Schematic of Severin's machine

one is in the filling position (the gob loading position) while the other has already turned the blank to the blow mould.

The double headed mould was first used by Andere Leistner who was awarded the following German patents, number 151 628 on the 20th October 1901 for it, and later numbers 183 524 and 183 525 nothing much is known about these later machines. It would seem that Severin took Leistner's idea, improved it and then incorporated it into his machine.

By 1909 Severin had machines working at Messrs Heye at Nienburg and at von Bochringer at Achern. Outside of Germany there were machines in Bohemia, France and Italy, i.e. at Hostomitz near Teplitz, St. Yorre, Marseille, Gouhenans, Bordeaux, Sao Paulo and Sansurinowka.

Figure 4,35 shows a drawing of a machine as used for patent applications while Figure 4,36 is a photograph of an actual machine.

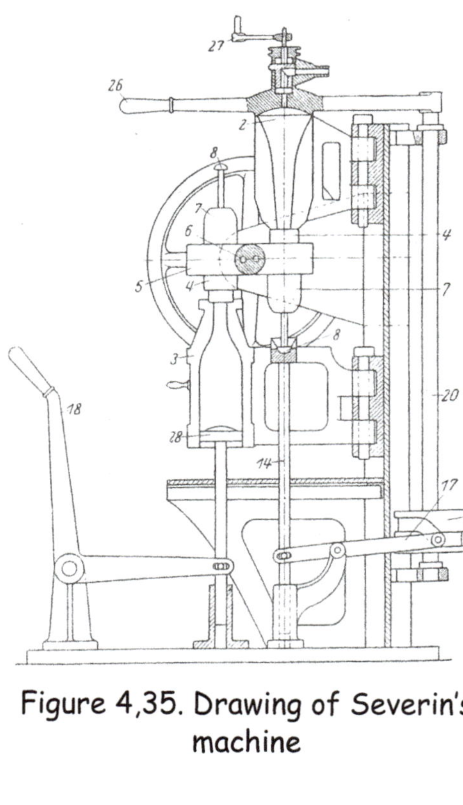

Figure 4,35. Drawing of Severin's machine

Another simple machine was that developed by A. Tourres of Graville-St. Honorine in France, its operation was similar to the Boucher and Severin machines. It turned the blank like the Severin machine but it only had one neck mould. Productivity claims as 180 bottles of 450 gram weight per hour.

All the machines listed so far inverted the blank, but in a press and blow machine this invertion was not required. First Proeger in America (Am Pat. 774 009, January

Figure 4,36. Photograph of Severin's machine

1902), and then Schiller (Ger. Pat. 157 520, 1st October 1903), who patented a simplified and improved version of Proeger's machine, where the parison mould and the blow mould were placed side by side on a table.

This arrangement enabled the 'presser' and the 'blower' to stand side by side at the machine which was advantageous to the machines operation. While the presser blows the parison in the parison mould the blower can blow the bottle in the blow mould.

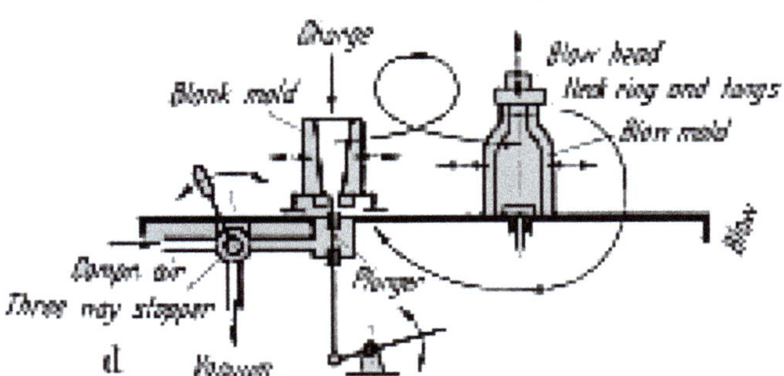

Figure 4,37. Schematic of Schiller's machine

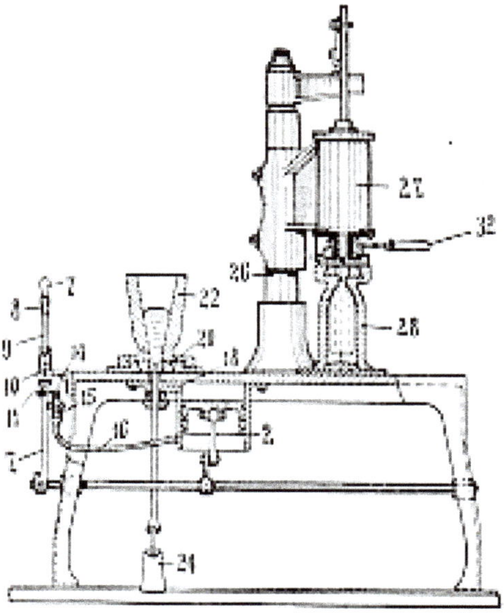

Figure 4,38 Schillers original machine

Figure 4,39. One of Schiller's original press and blow machines

Figure 4,40. Another of Schiller's press and blow machines. This is a later model

Figure 4,41. Schiller's blow and blow machine. First machine worked in 1907

The transfer from parison mould to blow mould is done manually.

In all previous of the machines the glass was allowed to flow into the parison or it was pushed down into it by covering the blank end with a lid and blowing compressed air through it. Schiller however sucked the fluid glass down into the neck mould by vacuum, probably the first time that this was done.

The first machine due to Schiller was constructed around 1905 and it went into operation in 1906 at Glashütte Westerhüsen (A. Grafe Nachfolger), near Magdeburg. His blow and blow machine (or suction and blow machine as some would call it) was constructed the following year and was first introduced at Sörnewitzer Glashütten A.G., Sörnewitzn near Meissen in 1907.

Another important customer was A.G. für Glas-Industrie, formerly known as Freidrich Siemens of Dresden who were supplied with 44 suction and blow machines. Over the years a number of modifications and improvements were made to the machines making them one of the favourite semi-automatic machines that were around.

Schiller made many of these machines some of which he used two blow moulds see Figure 4,42. There were two types and in BI he used the so called press on method. Glass filling the parison

Figure 4,42. Schiller's two mould machine

mould is not only vacuum sucked to the mouth part but it is also pushed up by a press plunger. A gatherer can normally gather for two machines.

Outputs for the two types are quoted as;

BII 160-180 average bottles per hour.

BI 150-170 average bottles or 120-140 one litre jars per hour.

During the war years 1914-1918 Schiller's machines were made in the United Kingdom and sold under the name of the 'Harlington' machine by Messrs. C. Melin and Co., London, which is shown in Figure 4,43. Whether or not any other machines where made outside of Germany it is not known.

This machine was suitable for the production of narrow neck ware. It consists of a table, on the left hand side was the parison mould, while on the right there was two finishing moulds carried

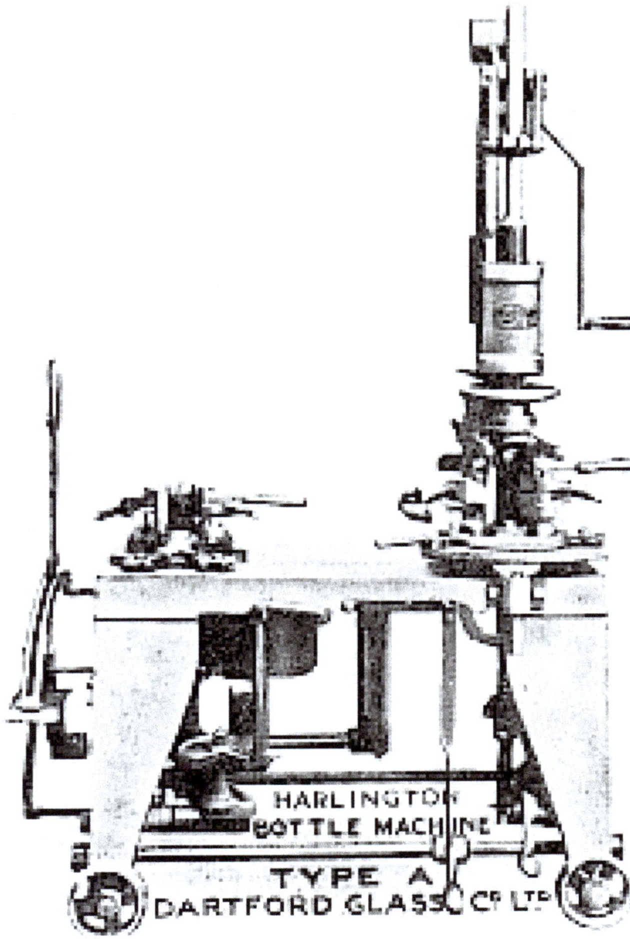

Figure 4,43. The Harlington machine

on a revolving platform supported by a central column. Below the table, near to the parison mould, is an air cylinder operated by a hand lever, which in one motion causes suction in the parison mould, but on reversal gives air pressure at the same point.

When the glass is delivered, the plunger is in the 'up' position, that is, projecting into the neck of the mould, but is so connected that at the end of the forward motion of the lever it is automatically withdrawn. The machine operates as follows; a supply of glass is delivered into the parison mould, the requisite quantity being cut off with the shears. By gravity, the molten metal runs downwards into the space between the neck ring and the plunger. It is assisted by suction, produced by pulling forward the lever. At the end of its stroke the lever releases the vacuum and withdraws the plunger, when a reverse movement of the lever admits air under pressure into the opening left by the plunger and blows up the parison to the shape of the parison mould.

The catch holding together the hinged halves of the mould is unfastened, the mould opened, and the parison lifted by means of the ring-mould and placed in an open blow mould, which is closed and locked. When the blow-head is raised the second finishing mould is rotated into position, the first opened, and the bottle removed.

In 1920 Schiller equipped some of his machines for small bottles with double or triple cavity moulds. A gatherer who had to work fast and skilful could service the two or three moulded machine. Schiller gave outputs of 6000 pieces per 8 hour shift for smaller bottles working triple.

	No. of Machines
Czechoslovalia	40
Roumania, Bulgaria, Servia, etc.	45
Italy	35
Spain	50
Portugal	15
Switzerland	35
France and Belgium	120
Holland	35
Poland	45
Russia	65
Turkey and Greece	30
Scandinavia	40
Great Britain	75
Germany	223

Table 4,5. List of Schiller semi-automatic machines in Europe to the end of 1932. Both press and blow and blow and blow

Schiller of the Glas-machinen-Industrie G.m.b.h. of Berlin had amongst his German patents numbers 157 530, 187 599, 210 027 and 277 873. Turner said that in 1932 there where 1150 Schiller machines of different types working throughout the world. The European countries and the number of machines working in each country is listed in Table 4,5.

In the United States no machines for 'narrow neck' bottle manufacture were made until 1908 when copies of an English designed machine were manufactured and licensed under British Patents. One widely used bottle machine was known variously as 'Johnny Bull', 'English' or 'United' and in 1909 there were 14 'United' machines in operation which by 1915 had risen to 114 out of a total of 265 'narrow neck' and 193 'wide mouth' machines.

The following extract from the 1944 Journal of the Society of Glass Technology states;

511 The Johnny Bull Semi-Automatic Forming Machine. - A Trail Blazer in Bottle-Making Machinery. (Anon. Glass Industry Feb. 1944, 25, 63-65, 78).

Deals with the early semi-automatic machines in the glass container. Among the comments are, "The Ashley machine is said to have met with some success in England for this purpose," and, of the Simpson (Simpson-Bradshaw ?) machine, "It seems, however, that this equipment was used only in the Simpson glass plant and its efficiency was somewhat questionable,"

The Pierpoint machine and "Johnny Bull" - from its English origin - was a 3-man machine, namely two operators and a gatherer, patented in Britain in 1902 and in the USA later (U.S.P. 795 304) It was exploited in the States by the United Bottle Machinery Co., who acquired the patents of F.H. Pierpoint and also of W.J. Demming from the P.B.M. Syndicate. On account of difficulties with the imported machines, some fifty "Uniteds" were built by the Fuches and Lang Manufacturing Co. of Rutherford, N.J., by January 1909, and soon fourteen of these were operating, licensed for seven years at a yearly rental of $500 each.

By the end of 1909 some 34 machines were in action, the

operators, when experienced, producing as many as 245 dozen pint beers per day. In 1911 there were 71 machines working, but it was obvious that they could not compete with the Owens fully automatic machine, and during 1912 only one more machine was installed.

The company designed a United Automatic Machine, mounting units on a circular rotating turret, but this suffered a number of setbacks, and although one was working at Swedesboro, N.J., it was not successful enough to be demonstrated. The development came to grief in 1918 on the problem of feeding glass to the moulds, after $212 429 had been spent on it.

One customer of these American developed Ashley type machines was J. B. Graham of Indiana and he further developed it into a large rotary machine that looked like an Owen's machine. His machine impressed the Owen's Machine Company and they bought him out and by 1925 turned it into probably the first continuous feeder fed rotating machine. Its life was short lived as it could not compete with the IS machine.

Meanwhile back in England the records of the Yorkshire Glass Bottle Makers United Protection Society began by this time to show some anxiety to the growth of glass bottle making machines. Rule 47, of date 1906, of the Society was as follows:-
"Bottle Making Machines"

Bottle making machines shall not be put into the bottle house amongst the bottle hands, and members shall not be displaced for the machines. If they be displaced they shall be put on Donation allowance as per scale.

Any member accepting a situation at machines or at a machine factory shall declare on 32nd Rule.

The inevitability of the machine was eventually recognised, but it was not until January 1931 that a Machine Section of the Glass Bottle Makers was established.

Glass Bottle Makers Society continued to issue its Quarterly Reports and one of those from 1907 states the names of the works where the Society members were employed which actually had machines in operation. The list, shown in Table 4,6, is not

Firm	Centre	No. of Machines	Amended Number
Yorkshire			
Breffit & Co.	Castleford	5	
John Lumb & Co.	Castleford	23	
Kilner Bros.	Thornhill Lees	?	+1
Turner & Co.	Savile Town	?	+2
Bagley & Co.	Knottingley	13	
Kilner Bros.	Conisborough	?	
Rylands Glass and Engineering Co.	Stairfoot, Barnsley	20	
Lancashire			
Cannington Shaw	St. Helens	7	
Forster & Co.	St. Helens	7	
Nuttal-Dixon & Co.	St. Helens	?	
Wales			
South Wales Glass Bottle Works	Newport, Monmouthshire	?	
Scotland			
Alloa Glass B. Works	Alloa	?	
Brown & Monise	Tollercross, Glasgow	?	

Table 4,6 List of firms operating bottle machines in 1907. Partially complete only. Report based on the 1907 Glass Bottle Makers Society Quarterly Returns

exhaustive and includes factories which have tried out machines but abandoned them after experimentation.

Elsewere in England other factory located engineers continued to design or install glass bottle machines. The first one is the Horley machine which was thought to have come from the Manchester area and was installed at T. Turner & Co., Saville Town, Dewsbury. In 1907 a party of American businesmen visited the factory and saw this machine in operation, as a consequence of this visit Mr. Roland Hirst went to America in 1908 to install the machine where he stayed from 1908 to 1916.

Prior to this he had served his apprenticeship in the mould shop of John Kilner of Wakefield. When he became twenty years old he went to the glass works of T. Turner & Co. in Saville Town, Dewsbury where he stayed for the next five years, recollecting during that period that all of the bottles made there were made by hand.

His next period of employment was at John Lumb & Co., Castleford where they were installing their Simpson and Bradshaw machines. He stayed there for two years after which when he returned to work at Turner's. This was about the time

Figure 4,44. The Horley machine. Installed at T. Turners & Co., Saville Town, Dewsbury, Yorkshire

when the Horley machine was being installed.

After taking the machine to America the Horley became the first narrow neck semi-automatic machine to be seen in that country. This statement may or may not be true as other people state that machines such as the John Bull or English machine where the first.

The first machine was installed at the Glenshaw Glass Factory at Wetmar, Pittsburg, Pennsylvania. Later, the patents were purchased by the United Bottle Machine Co. of Chamber St., New York who afterward removed to West Broadway in the same city; but the manufactury of the machine was undertaken by Fuchs and Lang in Rutherford, New Jersey.

While he lived in America, Hirst was employed by the United Bottle Machine Co. were he installed their machines at factories such as Charleroi, in Pennsylvania, Geneva in New York, Greenfield in Indiana, and also at Swedesborough, Terre Haute, Kansas City and many other places.

As it can be seen from Figure 4,44 that the Horley machine consisted of two independent units arranged back to back. The blow moulds were carried on the large cast iron brackets shown at the bottom of the picture, which swung away to allow the parison to be inverted. The parison mould, the neck ring and plug were carried on a bracket which was pivoted on a short spindle in the main column.

After the glass had been gathered and placed in the parison mould, the blown-down heads, shown on the long arms at 45 degrees at the top of the picture, were brought into position and the glass blown down into the ring. The blow-down heads were removed, the plug which was already in position was withdrawn and the parison lightly puffed.

The parison assembly was then inverted and the parison mould opened well back (the ends of the handles can be seen in the photograph). The blow mould was now brought into position and the bottle blown whilst still supported by the neck ring. Finally the neck ring was opened, allowing the bottle to be taken out when the blow mould was opened.

It was claimed that a gatherer, two blowers, a 'taker off' boy and a 'taker in' boy could produce 20–22 gross of half pint bottles per shift.

The next machine is the Knowles–Inman machine which was patented by Kilner Bros. and worked at both of their factories. Here I will include the next two machines because they were built by Messrs N. S. Kilner and Co. (Engineers), Ltd. of Leeds and Keighley (or Shipley) the glassworks contractor and I do not know if they are a related company. They may be a separate company with the same surname but it seems strange that a glass manufacturer is willing to sell his 'secrets' to his opposition. The source of the information is pre-1924 but the illustrations indicate that they most probably are after 1910, unfortunately there is no patent numbers. N. S. Kilner & Co. also marketed and sold the Simpson-Bradshaw machine.

The first one is the Kilner Machine No. 1, which you can see in

Figure 4,45. Messrs. N.S. Kilner's Number 1 machine

Figure 4,45, is a press and blow machine. The machine is suitable for making wide mouth bottles or jars. The parison mould is held in a slide on the left-hand side of the cast iron table. The cast iron mould is of less diameter than the finished mould but is so shaped that the glass moulded in it, when later blown out, readily conforms to the shape of the blow mould. For shaping the neck of the jar a neck-ring mould is used, comprising of two independent semicircular halves, hinged together and provided with a handle, and also a locking catch.

Behind the parison mould is a column supporting a plunger which is actuated by a hand operated lever, and set symmetrically over the opening in the parison mould. On the plunger arm an adjustable plate is secured by four long bolts and powerful springs. In operation this plate is brought onto the upper surface of the neck-ring mould, holding it in position, and, by its close fit against the plunger, preventing the squeezing out from the mould as the plunger descends.

On the right-hand side of the table are two finishing or blow moulds somewhat similar in design to those used in hand working. On a second pillar between and behind these moulds is a blow-head which can be swung over each mould in turn, when, by the operation of an air valve, a piston is caused to descend until a flat disc on the end the connecting rod of its accurately against the upper face of the neck ring, as the later lies in the mould. As soon as the disc seals the top of the mould, its action opens a secondary valve which allows compressed air to enter the mould for the blowing process. On closing the lever on the blowing-head, the piston once more rises and the secondary valve closed.

The machine can be operated by two youths and a boy working together the procedure is as follows - molten glass in sufficient quantity is gathered by the first youth on a gathering iron, dropped into the parison, and cut of with the shears. The second youth places on this mould the ring mould and pulls down the press lever, so causing the plunger to enter the molten metal and force it to the shape of the mould and also up into the neck ring. This ring being slightly cooler than the glass slightly chills it, and thereby allows the glass to retain its shape. The lever travelling back to its top position, removes the plunger, and the ring mould with the pressed blank is lifted from the mould and placed in one of the finishing moulds.

The boy now swings over the blowhead, and by turning the lever causes the piston to descend. When in position the compressed air valve, as we have seen, is opened and the bottle blown to shape. By again touching the lever the boy cuts off the supply of compressed air and raises the blow head. The finished bottle is then removed from the mould and the ring mould opened. Should a screw neck bottle be required a one-piece ring mould is employed, suitably cut so as to give the required thread on the glass. When the bottle is blown, the ring mould is screwed off before the bottle is removed form the blowing mould.

The second one is the Kilner Bottle-Making Machine, Turn-

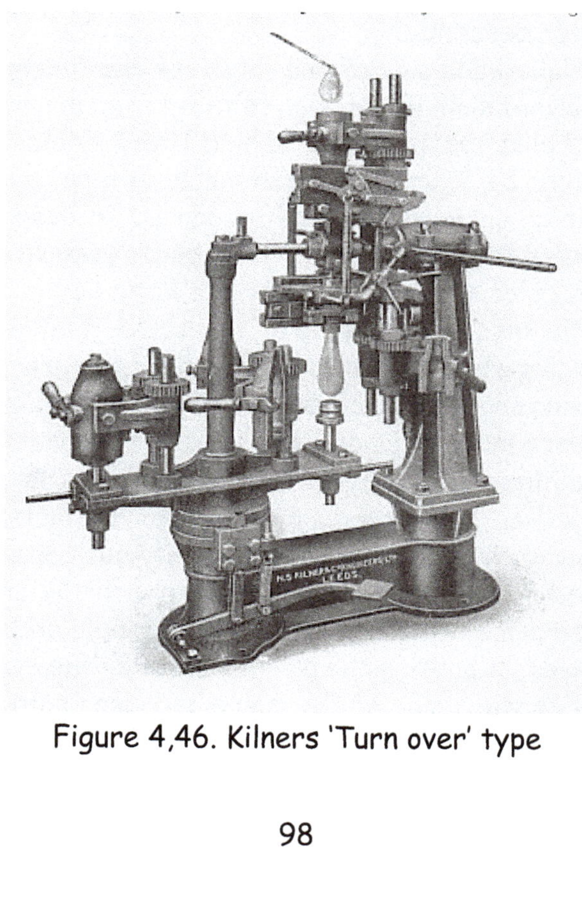

Figure 4,46. Kilners 'Turn over' type

over Type which is illustrated in Figure 4,46. This machine has been specially designed for the output of heavier bottles and jars, either wide mouth or narrow neck, such as are used for ale, wine, jam, pickles, etc. This machine comprises two parison moulds and neck-ring moulds placed mouth to mouth at opposite sides of a horizontal spindle, about which they can rotate.

Glass is placed into the upper, inverted mould and flows down into the neck ring under the action of gravity. The parison is then rotated through 180° and the second mould brought uppermost ready to receive a gathering of metal. Whilst the second one is being filled the first is opened and one of two blow moulds on a horizontal turntable brought in position, closed around the parison then locked.

Compressed air is now admitted and the bottle blown. On cutting off the air, the ring-mould is opened and the mould containing the bottle swung out of position, when it is opened and the bottle removed. As the bottle is removed the neck-ring

Figure 4,47. The Single mould Harrison machine

Figure 4,48. Harrison's Three head machine

mould and the parison mould are closed and reversed to receive a further charge, whilst the second parison and finishing moulds in turn come into position for completing the succeeding bottle.

Redfearn Brothers of Monk Bretton, Barnsley designed two machines, the single mould machine, see Figure 4,47, was patented in 1909, came into service in 1908 or 1909. This was followed a few years later by the three head Harrison machine, see Figure 4,48, or the four head machine. This machine nicknamed the 'joy wheel' was introduced during the 1914-1918 war. One of the advantages of these machines was their ability to make a wide range of products with very little adjustments required to the machine. They are reputed to be able to manufacture bottles from 2 oz to 70 oz.

In Europe Mainzer (D.P.R. 281 306, 292 167, and 211 077 dated March 13th 1901) tried to use one set of moulds with his fully automatic tilting machine, but it still had to be fed by hand. The machine also goes under the name Koenig machine.

Jean Wolf of Cologne (Ger. Pat. 176 511 granted 7th February 1905) developed a machine that was similar to Schillers except that it was more mechanised with most of the motions being pneumatically activated. The machine has three plungers and two blow moulds and the fluid glass is sucked into the mouth of the mould as in the Schiller's machine. The mould bottom is inserted from below, prior to blowing, to support the gob. The parison and blow moulds can be cooled intensively. The manufacturers gave outputs of 2400 600 gram bottles per eight hour with a three man shop. Wolf first machine went into operation in 1911.

Figure 4,49, D.R.P. 343 605, shows one of Wolf's later machines and Figure 4,50, an advert in a diary dated 1909. F.W.

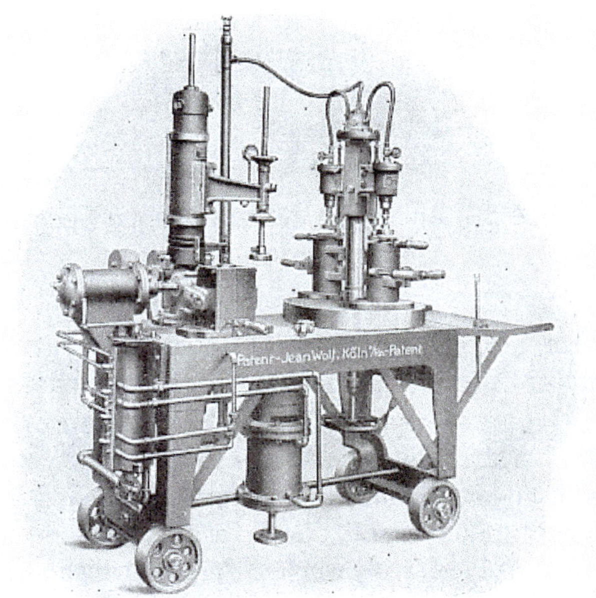

Figure 4,49. One of Wolf's later machines

Figure 4,50. An advertisement from a German calendar dated 1909

Kutschner had his first machine working in 1913. A reference to a German patent application is found in the abstracts.

This next patent was applied for by F. Sweeting and W. Hardman of Castleford in England. Up to 1914 Fred Sweeting, the manager, and William Hardman, the engineer, worked for the Aire and Calder Bottle Company in Castleford and after the company was bought out in 1914, they worked for the United Glass Bottle Manufacturers and in 1923 the submitted a patent application regarding their semi-automatic machine. This machine was nicknamed 'The Swan' an abbreviated version of the two makers names and was an attempt to improve upon the productivity of the semi-automatic machines.

It was a four head machine and it evidently filled a niche in the market as it was ideal to make small orders efficiently. It was sold in large numbers both at home and abroad. When the Castleford factory was closed down sixteen of the machines along with their operators were transferred to St. Helens and continued to work up to 1957. There were two other Sweeting-Hardman machines working at Davy and Moores in Brimsdown until 1981. One had been converted to press and blow.

Fred Sweeting chose to stay in Yorkshire joining Beatson Clark and later Bagleys. He was a founder member of the Society, a Vice President and its first Treasurer.

Figure 4.51 is a photograph of the Sweeting-Hardman machine, it was in competition with automatic machines and its

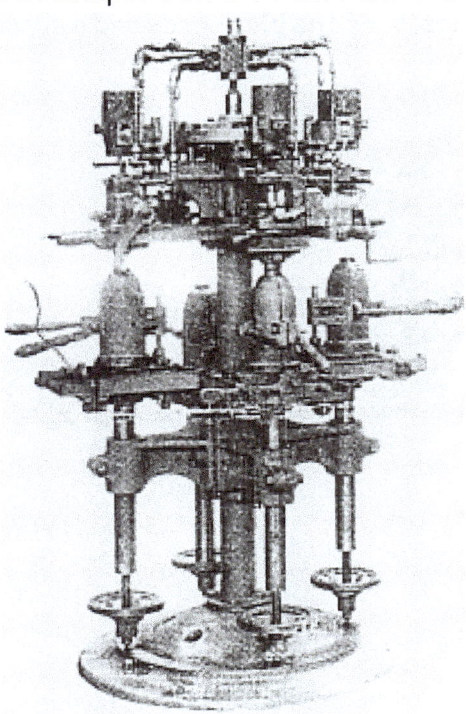

Figure 4,51. The Sweeting-Hardman machine

long life is a tribute to its designers.

There is an abstract in the Society of Glass Technology's 1925 journal regarding this machine, which is as follows;

#287. A Semi-automatic Bottle-making Machine. United Glass Bottle Manufacturers, Ltd., London, and F. Sweeting and C. Hardman, Castleford, Yorkshire (Brit Pat., No. 221 276 June 6th, 1923). - A semi-automatic bottle-making machine, of the kind in which the moulds a carried by a table intermittently rotated about a central column, comprises a sleeve adapted to rotate about a central pillar, the sleeve being slotted longitudinally and provided with vertically adjustable brackets on which the finishing moulds are mounted, so that the machine can be adjusted for making ware of different heights and sizes.

The central pillar, 2, is rigidly secured to a pedestal, 1, and carries a ball bearing, 3, on which a turntable, 4, revolves. The table the housings, 5, of the blowheads and a support, 6, on which

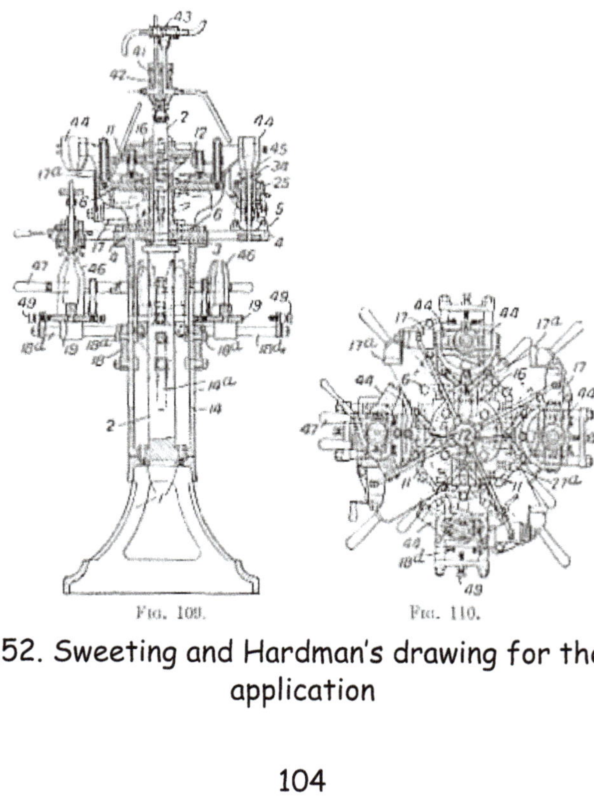

FIG. 109. FIG. 110.

Figure 4,52. Sweeting and Hardman's drawing for their patent application the parison, 44, are pivoted. The moulds are opened and closed by toggle links, 11, connected to studs, 12, carried by plates which slide in the guide on the support, 6, the studs engaging a cam, 16, fixed to the pillar, 2. Thus as the table rotates, the moulds will be opened and closed at the proper times by the cam, 16. A light sleeve, 14, of rectangular section is bolted to the underside of the table and is formed on each of its sides with a elongated slot, 14a. The table, 19, of each finishing mould, 46, slide on bars, 18d, projecting from a support, 18, which is clamped to the flat side of the sleeve, 14, by means of bolts, 18a, passing through a slot, 14a. By loosening the bolts, the support may be raised or lowered to adapt it to the height of the finishing mould.

The finishing moulds are opened and closed, and are moved on the slides, 18d, by means of handles, 47, stops, 49, secured to the bars, 18d, enabling the moulds to be positioned accurately beneath the blowheads. The ring moulds, 45, each consist of two sections sliding in guides, 25, carried by a turnover head, 17, which is inverted by means of a handle, 17a. The mould halves are closed by a spring and opened by means of a lever, 27a, pivoted on the slide.

A plunger, 34, for forming the opening in the parison, slides through the blowhead and is held in the ring mould by means of a spring, a lever being provided to enable the plunger to be withdrawn from the mould. A charge of glass is dropped into the parison mould and is drawn down into the ring mould by suction, air being withdrawn from the blowhead by means of an ejector, 43. The ejector is attached to a non-rotating bearing, 41, which is seated in a distributor, 42, that rotates with the table. The distributor cuts off the suction at the appropriate time and admits compressed air to the blowhead. After the parison mould is opened, the blowhead is inverted and the ring mould carries the parison down into the finishing mould. After the bottle is blown, the ring mould is opened and the head is inverted again, and the parison mould is closed to receive a new charge.

Now we come to the final machines in this semi automatic section and they were made by Johnson Radley a mould maker in Hunslet, Leeds, and sold them under the 'Cristelle' label. According to one of the family they only stopped selling them in 1963 when they closed the department down and the head office and part of the works were moved from Hunslet to Pudsey on the west side of Leeds. In the end they were being sold as a 'service to the customer', probably as a means to sell their moulds.

The semi-automatic single head Type 'AP' machine and the double head Type 'AS' machine both make high class perfumes, medicals, sauces, coffees, creams, milks, cordials, tabloids, oils, pomades, vinegars, inks, beers, minerals, syrups , etc. The ware range is 2 drams to 30 ounces for the 'AP' and from 1 ounce to 40 ounces for the 'AS'. There is no advantage to using the 'AS' machine at the lower weights but there is a decided advantage with the higher weights, one can be settling and cooling while the other starts a new cycle. Both machines work on the suck and blow principle.

Figure 4,53. The 'Cristelle' Type AP Machine. (B & B single)

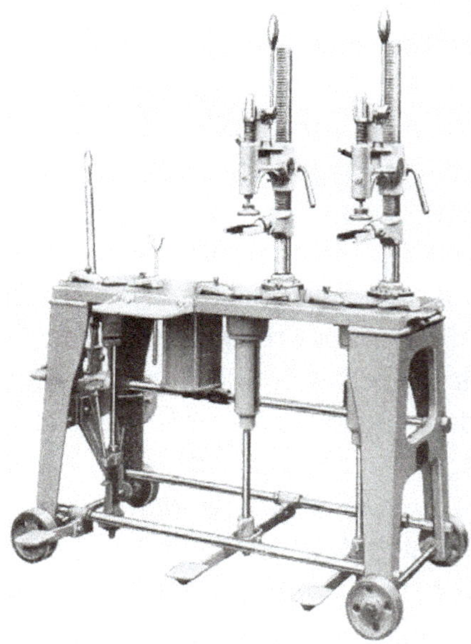

Figure 4,54. The 'Cristelle Type AS Machine. (B & B dual)

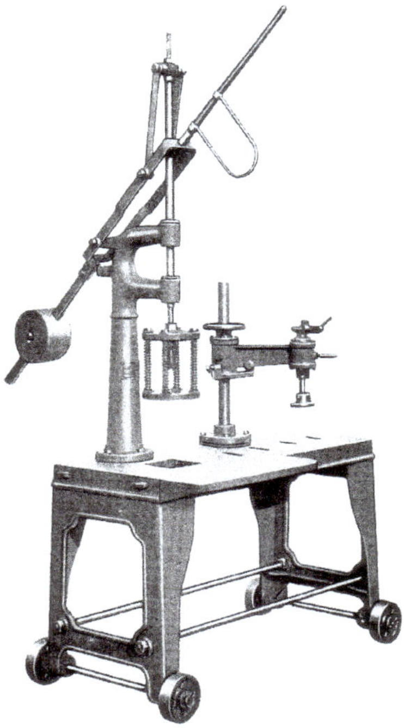

Figure 4.55. The 'Cristelle' Type BW machine. (Press & Blow)

The semi-automatic Type 'BW' is a press and blow machine, built for small and medium ware ranging from 2 ounces to 30 ounces in weight. It is particularly adapted for the production of meat pots, vaselines, pomades, face creams, pickles, jams, honeys, sweets, jellies, battery jars, lighting ware, domestic ware, tumblers, etc.

Type 'CG' for small to medium pressed ware - dishes, bowls, vases, candlesticks, powders, butters, ash trays, cream jugs, beer glasses, soda goblets, tumblers, trinket sets, stoppers, lighting ware, domestic ware etc.

Type 'CCP' for medium to large pressed ware - dishes, bowls, vases, quart jugs, trays, lenses, cathode ray tube fronts (up to 12" in diameter), etc. If required this press can accommodate smaller jobs, for example ash trays, tumblers, sundae dishes, ice plates, 10 oz tankards, etc.

Mould requirements.

Type	'AP'	'AS'	'BW'
Mould	1	2	2
Parison	1	1	1
Neck rings	3	3	3
Plunger	1	1	1

The Pressed Ware Specifications were;

Type	'AP'	'AS'	'BW'
Maximum height under bottle ring	12"	12"	12"
Minimum height under bottle ring	$\frac{7}{8}$"	$\frac{7}{8}$"	
Maximum diameter of body	$3\frac{1}{4}$"	$3\frac{1}{4}$"	6"
Maximum diameter of neck	2"	2"	4"
Maximum outer dia. of parison			$5\frac{7}{8}$"
Maximum stroke of plunger			12"
Approx. Nett weight c.w.t.'s	6	7	8

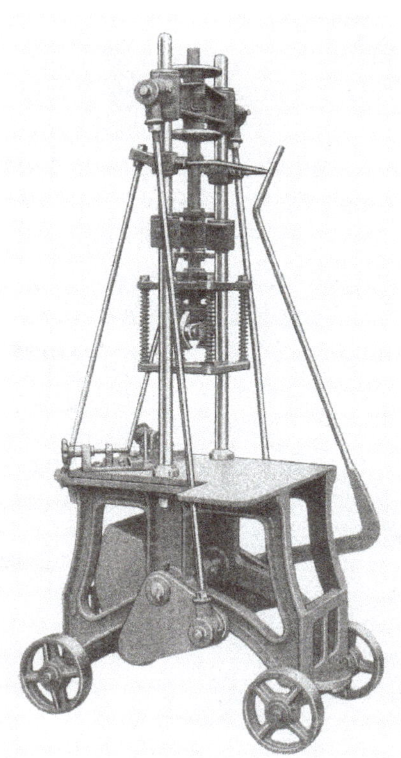

Figure 4.56. Cristelle Type 'CG' side lever spring cage glass press

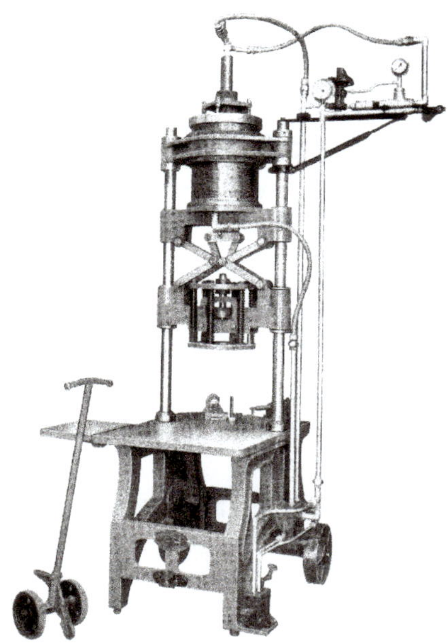

Figure 4.57. Cristelle Type 'CCP' pneumatic spring cage glass press

Figure 4.58. Cristelle Type 'CJ' rotary spring cage glass press

The ware range is quite extensive for example, the Type 'CJ' for small ware - could make dishes, plates, bowls, powders, butters, ash trays, sundae glasses, stoppers, eye glasses, glass lids etc.

Machine and pressing specifications.

Type	'CJ'	'CG'	'CCP'
Max. Dia. of mould base	8"	14"	16"
Max. height of mould	6"	12"	12"
Max. length of mould	15"	22"	
Max. stroke of pressing head	6½"	9"	10"
Max. dia. of spring cage	3¾"		
Max. dia. of plunger	4½"		
Max. clearance under cage			20"
Approx nett weight cwt.	7	14	25

Two other aids to the semi-automatic production of containers and pressed ware are a range of neck ring holders and a range of 'mechanical boys'. The neck ring holders come in four sizes 25 mm, 35 mm, 50 mm, and 75 mm. The 'mechanical boys' come in the following sizes:

	No. 1	No. 2	No.3
Max. height of mould	$7\frac{1}{2}$"	11"	12"
Max. dia. of mould	4"	6"	8"
Trough size	20×15×8	24×18×11	$24\frac{1}{2}$×20×$12\frac{1}{2}$
Water capacity galls	9	15	$18\frac{1}{2}$
Height of boy	21"	24"	$25\frac{1}{4}$"

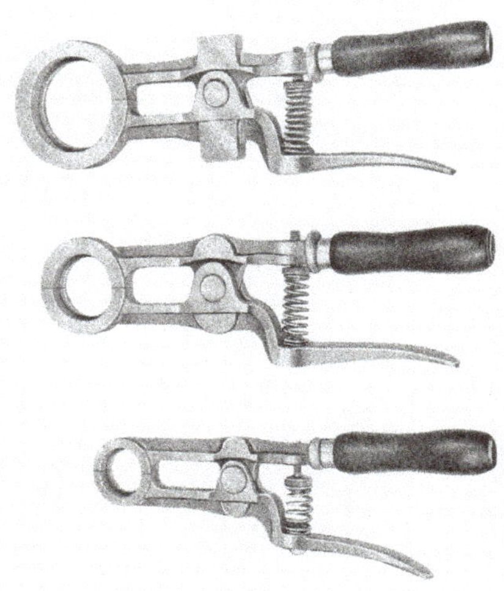

Figure 4,59, 'Cristelle' Neck Ring Holders

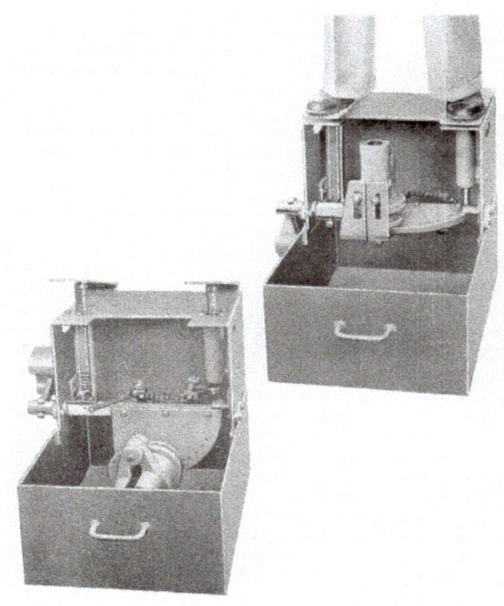

Figure 4,60. 'Cristelle' EP Mechanical Boy

Chapter 5

The rise and fall of the Suction machine

Figure 5,1.
Michael J. Owens

The first suction machine was shown to the public in 1904 and it was the first fully automatic machine to be developed. It designer was Michael Joseph Owens (1859-1923), who was the son of an Irish coal miner who had emigrated with his family to West Virginia in the United States of America. His father died when he was ten years old so he had to forgo his education and go out to work to help support his mother.

Mike Owens was born on 1st of January 1859 in Mason County, West Virginia, he started his employment when he was sent to work at the flint glass factory of Hobbs, Brockunier & Co. at Wheeling where he experienced the deplorable conditions which the glass blowers had to endure, especially the children to whom all of the 'dirty' jobs were delegated to. He passed through the various jobs for boys, as a 10 year old he started by shovelling coal into the furnace called "firing in the glory hole". At 11 he was a "carry in boy" taking newly made ware to the annealing lehr and a year later a "carry out boy" removing the ware out of the lehr. He also worked as a mould-boy opening and closing the mould for the master blower and as a gatherer skilfully collecting a gob onto a blowpipe. By the time he was fifteen years old he had learnt how to blow glass and soon became proficient at his art.

Deprived of his education through his family's misfortune he set out to become self-taught attributing much of his knowledge and success as coming from his membership of the local debating society. He joined the American Flint Glass Workers Union an soon became a forceful member and a leader of the branch. He was in fact one of the agitators who caused trouble at the New England Glass Co. of East Cambridge, Massachusetts, which had been established in 1818.

Figure 5,2. Edward Drummond Libbey

In 1878 W. L. Libbey, the General Manager of the New England Glass Company, acquired the lease of the factory. His son, Edward Drummond Libbey (1854-1925) who eventually became concerned with bottles, sheet glass and paste mould manufacture, joined the company as a clerk in 1874. W. L. Libbey died in 1883 and his son assumed control.

In 1888, following two strikes due to labour problems, he shut down the Massachusetts factory, and tempted by the cheaper natural gas prices that were available to the west of the Allegheny mountains moved with 115 of his workers to Toledo, Ohio, where with the support of local business men the W. L. Libbey & Son Glass Company with a capital of 100 000 dollars to "manufacture and sell rich cut and artistic glass." In 1892 E. D. Libbey changed the company's name to the Libbey Glass Company.

At first the new firm struggled but a strike by blowers at Cornings making bulbs for electric lamps allowed Libbey to take up the opportunity and move into their production in 1890. This he did initially at a works in Findlay, Ohio, then as the firm prospered moved them all to his Toledo factory.

Mike Owens started his association with Edward Drummand

when he went to work for him in August 1888. He started as a second ranking blower making chimneys and glass shades for oil lamps using paste moulds but within three months he had become foreman of the blowing department and two years later became superintendent of the factory.

There are many stories told about Owens one of them is "Owens strode into Libbey's office one day and announced that he was quitting work because the plant superintendent was not properly disciplining his workers. He further predicted that the company would be bankrupt within a few years. Libbey agreed with him and asked Owens if he thought that he could restore order in the furnace room. When the latter emphatically affirmed that he could, Libbey offered him the superintendency.

Owens accepted, but only upon the condition that neither Libbey or any other member of the front office would interfere in the least with his duties. He then proceeded to discharge the entire labouring force and to re-hire only those men whom he considered both capable and manageable".

Whether this story is true or not is open to debate, but it is true that Owens rose through the ranks primarily because he possessed remarkable imagination, ingenuity, initiative and will power. He had charge of workmen blowing bulbs at a branch shop in Findaly in 1891-1892 and superintended Libbey's glass exhibit at Chicago's Columbian Exposition in 1893.

While supervising the exhibition at Chicago Owens had sufficient 'free time' to enable him to sit back and take stock of Libbey's glass making activities and to think up ways of how to improve them. He had sufficient leisure time to formulate how glass could be blown mechanically. In 1894 he obtained two patents covering apparatus for mechanically operating paste-moulds in glass manufacture; these he assigned to the Libbey Glass Company for use in making light bulbs only.

This part of the Owens life history is important in so much as to its development of the automatic bottle machine. In 1895 the formation of the Toledo Glass Company was made, to which Owens and Libbey assigned their remaining rights in the paste mould machine and agreed to turn over all of the patents that they may obtain in the next seventeen years. Very soon they sold off their rights to the tumblers and lamp-chimney manufacturers and diverted their attention from manufacturing those to bottle making machines.

Owens claimed that he conceived the basic ideas in 1898 when he started to carry out trials to make bottles automatically. He was conversant with the idea of machines being multi-headed due to his inventions with paste mould machines, what was needed was a means of automatically loading the glass to the machine. What followed was five years of patient development work, often causing machines to be built and then broken up because they were going down a blind alley.

During these years Owens was fortunate to have Libbey as a financial partner, according to W.C. Scoville, Libbey personally met all the deficits of the Toledo Glass Company while Owens was making his experiments.

William S. Walbridge, who had been a colleague of Owens while at the American Lamp Chimney Company, became an official and stock holder of the Libbey Glass Company and treasurer of the Toledo Glass Company, wrote an account on the development of the Owens machines, of which he was an eye witness.

He said that Owens first commenced by sucking up molten glass with a 'hand-gun', an instrument like a large bicycle pump. This was based upon an apparatus called the 'tally-ho' or 'sucker-upper' which was used in the Libbey glassworks between 1890 and 1895. [Croskey and Locke patented a similar device in 1897 but failed to bring it to a successful conclusion]. The 'tally-ho' device had not been patented and Owens succeeded in obtaining

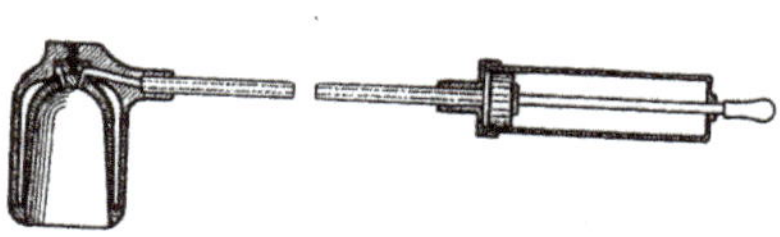

Figure 5,3. Croskey and Locke "tally-ho" device

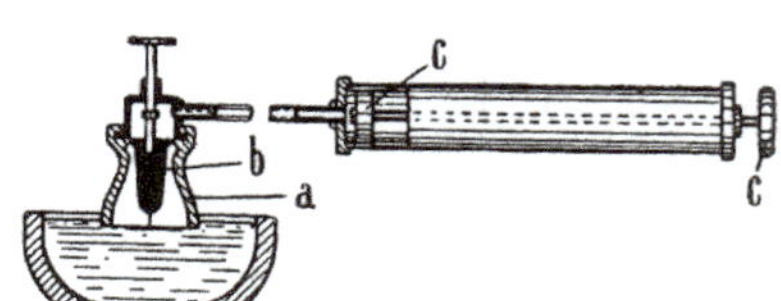

Figure 5,4. Owens' "gun"

a patent on his 'gun' in 1904 (US Patent No. 759 742), the patent was applied for in 1899.

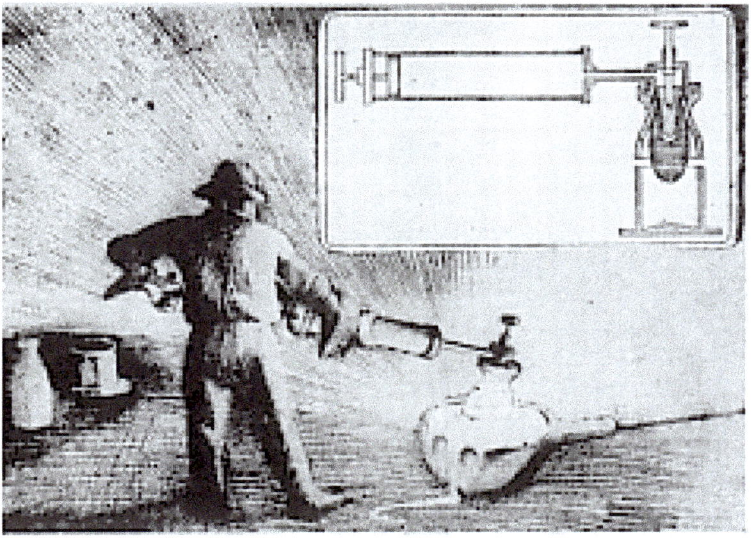

Figure 5,5. Owens first experiment apparatus

In Figure 5,5 the 'gun' is shown sucking a charge from a pot of molten glass into a two part iron milk bottle mould. The upper part of the mould formed the neck and lip around a plunger, the lower mould was opened and the parison, suspended by the neck in the upper mould, was placed in the finishing mould. Then the piston in the 'gun' was reversed and air was forced in to the parison which expanded to fill the blow mould. It took him many months of

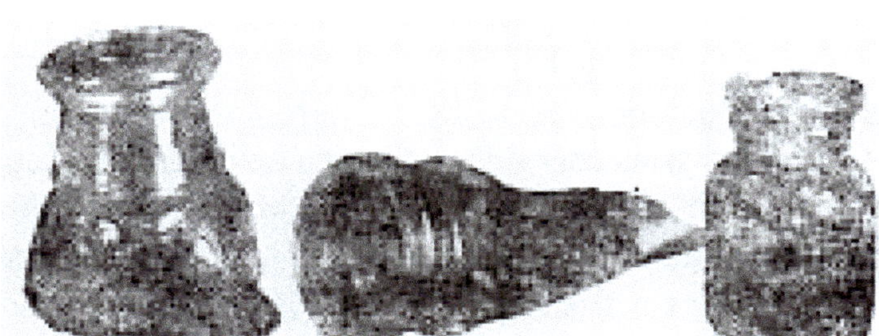

Figure 5.6. Some of the first attempt to make bottles by the vacuum device

Figure 5,7. Owens second experimental machine

experiments before he could make a bottle successfully, Figure 5,6 shows some of his failures.

A second experimental machine was built, see Figure 5,7, this time it included a small tank furnace to supply it with glass. This time the 'gun' was mounted on a column and provided with wheels to move it to and from the tank.

With this machine some bottles were made but it was not automatic so Owens continued and proceeded to develop his ideas of an automatic machine further.

N. B. The experimental machine terminology is by Meigh in his book 'The Development of the Glass Bottle Machine'.

In all of this work he was ably assisted by his engineering staff of William Emil Bock and C. William Schwenzfeier and from 1901 by Richard La France.

The third machine was to be an automatic rotary machine having a series of arms carrying gathering and finishing moulds. In its test position it only had one complete gathering and finishing mould, when tested it showed that it was impossible to continue to gather glass from the same position (same area of glass) as the glass became too chilled in the process. This experience was invaluable because it led to the conception of a changing surface, which was achieved by an auxiliary rotating tank (or revolving

Figure 5,8. Owens experimental machine No. 3

Figure 5.9. Owens experimental machine No. 4

pot) for which Owens was granted a patent on the 8th November 1904 (US Patent No. 774 690).

Machine No. 4 (Figure 5,9) had six arms each carried a gathering and finishing mould. For this trial the Toledo Glass Company built a glass factory with a furnace and a revolving pot. This demonstration proved without doubt that bottles could be made automatically. This led to changes in the organisation, in 1903 the Owens Bottle Machine Company was formed with a capital of three million dollars. The new company took over the plant of the Toledo Glass Company and the American rights on all inventions with the object of manufacturing some glass articles with the machine, but the main function of the company was to licence others to use the Owens machine.

The manufacturing experience gained from the fourth machine soon led to vital and radical improvements being made which were included in the development of the fifth machine (Figure 5,10), which became the foundation of all of the machines licensed in 1904-1906.

There are two another versions about the fourth machine. According to the American Society of Mechanical Engineers when they designated the 'AR' machine as an International Historic Engineering Landmark (17th May 1983) they stated that the fourth machine had five arms ('bicycle pumps' as they say they where termed.) U.S Patent Number 766 768. It had a revolving pot and it was ready for production in 1903.

They also say that each of the arms dipped individually to gather the glass, not like the first commercial models where the whole machine worked intermittently, the whole machine was lowered and then lifted for each gather while stood. In its first test the Number 4 machine made eight pint beer bottles per minute, another thing (in the USA) was that the beers were 'narrow neck', previously all semi-automatic machine made only 'wide mouth' containers.

Cable in his paper 'The Centenary of the First Automatic Bottle-Making Machine' states that there were patents for two five arm machines, the first machine was driven by a Geneva drive making its rotation intermittent and that it gathered glass from a stationary pool of glass. This highlighted that the 'chilled glass' at the gathering point was causing problems and it must have been serious because another patent was applied for eight months later.

Figure 5,10. The Owens fifth machine

They solved the problem by building a newly designed furnace and revolving pot along with a newly designed second five arm machine where each of the arms dipped individually into the pot (US Patent 774 690).

The fifth machine had a series of six fixed arms carrying the gathering moulds. The entire machine was lowered and raised by counterweights in the process of gathering. This stopped the need for the machine and the revolving tank to be halted while gathering took place. Instead the gathering mould dipped into the glass while revolving, thus allowing for an increase in speed and hence higher output.

When considering the out come of this successful machine one can only admire a man who accepted that a machine weighing ten tons or more needed to be lowered and raised every time a charge of glass was gathered. This No. 5 type machine was licensed to such important customers as the Thatcher Manufacturing Company, the Ohio Bottle Company and the American Bottle Company during 1904–1906.

Owens was not satisfied with the Number 5 machine and soon produced Number 6 (Figure 5,11) with improvements leading to increased efficiency. This became the A type machine and from 1906 to 1911 was licensed to the Hazel-Atlas Company, the Charles Boldt Company, the Illinois Glass company, and the Whitney Glass Works. By the end of 1911, 103 machines had been installed with a capacity of over four million gross of bottles per year.

Figure 5,11. Owens experimental machine No. 6

Owens had already obtained a patent in August 1904 (US Patent No. 766 768) for a five arm machine (see Figure 5,12) utilising the dipping head principle it was driven by a Geneva drive which made it an intermittent rotating machine.

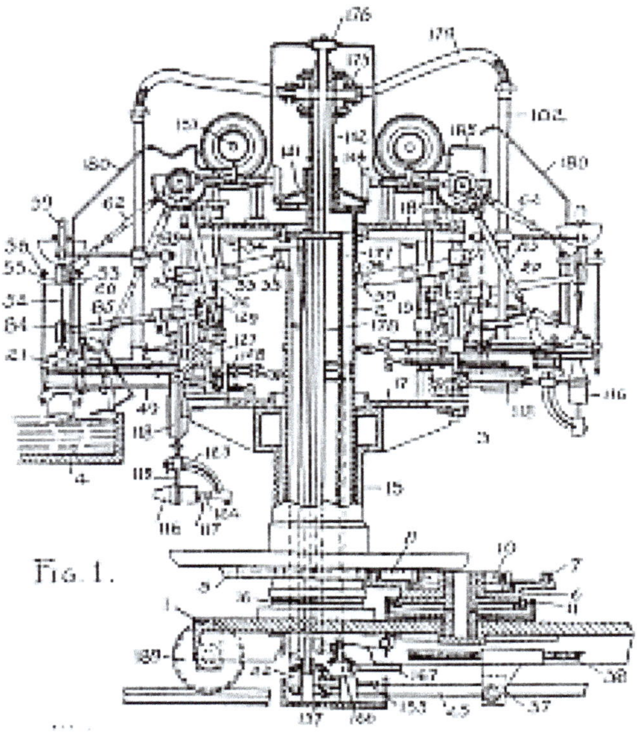

Figure 5,12. Owens patent drawing for his five arm dipping head machine

In 1911 he adopted the dipping head gathering principle, and started to sell these machines in 1912 the first models being the two ten arm machines types AN and AR. This was considered to be a major improvement and it made life much easier as the machine did not have to be lowered and lifted for each gather of glass.

The 15 arm AQ came next in 1912, followed by the 6 arm AT, for making carboys, in 1914, then another 15 arm machine the AV in 1917. The final machines were the CA and CB machines with

10 arms each with two heads making them the equivalent of a 20 arm machine, introduced between 1920 and 1923.

The complete range of their products, the year or years when they were introduced, the number of arms each machine had, the capacities that they could make and the production speeds can be found in the following Table 5,1.

Type	Number of arms	Year introduced	Capacity US Fl. Oz.	Production per minute
A*	6	1903-4	4-40	10-20
AC*	6	1908	4-40	11-22
AE*	6	1909	4-64	12-24
AD*	10	1909	4-64	24-26
AL*	8	1912	16-160	8-12
AN	10	1912	0.25-11.5	20-60
AR	10	1912	4-80	12-45
AQ	15	1914	4-80	18-60
AT#	6	1915	0.75-15.25 (galls)	6-7
AV	15	1917	0.125-11.5	30-90
CA	10 (x2 heads)	1920-23	max 8	80-320
CB	10 (x2 heads)	1920-23	8-32	80-160

* Fixed arm machine # Carboy machine

Table 5,1. The family of machines built by the Owens Machine Bottle Company

N.B. You will notice the discrepancy with the dates of introduction

In 1915 the Owens Bottle Machine Company built their biggest machine yet, it was for the manufacture of carboys. It stood over 15 feet high, and more than 15 feet in diameter and weighed more than 100 tons and required 100 horse power for its operation. It could produce five to six 5 gallon bottles per minute each weighing about 12 pounds, and the Illinois Glass Company obtained an exclusive license to use it. Figure 5,13 gives an indication of its size, the man placing equipment on it is dwarfed by its height.

Owens' first machine was the 'A' type, see Figure 5,14, it had six fixed arms and it released the finished bottles 40 degrees

Figure 5,13. The Owens carboy machine

before gathering point. At this time the bottles had not cooled enough to be free standing and they were transferred to an auxiliary machine to cool them and fire-finish the closures.

His next improvement was to make the delivery point some 60 degrees after the gathering point which gave the bottles adequate time to cool and set. This was again a fixed six arm

Figure 5,14. Owens 'A' type machine

Figure 5,15. Owens first machine where the bottle finished after the gathering point

machine, its designation was the 'AE' type although the 'AC' could have had these improvements as well. The machine is illustrated in Figure 5,15.

The next improvement was to make all of the heads capable of dipping into the glass surface individually, this meant that

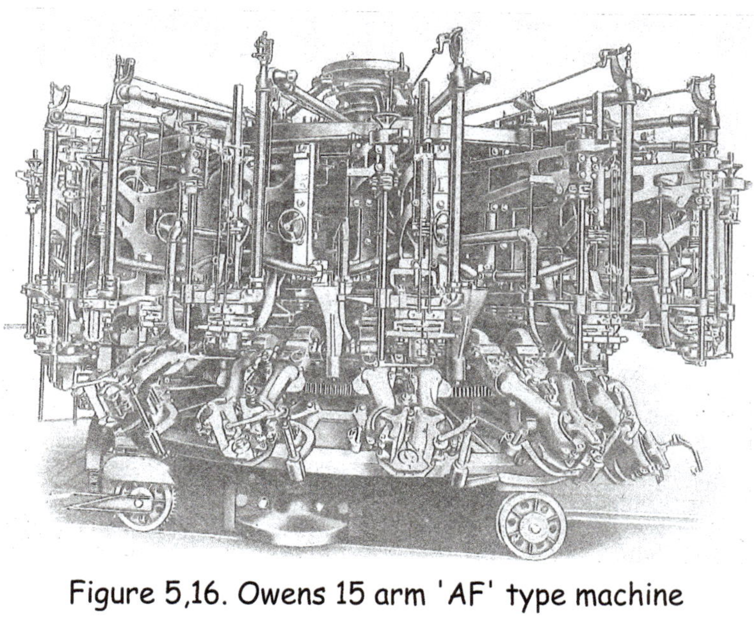

Figure 5,16. Owens 15 arm 'AF' type machine

the machine did not need to be raised and lowered as each head passed over the rotating pot. This must have had a major impact on the amount of maintenance that the machine required. The first machine to be marketed with this new modification was the 'AF' machine, every Owens machine after this model included this feature.

Many of the machines could work with plural moulds either dual or triple, the two 15 arm machines were examples of this but this was not good enough for Owens.

To increase the production even more was made when with the introduction of the 'CA' and 'CB' type, see Figure 5,17. They only had 10 arms but each arm carried two sets of moulds meaning that for one complete revolution of the machine 20 heads would need to be charged with glass. The number of moulds needed filling would have been 20, 40, or 60 depending on the bottle size.

Figure 5,18 gives a close up view of the 'C' type machine showing the triple moulds at the mould opening position.

Figure 5,17. An Owens 'C' type machine

Figure 5,18. A close up of the 'C' type machine showing the triple moulds

A down side to these new machines was the need to increase the diameter of the revolving pot with its increase in fuel consumption. The type 'CA, and 'CB' required a 13 foot diameter pot.

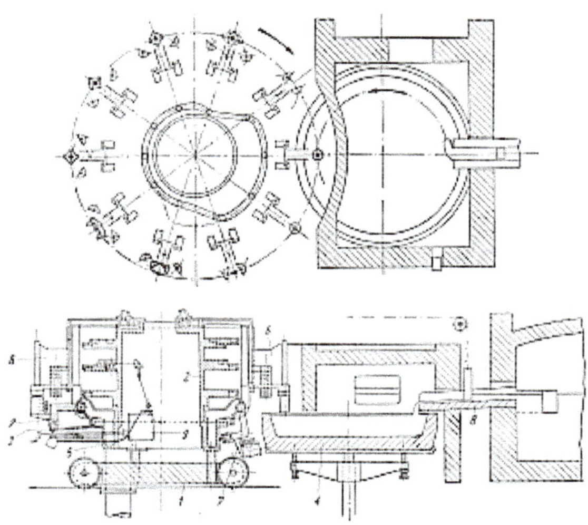

Figure 5,19. Schematic view of a 'revolving pot'

The following photographs, Figures 5,20 and 5,21 showing an actual machine from the pot entry side of the machine and at the blank transfer position.

Figure 5,20. Machine looking from pot entry side

Figure 5,21. Machine at blank transfer position

Meanwhile at the same time that Owens was developing his machines, Owens and Libbey were trying to license these machines throughout the industrialised countries at great advantage to

Year	No. of Machines	Production Capacity (1000 gross) †	Total Production of Bottles & Jars in the USA (1000 gross)	No. of Hand and Semi-automatic Machine Operators †
1905-06	8	----	----	8000
1906-07	18	----	----	8559
1907-08	34	----	----	9627
1908-09	46	----	----	8822
1909-10	61	1500	12 313	7005
1910-11	103	2500	----	7634
1911-12	124	4200	----	7477
1912-13	152	7000	----	7703
1913-14	164	8000	----	7703
1914-15	176	11 000	19288	6868
1915-16	187	----	----	4941 ‡
1916-17	200	----	----	5314 $

Table 5,2. The development and use of glass-making machinery. Number of Owens Automatic Bottle-Making Machines in the USA*
* Glass Bottle Blowers Association Reports, USA. † This data for the calendar years 1905, 1909, 1910, etc. ‡ Includes also 500 Owens machine attendants and unskilled workers of the American Bottle Co. $ Includes also packers in general employment.

themselves. They organised the Owens European Bottle Machine Company in October 1905 with the intention of selling the patent rights to the European bottle makers.

They decided that the best way forward was to build a factory to demonstrate their machines. The location chosen was England, and a site at Trafford Park in Manchester was obtained in May 1906, and by April 1907 they were producing bottles successfully on a tank furnace that supplied glass to two 'A' type Owens machines.

The European glass industry saw the machines being demonstrated and were quite alarmed with the threat when they were told that the factory would stay in competition and possibly be enlarged unless the European rights were not purchased. The purchase price was 12 000 000 gold marks. This caused much consternation especially with the German manufacturers due to competition in Germany by a new supplier and a cartel was soon formed.

The negotiations were carried out by three German Manufacturers namely A.G. Gerresheimer Glashüttenwerke vorm Ferd. Heye of Düsseldorf, Freidr. Siemens of Dresden and H. Heye Glasfabrik of Schauenstein acting for a large group of European manufacturers. The American side of the negotiations were the Owens European Bottle Machine Company of Toledo, Ohio USA and the Toledo Glass Company of Toledo, Ohio, USA.

By the 28th November the already formed company the Europaischer Verband der Flaschenfabriken generally known as the E.V. met with the Americans to discuss the agreement known as the "Option Agreement of 5th August 1907" but there was no first class German Bank who could offer the bank guarantees. The agreement date was changed from the 2nd December 1907 to the 1st March 1908 with a new schedule of payments. First payment to be 3 000 000 marks followed by the rest paid over the period of 1909 to 1917.

Guarantor	**Amount guaranteed with interest over period 1909-1917**
The Bömische Union-Bank, Prague for Austro-Hungary	1 150 000 marks
The German Association of Bottle Manufacturers.	3 840 000 marks
The British Association of Glass Bottle Manufacturers.	2 400 000 marks
Surte Glassworks and Liljedals Glassworks in Sweden and Laurvig Glassworks in Norway Johnsen & Jörgensen also gave same guarantee.	256 000 marks
Aktieselskabet Kastrup Glasvaerk	156 000 marks

The 'E.V.' agreed to increase its original capital to 6 000 000 marks.

Table 5,3. List of E.V. guarantors

The formation of the 'E.V.' was made on the 15th November 1907, its objective was to form an international organisation to finance the purchase of the rights to the Owens patents and operate them in a co-operative way. The original countries were; Germany, France, Great Britain, Austro-Hungary, Belgium, Holland, Norway, Sweden and Denmark.

With Germany being the largest producer of bottles it was decided to make the head office in Berlin, later moved to Düsseldorf. (See paper by R. S. Biram obtained from the SGT. It is a well documented record of the negotiations, etc.) Between 1907 and 1914 some 60 machines were licensed and installed, the types and the countries are listed below.

Of the sixty machines licensed in Europe some were made in the USA, some made by Berliner Maschenenbau Aktiengesellschaft vormals L. Schwartzkopff and some were ordered from the Linotype Machinery Company of Birmingham, England, but owing to the First World War none of the latter were delivered. In addition to the sixty another 15 were sold to "outsiders" such as Apollinaris in Germany, Companhia Vidraria in Rio de Janero, S.A. Per L'Industria in Italy, Sourse Perrier in Aigues Mortes, France and Nueva Fabrica de Heilo in Havana.

Partners			**Types**		
	A	**AE**	**AR**	**AN**	**Total**
Austro-Hungary	8	---	---	---	8
Denmark	2	---	---	---	2
Germany	23	5	1	1	30
Gt. Britain	6	3	1	---	10
Holland	1	6	---	---	7
Sweden and Norway	1	2	---	---	3
				Total	60

Table 5,4. Owens machines installed by country 1907 to 1914

The dates when the first Owens machine was installed in the various countries are in Table 5,5.

The war caused all sorts of trouble regarding the imports of machines to Great Britain but after 1916 some of the difficulties had been ironed out. A list of machines working or ordered in Great Britain in 1919 is shown in Table 5,6.

Country	Original Installation	In 1913 No. of Machines in Use	In 1913 No. of Firms with Rights of Use
England (original)‡	1906		
Germany	1907	24	7
England (first re-installation)	May 13th, 1908	4	7
Holland	June 11th, 1908	2	2
Austria	July 14th, 1908	6 †	5
Sweden	Aug. 23rd, 1910	2 Ж	1
France	Oct. 1910		
Denmark	July 26th, 1912	2	1
Italy	Aug. 7th, 1912		
Norway	Jan. 1913		
Hungary	May 19th, 1914		

Ж Operations began March 10, 1911. ‡ The original two machines from Trafford Park were transferred to, Alloa Glass Works in Scotland and to David Wright & Co., of Belfast, Ireland. † Four of these were in Bohemia.

Table 5,5. Dates of machines first installed by country

In 1907 the matter of consternation, mainly, to the German manufacturers was that the Owens European Bottle Company had granted a license to the Rheinahr Glasfrabik company, a company owned by the Apollinaris Company (Apollinaris & Johannis Ltd.), a British company. They had built a new factory at the mouth of the River Ahr, near to the town of Sinzig, to manufacture their own bottles under license using Owens machines.

Meigh states that "The Apollinaris directors had achieved a brilliantly clever financial arrangement which makes for interesting reading. Having desired for a long time to make their own bottles because of constantly increasing prices charged by the established German producers, and having heard of the success of the Owens machine in America, Mr Julius C. Prince, the Managing Director, went to the United States in January

Works	Type	No.
Wright's Belfast	A	1
Alloa, Scotland	A	1
Nuttall's St. Helens	A	1
Cannington Shaw, St. Helens	A	2
Bagleys, Knottingley	AE	1
Cannington Shaw, St. Helens	AE	1
Bagleys, Knottingley	AR	1
Cannington Shaw, St. Helens	AR	1
Irish Glass Bottle Co., Dublin	AR	1
Nuttall's, St. Helens	AR	1
	Total	11
Owens machines in the UK at July 1919 but not installed.		
Rylands, Stairfoot	AR	1
Kilner's, Connisbrough	AR	1
Wood's, Portobello	AR	1
Redfearn's, Barnsley	AR	1
U.G.B., Charlton	AR	2
U.G.B., Charlton	AN	2
	Total	8
On order in USA		
Rylands	AR	1
Rylands	AN	2
Bagleys	AR	1
U.G.B., Charlton	AR	1
U.G.B., Charlton	AN	1
Kilner's	AR	1
	Total	7

Table 5,6. Owens machines in the UK July 1919

1907 where he saw thirteen Owens machines working at three different factories. He arranged for the Apollinaris Company to subscribe for 1410 one hundred shares in the Owens European Bottle Machine Company Limited. The Americans were glad to be able by this means to enlist the support of the Apollinaris Company to influence their proposed European deal.

It turned out to be a remarkably good investment for Apollinaris because as owners of about one-fifth of the Owens European Bottle Machine Company, for which they paid £29 040, their share of the 12 million marks paid by E.V. amounted to £121 520, thus giving them a gross profit of about £92 480 irrespective of interest. The cost of the Rheinahr factory was about £112 000 so Apollinaris had only to find some £20 000 from their reserves. They had acquired a modern factory using the latest automatic bottle machines almost for nothing, a remarkable brilliant performance which had a powerful effect in causing the established German glass manufacturers readily to accept the American terms".

By 1950 the E.V. company was losing its power and was gradually changing into a company that acts as a link between all of the original E.V. members and it was concentrating its efforts with Schwartzkopff the German manufacturer of the Owens machines and spare parts. In 1952 they took out an agreement with them lasting for ten years, Schwartzkopff under the name of Berliner Maschinenbau Aktiengellschaft was one of the three original glass machine manufacturers for the manufacturing of Owens machines in the E.V. alliance.

In 1954 Schwartzkopff put on the market their Owens-Schwartzkopff type 'OS' machine and in 1958 the much smaller KS 6/60 machine. The 'OS' was a six head machine based on the types 'A', 'AC' and 'AE' machines but with the dipping heads of the 'AR' and 'AN' machines. With the additional facility of being able to change the dip stroke of each head up to 50 mm individually which could be carried out without stopping the machine. The machines were also much quieter due to the removal of the counterweights and replacing them with springs. The drive is now a stepless P-I-V.

It is quite easy to change from single gob to double gob moulds and vice versa. The double mould operation is called the DS (Double Mould Schwartzkopff), the most important data for the 'OS' and 'DS' machines is length 4050 mm, width 2840 mm, and height 3400 mm. Number of heads six. Figure 5,22 shows an 'OS' machine in the single model

The KS 6/60 machine automatic blowing machine is an innovation of the Berliner Maschinenbau AG. formerly L. Schwartzkopff, Berlin.

Figure 5,22. The Owens-Schwartzkopff type 'OS' machine

Figure 5,23. The Berliner Maschinenbau AG. KS 6 Schwarkzkopff machine

The machine produces flacons, medicine, container and beverage bottles, the production range is 25 to 220 mm in height (includes the finish) and 5 to 420 g in weight.

Other dimension required for the installation of this machine can be seen in the schematic Figure 5,24.

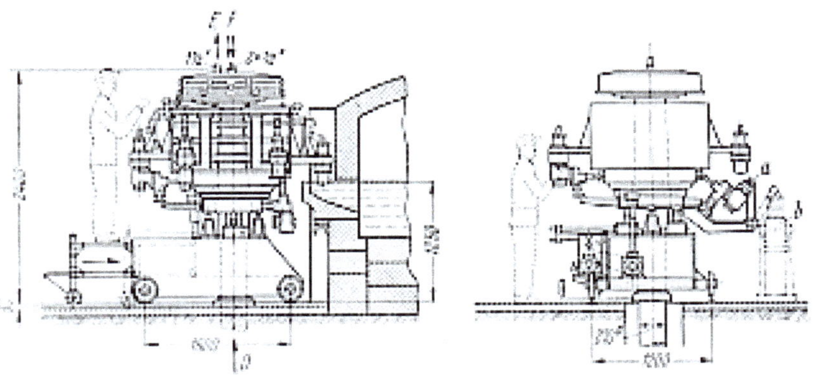

Figure 5,24. Schematic of a KS 6 machine

Rates of production;
'OS' machine. Rate by supplier. Fastest rate 42 heads/min.

Single moulds,
12 000 ale bottles per 8 hours at 640 g wt. and 650 cu. cm contents.
12 000 cognac bottles per 8 hours at 690 g wt. and 700 cu. cm. contents. 10 500 water bottles per 8 hours at 800 g wt. and 1000 cu. cm. contents.

Dual moulds,
18 000 to 19 000 Vichy crown cork bottles per 8 hours for 1350 gm. wt. and 330 cu. cm. contents.
18 000 to 19000 beer bottles per 8 hours at 380 g wt. and 350 cu. cm. contents.

'KS 6' machine. Rate by supplier.

	Wt. (g)	Production (Bottles/day)
Single moulds.		
	50	65 000
	100	50 000
	200	35 000
	400	25 000
Dual moulds	20	114 000

Between the years 1905 and 1926 317 Owens Bottle machines were put into production including 64 shipped abroad. This included models "A", "AC", "AD", "AE", "AL", "AN", "AQ", and "AV". In addition, 119 "AR" machines were manufactured between 1912 and 1941, with 28 being shipped overseas.

In 1923 just 20 years after the successful trial of the Number 4 machine the National Association of Bottle Manufacturers said that 94 out of every 100 bottles in the U.S. were being made by machinery - either semi-automatic or automatic. The last two Owens machines in production, 15 arm "AQ" models, were operated at Gas City, Indiana, until 17th December 1982.

Figure 5,25. Michael Owens with some members of his original organisation

The photograph, Figure 5,25, is of Michael Owens with other members of the original organisation. Owens is fifth from the right.

To quote Meigh, "In Mike Owens the glass industry had a genius who had come up in what could be called 'the hard way', struggling through a carrier in glass factories, from the hardships of early boyhood to the success of important directorships and financial influence. He had the failings of many men of outstanding ability; possessing a dominating personality he was impatient of advice from his subordinates especially if they seemed to know more than he did about some practical problem, and he treated some of them badly."

He despised the work of other inventors in the same field including that of Peiler, to the great disadvantage of his own company, although, as it turned out, to the ultimate benefit of the industry. Nevertheless, despite these shortcomings, he improved the conditions of work in glasshouses, and by his 'wonderful mechanical creations' led the way to automation in the glass world and gained the respect of the engineering world outside. He was awarded the Elliott Cresson Gold Medal in 1915 by the Franklin Institute, a very high honour.

Michael Owens died in his office in Toledo on the 27th December 1923, five days before his 65th birthday, where he was vainly urging his co-directors to agree to the Illinois Glass factory being equipped with 'CA' and 'CB' machines. Many tributes were paid to him as a public benefactor, one such being: "His name will be associated with bottle manufacture as long as civilisation remembers its benefactors" - a fitting epitaph.

In 2007 Michael Owens was inducted into the American National Hall of Fame.

Owens had numerous patents to his name; some of his patents take out in his own name, some taken out with others and patents taken out by his subordinates, i.e. up to his death are listed in the following tables.

The man left the foundations to what is now the largest glass bottle manufacturer in the world Owens-Illinois.

Owens, M.J.	1892	482526	Apparatus for mechanically operating paste-glass moulds.
Owens, M.J.	1893	489543	Apparatus for mechanically operating paste-glass moulds.
Owens, M.J.	1893	506719	Receptacle for leers or annealing ovens.
Owens, M.J.	1894	515838	Device for removing moils from blowpipes.
Owens, M.J.	1895	534840	Apparatus for blowing glass.
Owens, M.J.	1895	548587	Machine for blowing glass.
Owens, M.J.	1895	548588	Machine for blowing glass.
Owens, M.J.	1896	570879	Mechanical glass blower.
Owens, M.J.	1897	588926	Glass-blowers mould.
Owens, M.J.	1899	625513	Glass blowing machine.
Owens, M.J.	1899	635344	Machine for blowing glass.
Owens, M.J.	1899	682595	Art of making glass articles.
Owens, M.J.	1899	759742	Method of blowing glass. (Granted in 1904)
Owens, M.J.	1902	761178	Fire finishing machine. (Granted in 1904)
Owens, M.J.	1904	766768	Glass shaping machine.
Owens, M.J.	1904	774690	Glass melting and shaping device.
Owens, M.J.	1904	774691	Machine for filling leers.
Owens, M.J.	1904	744692	Leer.
Owens, M.J.	1904	840245	Glass tank. (Granted in 1907)
Owens, M.J.	1907	854836	Fire glazing machine.
Owens, M.J.	1908	877729	Process of making sheet glass.
Owens, M.J.	1912	1364514	Glass Gathering and Shaping Machine (Granted in 1921)
Owens, M.J.	1914	1322726	Bottle-blowing machine (Granted in 1919)
Owens, M.J.	1921	1547439	Glass forming machine (Granted in 1925)
Owens, M.J.	1923	1600483	Mechanism for transferring ware to leers. (Granted 1926)

Table 5,7. Owens patents made in his own right

Owens, M.J. & Colburn, H.J.,	1897	576074	Glass blowing machine.
Colburn, H.J. & Owens, M.J.,	1897	594343	Glass blowing machine.
Owens, M.J. & Wilson, A.R.,	1899	682595	Machine for fire finishing glass articles (Granted 1901)

Table 5,8. Owens' joint patents

Bock, W.E.	1902	761139	Fire finishing machine. (Granted 1904)
Bock, W.E.	1900	792581	Machine for forming parisons for glass blowing. (Granted 1904)
Bock, W.E.	1905	796410	Glass tank.
Bock. W.E.	1907	852097	Machine for shaping glass.
Bock. W.E.	1907	852098	Method of shaping glass articles.
Bock. W.E.	1907	852099	Mould for glass shaping machine.
Bock, W.E.	1907	854793	Fire polishing machine.
Bock. W.E.	1907	870664	Machine for gathering and shaping glass.
Bock. W.E.	1909	1063108	Glass blowing machine. (Granted 1913)

Table 5,9a. Patents of Owens subordinate W.E. Bock

La France, R.	1913	1185687	Glass gathering and shaping machine. (Granted 1916)
La France, R.	1913	1296818	Automatic Take-in.
La France, R.	1920	1413742	A Vacuum Capacity Device for Glass Machines. (Granted 1922)

Table 5,9b. Patents by Owens subordinate R. La France up to the time of Owens' death

Haddan, A.J.H.			Conveyor for Bottles.

Table 5,9c. Patents by Owens subordinate A.J.H. Haddan up to the time of Owens' death

Soubier, L.D.	1920	1356174	Means of Transferring Glass to Moulds.

Table 5,9d. Patents by Owens subordinate L.D. Soubier up to the time of Owens' death

The next engineer and inventor we will talk about is the Frenchman Emile Roirant (1882-1955), he made many machines but none where installed in the United States. According to an abstract taken from Glass in the 1936 edition of the J.S.G.T. (Abstract Number 940) "the first Type 'B' machine was installed in the Puy Guillaume factory of the Vichy Company in France. His Type 'F' was marketed in 1929 which had a maximum output of 12 000 bottles per day and his 'M8' machine was developed almost at the same time, the M8 was an 8 mould type, weighing 4.5 tons and measuring about 1·5 metres in diameter. This machine has been working satisfactory for the last five years with an average daily output of 38 000 litre bottles per day of the Vichy type. The 'A-6', a six arm machine came next along with a revolving pot installed in the refining end of the furnace thus saving the expense if firing a separate chamber. His latest development was the 'BB' type designed to make large capacity (5-60 litre or 2·25–26·5 gallons) ware, which has 2 parisons and 2 blow moulds". We are going to discuss his suction type machines here and his flow machines later.

Around 1922 Roirant succeeded in developing an automatic machine for the fabrication of smaller orders by suction, feeder fed machines will follow later. He called it the Roirant Type 'B' mainly because its market was aimed at the bottle, or bouteille, trade. It covered the weight range of 250 to 1500 g, and the Roirant Type 'F', which came along a little later was aimed at the smaller bottles or 'flacons'. These covered the weight range 40 to 350 g.

This description of how it works is essentially for the type 'B' but the types resemble each other. The machine takes the glass

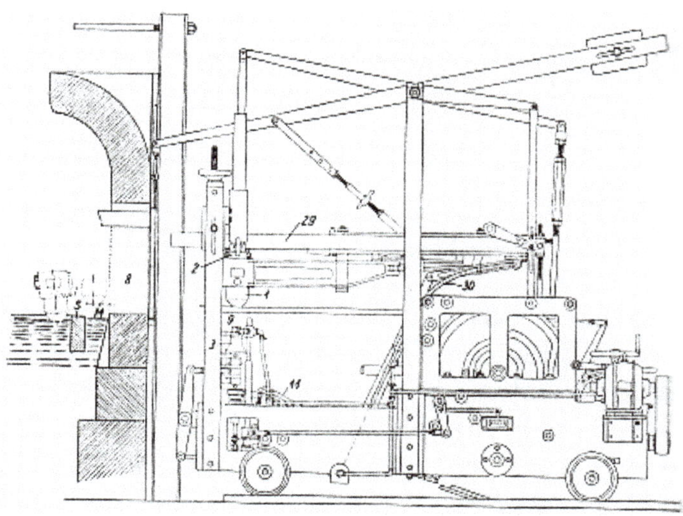

Figure 5,26. Schematic of the Roirant 'B' machine

directly from the tank; for this purpose the suction and neck ring move horizontally between their place above the blow mould and the suction place, dipping into the glass in the furnace for the filling operation. The glass is sucked from a boot while the filled parison mould withdraws, excess glass is cut off while still within the furnace. The cutting falls into the tank between the boot and the tank block and so is kept away from the suction position.

During the transfer of the parison the parison mould stands vertically above the blow mould; the parison hanging from the neck ring drops, passing the opening parison mould, and enters the blow mould. After closure of the blow mould the neck ring opens and the parison hangs free in the blow mould. The neck ring then moves back to its original position, above the parison mould. The finished bottle is then taken out by hand or mechanically. Figure 5,27 shows a 'B' type machine.

Figure 5,27. The Roirant type 'B' machine. French Pat. Nos. 575 544, 589 018, and 589 805

Emile Roirant designed the machine but they were sold by the Société Anonyme D'Études et de Constructions D'Appareils Mécaniques pour La Verrerie, initially from 3, Rue de Castellane, Paris but later, sometime between 1934 and 1936, from the Société Anonyme D'Études et de Constructions D'Appareils Mécaniques pour La Verrerie S.A. of Sternengasse, Basel, in Switzerland. All of the machines were as far as I know made by Ateliers J. Hanrez, in Belgium at Monceau sur Sambré and later from Charleroi a nearby city but maybe this is just the postal address. How Wilzin fits into the equation I do not know.

It took Emile Roirant a number of years to perfect his machine, as shown in abstracts from the J.S.G.T. records. The patent applications are for a group of individuals but they all seem to be connected, the applicants were; Emile Roirant, A. Wilzin and the Societie Anon. D'Etrudes et de Constructions D'Appareils Mécaniques pour La Verrerie.

The 'F' type machine can be seen in Figure 5,28. Between 1925 and 1950 more than 500 type 'B' and 300 type 'F' machines were sold. Unlike the 'B' type machine which has a control wheel

Figure 5,28. Roirant's type 'F' machine

to regulate the working times and the production speed, the manufacturing speed of the 'F' type machine is regulated by the diameter of the pulleys, the relationship is as shown in Table 5,10. The motor shaft speed is 1.420 rev/min. (1500 rev/min. theoretically) and the distance between the pulley axes is 450 mm. The same pulley belt can be used for each set of pulleys.

Manufacturing speed (s)	No. objects made per minute	Dia. of motor pulley (m/m)	Dia. of machine pulley (m/m)
14.8	4.06	95	144.5
13.5	4.43	100	139.5
12.5	4.80	105	135.0
11.5	5.23	110	130.0
10.5	5.76	115	125.0
9.7	6.17	120	120.0
9.0	6.70	125	115.0
8.2	7.30	130	110.0
7.5	7.93	135	105.0

Table 5,10. Manufacturing speed as a function of the pulley diameters

The machines have some control over the speeds quoted above, the theoretical and actual manufacturing rates for the 'B' and 'F' types along with bottle sizes can be found in the following two Tables 5,11 a and b, and 5,12 a and b.

Bottle weight (g)	Total (Pcs./min.) Theoretical	Actual
300	4.6	
400	4.3	at 350 g 4.7
500	4.0	at 500 g 4.2
600	3.7	
700	3.5	at 700 g 4.0
800	3.3	at 850 g 4.0
900	3.1	

Table 5,11a. The maximum working rate of the Type B machine was 4.7 pcs/min

Limiting bottle dimensions	
Height with finish	140-360 mm
Max. outside diameter	120 mm
Finish Diameter	60 mm
Weight	250-1500 g
Max. height of bottom pushup	50 mm
Plunger max. diameter	49 mm

Table 5,11b. The operating parameters of the Type 'B' machine

Tables 5,12a and 5,12b showing various operating parameters are shown below.

Bottle weight	**Total (Pcs. / min.) Theoretical Single mlds (EF)**	**Dual mlds (DF)**	**Actual**
50	8.0	14	at 60 g 13 DF
100	7.1	12	at 110 g 9.0 EF
150	6.3	11	
200	5.7		at 200 g 6.5 EF
300	5.8		
400	4.2		at 400 g 5.0 EF

Table 5,12a Showing the theoretical and actual speeds for both single and dual operation. The maximum rate for single moulds (EF) was 9.0 pcs/min, and for double moulds (DF) 16.0 pcs/min

Limiting bottle dimensions	**Single mlds.**	**Double mlds.**
Height with finish	50-220 mm	50-220 mm
Outside diameter	to 100 mm	to 45 mm
Finish diameter	to 55 mm	to 25 mm
Weight	40-350 g	40-150 g

Table 5,12b. Dimensional ranges for bottles that could be made on the Type 'F' machine

In a anonymous abstract from the National Glass Budget which appeared in the 1936 edition of the J.S.G.T., Abstract Number

722 the following average outputs for a variety of bottles were quoted as;

	Weight oz.	Capacity oz.	Glass Colour	Ave.prod. Per 24 hrs Gross
Model 'B' Machine				
1 Qt. Crown cork mineral	25.33	35.0	Green glass	24*
1 Qt. Whiskey	18.00	36.0	Flint	26
1 Qt. Beer, swing stopper	30.00	30.0	Green	26*
1 Pt. Beer, swing stopper	18.50	16.0	Green	31*
0.2l. Rhine wine	26.50	26.5	Green	25*
0.2l. Burgundy	25.33	26.5	Pale green	28*
Model 'F' Machine				
0.5 Pt Crown cork mineral	11.25	8.5	Green	49*
Sample brandy	4.25		Pale green	80
Milk bottle 45 mm capacity	11.25	8.5	Flint	45*
Square perfume	3.625	1.75	Pale green	90
Lemonade	15.75	15	Green	47*

Total marked with a * are goodware packed.

Table 5,13. Production rate of the Roirant B and F Type machines

Wherever possible the 'B' and 'F' types used interchangeable ancillary equipment. Illustrations are shown in the following diagrams.

SHAPING MOULD HOLDER, Model A

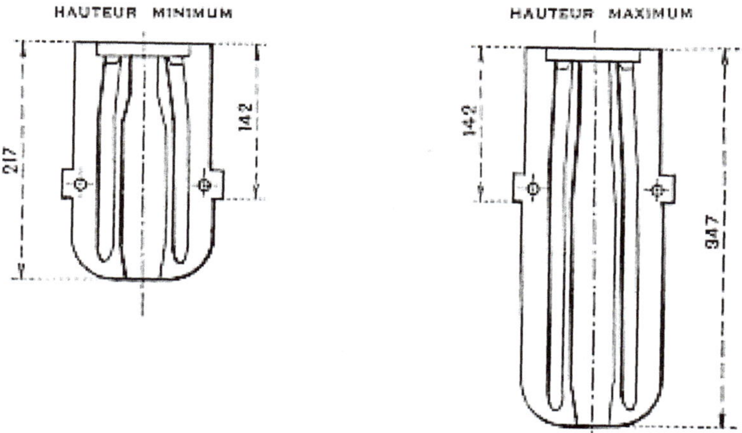

SHAPING MOULD HOLDER, Model B

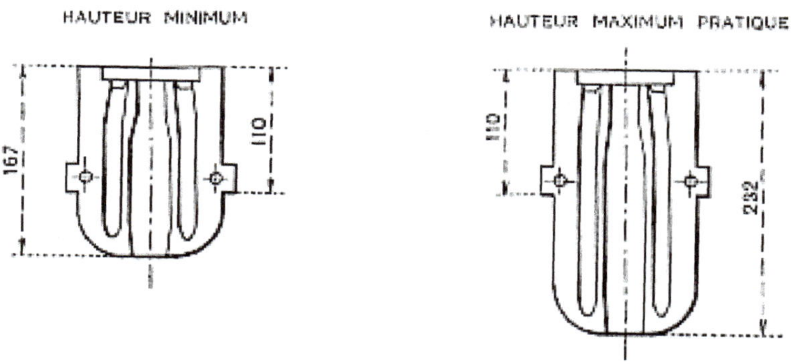

SHAPING MOULD HOLDER, Model D

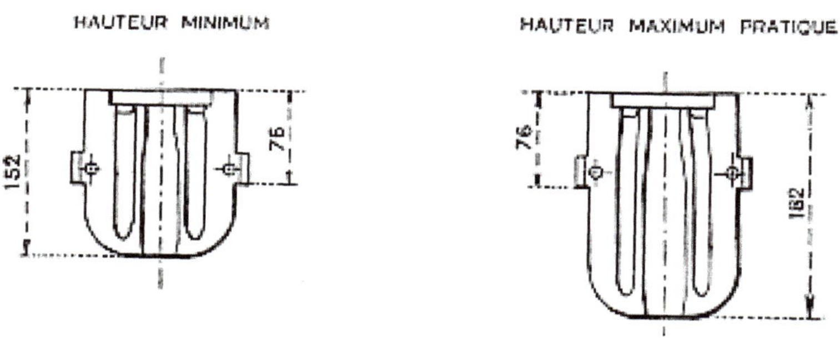

Figure 5,29. Limit dimensions for Shaping Mould Holders, Model C, not shown here, is for large diameters up to 130 mm

FINISHING MOULD HOLDER, Model A

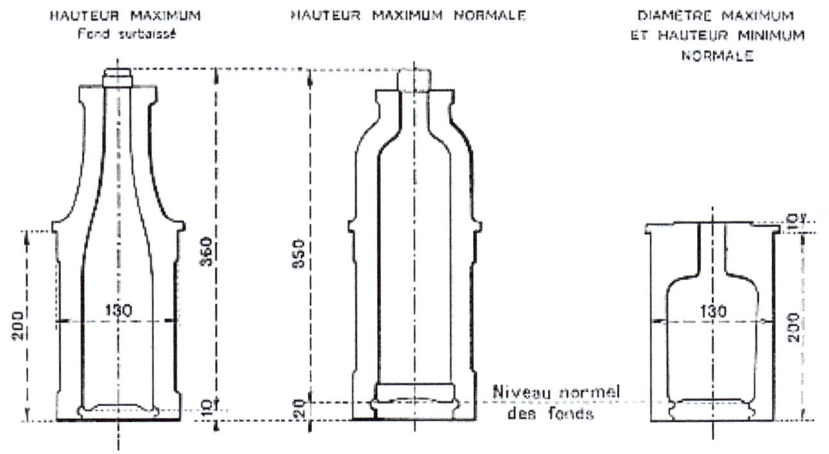

FINISHING MOULD HOLDER, Model B

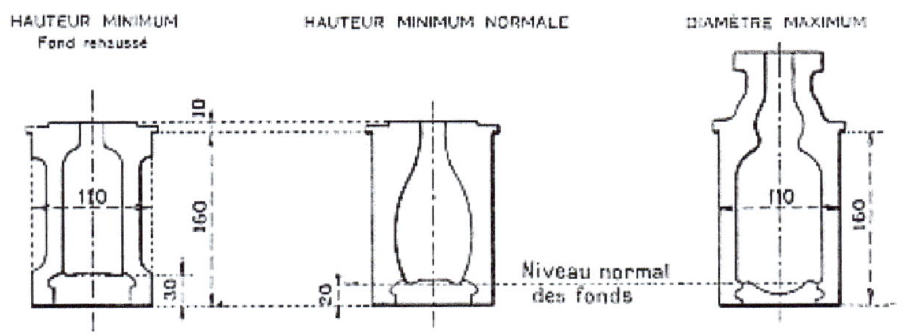

FINISHING MOULD HOLDER, Model C

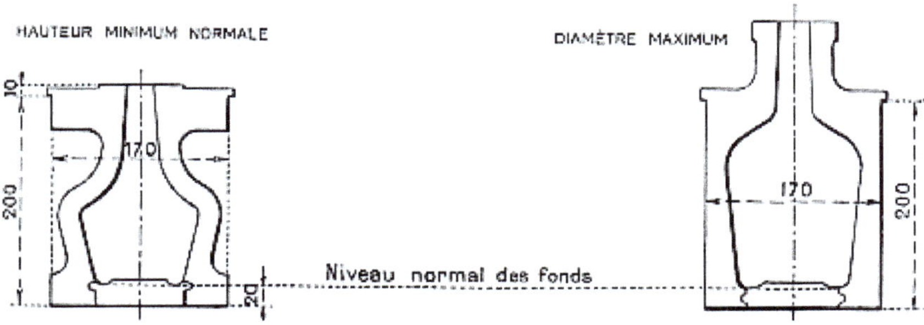

Figure 5,30. Limit dimensions of Finishing Moulds

RING MOULD HOLDER, Model A

MOULE DE BAGUE EN 2 PIÈCES
ET MANDRIN RAPPORTÉ

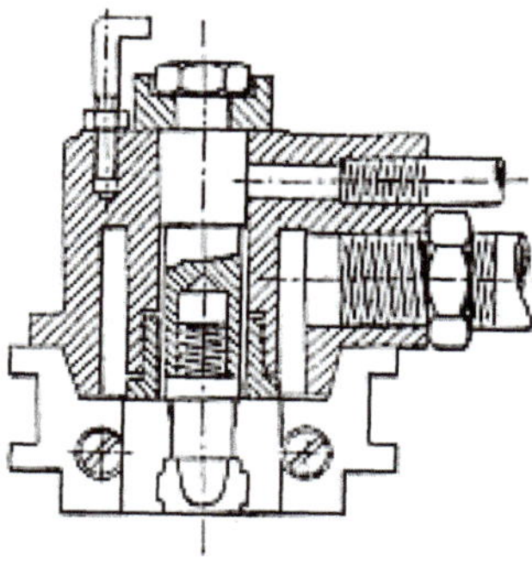

MOULE DE BAGUE EN 3 PIÈCES
DIMENSIONS MAXIMUM
DES BAGUES

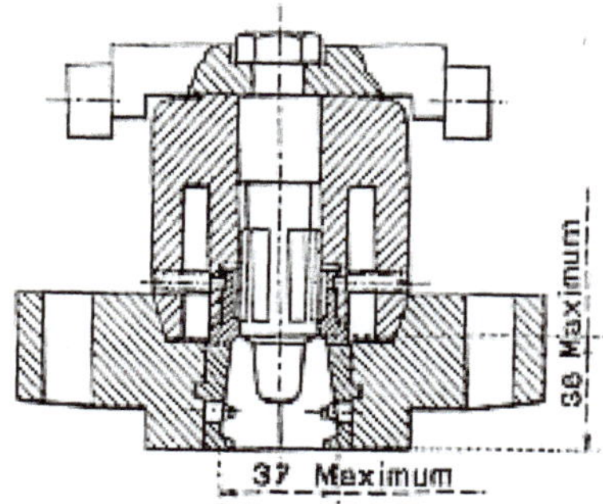

RING MOULD HOLDER, Model C

MOULE DE BAGUE EN 2 PIÈCES

MANDRIN A AILETTES

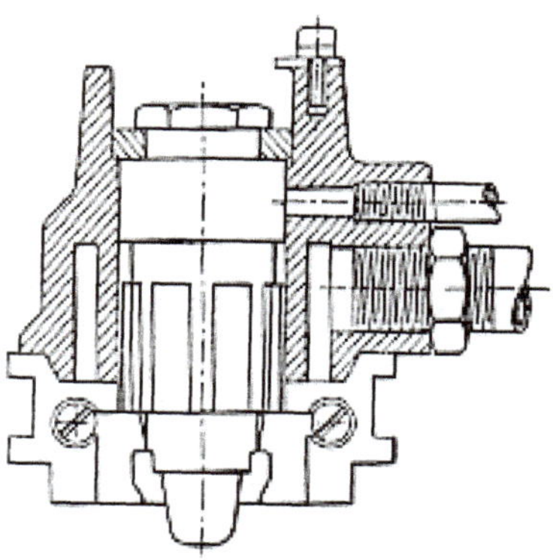

MANDRIN A AILETTES
AVEC REFROIDISSEMENT INTÉRIEUR

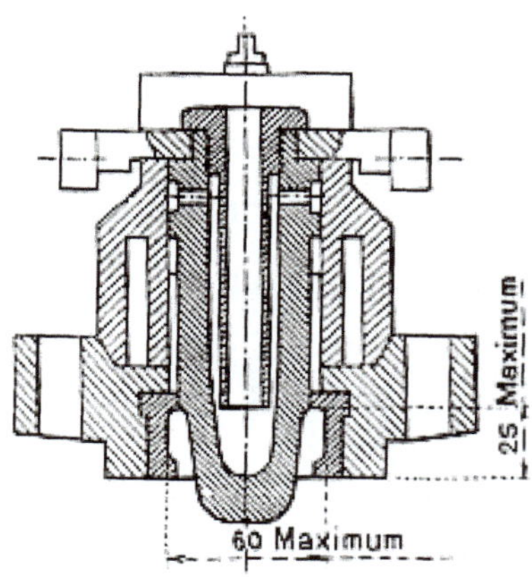

Figure 5,31. Limit dimensions of Ring Mould holders

SIDE-ARRANGEMENT OF THE FURNACE

AT THE DRAWING-POST

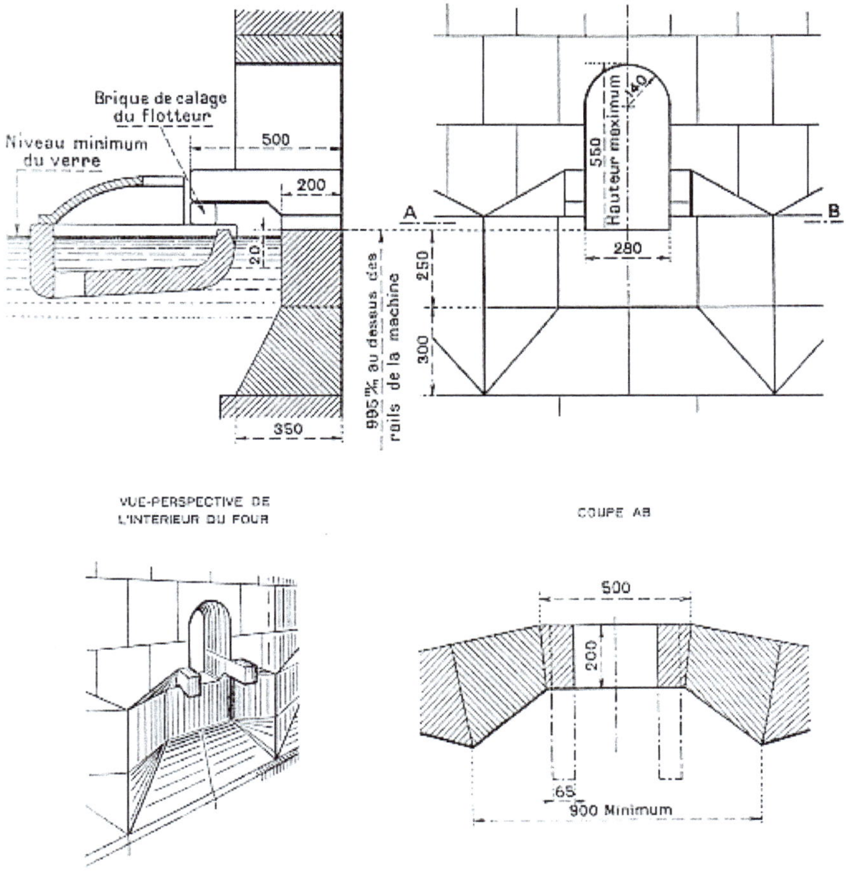

Figure 5,32. Details at the Drawing Post, showing the access hole and the floating boot

The Drawing Post is the name given to the furnace opening and the floating boot, it shows the neatly tailored opening that was required for the machines. The two small arms that protrude out from the furnace wall are used to keep the floater in position by buoyancy.

The position of the floater is critical to the operation, it needs to be placed so that the gather always comes from the centre, if the gather comes from the side then the thermic homogeneity will be wrong and, if too far back the sheared glass will fall back into it rather than between the floater and the end wall.

The floater size must be adequate for the required weight range. The level of the glass surface inside the floater must not vary too much, the changes to the level are brought about by the speed of glass extraction being greater the speed of glass replacement.

ONE-PIECE SCREEN

(Fig. 2)

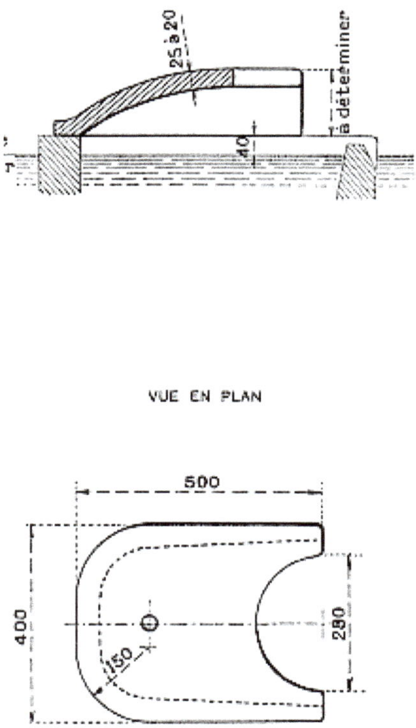

Figure 5,33. Drawing of one piece cover

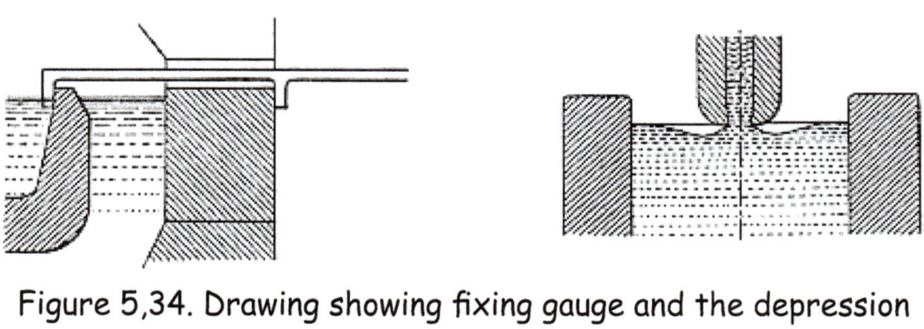

Figure 5,34. Drawing showing fixing gauge and the depression caused to the glass surface when making a gather

Periodically it is required to reposition the floater, a gauge such as the one shown in the left hand drawing is very helpful.

Around 1928 Roirant brought out a new, higher productivity, suction machine designated the Roirant Type 'A6'. It had six heads and rather than draw glass directly from the working chamber, (or working end), of the furnace it used a small revolving pot. He sold many of these machines and around 1970 there where about 60 of these machines still working. Once again no A6's were sold in the United States. The machine had twin tables each with six moulds, each station having one parison and one blow mould, it produces glass in the weight range 300 to 1500 g. The parison dipped into the glass each time it made a gather and being a suction machine it was possible to make within reason different bottles of different weights at each station.

Legend for Figures 5,35 and 5,36

1. Revolving pot wt. approx. 1500 kg
2. Roirant machine 'A6' wt. approx. 5100 kg
3. Melting tank
4. Overall length 2400 mm
5. Approximate height 2500 mm
6. Space for moving machine back 2000 mm
7. Rails
8. Overall width 1950 mm
9. Distance between centre line of machine and centre line of damper 105 mm
10. Approximately 1095 mm
11. Furnace front
12. Spout
13. Control of damper
14. Damper
15. Height of revolving pot 1370 mm
16. Diameter. of revolving pot 1500 mm
17. Height from rails to glass level 1660 mm
18. Glass level
19. Bottle chute

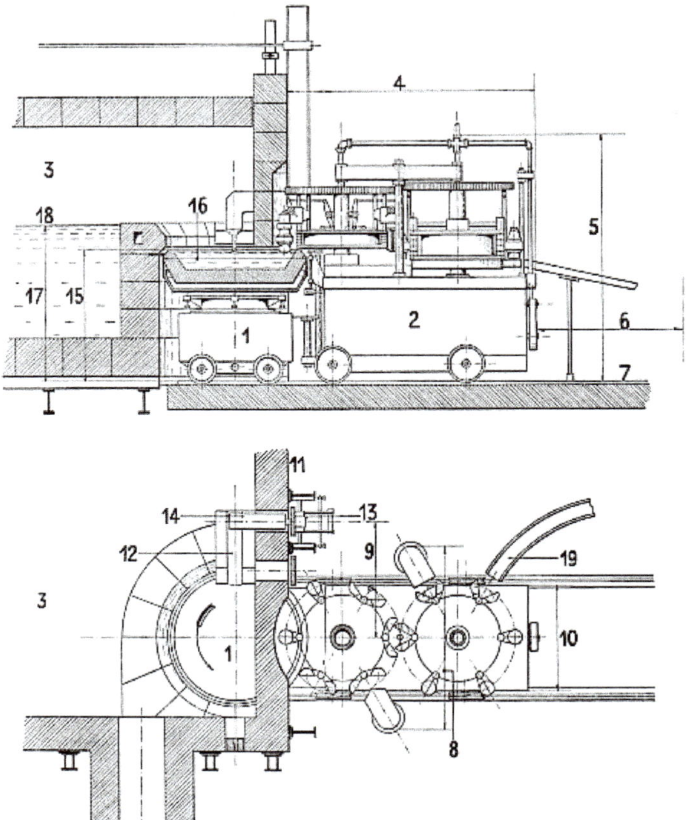

Figure 5,35. Layout of an A6 machine

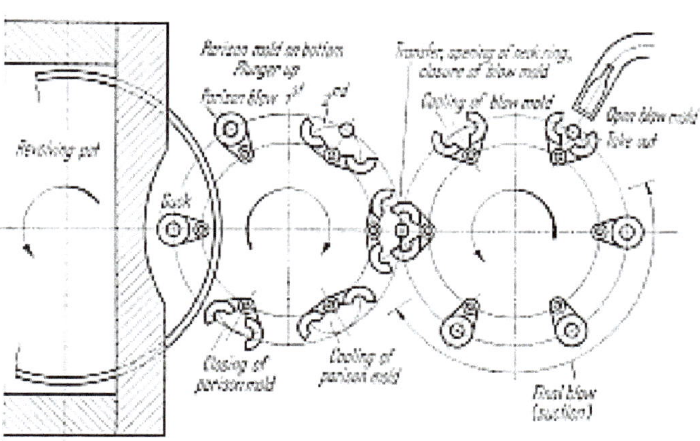

Figure 5,36 . Operational sequence

The working principle;

Station 1.

Dipping of parison mould into glass mass of revolving pot, suction, lifting and cutting, and, on the way to the next station, plunger withdrawal;

Station 2.

Dropping the filled parison mould onto a hollowed shaping block, first parison blow, lift while turning, opening of parison mould;

Station 3.

Dropping and ramming of parison, second parison blow by a small blast, lifting;

Station 4.

Dropping, closure of the blow mould, opening of head mould, lifting;

Stations 5-7.

Blowing of parison in blow moulds, by vacuum;

Station 8.

Opening of blow mould and bottle ejection.

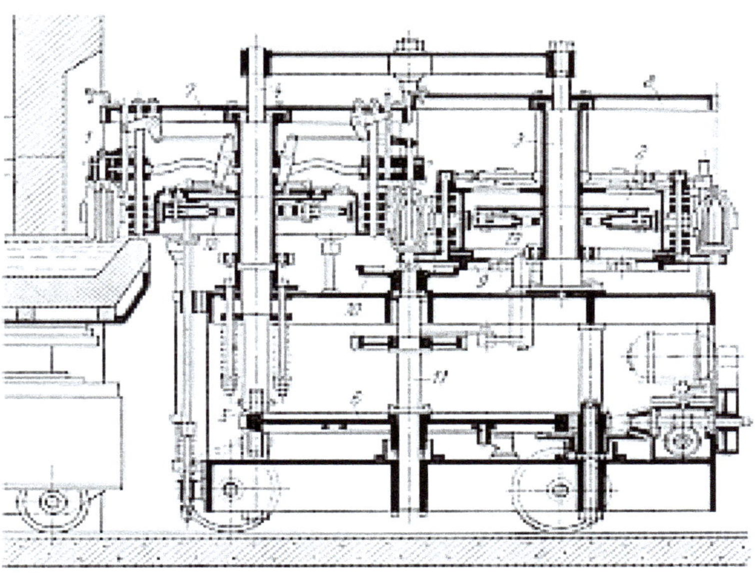

Figure 5,37. Construction of a Roirant A6 machine

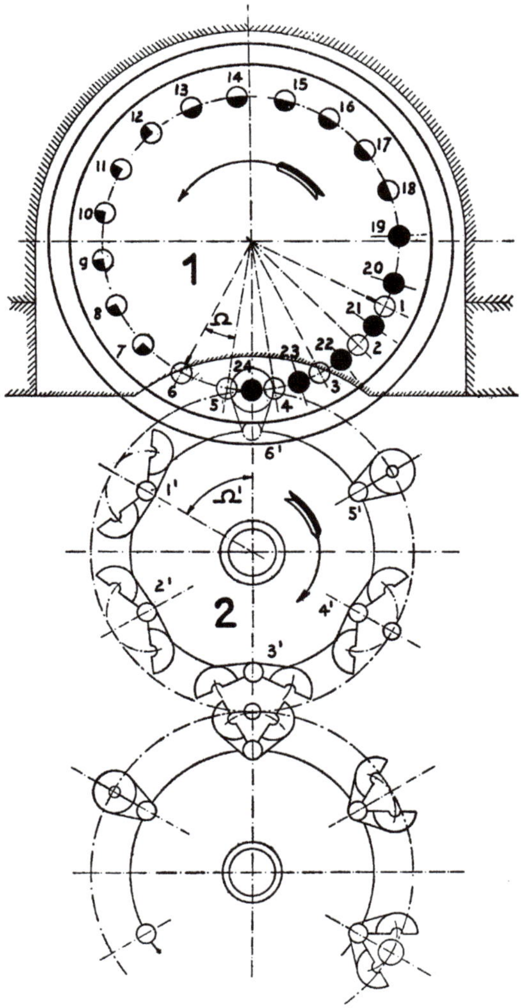

Position of the 4x6=24 drawing spots 1 to 24 is the revolving pot after four complete revolutions of the parison mould table.

- ○ suction spots 1 to 6 - first revolution of the parison mould table 2
- ◔ suction spots 7 to 12 second revolution of the parison mould table 2
- ◒ suction spots 13 to 18 - third revolution of the parison mould table 2
- ● suction spots 19 to 24 - fourth revolution of the parison mould table 2

Figure 5,38. Diagram showing the position of the dips in the pot

ROIRANT
TYPE „A. 6" MACHINE

The photograph illustrates the first machine at the trial tank.—
Position when drawing the glass from the revolving pot.

SOCIÉTÉ ANONYME D'ETUDES ET DE CONSTRUCTIONS
D'APPAREILS MÉCANIQUES POUR LA VERRERIE
3, rue de Castellane, PARIS 8

Figure 5,39. Roirant's prototype A6 machine from a 1934 brochure. Note the address

The rotational speed of the pot and the machine are adjustable but they are linked together in such a way that allows the machine to dip the second time between two of the dips

ROIRANT
TYPE "A 6" MACHINE

SOCIÉTÉ CONTINENTALE D'APPAREILS MÉCANIQUES
POUR LA VERRERIE S. A.

Sternengasse, BASEL (Switzerland)

Figure 5,40. Copy of 'A6' machine from 1936 brochure. Note the address

of the first rotation. See Figure 5,38 which explains this in a schematic form.

Figures 5,39 and 5,40 show both the prototype A6 machine and the commercial machines.

In terms of output and bottle specifications of the A6's figures are shown in Tables 5,14 and 5,15.

Bottle weight (g)	**Total (Pcs/min)**	
	Theoretical	**Attained**
300	24.0	At 340 g 27.0
400	22.0	
500	20.5	
600	18.7	At 550 g 21.0
700	17.7	At 700 g 19.4
800	16.6	At 820 g 18.5
900	15.6	At 870 g 17.5-18.0

Table 5,14. Pieces per minute for various bottle weights

Height with finish	150-360 mm
Outside diameter	to 120 mm
Finish diameter	36-60 mm
Weight	300-1500 g
Plunger diameter	53 mm

Table 5,15. Range of bottle dimensions

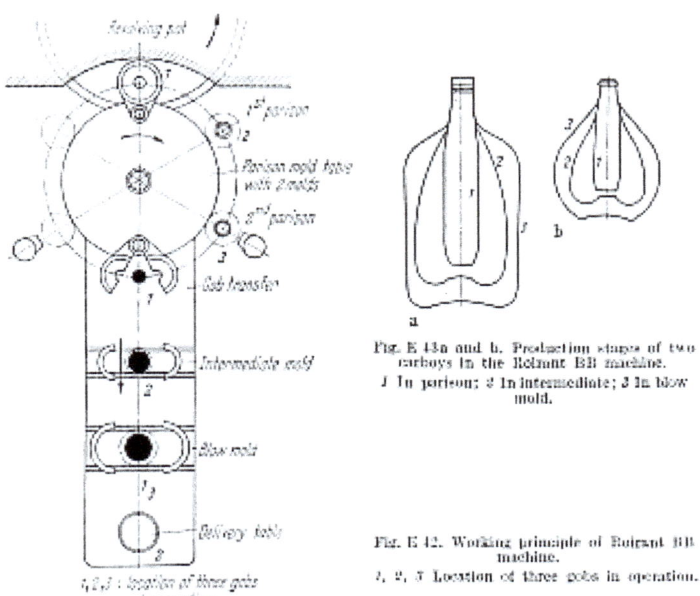

Figure 5,41. Roirant's Type 'BB' machine

The A6 permits the making of different shapes and sizes at the same time, with weights variable by as much as 150 g. The economical limit is probably by two seconds per bottle in fabrication times, i.e. 20-22 s/bottle, or 3-2.7 bottles/minute.

In 1935 Roirant succeeded in the manufacture of large ware automatically and marketed his machine as the Roirant Type 'BB'. It is the only machine in the world to make containers that hold between 10 to 60 litres in contents, and there is about 15 machines working in the world (circa 1965).

At station 1, the parison sucks glass up from the surface of a revolving pot, just like the A6. At stations 2 and 3 the parison is blown. At station 4 after opening the parson it is transferred by a transporter to an intermediate mould at station 5 (marked as intermediate mould on diagram) and vacuum blown to an intermediate shape. Finally the parison is transferred to station 6, the blow mould, where it is blown to size. The carboy of large bottle is then manually extracted and transferred to the lehr.

The output and dimensions of the 'BB' machine are shown in Tables 5,16 - output and 5,17 - dimensions;

Bottle weight g	**Total (pcs/min)**	
	Theoretical	**Actual production figures**
1500	2.4	at 2500 g 2.4
5000	1.6	at 4800 g 1.7
		at 6500 g 1.3
7500	1.2	at 9000 g 1.0

Table 5,16. Actual and theoretical speeds for different bottle weights

Max. head diameter	80 mm
Max. plunger diameter	59 mm
Max. body height	465 mm
Max. total height	320-640 mm
Capacity	10-60 l

Table 5,17. Maximum dimensions and capacity

Diameter (mm)	Height (mm)	Weight (kg)	Capacity (litres) Cylindrical	Spherical
275	400	2.5	---	10
375	500	5.0	25	30
450	600	7.5	30	50

Table 5,18. Approximate dimensions of commercial carboys

Figure 5,42. Advertisement from the 1934 Edition of the Directory for the British Glass Industry

Next is the Redfern machine. It was a project aimed at designing a machine that followed the Owens production principles but avoided the patent restrictions applied by Owens. The design of the machines was given to Robert Frederic Hall and Sidney Hunt of Mosley, Birmingham, England the finished machines to be used by John Lumb & Co. Ltd a wholly owned subsidiary of John Walker Ltd of Kilmarnock, Scotland.

They only worked at the Castleford plant in Yorkshire but two 15 arm machines where sold to Mitsubishi Shoji Kaishy Ltd on behalf of the Kirin Beer Krishna of Japan. A 12 arm machine, or possibly two, were later sold but these were lost at sea during transit. The machines were named Redfern after the then Managing Director of John Walker's, Francis Redfern.

The abstracts show that the first patent were applied for on the 10th July, 1917, meaning that work was started most probably in 1916 or even before. This was well before the feeder became a commercial proposition so that any automatic machines developed had to be of the suction type. John Lumb's in 1914 had 38 machines of different types, all were semi-automatic machines so the step from hand gathering to automatic gathering would have seemed to be a logical step forward in the evolution of bottle making machines.

The designs were for four types of machine, a 6 arm, a 10 arm, a 12 arm and a 15 arm, moulds to be single, double or triple according to the bottle size, the machines were designed to make bottle and jars by both the press and blow and the blow and blow processes. Only the 10, 12 and the 15 arm machines were made.

The first machine, a 10 arm, started production around 1920, this was followed by the much improved 15 arm version which was put into production on the 17th February 1927. So pleased was Francis Redfern about his new machine that he gave a paper to the Society of Glass Technology on the 18th of May 1927.

The following are extracts mainly from his paper, but some are also from Dralle. Ideally they need to be read in conjunction with the patent applications as they describe how each works. The machine consists of two elements, a centre one that is fixed and an outer on that moves, see Figures 5,43 and 5,44 for a 10 arm machine.

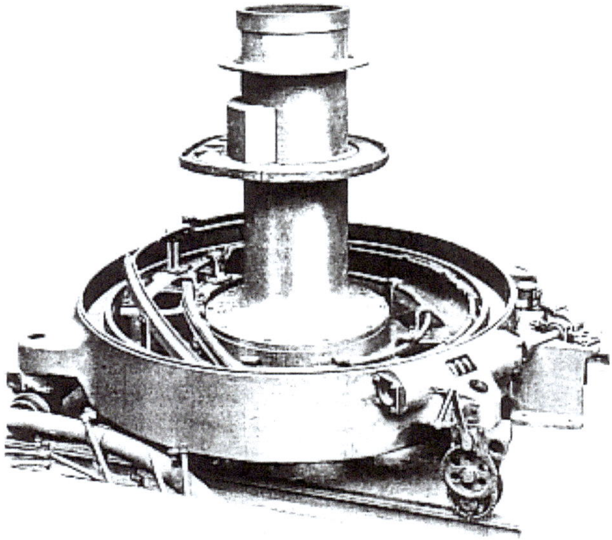

Figure 5,43. 10 arm Redfern centre column

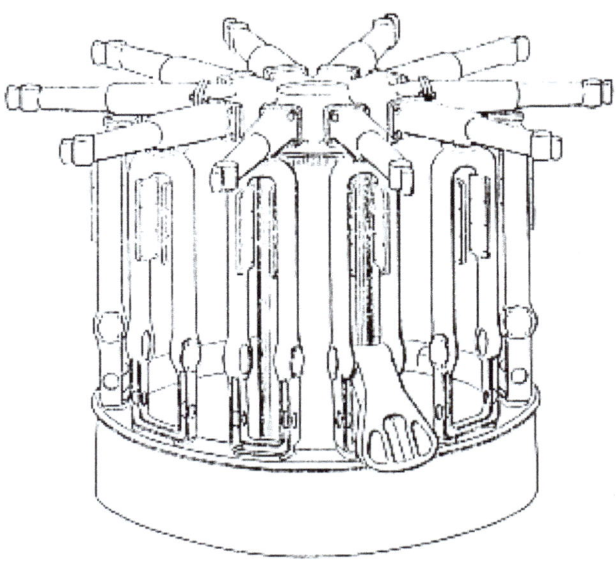

Figure 5,44. Part of machine that rotates around the centre column

The cams in the base of the machine were experimental and have now been superseded. In essence this element consists of a circular trough mounted on four wheels to run up to and from the furnace on a set of rails. In the base of this element lie the cams which raise and lower the tables carrying the finishing moulds and also open and close them.

A hollow tubular structure rises from, and is attached to, the centre of this base to which all of the cams operating the moving parts of the machine are fixed, other than the two aforesaid cams that have just been mentioned. A case is bolted to the outside of the trough housing a worm which is driven by a motor and let us imagine that the trough is partly filled with caster oil for lubrication.

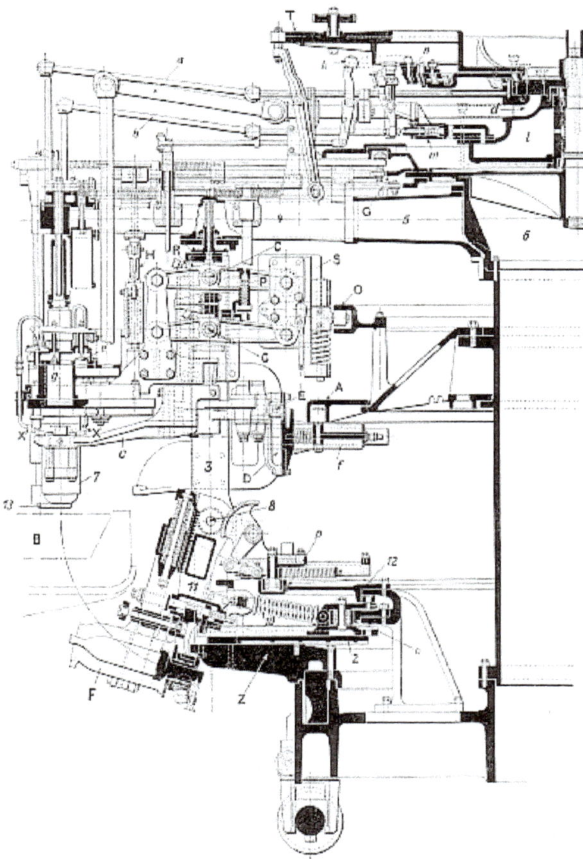

Figure 5,45.
A line drawing through a 15 arm machine

Considering the movable element, from the bottom upwards there is the ring, the lower edge forms a bearing. On the upper horizontal face of the ring are built a number of bifurcated columns or elongated loops. On top of these bifurcated columns are an equal number of radial arms, all connected to, and communicate with, a central chamber. The bifurcated columns, the radial arms, and the central chamber are hollow to allow for the passage of cooling air.

The moveable part - the outside element - revolves around the column driven by the worm into which a toothed part of the movable base fits. All of the units or arms bolt on to this outer element. Figure 5,45 shows a line diagram of one of the units, from the cam roller 'A' to the nose of the parison mould dipping into the pot at point 'B' there are two parallel levers connecting both of which pivot on the bifurcated column. As the roller travels round the stationary cam, the nose of the parison is lowered or lifted from the glass in the pot. Figure 5.45 also shows how the parison is counter balanced while the blowing mould is not.

Figure 5,46. 15 arm Redfern with the blank mould in the dipping position as it passes under the pot

Figure 5,46 shows all of the important cams which can give easy motion up to speeds of 6 or 7 revolutions per minute. It also shows the hollow radial arms which lead cooling air from the central column down the hollow bifurcated columns through the hollow tables to the moulds.

All cooling exits are controlled by self-locking valves. By these means not only are the moulds, the knife, and other parts kept cool, but the whole of the framework of the moving element of the machine. The framework can be tapped into at any point thus ensuing that the machine is not covered with a nest of cooling pipes. In case of damage a unit can be exchanged in 45 minutes, the new unit being ready for bottle making immediately.

The next photograph, Figure 5,47, shows the unit from the other side. Here you can see the arm reaching out from the bifurcated column carrying the knife spindle. This arm is hollow

Figure 5,47 15 arm machine from the outlet side of the pot

allowing cooling air to pass from the bifurcated column through the hollow casing which carries the knife spindle bearing - so keeping the bearing and spindle cool. This figure also gives an indication of the dipping section of the cam which controls the position of the parison head.

And Figure 5,48 shows the position in which the bottle is blown. The arms carrying the parison mould have opened to permit the finishing mould table to come up to the horizontal bottle making position and the two halves of the finishing mould have closed around the parison, which as it hangs like a skittle from the ring mould, and so, in this position the bottle is blown.

Figure 5,48. 15 arm Redfern with mould in the blowing position

Figure 5,49 shows two units mounted on the machine. The one on the left is in the dipping position and the one on the right is in the blowing position. At the top of each of these units is a tank from which oil is fed to the moving parts by gravity. Most of the rollers operating on the cams are ball-bearing rollers.

The most important bearings on the machine are made by Timken, both the roller and the ball types, they are sealed from the dust and filled with grease, occasionally they need replenishing using a grease gun.

Figure 5,49. 15 arm Redfern with two units mounted. One in the dipping position and the other in the blow position

Figure 5,50 shows the moulds in position with the knife. The knife can be adjusted under the bottom of the parision mould without the use of a spanner, the operator can make the adjustment while the machine is moving. Adjustments available are to raise and lower the knife underneath the head, to raise and lower the head from the knife after cut-off or to do both can be done simultaneously. The object is to avoid scarification of the bottom of the parison by the knife.

An adjustable cam enables the position to be varied when connection is broken between the bottom of the parison mould and the knife. Similarly an adjustable cam enables the position to be varied at which the plug is lowered and raised.

Figure 5,50. The mould system and the knife

The two halves of the parison mould are carried on arms that can be detached from the main bearing from which they hang. The two halves of the finishing moulds rest on by their own weight on a horizontal plate resting on the finishing table. They are opened and closed by hinges located at the mid point of the moulds.

Figure 5,51 shows the two halves of a parison mould with half of a ring mould set between the two. These are shown dismounted from the machine, but when mounted on the parison arms each half of a parson mould is carried on three points in such a manner that as the two halves come together they automatically find their correct relationship one to another; and since they close around a register on the ring mould carriers located in the blowing head, they automatically find their correct relationship with the ring moulds. As is known the plug sits on top of the ring moulds and the suction comes through the space left between the plug and the ring and across the meeting faces of the parison mould from the channel which is shaped to form an embryo or parison of the bottle.

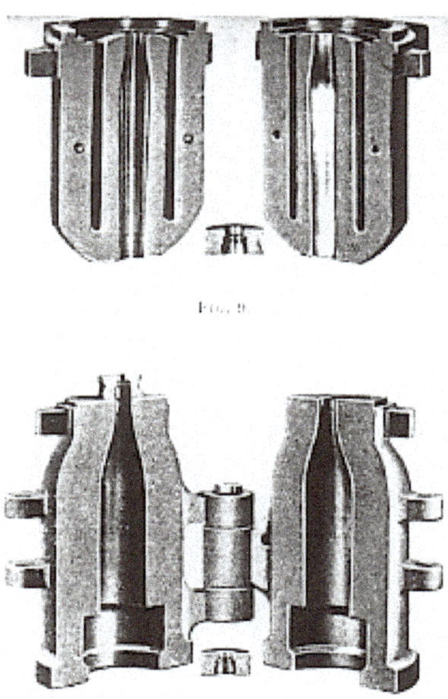

Figure 5,51 Mould equipment

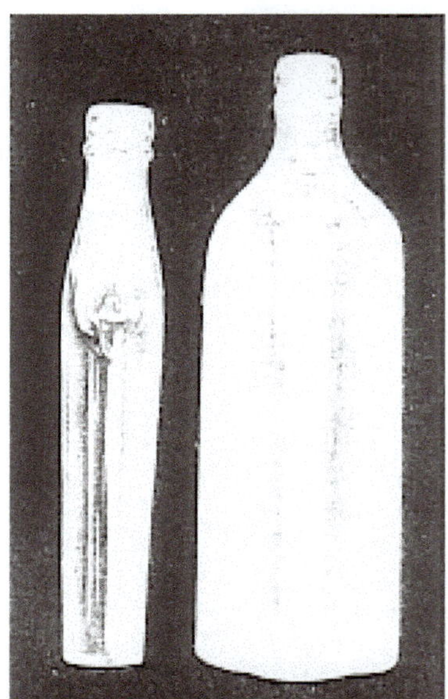

Figure 5,52. Parison and bottle blown from such

The bottom half of Figure 5,51 shows the two halves of a finishing mould dismounted from the machine. The left half shows a half ring mould on top of the finishing mould. This illustration does not show the hinge post but the latter is fixed to the table in such a way that it is easily removed when changing moulds.

The adjacent Figure 5,52 shows a parison of a square quart bottle alongside a bottle blown from one such parison. Figure 5,53 is a view from the top of the machine and shows in a circle the blowing cams. These blowing cams are in small segments and can be adjusted outwardly from the centre of the machine and inwardly towards the centre by hand whilst the machine is going, so that it is unnecessary to stop the machine to adjust the control of the bottle blowing.

There are 10 tiers of valves which can be adjusted and by shaping each set of cams it is possible to make 10 different bottles at anyone time.

There are approximately 40 tons of moving parts mounted on the stationary element giving a total weight of about 60 tons. The machine will revolve up to 6 or 7 revolutions per minute. At 6

Figure 5,53. Oiling position 15 arm machine

revolutions per minute the machine can hardly be heard to move, at this speed utilising an electro-mechanical brake the machine can stop within two and a half feet. The power consumption at 6 revolutions is 11.5 horse-power.

The range of bottles that can be made are from the neck ring parting line to the base is from 2.5" (64 mm) to 15" (382 mm), with a maximum blow mould diameter of 9" (228 mm) and of parison mould up to 7" (178 mm) maximum. Wide mouth ware can be manufactured in which the outside diameter of the neck mould is 5.25" (133 mm) and the maximum length of plunger which can be entered into the bottle is 2.875" (73 mm). For dual moulds diameters of up to 2.5" (64 mm) can be made on the machine for example a reputed pint bottle. Smaller ware can be made on triple moulds.

The 10 arm machine has produced 1678 gross of merchantable quart bottles per week, these are for reputed square quarts, round quarts would be in excess of this number. Castleford has produced 2042 gross of merchantable square pints in a week working single moulds. We estimate based on our experience that on a 15 unit machine working dual moulds it is capable of making 120 pints per minute or say 7500 gross of merchantable bottles per week.

"Some people mistakenly believe that mixed jobs cannot be made on a suction machine. With a six unit machine it is within the mechanical limits of adjustability it is possible to make six different jobs with height differences of up to 3.5". The variations in weight would depend upon glass conditions, temperature, and the skill with which the mould equipment is designed. At Castleford with a 10 arm machine we have made bottles varying in weight by 4 ounces."

Machine	**Reputed Sq. Quart (0.756 l)**		**Reputed Sq. Pint (0.378 l)**	
Size	**Single (gross)**	**Dual (gross)**	**Single (gross)**	**Dual (gross)**
6 Arm	1120	1680	1473	2240
10 Arm	1866	2800	2489	3733
12 Arm	2240	3360	2987	4480
15 Arm	2800	4200	3733	5600

Table 5,19. Estimated rates of output from Redfern

Redfern states that the machine can be produced in three sizes; 6 units, 10 units and 15 units whereas Dralle quotes that 6, 10, 12 and 15 arm machines can be made. He gives rates of output for Reputed Quarts and Pints being made on all four machines as in Table 5,19 above.

Other output figures from Wendel state the production in pieces/24 hours;

No. of arms	0.756 l		0.378 l	
	Single mould	Dual mould	Single mould	Dual mould
6 arms	26880	40320	35832	53760
15 arms	67200	100800	89600	134400

Table 5,20. Estimated output in pieces/24 hours from Wendel

Finally Figure 5,54 shows a part assembled 10 arm machine where you can see by comparing it with Figure 5,44 how much the

Figure 5,54. 10 arm outer section part assembled

Figure 5,55. The 15 arm Redfern machine

moving element has been tidied up all of the heavy pipe work has been simplified. Figure 5,55 shows a complete 15 arm machine.

The following are some of the German Patents' the patent numbers are;

443 469 vom 8. VII. 1920, priority, vom 10. VII. 1917
447 427 vom 5. XI. 1920, priority vom 8, XI. 1919
470 901 vom 2. VII. 1920, priority vom 2. VII. 1920
474 752

The **Hillman** machine, see Figure 5,56 was in development from 1920 to 1928 and it too was an improvement to the construction of the Owens machine using several new ideas. However its final perfection in practice was prevented and not pursued due to lengthy patent litigation problems.

The chief purposes of this 10 unit machine when designed was the simplification of the machine and the furnace construction.

Figure 5,56. Hillmann ten arm machine

New ideas compared to the Owens were, transfer from the parison mould (2), to the blow mould (1) by lowering the neck ring (4) instead of lifting the blow mould. The neck ring, the parison and the blow mould halves all close using springs, this gave troubles due to the release springs yielding a little during pre-blow.

As all of the cams where in the base this led to the machine being able to manufacture bottles of different shapes and dimensions. It also allowed for the vacuum and air to be controlled automatically by an electric control system. (Ger. Pat. 552 092 inventor Hillman). The control drum was situated at the top of the machine.

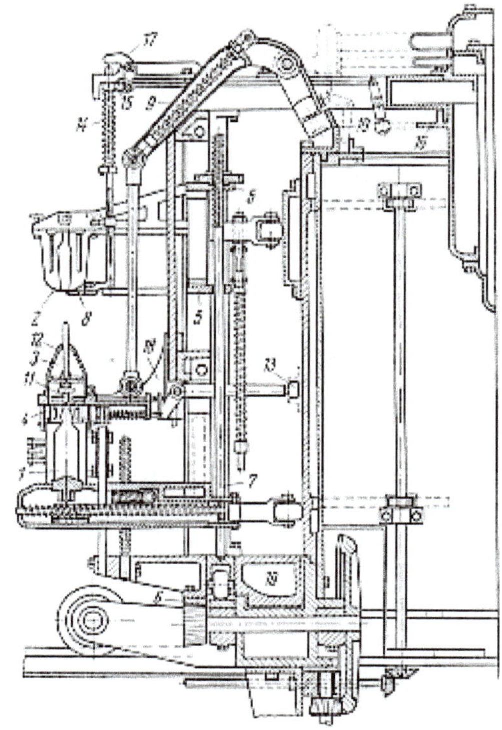

Figure 5,57. Schematic of the Hillman machine

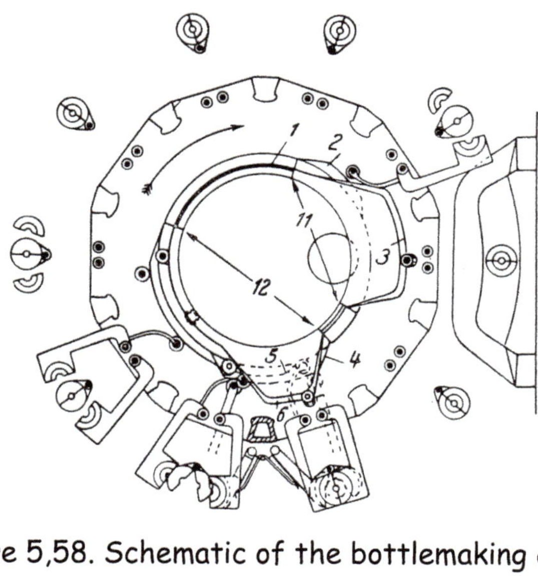

Figure 5,58. Schematic of the bottlemaking cycle

To eliminate the revolving pot Hillman provided a fixed basin-like protrusion of the tank for the dipping of the parison moulds. This worked successfully for colourless glass but not for green. To remove the cut off scars (knife cuts) from the glass surface either skimmers preceding the parison mould or guiding walls which adjusted the convection currents to give a new glass surface for each dip.

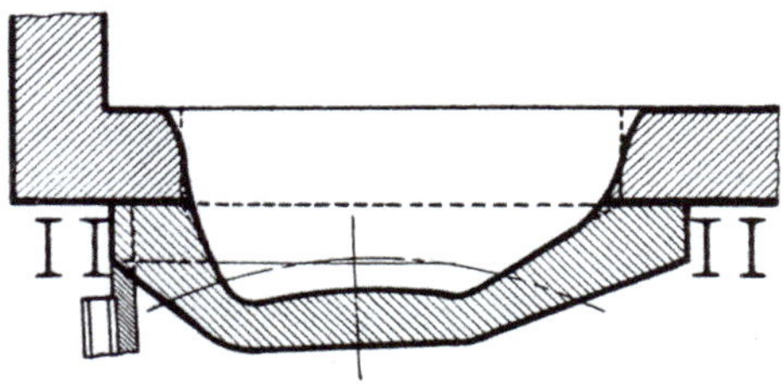

Figure 5,59. Schematic of the gathering position

For colourless glass the production figures were;

300 g 250 cm^3:	1250 pieces per hour
450 g 500 cm^3:	1050 pieces per hour
850 g 1000 cm^3:	900 pieces per hour

Table 5,21. Production figures for the Hillman machine

Unfortunately after a promising start the range of production was curtailed due to problems with creating a new surface at the dipping point. Further development work could not solve the problem.

There are British patents applied for in December 21, 1921 (172947) and in March 10, 1927 (287067) and a further patent for Hillman was DRP 478 817 dated 27th June 1929.

Next comes the **O'Neill** suction-blow machine British Patent 315 154 10th October 1928. This machine can have three, four, six or eight arms, and it was redesigned several times. As with the majority of O'Neills machines it was a twin table model and the model number for the 8 mould machine was 80, which possibly indicates that the series were the 70s.

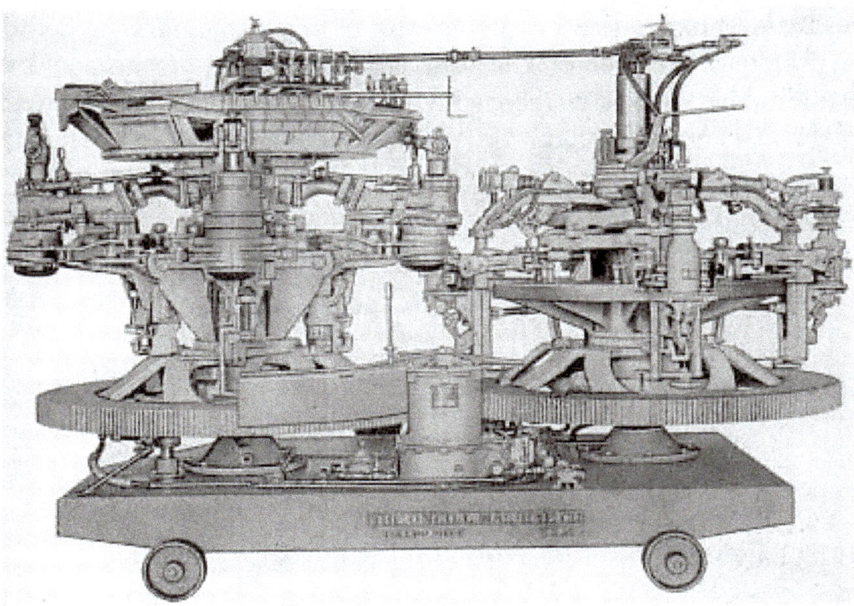

Figure 5,60. O'Neill four mould suction & blow machine

Figure 5,60 shows a four mould, electric driven machine. It accepts glass from a fixed forehearth or bay attached to a tank, the dipping mould is on the left of the photograph. Both of the tables rotate each in opposite directions to the other, they are driven by an electric motor through large gear wheels. The parison moulds are guided in due order over the forehearth. There is no special stirrer for the equalisation of the glass mass contaminated by the shearing of the glass left after the parison has been loaded although a certain amount of homogenisation occurs at the point of suction due to the dipping of the parison moulds into the metal.

Various adjustments can be made while the machine is running, namely the adjustment of suction, of pre-blowing, the movement of the plunger and the dwell time of the parison while the parison is closed. Important when making the heavier weight jobs. If jobs of more than one height are made at the same time, then the varying heights of the different bottles are compensated by using neck rings that vary in height.

Typical production figures supplied by two users of the 8 mould model 80 are;

Wt. (g)	Vol. (cm^3)	Ht. (mm)	Dia. (mm)	Fin. Dia. (mm)	Output (p/h)
150	200	125	60	30	2000
190	215	165	62	28.5	1500
320	265	200	60	30	1500
650	1030	300	100	75	800
700	1000	320	84	31.5	900

Table 5,22. Production figures for an 8 mould Model 80 machine

This machine is often referred to as the Knox-O'Neill machine, it was developed as a joint venture which started in 1929. By October of that year they had sold 95 machines to such companies as, the Carr-Lowery Glass Company, the Whitall-Tatum Co. and Owens-Illinois, while Knox used two themselves.

As the British patents were applied for or granted in October the previous year (1928) and the American ones must be from an earlier date the 4 mould machine and the Knox-O'Neill machine may be different, although the speed that they took to get it into production could indicate that O'Neill already had a machine and he needed someone to finance the building of it and give it proving trials which were supplied by Knox.

A 1937 catalogue of parts issued by the O'Neill European Machine Co. of Montreal, Canada for the No. 93 and 96 Vacuum Machines. This is not conclusive evidence of the series numbers, the O'Neill Co. had a history of naming an improved series starting 20 higher than the original, for example the Series 20 were replaced by the Series 40 when they were upgraded. Series 70 could quite easily be the original model. Figure 5,61 is an example of a 93 or 96 machine.

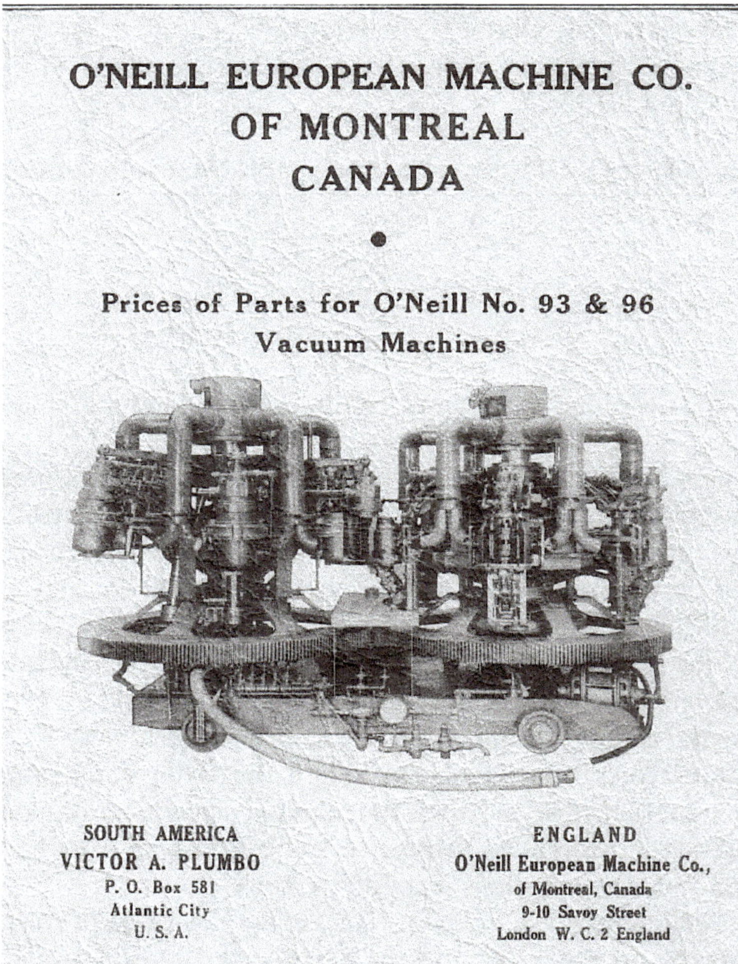
O'NEILL EUROPEAN MACHINE CO.
OF MONTREAL
CANADA

Prices of Parts for O'Neill No. 93 & 96
Vacuum Machines

SOUTH AMERICA
VICTOR A. PLUMBO
P. O. Box 581
Atlantic City
U. S. A.

ENGLAND
O'Neill European Machine Co.,
of Montreal, Canada
9-10 Savoy Street
London W. C. 2 England

Figure 5,61 A 1937 Parts list for the 93 and 96 models

The Wyss AW suction and blow machine

This one mould machine by A. Wyss is a continued development of the Roirant 'B' machine. The most important development is a heat-protecting hood made of cast steel which constantly seals the working hole it is provided with a refractory plate that closes the hole on the outside face. When the parison is in the suction position the hood repositions itself forming an effective shield from the radiant heat from the working hole. This allows for the cooling air of the suction members to be reduced considerably.

Mould changes can be made with minimum downtime, height adjustment of parison mould and the adjustment of neck ring to parison mould is performed by adjustment of a rotating spindle, the blow mould is not moved. The bottles can be transferred directly from the machine to the lehr, this is achieved by each blow mould having several bottoms on which the blown bottles can be transported. Improperly blown bottles and fallen parisons do not jam the moulds on closing and can be removed quite easily, these improvements result in higher productivity.

Production limits are:

Height of finish mould	160-320 mm
Diameter of bottle bottom	105 mm max
Finish diameter	40 mm max

The Lagena machine another machine that resembles the one arm Roirant. It was developed by the Mühlig Union Company of Usti (Aussig), of Czechoslovakia. This machine did not travel outside of its domestic region but they have been excellent results and were still in operation up to the 1960s.

Production limits are:

Weight	10-1050 g
Height	42-362 mm
Diameter	12-90.4 mm
Finish (interior)	8.5-34.4 mm
Total power requirements	1104 kW

The machine weighs about 2200 kg and has occupies a footprint of 2.8 m in length by 0.68 m in width, its height is 3.38 m.

Production rates are:

Bottle weight (g)	Pieces/h
1050	105
840	135
500-550	210
10.5	1000

Table 5,23 Production rates of the Lagena machine

A schematic of the machine is shown in Figure 5,62, unlike the Roirant 'B' and 'F' machines the parison mould does not reciprocate on a straight tiltable slide for the suction dip. The parison mould 11 is guided along a flat arch of radius R from the working position into the furnace. From position C to C". Because of the different ways the plunger movement is coupled to the head the Lagena can start to blow the parison earlier than the Roirant hence increasing the operational speed.

The final blow is done while the bottle head still sits in the neck ring mould B so that the arrival of the next parison cannot be transferred until the final blow has been completed and the neck ring has been returned to the suction head.

The close similarity that the machine has to the Owens means it has the advantage of being able to use Owens mould sets.

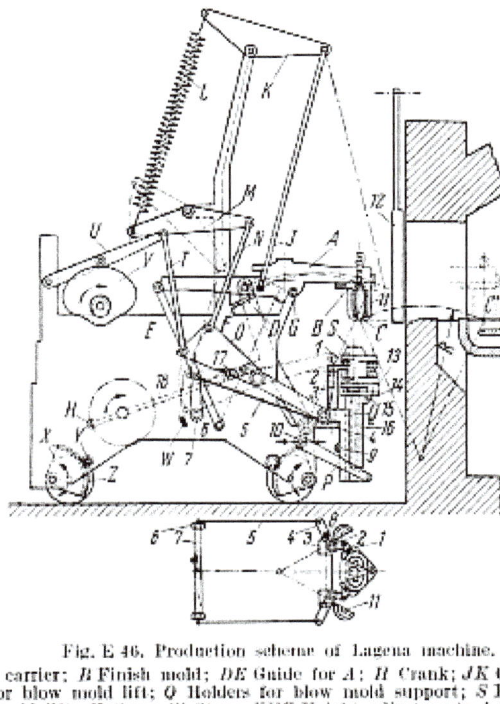

Fig. E 46. Production scheme of Lagena machine.

A Parison mold carrier; *B* Finish mold; *DE* Guide for *A*; *H* Crank; *JK* Control gears; *L* Spring; *MNOP* Gears for blow mold lift; *Q* Holders for blow mold support; *S* Blow mold; *TU* Control gears for blow mold lift; *V* Cam; *W* Stop; *XYZ* Height adjustment; *1* Connection; *2*–*7* Gears for blow mold opening; *11* Parison mold; *12* Work hole baffle; *13*–*16* Blow mold supports; *18* Push rod for parison mold support.

Figure 5,62. Schematic of the Lagena machine

The Monish machines were marketed by the Monish Glass Machine Company of Perth, Scotland, the name derives from the producer John Moncrieff Ltd. of Perth and the inventor A. MacNish; there were four types, the single mould, the three head Major and Minor and the Twin Table machines. They are all illustrated in Figures 5,63 to 5,67.

The first Monish machine came onto the market around 1928 (British Patent No 320 024 June 1928) it was a single mould machine. The machine had two tables a parison mould table and a blow mould table. The parison mould table travels through an angle of 90° from the forehearth to the blow mould table, which is positioned outside the circle of travel of the parison mould.

The parison is transferred to the blow mould by a forward movement of the plunger bracket. After the bottle is blown in the blow mould, the neck ring is opened and the plunger bracket returns to its original position before being moved to the gathering position for another gather.

Whilst the parison is travelling from the forehearth to the transfer position, a cover moves over and covers the opening in the forehearth in the gathering position, thereby keeping the glass hot for the next gather during the time the parison is being transferred and the bottle blown in the blow mould.

The Monish Minor is a scaled down version of the Major, they both have three blow moulds and three parison moulds. Both machines can if wished

Figure 5,63. The single head Monish machine

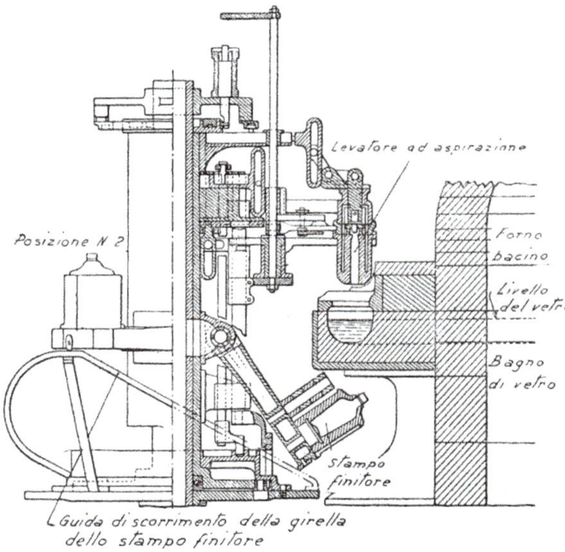

Figure 5,64. Schematic of a Monish machine showing the "forehearth" gathering position

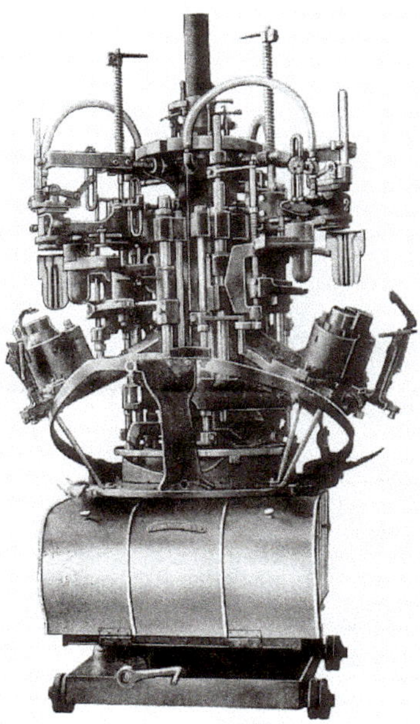

Figure 5,65. The three head Monish Major machine

Figure 5,66. Three head Monish Minor machine at a furnace

make three different articles which are of slightly different weights, heights and shapes. The technical data for the machines is shown in Table 5,24.

	The Monish Machine		
	Single mould	**Major**	**Minor**
Motor power and auxillaries	17.25 HP	19.0 HP	15.5 HP
Power comsummed	14.0 HP	15.0 HP	11.0 HP
Output per minute	Up to 8	12	20
Capacity Min.	0.5-80	5-80	0.5-5
(contents) Max.	ounces	ounces	ounces
Height of bottle	1.25-15	4-12	1.5-6
to bottom of N.R.	inches	inches	inches
Max. O.D. of neck of Std. Machine.	2.25	2.25	1.375
With Wide mouth heads.	inches	inches	inches
	--------	4.25	2.50
		inches	inches
Nett Weight	18 cwt.	4.25 ton	2.27 ton

Table 5,24. Technical data for Monish machines

Figure 5,67. Monish Twin-Table Two Mould machine

Two quite different bottles in shape and size can be manufactured at the same time without too much difficulty. The tables are moved by air cylinders and gears, and the machine is timed throughout by an electric timer.

Production data and power requirements;

Motor power required including all ancillary equipment		26.5 HP
Approximate power consumed		23.0 HP
Range:	Capacity	from 2 oz to 80 oz
	Speed	up to 15 bottle/min
	Height under ring	from 2 to 15" high
	Nett weight	3 ton 5 cwt.

For smaller glassware under 3 inches high and wide mouth ware from 2.25 inches to 5 inches outside diameter of neck, and for producing internal screw necks, special devices can be supplied.

The following advertisment from company literature dated 11/1938 and 1/1947 shows a quite large and diversified range of articles that can be made on the Monish machines.

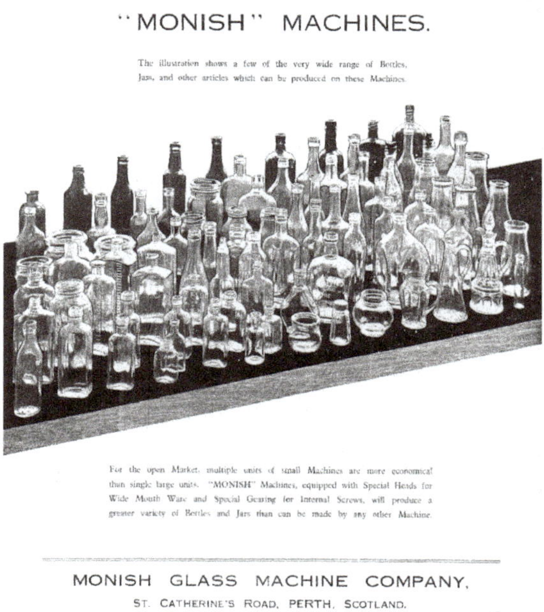

Figure 5,68. Advertisement from 'Monish' catalogue

The dimensions for the Major and Minor machines are 2.0×1.55×3.075 m for the Major and 1.1×1.0×1.765 m (L×W×H) for the Minor.

Bottle weight (g)	Pieces/h Major	Pieces/h Minor
800-1000	240-300	
400-600	360-480	
200-300	600-730	
150-200		600-730
100-150		730-840
40-60		960-1080
20-40		1080-1200

Table 5,25. Manufacturers production details for the Monish Major and Minor Machines

Figure 5,69. The Pöting AM 6/D machine

Now we turn to the last group of machines in this section on suction-blow machines, they are the Pöting suction-blow machines. These machines were designed and made in Germany for the production of 10 to 250 ml content medicine bottles and containers. The two continuously rotating one table machines were developed and marketed in 1940/41 as the Pöting AM 8 and the PD, later AM 6/A with 6 and 4 arms.

The use of the AM 6/A spread considerably and has recently (late 1960s) been updated and improved as the AM 6/D shown in Figure 5,69. This occurred because it can be operated in a simple manner and gives good results.

Over a small bay that protrudes from out of the tank it sucks up glass directly from the surface of the furnace chamber. In it is located a stirring mechanism for the homogenisation of the glass at the point of loading (German Patent No. 880 878). The apparatus comprises of a refractory disc fixed to a water cooled shaft, the refractory end is immersed into the glass.

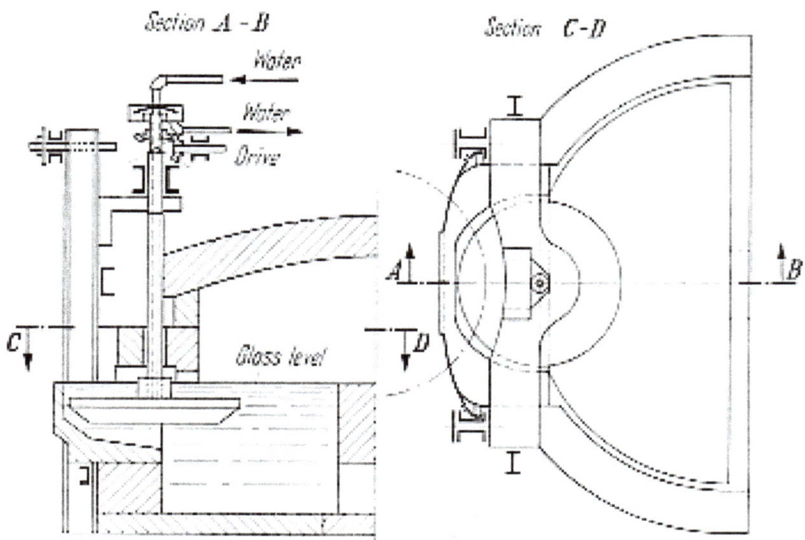

Figure 5,70. Schematic of the stirring unit for the Pöting AM 6 machine

The unit rotates and its speed of rotation is adjustable, likewise the speed of the machine. By adjustment to the height of the parison mould holder bottles of varying heights and weight can be produced, length of parison variation ±10 mm as in the Owens machine. During dipping only the parson mould holder moves, with pneumatic damping.

The machine can be adjusted in height (dipping depth) by a hand wheel. Experiments with double moulds for small bottles are underway (late 1960s).

The AM 6/D has 6 arms and weighs 3600 kg, it has a mould circle diameter of approximately 1.75 m and the height is 2.3 m. It requires 1.5 kW for the machine drive. The air requirements are;

Vacuum 95%:	150 m^3/h
Compressed air 3 atm.:	60 m^3/h
Cooling air 400 mm. water col.:	7200 m^3/h

Total power requirements are around 30 kW

Production limits are;

Height (including finish)	40-200 mm
Max. diameter;	
Standard blow mould holders	75 mm
Large blow mould holders	85 mm
Max. finish diameter	40 mm
Weight	5-300 g

Production figures are;

Contents (cm^3)	10	30	50	100	150
Pieces/min	44	41	38	33	27

Chapter 6

The development of the Feeder and the Forehearth

The development of the forehearth did not finally take place until around 1916 to 1917 when the then commercial offerings could now start to become recognisable as the forerunners of today's products, and that, most importantly, was they could feed a machine with a continuous stream of glass gobs.

As soon as the bottle making machine was invented people had tried to develop a means of feeding them automatically, to do away with the skilled man the 'gatherer', and in so doing make the machine an 'automatic' one. The solving of this process started with the various press machines that where invented first and later these were followed by the press and blow machines and then by the blow and blow machines. The problems that they all encountered when they started out was; how to get individual lumps of glass that were consistently of the same weight to fall into a machine with the same time interval between each gob.

This was difficult with the single unit machines but once the table machines, both the single and the twin, became more prevalent, then progress was made much more quickly. By 1900 many applications had been made for patents for feeding glass mechanically to bottle making machines. Some of the various ideas can be seen later in the abstracts.

The obvious method to try was to allow a stream of liquid glass to flow out through a hole in the bottom of a furnace forehearth, see Figure 6,1. But many failures occurred in trying to find a means of interrupting the stream when one mould was filled and then had to be moved away to make room for another.

Figure 6,1. Glass falling from an orifice

Historical Survey or a 'cull' of the Literature
These have mainly been taken from the paper given by G. Dowes and E. Meigh at the London meeting of the Society of Glass Technology on April 20th 1921. Here I have used their division of the time periods. The development in the art of delivering molten glass to a machine automatically falls into three distinct periods chronologically:-

A. The Early Period (up to 1900)
Though the machine came into effective used by 1890 there was no demand for a feeder, most likely due to a forming machine that was suitable for, or one that required one, a feeder was not in existence or was not efficient. This did not stop the glass works engineers looking and designing one. The early endeavours fell into one of two types; Type I a simple siphon or tap and Type II a reciprocating slide to cut off and deliver measured charges.

Type I. The simple siphon or tap.
In 1885 a patent was granted to Dan Rylands for a device for siphoning off the metal (glass) from a tank. (British Patent 14 573/1885) and another granted to Rylands and B. Stoner in 1887 (British Patent 354/1887). In the latter case the outer end of the siphon was to be closed by a plug which could be manipulated to deliver charges of molten glass to a machine. This idea of using a

siphon to deliver charges to a machine seamed to fascinate many of the earlier inventors, this included Owens who used a siphon in his early experiments.

Type II. A reciprocating slide to cut off and deliver measured charges.

An example of this type is shown in Figure 6,2, it was patented in 1900 by A. Sowden (British Patent 14 909/1900).

The device worked as follows; a slide (4) having a measured chamber (6) is reciprocated horizontally below a tank extension (2), and during its traverse the chamber (6) becomes filled with molten glass which passes through an opening (3) in the bottom of the tank. The charge of metal in the chamber (6) is carried forward by the slide, and, simultaneously, another slide (5) is withdrawn from beneath the chamber and a plunger (22) descends and forces the charge of metal into a ladle (27), by which it is carried to a mould.

This method of feeding measured charges of metal to the bottle-making machine became quite popular and a number of patents were registered. In some of these a rotating disc having a broad edge which closed an outlet to the tank and was formed with pockets replaced the reciprocating slide. As the disc revolved the pockets became filled with glass as they passed the opening so that the device delivers a succession of measured charges. One such device was patented in 1895 by Eunson (British Patent 585/1895), and it re-appeared in America round about 1920 (USA Patent 1 256 979).

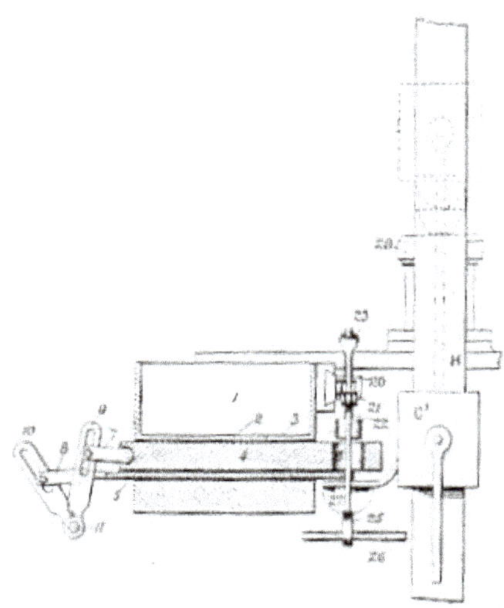

Figure 6,2. Sowden's feeding device

B. The Middle Period (1900–1914)

This era starts off with Homer Brooke who was granted a patent in 1901 (British Patent 24 324/1901) for a 'flow feeder'. This will be discussed in more detail a little later.

As more and more engineers became involved the development of feeders became more intense, many patents were obtained in large numbers and even if the devices were not themselves successful they showed the application of ideas which later became a success. At this time the devices fell into a number of distinct types, in the first type a plug reciprocating inside the metal in a forehearth or boot periodically opens or closes an outlet, thus permitting the stream either to flow and then stopping it. Numbers of patents have been issued for variations on this theme.

It is quite obvious that the increase in flow when the plunger pushed the glass out led the inventors to develop a second type of apparatus the 'impulse type'. An early example of this is by Morrison and Fischer is shown in Figure 6,3, the patent granted was British Patent number 27 227/1905.

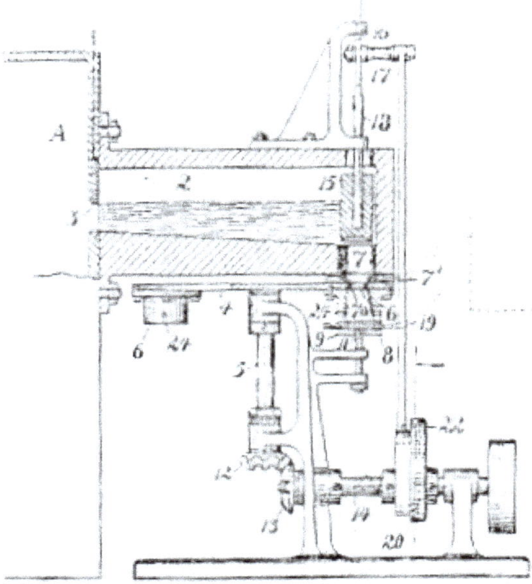

Figure 6,3. Morrison and Fischer's 'Impulse' feeder

The device works like this; a reciprocating plunger (15) is used to force the metal through an outlet (7) into a measuring and cutting off receiver (6) rotating below the outlet. An interesting fact arises here that if the oscillating or reciprocating plunger in the glass is located near to the outlet then the plunger forces out the glass on the down stroke but draws it back up into the feeder on the up stroke. This idea was developed by many others which we will discuss later.

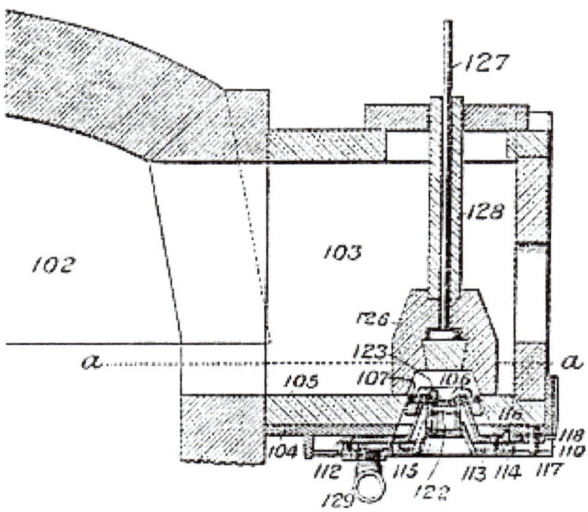

Figure 6,4. Dixon and Marsh's feeder

Another type of feeder in which an altogether different method of controlling the stream is used was patented in 1906 by Dixon and Marsh (British Patent number 5 040/1906, American Patent 814 773).

The device shown in Figure 6,4 consists of a tank having an outlet (122) which can be opened or closed by a plug (126). The plate surrounding the outlet is provided with a number of perforations or grooves. When the flow is to be interrupted a blast of high-pressure air is directed through the grooves onto the stream of metal, and the construction of the grooves is such that after the stream is severed low-pressure air is emitted from the grooves to support the metal until the filled mould is removed and another takes its place.

This idea, of supporting the cut end of the stream by an air blast, has recently (1920), been applied with success by William Miller, British Patent Number 112 947, and Tucker and Reeves British Patent Number 109 782. The former uses an air blast to uphold the cut end of the stream, while the latter uses a blast of hot gas.

Figure 6,5 (a) to (c) shows a Tucker, Reeves and Beatty feeder arrangement.

(a) Showing the gob after being sheared.

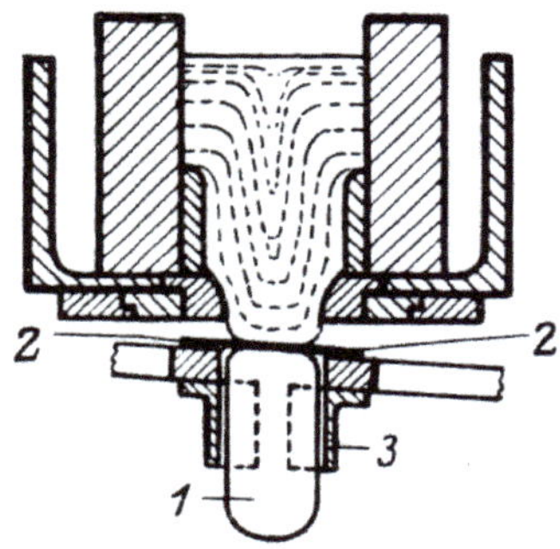

(b) Showing gases holding the glass back.

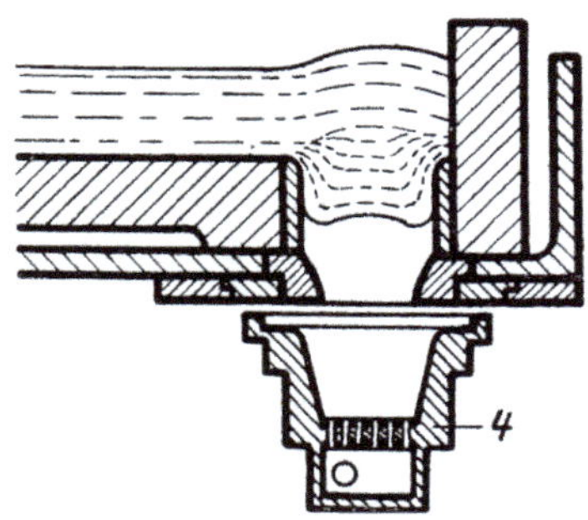

(c) Holding gas turned off and the glass resumes falling to form the gob.

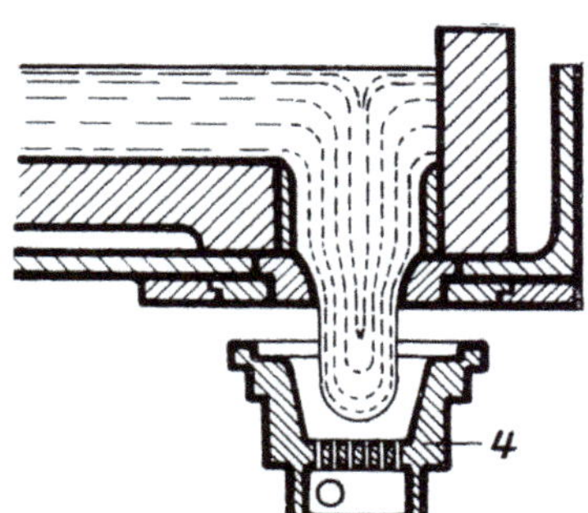

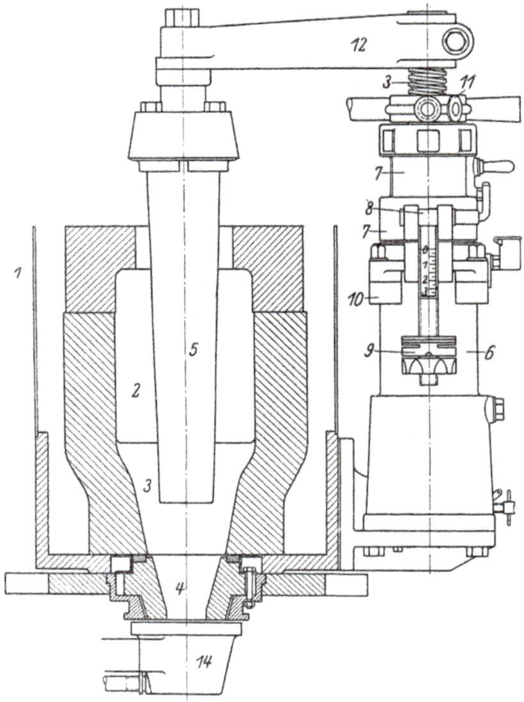

Figure 6,6. Schematic of Turner, Reeves and Beatty feeder

Various other feeders were proposed during this period such as an automatic ladle that dipped and scooped a charge from the tank and carried it to the parison mould or the automatic gathering on a punty rod such as Peiler's which will be discussed in more detail later. There were also feeders patented by Severin in 1908 (British Patent Number 9 69-89/1908 and USA Patent 892 013) and one by Simpson in 1913 (British Patent 20 863/1913) in which the metal is forced upwards through a tube into a parison by means of a plunger or by compressed air. These feeders were used by Severin and Simpson to demonstrate their machines in this country before the war (World War I), but were abandoned owing to various difficulties. The Simpson patent could possibly have been made by the originator of the Simpson and Bradshaw machines which were originally made by John Lumbs & Co. Ltd. in Castleford, England.

C The Modern Period (1914-1921)

During this final period of development there was a deluge of patents applied for and granted for feeders. This was due to the development of the semi-automatic twin tabled machines mainly in America, the shortage of boy labour which led to the elimination of the transfer boy and the development of the 'no boy' machine. The gatherer was now the only person between the furnace and the taker in, and it is a logical step to replace him by some mechanical means. It was also a fact that for these machines to be used to their best advantage they had to be worked at a higher rate of output than for hand working, this in turn made for means to keep the man cool and also give him a rest for 20 minutes in every hour.

The feeders that were now being invented were starting to fall into well defined types.

Type 1. A stream with shearing devices.

Notable examples of this type are those patented by Proeger (British Patent Number 14 734/1914) and Drey (British Patent Number 15 793/1915).

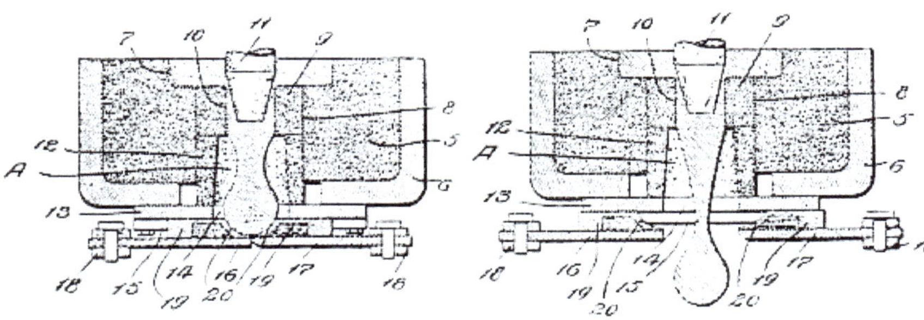

Figure 6,7. Drey's feeder

It had been found that if a stream is allowed to flow into a mould and then severed, the coiling of the stream which takes place in the mould mars and weakens the finished ware. Inventors had begun to realise that it was necessary to gather up the glass and deliver it to the machine as a whole homogeneous gob rather than one with a long tail attached to it. Proeger allowed the stream to gather on top of the closed shears and then opened and closed these shears quickly to sever the stream in the narrow neck formed behind the gob when allowed to fall. A drawing of Drey's device is shown in Figure 6,7.

In this feeder the stream is gathered into a gob on cupped shears which form the lower part of a chamber, A, heated by radiation from the metal. The lower figure in the diagram shows how the stream forms into a narrow neck behind the falling gob. Several feeds of this type are in use commercially (circa. 1921).

Type 2. The cut end of the stream is upheld by a blast of gas. Reference has already been made to this type, of which the feeders of Miller and Tucker are examples. Other Tucker patent applications can be seen in the abstracts.

Type 3. The cut end of the stream is withdrawn back into the tank by suction.
A example of this type is one designed by McCauley (British Patent Number 113 665). R. E. McCauley made a number of patent applications which can be seen in the abstracts.

Type 4. The reciprocating plunger type.
The reciprocating plunger type. This type, in which a plunger reciprocates in the body of the metal, has been mentioned above.

Type 5. The stream is supported, after severance, by lateral support.
Graham British Patent 127996, and Wadsworth USA Patent 1311474, have produced of this type in recent years.

Type 6. The 'paddle impulse' type.
The most famous of these is the one patented by Peiler for the Hartford Fairmont Company which will be dealt with later. A number of these were sold commercially.

The next development was to try and remove the shear mark and then the shaping of the gob so that besides being constant in weight they would all be of the same shape, ideally the same as the parison, which would then allow equal chilling when the gob enters the parison and the subsequent equal distribution of the glass. The best example of this type is the Hartford Fairmont 'Paddle Needle' feeder. British patent number 142 786.

In 1921 Meigh and Dowes listed what they thought the requirements of a gob should be and gives a comparison between the flow fed feeder and the paddle-needle feeder, the so called "modern design" feeder did not come on to the market until the following year. To quote they said;

"The perfect 'gob' is an embryo bottle or jar having weight and shape nicely calculated to suit the dimensions of the parison mould. The art of 'gob' feeding is being developed today for sound physical reasons, it is immediately apparent that if mould charges can be pre-formed accurately to fit the internal contours of the mould then the serious defects caused by folding or lapping of the glass or and trapping of the air might be avoided. Further the process becomes a much wider application enabling a greater variety of glassware to be produced".

Experience has already shown that 'gob' feeding (e.g. Peiler) possesses considerable advantages, over the cruder 'flow feeds' (e.g. Brooke), the chief of which may be summarised as follows;

1. The glass can be worked in a more viscous condition (as in hand gathering), and consequently the mould charge or 'gob' remains compact and cohesive, is easily controlled, and is sensitive to adjustment.
2. Since 'flowing' properties are not alone relied upon for locomotion of the glass from furnace to forming machines, the temperature of the glass need not be regulated to suit the feeder but to suit the forming machine and the moulds.
3. When for any particular forming machine and for any particular article being made and the highest permissible speed has been determined, this can be maintained constant, irrespective of temperature fluctuations.
4. The weight of individual 'gobs' can be pre-determined and maintained constant within remarkably close limits.

5. The temperature and viscous conditions of the glass and the method of pre-forming the charge confer upon the 'gob' greater resistance to trapping or entanglement of air, and consequently the ware does not suffer from 'blisters' to the same extent as in a gravity flow process.
6. In 'gob' feeding the moulds and other handling or shaping instruments are less heated than in other automatic methods and a shorter time is required for the ware to acquire its permanent set after shaping.
7. On account of the 'gob' having approximately the configuration of the parison mould, uniform cooling takes place and regular distribution is thus possible.
8. The mould charge is in the suitable physical condition for multiple distribution, and the rate of production per unit feeder can accordingly be increased by diverting gatherings to two, three or four machines.
9. 'Gob' feeding is not limited in application to the manufacture of wide mouth ware, but is equally applicable to narrow neck production, and compares very favourably in quality of production, with any other method of making narrow neck ware.

We will now discuss in more detail some of the different versions of gob feeding and the people that invented them starting with Homer Brooke. Of the very few stream feeders to reach a measure of commercial success the most famous was that invented by Homer Brooke in 1903. Homer Brooke was born in Yorkshire, England and while still small his father William Brooke with his family emigrated to America in 1849. There William became employed by a glassworks near

Figure 6,8. Homer Brooke

to New York City but later he must have branched out on his own because in 1863 when he died his business of making moulds and glass machinery was carried on by his son Homer.

His patent is shown in Figure 6,9, and Figure 6,10 describes how the stream feeder worked.

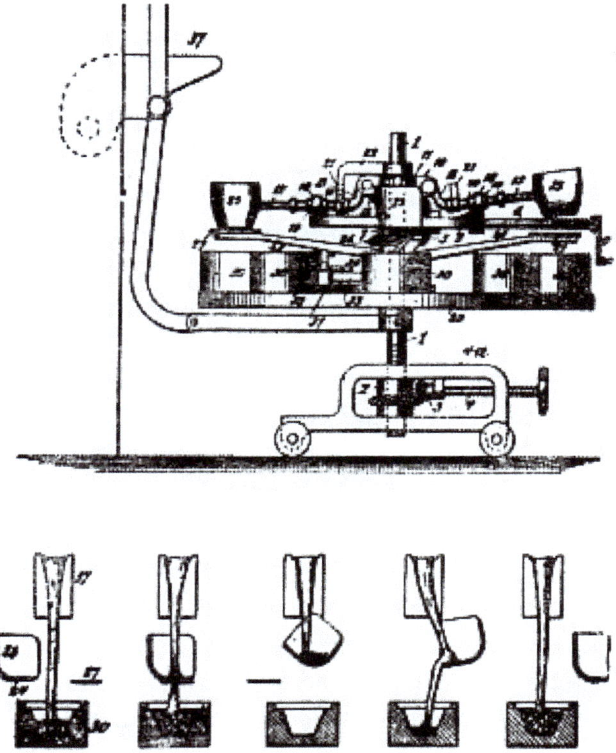

Figure 6,9. Homer Brookes stream feeding device

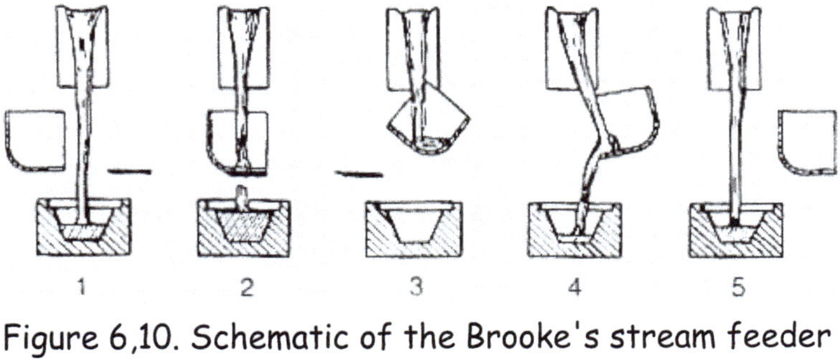

Figure 6,10. Schematic of the Brooke's stream feeder

The operation of a Brooke stream feeder works like this;

1. Continuous stream of glass falls from the orifice into the mould.
2. The mould becomes full so the cup and the knife sever the stream.
3. The cup is tilted to collect the still flowing stream while a new mould moves into place.
4. The cup is tilted further and empties into the mould.
5. The cup moves aside until the mould is full and the cycle starts again.

His feeder used a steady stream of relatively fluid glass, when the mould was full the stream was sheared and gathered in a cup until a new mould was moved into position and the contents were dumped into the new mould. The machine had four sets of cups and shears that were mounted on a rotating table. The main problem was the poor thermal homogeneity of the glass in the parison mould caused by the chilling of the thin stream of glass coming into contact with the cup.

Even the best continuously flowing feeder caused defects in the glassware due to the skin formed as the thin stream passes through the cool air and as it coiled and folded on entering the mould. Laps and creases were formed in the cup and the air bells or blisters were entrapped and showed as blemishes in the finished article. In spite of the low quality glass that resulted it was acceptable enough for firms such as Hazel Atlas and Ball Brothers who used it to make cheap pressed jars and lids in large quantities. They both took out licenses under the Brooke patents paying out substantial royalties.

In 1915 to safeguard the feeder fed Graham's machine, which they recently bought, the Owens Bottle Co bought out all of the Brookes patents and licenses which they did not already have. Hazel Atlas persevered with the stream feeder taking out a patent in 1925 for a similar device but with shears above the cup.

By 1920 they began to use Hartford feeders and gradually added to their number such that they became one of Hartford's important licensees. Mike Owens made a great mistake in either

1915 or 1916 when Hartford offered to sell their device and patent rights which Owens refused. In 1924 just after Owens death there was an agreement made between Owens and Hartford Empire ensuring close co-operation between the two.

This is a biopic of Peiler's investigations which led to the forehearth of today. It is followed by the efforts put in by Hartford Empire, and later Emhart.

Figure 6,11, Karl E. Peiler

One of the first steps in the evolution of the glass industry was the move made by the Beech-Nut Packing Company of Canajohaire, N.Y. in America, they had contemplated packing some of their products in glass jars but the quality of the jars as supplied by their suppliers was not good enough for vacuum sealing. Their President was the very able and enterprising Mr Barlett Arkell who consulted with Messrs W. H. Honiss & W. A. Lorenz, mechanical engineers and patent attorneys of Hartford, Connecticut.

At this time Honiss & Lorenz had working for them a brilliant young engineer called Karl Ernest Peiler who had graduated in engineering in 1904 from the Massachusetts Institute of Technology. In 1911 he had been assigned the specific research on the automatic feeding of glass forming machines. He had never been into a glass works so he had not any preconceived ideas of how to solve the problem. He was an 'academic' rather than a 'glass man' such as Michael Owens.

This meant that his research proceeded along scientific lines rather than by technological improvements. He soon tossed aside the relatively high temperature method of feeding used by the 'stream' and the 'suction' methods and concentrated on feeding and working glass at temperatures that were around those used by the 'hand workers'.

He took two different approaches, the first being his 'punty feeder' and the second his 'paddle feeder' which was soon modified to become his 'paddle and needle feeder'. The 'punty feeder' was an attempt to automate the gathering of glass by hand. He designed and patented the apparatus shown in Figure 6,12.

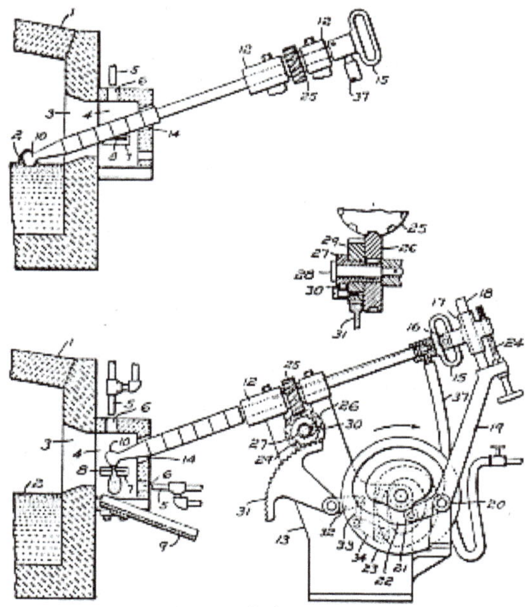

Figure 6,12. Peiler's Punty Feeder.
(After US Patent Number 1 324 464, 1919)

The results were disappointing, the gathering iron rotated continuously while inside the furnace but it stopped when it was withdrawn, allowing a gob of metal to be formed under gravity it then was cut by shears and the gob then fell down a chute into the mould.

The problems were that amount of the glass mass gathered on the rod varied and due to the shear cut being timed, i.e. there was no adjustment, both of these together gave rise to a varying amount of glass being fed to the machine continuously, which was a far from satisfactory situation.

The following lines was penned, after the disappointment, by one of his colleagues and given by him to Edward Meigh testifying;

Peiler's Lament.

My punty lies over in Glenshaw,
My punty ain't done so damm well,
My punty is sick to the stomach,
And W------ is chucking like hell.

Foul curses on the power line.
That fluctuated so,
And made my punty feeder feed,
Not handsome gobs which we did need,
But chicken entrails, crude indeed,
A truly cruel blow!
(Anonymous)

The next step was the 'Paddle' feeder, See Figure 6,13, here the glass was pushed out of the forehearth and allowed to hang from the lip, after a predetermined time is was then severed and allowed to drop to the machine.

For the first time a gob of glass of predetermined weight, and to some extent shaped, and severed in suspension, was delivered to the machine.

The paddle feeder was first used by the Monongah Glass Company in 1913 and by 1915, thirty of these paddle feeders were installed in the United States and several in other foreign countries.

The Hartford 'paddle feeder' came into commercial use in conjunction with a twin-table milk machine and a twin press machine that had

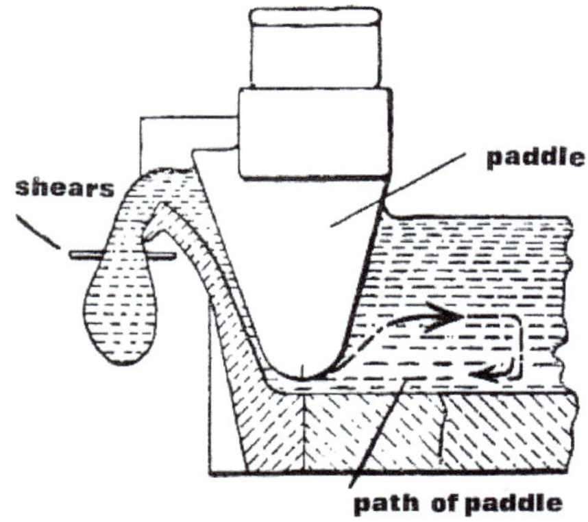

Figure 6,13. Schematic view of the 'Paddle' feeder

been designed by Honess & Lorenz. The pear shaped gobs were guided into the moulds by means of a swinging trough delivering glass to two fixed chutes. The chutes had fine water sprays on them enabling the gobs to fall on a bed or cushion of steam, the fine droplets had assumed spherical shape under the heat from the glass. Large numbers of disc-mouthed milk bottles and pressed ware such as tumblers where made by the combination of the paddle feeder and the Hartford milk bottle machine and the Hartford press. These feeders allowed the machines made by O'Neill and Lynch to become automatic from around 1916.

Peiler realised that the paddle feeder was not ideal to manufacture narrow necked bottles. It had limited gob-shaping possibilities and was unequal to the task of producing a satisfactory range of narrow neck ware so he returned to his plunger research. His solution was to put a bowl after the lip and used a thin needle to punch out glass from a hole situated in the bottom, the hanging glass was severed by a timing device. See Figure 6,14.

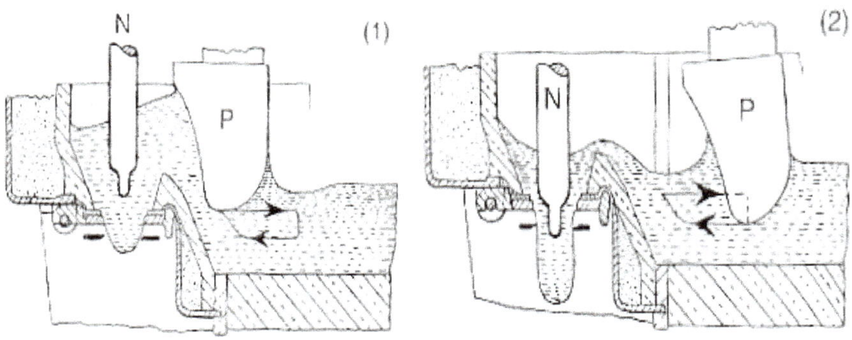

Figure 6,14. Peiler's Paddle and Needle feeder

Its operation is shown in Figure 6,15, working from the left, the paddle is drawn back, then it dips into the glass and moves back towards the bowl. The already raised needle (plunger) is lowered pushing the glass out through the hole in the base of the bowl. Then after sufficient time to allow the gob to form the glass is sheared and the gob goes to the machine or machines. The cycle then starts again.

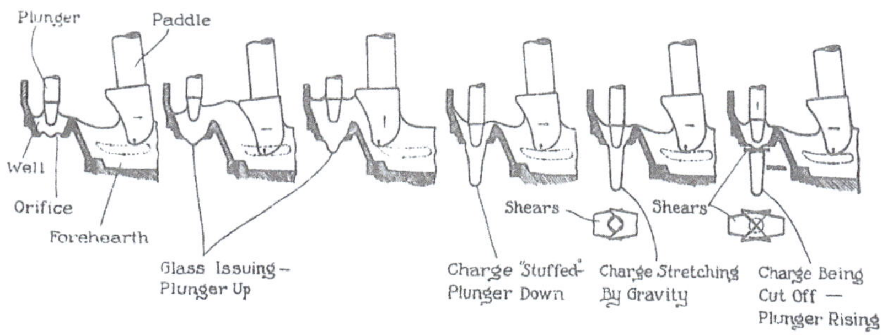

Figure 6,15. A schematic showing the operation of Peiler's P & N feeder

In order that the quantity of glass swept over the dam may be regulated while the feeder is in operation, the paddle is provided with three means of adjustment affecting, the length of the paddle stroke, the depth that the paddle dips into the glass, and the distance at which the paddle works behind the spout dam. The third of these adjustments is used more particularly in connection with the temperature that the glass is being worked; the paddle generally being adjusted to work near to the spout dam if the glass is being worked 'hot' and further away if being worked 'cold' - the terms 'hot' and 'cold' being used relatively.

Meigh goes on to say that, here it might be emphasised that the permissible variation in the working temperature of the glass is not determined by the feeder, for the latter can be operated to deliver the required gob at the speed and temperature best suited to the type of forming machine, moulds, and particular article being manufactured.

After the glass is paddled over the lip of the spout it is not allowed to flow freely by gravity through the orifice ring, but is controlled by the vertical action of the secondary impeller or 'needle'. It has already been noted that in order to prevent any folding or lapping of the glass in the mould with the consequent defects, it is necessary to pre-form the gather so as to conform as nearly as possible to the shape and size of the mould to be fed, and thus minimise 'settling'. The types of gathering required for the manufacture of ordinary bottles or jars will usually fall approximately under one of the following headings;

1. A spheroidal-shaped gathering for pressed tumblers, etc.
2. A cylindrical gathering with blunt ends for wide-mouthed bottles
3. A gathering with straight sides, but tapering down at the lower ends for most press and blow products.
4. A gathering with cylindrical body, but having distinct shoulder with reduced diameter at the lower end used for narrow-neck bottles.

By the proper use of the needle, and the means for its adjustment and in conjunction with the shears severing the gather a variety of gobs can be made. See Figure 6,16 for a schematic showing some of the different shapes that can be formed for different types of machines.

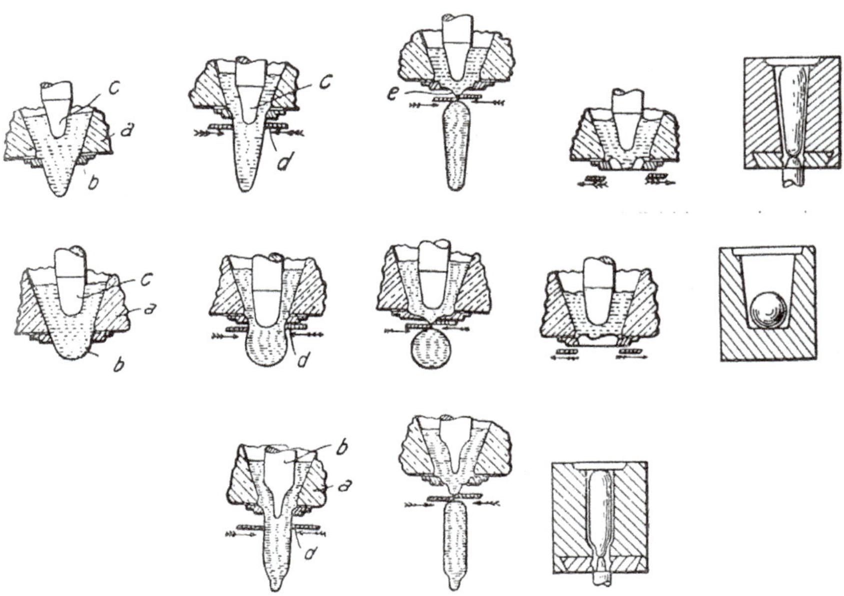

Figure 6,16. Different mould charges.
Top, blow and blow and press and blow gob formation.
Middle, gob formation for press.
Bottom, gob formation for narrow neck ware

The needle is operated by cam with a removable lobe for the rise and fall of the needle along with provision to adjust the amount of dwell time between the rise and fall of the plunger. There is also means for adjusting in operation;

1 The height of the needle stroke in relation to the orifice ring, by sliding the connecting-rod pin in the stroke cam.
2 The commencement of the needle up-stroke relative to the time of the cutting action of the shears, by means of a differential gear.

The P & N feeder came into commercial use in March 1918, more than 120 of these P & N feeders were installed in the United States and in several foreign countries.

The next two figures show Peiler's two Patent applications, Figure 6,17 for the Paddle feeder and Figure 6,18 for the Paddle and Needle Feeder.

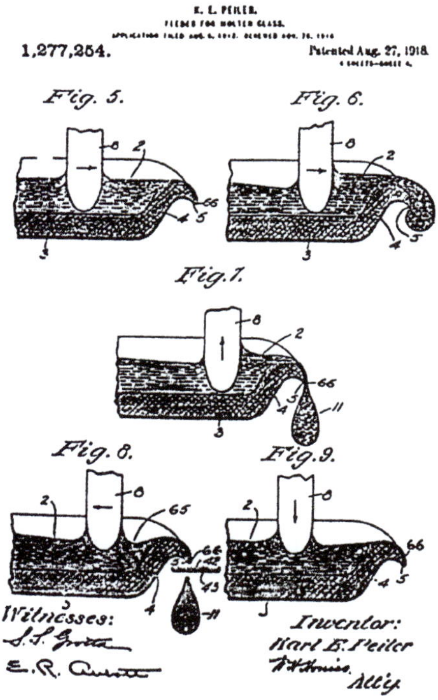

Figure 6,17. Peilers Patent application drawing, (US Patent Number 1 277 254, dated 27th August 1918)

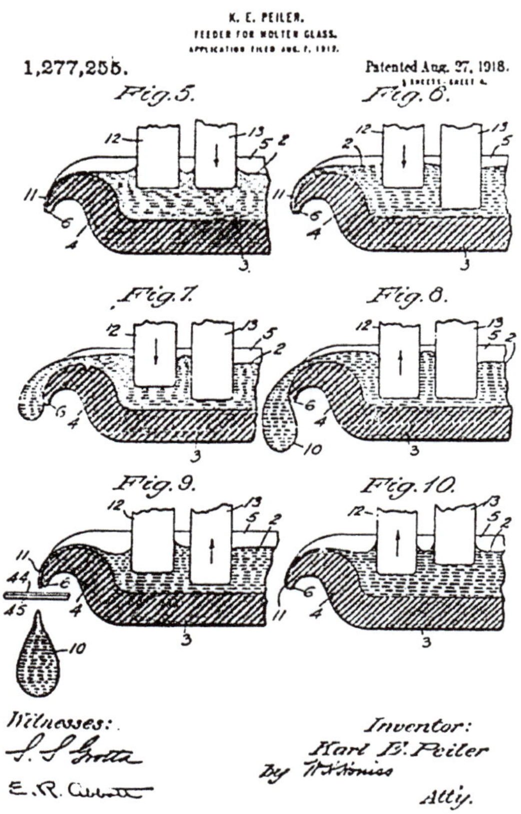

Figure 6,18, Peiler's Drawing for his Patent application for the 'Paddle and Needle' Feeder, (US Patent Number 1 277 255, dated 27th August 1918)

Distribution of mould charges.

Meigh continues - the subsequent history of the 'gob' depends upon whether the feeder is for high production or for low production. In the latter case the gob drops directly into the mould positioned vertically beneath it; this is the slow-speed single-unit feeder, which is most useful where a large variety of small quantities are required. For high speed production a distribution system is preferred. This method is usually employed in the Hartford-Fairmont process.

Figure 6,19. 'P & N' feeder from the rear

Figure 6,20. 'P & N' feeder from the front

Figure 6,19 gives a view of the rear of the paddle feeder while Figure 6,20 gives a front view, (the quality of the photographs leaves much to be desired), At first sight the use of troughs would tend to spoil the ware by uneven gob temperatures caused by the cooling from the troughs. This is prevented by the use of a film of vapour created by spraying water upon the trough. The 'hot' gob causes the water droplets to assume the spheroidal state and thus creates a cushion of steam which conveys it along the trough, and at the same time acting as a nearly perfect insulator. The shapes of the troughs are carefully designed to counteract the tendency of the gob to become distorted. The unit may consist of two, three or four troughs according to the type of forming machine and the class of ware being produced.

Figure 6,21 shows the installation of a Hartford 'P and N' feeder along with an unnamed bottle forming machine. It is difficult to see but it looks as if it may be feeding two machines.

Figure 6,22 showing a Hartford 'P & N' feeder feeding a

Figure 6,21. Two views of the Hartford Paddle and Needle feeder. Installed at Alloa Glass Works Co. Ltd, Glasshouse Loan, Alloa, Scotland

Hartford bottle machine and a Lynch LA type which is obscured. The installation was completed in 1919 at the Greenford factory of Rockware.

The development work continued and in 1922 Peiler in the shape of Hartford put the now familiar 'Needle Feeder' on the

Figure 6,22. Rockware Glass Plant Greenford showing a 'P & N' feeder, feeding two machines

market, see Figure 6,23. This, like the IS machine has had a long life, it has withstood improvements and modifications, all of which were carried out in the name of progress, i.e. to ensure that the machine to which it was feeding received an adequate and consistent stream of glass which in most instances was to a bottle forming machine.

The feeder works in the following manner, left to right in Figure 6,24,

- first the glass flows out of the orifice under gravity
- secondly the glass is accelerated by being pushed down by the plunger
- then the plunger reverses its downward motion and starts to pull the glass back into the body of the feeder

- finally a set of shears set just below the orifice severs the falling stream of glass

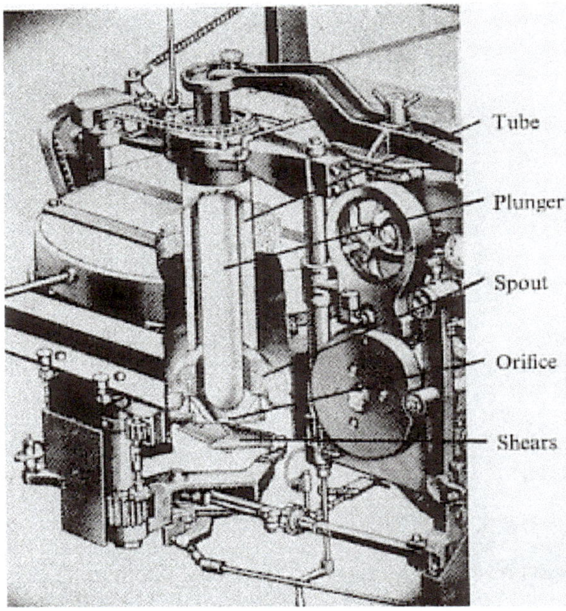

Figure 6,23. The Hartford Empire Single Feeder, Type 144

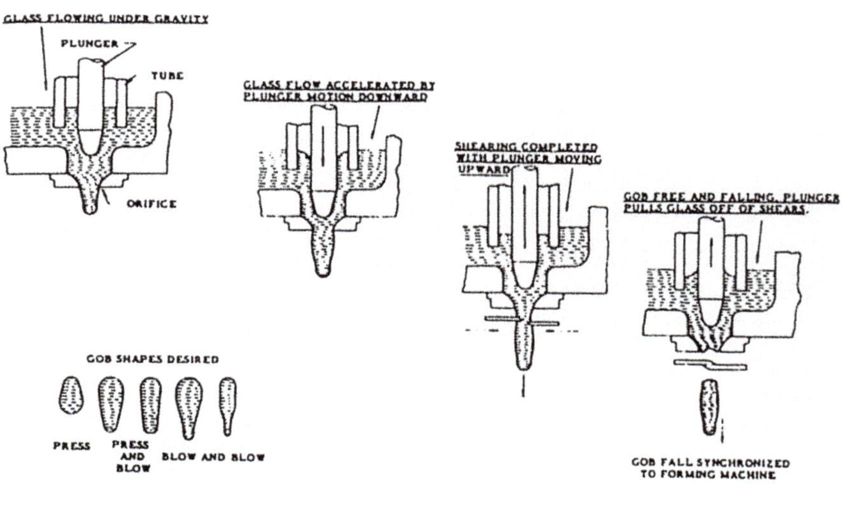

Figure 6,24. The operation of the Single Feeder

This feeder went through a number of small modifications with types A, B, C and D being on offer until it eventually became the Hartford Type 144 feeder.

It can be seen from the attached letter or memorandum which listed all of the major company salaries in February 1923, that the letter heading also lists all of company owned or licensed by Hartford Empire. The only one which specifically mentions feeders are the ones designed by George E. Howard, a man who now worked for Hartford Empire according to the salary listing.

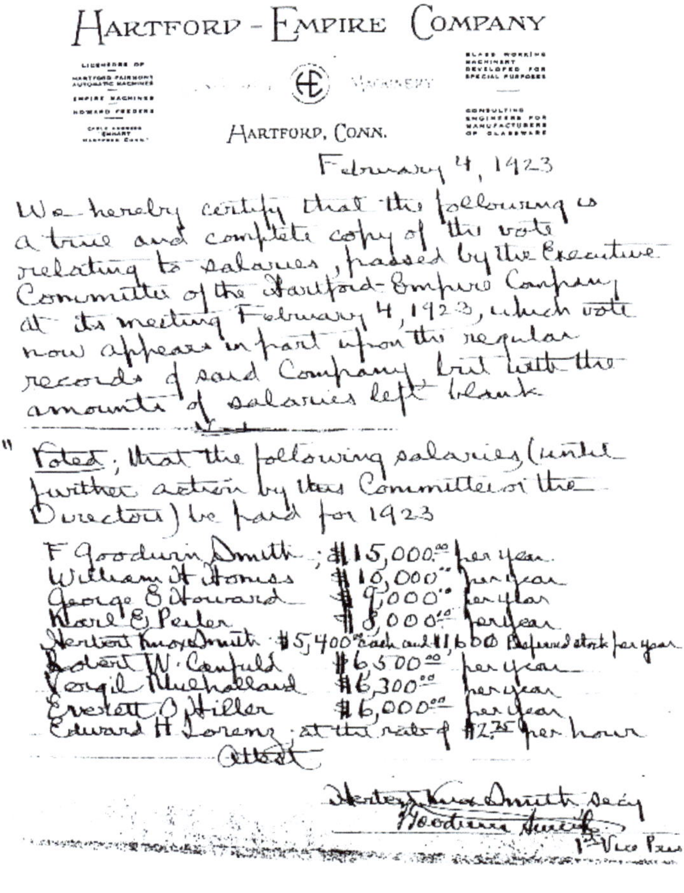

HARTFORD - EMPIRE COMPANY

Licensees of Hartford Fairmont Automatic Machines — Empire Machines — Howard Feeders

Glass working machinery developed for special purposes — Consulting engineers for manufacturers of glassware

HARTFORD, CONN.

February 4, 1923

We hereby certify that the following is a true and complete copy of the vote relating to salaries, passed by the Executive Committee of the Hartford-Empire Company at its meeting February 4, 1923, which vote now appears in part upon the regular records of said Company, but with the amounts of salaries left blank.

"Voted; that the following salaries (until further action by this Committee or the Directors) be paid for 1923

F Goodwin Smith ; $15,000.00 per year
William H Honiss $10,000.00 per year
George E Howard $9,000.00 per year
Karl E Peiler $8,000.00 per year
Herbert Knox Smith $5,400.00 cash and $1,600 Preferred stock per year
Robert W. Canfield $6500.00 per year
Virgil Mulholland $6,300.00 per year
Everett O Hiller $6,000.00 per year
Edward H Lorenz at the rate of $2.75 per hour

Attest

Herbert Knox Smith Secy
F Goodwin Smith 1st Vice Pres

Figure 6,25. Record of salaries paid to Directors in February 1923

The next innovation was the Type 81 Turbo feeder introduced in 1950, see Figure 6,26, just like the Type 144 over time the 81 gave rise to many offshoots all of which were mainly to increase the tonnage, increased machine speeds, and increased ranges for single, double, triple and quadruple gobs.

In 1951 the 81 was followed by the Type 113 with the HE High Torque air shears used in the making of television tubes, and the Type 115 feeder for an extended range of double gob operation. During the next decade the types 503, the 513 and the 515 were developed to be used with the wider forehearths to give tonnages up to 150 tonnes per day.

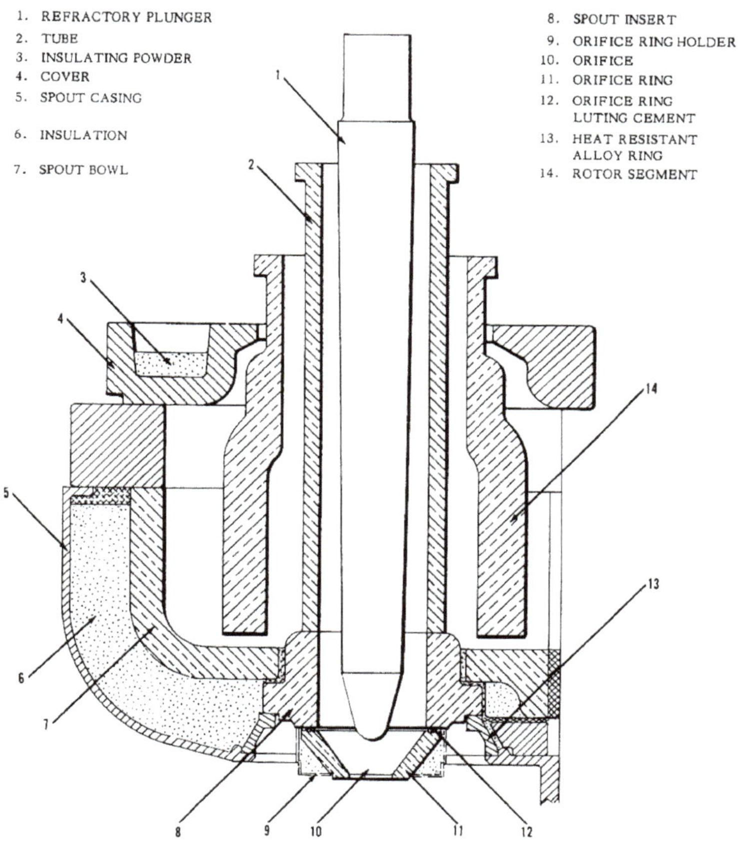

Figure 6,26. The Hartford 81 Turbo Feeder

In 1962 the Type 194 entered the market, followed in the 1970s by the Type 540 advanced forehearth system. Here it provided an improved cooling and heating system along with more sophisticated temperature control giving rise to greater temperature homogeneity. In 1978 Straight Line Shears were introduced.

In 1985 the Type 555 was introduced. This was microprocessor controlled with a servo-plunger mechanism and a servo-electric parallel shearing mechanism. This made for many applications to be computer controlled leading to the Flex-line system.

In the early 1990s the Advanced 740 Forehearth Control System was put on the market. On the 1st June 2006 Emhart ceased making forehearths passing on all of their customers to Sorg of Germany.

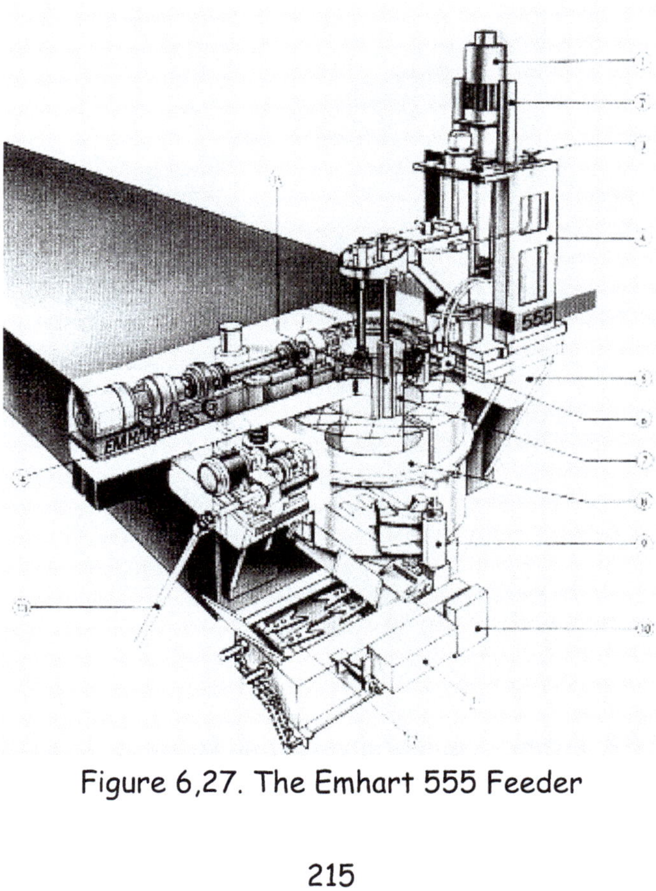

Figure 6,27. The Emhart 555 Feeder

Let's now go through the early development of the forehearth. The first feeders connected directly into the working end making them susceptible to direct interference from any changes to its temperature, see Figure 6,28, which is shown in an example of a patent filed by Tucker and Reeves. The USA Patent Number was 1 512 566 granted Oct 21st 1924 and 1 516 220 granted Nov. 18th 1924. The application was filed on Aug 12th 1918 Number 249 421.

The sketch shows that in this application that the heat input is controlled by burners or cooling pipes placed at the positions marked 9 and 10. The chimney No. 6 is used to regulate the heat input from the furnace.

This was cured by placing a refractory damper between the wall of the working chamber and the entrance to the forehearth,

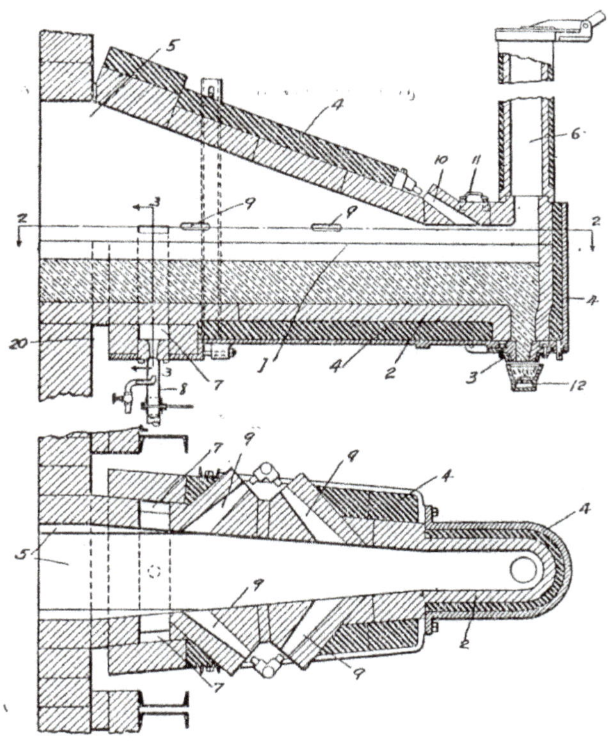

Figure 6,28. Patent application showing forehearth with direct access to the working chamber

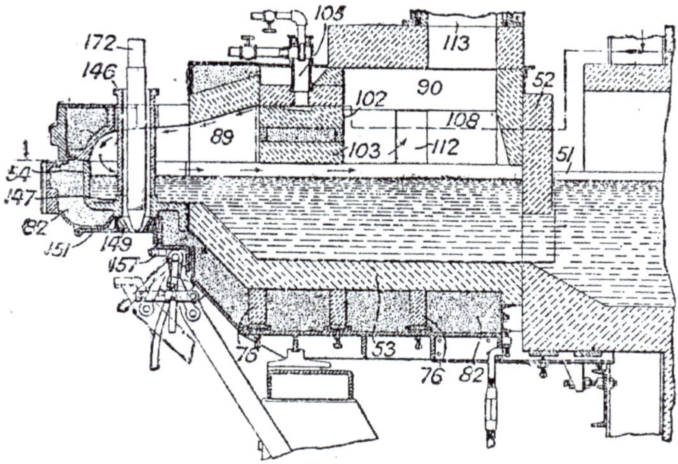

Figure 6,29. Part of Peiler's Patent application

the damper entered the glass to make the seal, see Figure 6,29. This is the first example that I have found regarding this problem, it was filed by K. Peiler at the British Patent Office on the 11th June 1924, being American in origin we can safely assume that the USA patents had already been filed.

The application to the British Patent Office comprised of two applications numbered British Patent Numbers 227 078 and 227 079. In the first one the abstract states, "In a feeder comprising a forehearth through which the glass is conveyed from the tank to a discharge outlet, means are provided for completely isolating the foreheath from the tank above the level of the glass. ------."

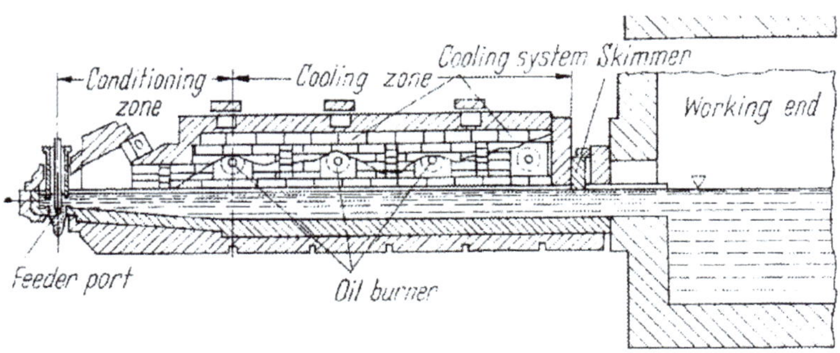

Figure 6,30. An oil fired 'Z' Type oil fired forehearth

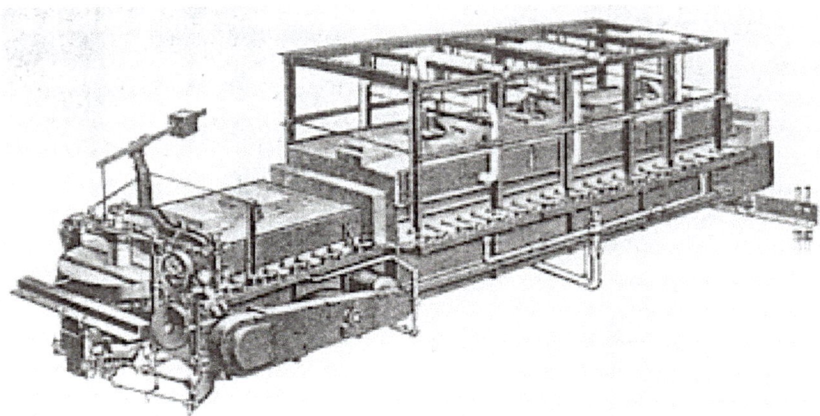

Figure 6,31. View of the 'K' Type gas fired forehearth

As the forehearths lengthened due to tonnage requirements so did the need for more control of the glass conditioning. Figure 6,30 shows a longitudinal section through a 12 foot 'Z' Type oil fired forehearth, it had four oil burners per side and one heating the spout area and Figure 6,31 shows the more familiar gas fired 'K' type forehearth.

Figure 6,32 shows a longitudinal section through a gas fired forehearth, while Figures 6,33 and 6,34 show the conditioning or equalising section of a conventional forehearth, and one with a deep spout. Figure 6,35 shows a Type 'KU' forehearth.

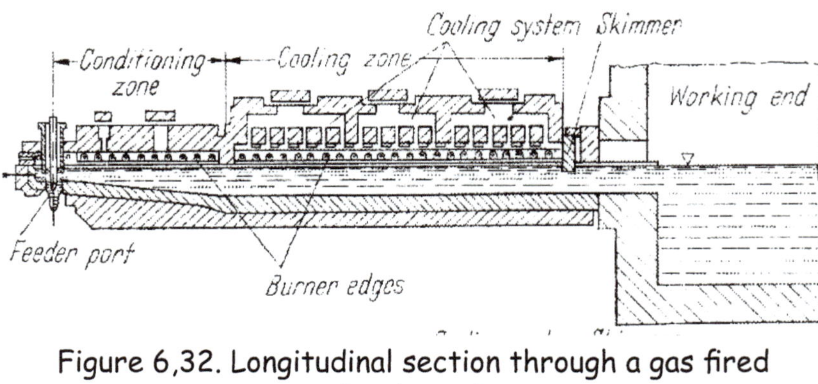

Figure 6,32. Longitudinal section through a gas fired forehearth

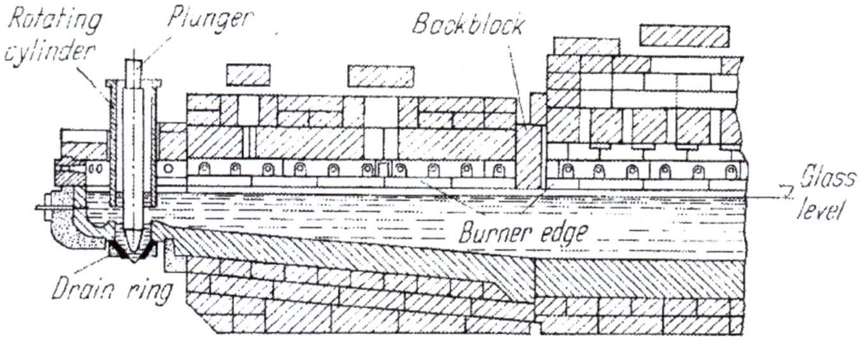

Figure 6,33 Section through the equalising zone, showing the shallow or standard spout

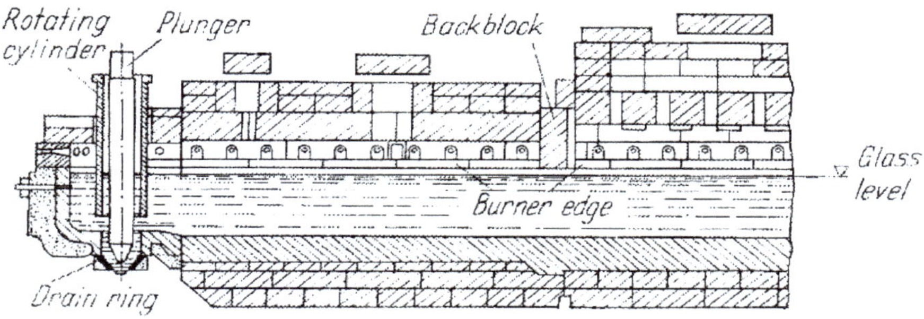

Figure 6,34. Section through an equalising section showing the arrangement for a deep spout

These were circa 1960s, three of the original spout configurations came to be standard with the glass depth being accommodated in the design of the equalising section. Using the standard channel depth of 7 inches then the shallow is 6 inches (glass depth altered in the last 4 feet), the standard is 7 inches (the forehearth bottom is a constant depth) and the deep is 10 inches (the forehearth is level as in the standard with a cascade down to the spout, or the forehearth is deepened over the last 4 feet. These are channel sizes the glass depth is one inch less.

Nowadays the channel size depends upon the feeder mechanism; using Emhart's terminology;

Make	Spout entrance	Equ. channel width	Spout depth*
144	14"	16"	6", 7" and 10"
		22"	6", 7" and 10"
		26"	6", 7" and 10"
81	16"	22"	7" and $9\frac{1}{2}$"
	18"	22"	10"
		26"	10"
		36"	10"
194	18"	26"	7" and 10"
115/503"	26"		11 and 5/16"
		36"	11 and 5/16"
		48"	11 and 5/16"
Type 1100 mm	22"	1100 mm	11 and 5/16"

* these are for level spout bottoms, for glass depth subtract one inch from spout depth plus the extra loss if using a raised tube seat.

Table 6,1. Various spout and equalising block combinations

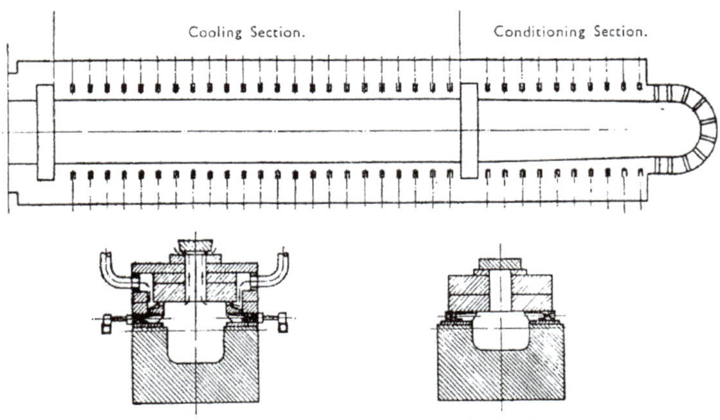

Figure 6,35. Sections through a Type KU forehearth

The burners shown in Figure 6,35 are arranged at equidistant centres, except where a zone baffle of some similar arrangement that needed a break in the superstructure. For a 14 foot forehearth there would be 66 burners at $4\frac{1}{2}$" centres plus 10 more around the spout.

TYPE "KU" FOREHEARTH FUEL AND AIR REQUIRED

MAXIMUM (6" or 15.2 Cm. of water MANIFOLD PRESSURE) and AVERAGE ($1\frac{1}{2}$" or 3.8 Cm. of water MANIFOLD PRESSURE)

Type Feeder	Forehearth Length Feet	Forehearth Length Meters	Sections	No. Burners	Amount Fuel Consumption	Forehearth Fuel (Gas) Δ City Gas C.F.H.	City Gas C.M.H.	Natural C.F.H.	Natural C.M.H.	Propane C.F.H.	Propane C.M.H.	Combustion Air* L.P. Fan Air C.F.M.	L.P. Fan Air C.M.M.	Compressed Air C.F.H.	Compressed Air C.M.H.
81	4	1.22	COOLING OR REAR SEC.	14	Max.	300	8.50	150	4.25	60	1.70	25	0.70	375	10.60
OR		1.22			Ave.	150	4.25	75	2.10	30	0.85	13	0.35	190	5.30
144	6	1.83		24	Max.	500	14.15	250	7.10	100	2.85	42	1.20	625	17.70
		1.83			Ave.	250	7.10	125	3.55	50	1.40	21	0.60	315	8.85
	8	2.44		36	Max.	750	21.25	375	10.60	150	4.25	63	1.75	940	26.55
		2.44			Ave.	375	10.60	190	5.35	75	2.10	31	0.90	470	13.30
	10	3.05		46	Max.	960	27.15	480	13.60	190	5.40	80	2.25	1200	34.00
		3.05			Ave.	480	13.60	240	6.80	95	2.70	40	1.15	600	17.00
	12	3.66		56	Max.	1150	32.60	575	16.30	230	6.50	96	2.70	1440	40.70
					Ave.	575	16.30	290	8.15	115	3.25	48	1.35	720	20.35
144	4'$7\frac{1}{2}$"	1.41	Cond.	30	Max.	620	17.55	310	8.80	125	3.55	52	1.45	775	21.95
		1.41	Sect.		Ave.	310	8.80	155	4.40	60	1.75	26	0.75	390	11.00
81	4'$8\frac{1}{2}$"	1.44	Equal.	30	Max.	700	19.85	350	9.90	140	3.95	58	1.65	875	24.80
		1.44	Sect.		Ave.	350	9.90	175	4.95	70	2.00	29	0.80	440	12.40
144	14'$7\frac{1}{2}$"	4.46	KU	76	Max.	1580	44.70	790	22.35	315	8.95	132	3.75	1975	55.95
		4.46	KU		Ave.	790	22.35	395	11.20	155	4.45	66	1.85	990	28.00
81	14'$8\frac{1}{2}$"	4.49	KU	76	Max.	1660	47.00	830	23.50	330	9.35	138	3.90	2075	58.80
		4.49	KU		Ave.	830	23.50	415	11.75	165	4.70	69	1.95	1040	29.40

(Δ Gas heating values: City 500 B.T.U./Cu.ft. or 4450 Kg. Cal./Cu.M.; Natural 1000 B.T.U./Cu.ft. or 8900 Kg. Cal./Cu.M.; and Propane 2500 B.T.U./Cu.ft. or 22,250 Kg. Cal./Cu.M. If actual heat content much different, adjust fuel consumption figures.)

C.F.H. = Cubic Feet/Hour
C.M.H. = Cubic Meters/Hour
C.F.M. = Cubic Feet/Minute
C.M.M. = Cubic Meters/Minute

(* L.P. Air $1\frac{1}{2}$ Pounds/Sq.in. or 105 Grams/Sq.cm. Fan.; Compressed Air 30 Pounds/Sq.in. or 2110 Grams/Sq.cm.)

Table 6,2. Fuel and Air requirements for a 'KU' forehearth

TYPE "KW" FOREHEARTH FUEL AND AIR REQUIRED
MAXIMUM (6" or 15.2 C.M. M.P.) and AVERAGE (1-1/2" or 3.8 C.M. M.P.)

Type Feeder	FOREHEARTH LENGTH Feet	Meters	Sections	No. Burners	Amount Fuel Consump-tion	FOREHEARTH FUEL (GAS)** City Gas C.F.H.	City Gas C.M.H.	Natural C.F.H.	Natural C.M.H.	Propane C.F.H.	Propane C.M.H.	COMBUSTION AIR* L.P. Fan Air C.F.M.	L.P. Fan Air C.M.M.	Compressed Air C.F.H.	Compressed Air C.M.H.
81	8	2.44	↑ REAR OR COOLING ↓	36	Max.	1035	29.25	515	14.60	205	5.85	86	2.45	1290	36.55
144	12	3.66		56	Max.	1605	45.50	805	22.75	320	9.10	134	3.80	2010	56.85
		3.66			Ave.	805	22.75	400	11.35	160	4.55	67	1.90	1005	28.40
144	14	4.27		68	Max.	1950	55.25	975	27.60	390	11.05	163	4.60	2440	69.30
		4.27			Ave.	975	27.60	490	13.80	195	5.50	81	2.30	1220	34.65
144	16	4.88		78	Max.	2240	63.35	1120	31.70	450	12.65	187	5.30	2800	79.20
		4.88			Ave.	1120	31.70	560	15.85	225	6.35	93	2.65	1400	39.60
144	18	5.49		90	Max.	2585	73.10	1290	36.55	515	14.60	215	6.10	3230	91.40
		5.49			Ave.	1290	36.55	645	18.25	260	7.30	108	3.05	1615	45.70
144	20	6.10		100	Max.	2870	81.20	1435	40.60	575	16.25	239	6.75	3590	101.50
		6.10			Ave.	1435	40.60	715	20.30	285	8.10	120	3.40	1795	55.75
144	4'7-1/2"	1.41	144	30	Max.	620	17.55	310	8.80	125	3.55	52	1.45	775	21.95
		1.41	Cond.		Ave.	310	8.80	155	4.40	60	1.75	26	0.75	390	11.00
81	4'8-1/2"	1.44	81	30	Max.	700	19.85	350	9.90	140	3.95	58	1.65	875	24.80
		1.44	Equal.		Ave.	350	9.90	175	4.95	70	2.00	29	0.80	440	12.40
144	14'7-1/2"	4.46	KW	76	Max.	1940	54.90	970	27.50	390	11.00	162	4.55	2425	68.65
		4.46			Ave.	970	27.50	485	13.75	190	5.50	81	2.30	1215	34.35
81	14'8-1/2"	4.49	KW	76	Max.	2020	57.20	1010	28.60	405	11.40	168	4.75	2525	71.50
		4.49			Ave.	1010	28.60	505	14.30	200	5.75	84	2.35	1265	35.75
144	20'7-1/2"	6.29	KW	108	Max.	2860	80.90	1430	40.50	575	16.20	239	6.75	3575	101.15
		6.29			Ave.	1430	40.50	715	20.25	285	8.10	119	3.40	1790	50.60
81	20'8-1/2"	6.32	KW	108	Max.	2940	83.20	1470	41.60	590	16.60	245	6.95	3675	104.00
		6.32			Ave.	1470	41.60	735	20.80	295	8.35	122	3.45	1840	52.00

C.F.H. = Cubic Feet/Hour C.F.M. = Cubic Feet/Minute
C.M.H. = Cubic Meters/Hour C.M.M. = Cubic Meters/Minute

(** Gas heating Values: City 500 B.T.U./Cu.ft. or 4450 Kg.C./Cu.M.; Natural 1000 B.T.U./Cu.ft. or 8900 Kg.C./Cu.M.; Propane 2500 B.T.U./Cu.ft. or 22,250 Kg.C./Cu.M. If actual heat content much different, **adjust** fuel consumption figures.)

(* L.P. Air 1-1/2 Pounds/Sq.in. or 105 Grams/Sq.CM. Fan. Compressed Air 30 Pounds/Sq.in. or 2110 Grams/Sq.CM.)

Table 6,3. Fuel and Air requirements for the 'KW' type forehearth

Tables 6,2 and 6,3 are tables showing the amounts of gas and air that was required for combustion. These are from Emhart literature which was probably from around the 1950s or 1960s, there is a drawing amongst this literature that indicates that it was before 1956 as it shows 5 section machines. The 6 section machine was not introduced until 1954 and then there would be a time lag on their implementation.

Figures 6,36 and 6,37 are schematics for the gas systems, Figure 6,36 is for a low pressure system and Figure 6,37 for the high. These are from the same Emhart 1950s literature. Figure 6,38 shows cross sections through a KW cooling section and a Type 81 Equalising section. While Figure 6,39 shows a common arrangement for a feeder connecting block circa 1950s-1960s.

Current practice would be for an electrocast block of high alumina content, (no zircon), or of other high grade materials to form the back cover block (lintel or arch) which is then built

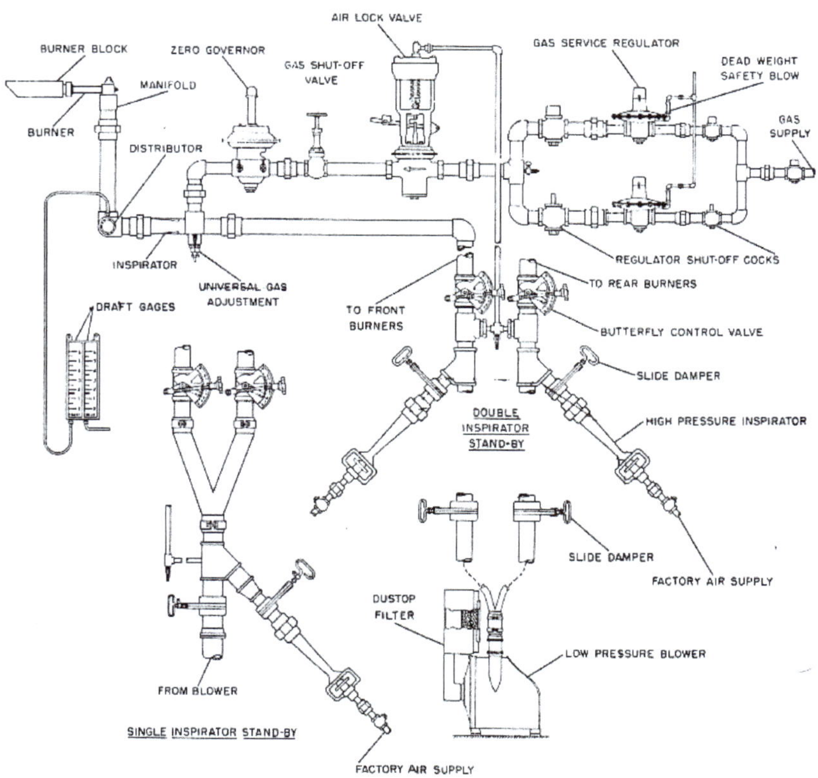

Figure 6,36. Schematic of L.P. gas firing for 'K' type

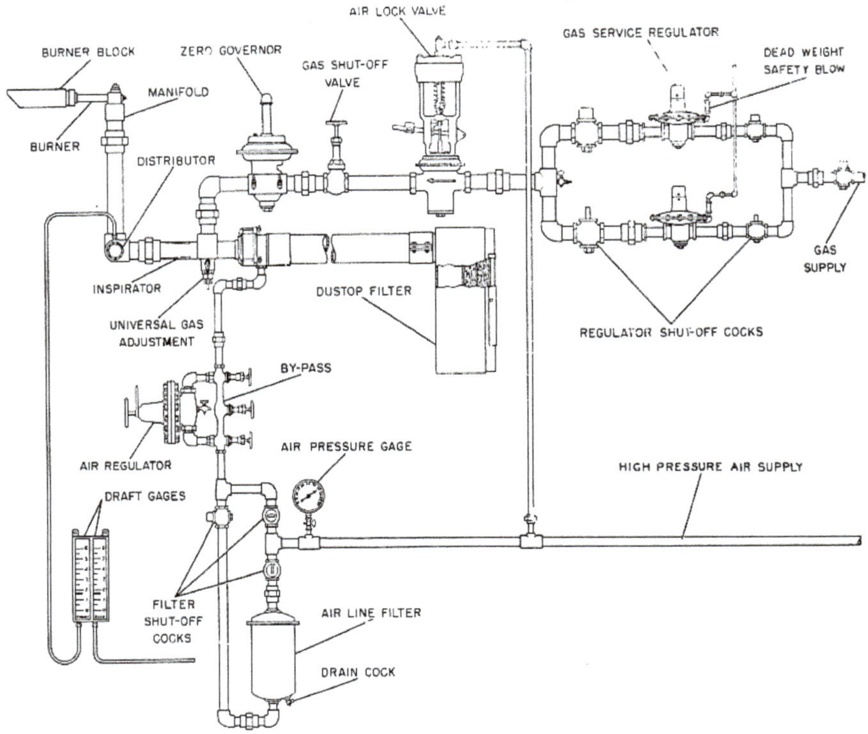

Figure 6,37. Schematic of H.P. gas firing system

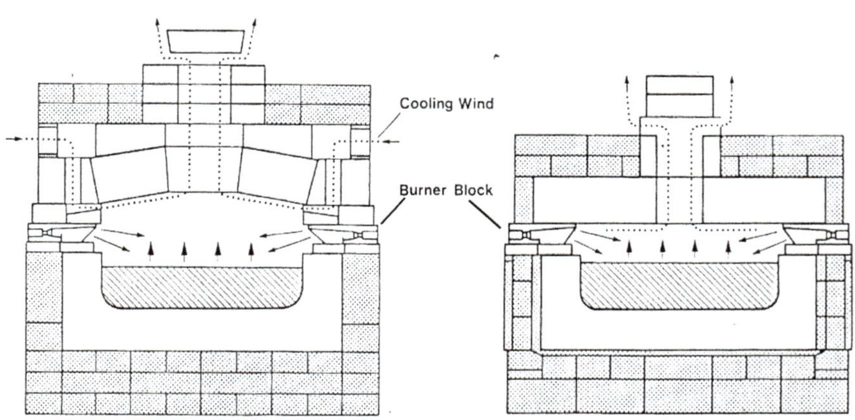

Figure 6,38. Cross sections through a KW cooling section and an 81 equalising section

into the furnace superstructure. If the furnace did not have a working end it had a distributor instead then the comment about built-in lintel blocks does not apply, more about distributors later.

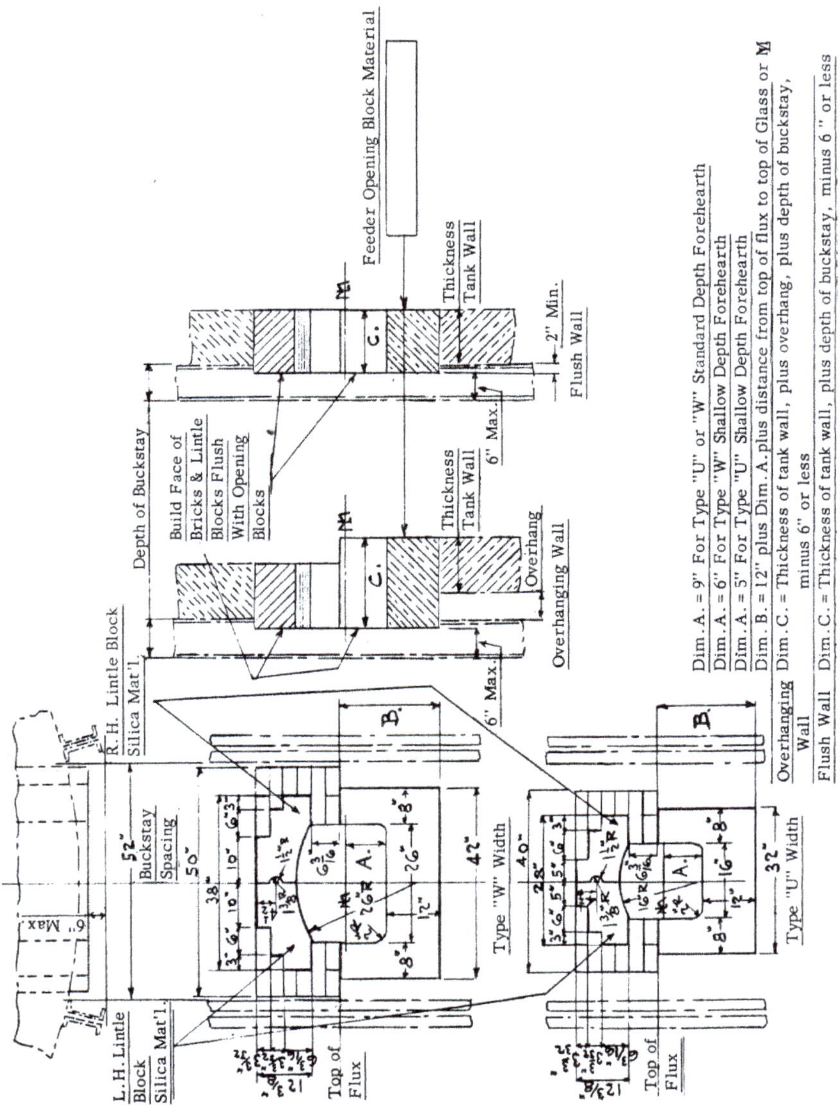

Figure 6,39. Arrangements for a KU and KW feeder connecting block

Of course the feeder/forehearth needs some form of control, originally there was no forms of temperature measurement the temperatures were judged by eye. Thermocouples were then used for temperature measurement, firstly the thermocouples were suspended into the glass from above, this was followed by horizontally mounted ones inserted into the glass through holes drilled through the spout and/or the forehearth sidewalls. These in some instances became a source for the origin of bubbles and blisters so the next step was to use radiamatic heads, which had a much longer life but the downside was they only measured the glass surface temperature.

Nowadays the temperature sensor is either a single thermocouple or a triplex thermocouple or a combination of both. The thermocouples hang from above making the maintenance of them simple, the life of a thermocouple is not as long as that of the forehearth so they need to be changed during the run.

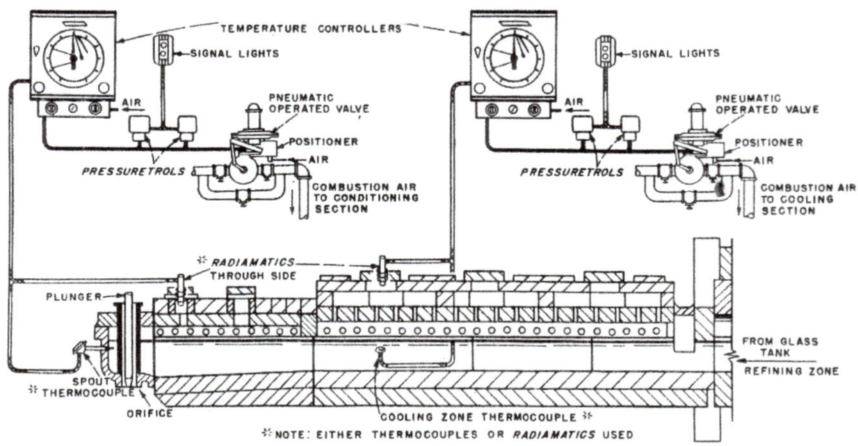

Figure 6,40. Control system for a 'K' Type forehearth, circa 1960s

In this case the instruments are either Brown or Honeywell Brown, the control system would most likely be the Air-o-line pneumatic controller. These instruments later gave way to the digital system such as that supplied by T.C.S. Instruments. The schematic shows the position of both thermocouples and radiation heads.

Many innovations have taken place to try and make the forehearth more controllable such as the 'side to side' gas firing system where each side of the forehearth can be adjusted separately. The cost of modern sensors and recording equipment is much more expensive than radiamatics but the benefits are worth the extra cost. The configurations of thermocouples are as follows:

1. A minimum system would be one single thermocouple at the forehearth entrance (possibly through a hole situated in the mantle block) and one at a typical distance of 9" back from the entrance to the spout, both would be on the centre line of the forehearth with the temperature being measured at a vertical position, distance from the bottom at what ever the company decides - better than nothing but not much use.
2. Same as above but using two triplex thermocouples positioned from the bottom of the forehearth of a 6" depth of glass at 1", 3" and 5". More information than before but again not much use.
3. Same as above but with an additional two single thermocouples located each side positioned at half way between the centre and the channel edge, thermocouple position 1" up from the bottom. Much more information gained especially the relationship between the centre and the edges.
4. Considered the minimum installation one tri-level thermocouple at the entrance to the forehearth and three tri-level thermocouples at around 9" back from the spout entrance. This gives the best information enabling the adjustment of the firing and cooling of the various zones to give the best conditions at the spout entrance.

A well used measure of the forehearth's efficiency is a formula which was developed by Owens-Illinois in the late 1970s, it is used by several glass companies worldwide and it is based upon the readings from three triplex thermocouples spaced equally across the forehearth with each giving information at three depths. Looking from the spout, they are arranged like this;

UL	UC	UR
ML	MC	MR
LL	LC	LR

and the formula is;

Thermal Homogeniety Efficiency (%) = $[1-(\Delta H+\Delta V)/MC]x100$

The reference temperature is taken as the value of MC as variations to this temperature do not affect the final calculation. Other variations of this calculation are used the main one being the use of five measuring positions arranged like this;

	UC	
	MC	
LL	LC	LR

The triangle of glass defined by the boundaries LL - UC - LR is considered to be where the primary glass comes from to make the gobs. One other item worth a mention is that the rotating tube can cause a build up of 'cold' glass on the exit side of the tube rotation and effect the temperature readings, i.e. on the side were the flow is opposite to the main direction of flow. To keep this within acceptable limits a tube speed of 3 revolutions per minute should be considered as maximum.

So far we've discussed the permanent refractories that are used, now we will look at the replaceable ones or the 'expendables' as they are known and what delivery equipment sometimes need changing when the jobs are changed. Covered in this category are; the spout and it's ancillary equipment such as the orifice, the plunger or plungers the tube and rotors, others are stirrers, zone baffle blocks and skimmers. The spout, the zone baffles and the skimmers although classed as expendables usually have a long life of at least a year depending upon refractory choice.

The non-refractory variable equipment that may need changing are the drop guides and the delivery equipment - this is for an IS machine.

We will look at the non-refractory items first. First we start with the glass weight to determine the sizes, the ideal gob shape is one that has a gob length that is three times its diameter, the relationship can be seen in Figure 6,41. This size is the ideal one

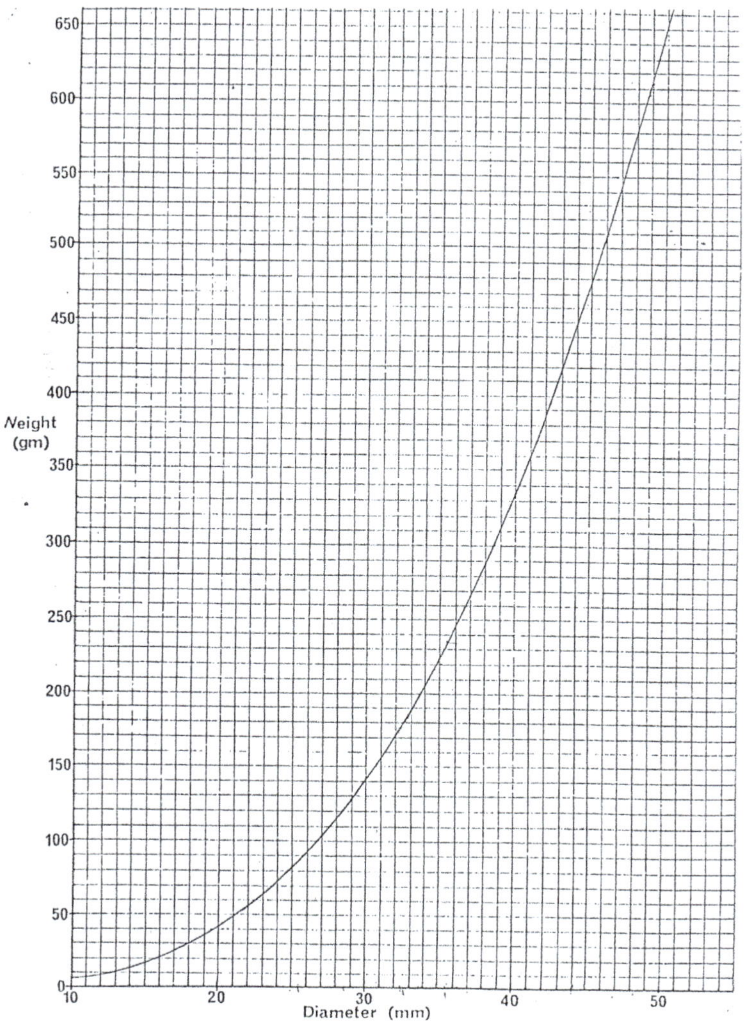

Figure 6,41. Ideal gob for parison design

not the actual one - sometimes they have to be shaped to match the blank shape. Once determined this gives the orifice diameter, for bottle manufacture the relationship between bottle weight and orifice size is as follows again these are just a guide:

A half ounce gob will pass through a $\frac{5}{8}$ths inch orifice
A 2 ounce gob will pass through a 1 inch orifice
A 4 ounce gob will pass through a $1\frac{1}{4}$ inch orifice
An 8 ounce gob will pass through a $1\frac{9}{16}$ inch orifice
A 16 ounce gob will pass through a $2\frac{1}{8}$ inch orifice
A 32 ounce gob will pass through a $2\frac{5}{8}$ inch orifice
A 64 ounce gob will pass through a $3\frac{3}{8}$ inch orifice

Once the diameter has been determined then the drop guides if being used may need changing. Estimates of the required sizes are;

- For single guides the selected size should be from one $\frac{1}{4}$ to $\frac{3}{4}$ of an inch larger than the gob diameter.
- For two drop guides the selected size should be either one $\frac{1}{4}$ or a $\frac{1}{2}$ inch larger than the gob diameter.
- For three drop guides the selected size should be either one $\frac{1}{4}$ or a $\frac{1}{2}$ inch larger than the gob diameter.

In all cases it is usually satisfactory to choose one size larger than the gob diameter. Finally the delivery equipment may need changing the size required is governed by the gob dimensions.

Figures 6,42 and 6,43 show expanded versions of the refractories the make up the feeder assemblies. Figure 6,42 is for a 144 feeder assembly and Figure 6,43 is for an 81 feeder assembly, including some of the restraining ironwork. Other things that sometimes need changing at a job change besides the orifice are the plunger (if the change in weight is significant) and the drop guides if they are being used (changes in weight alter the gob diameter).

The plunger selection depends on its shape and the number

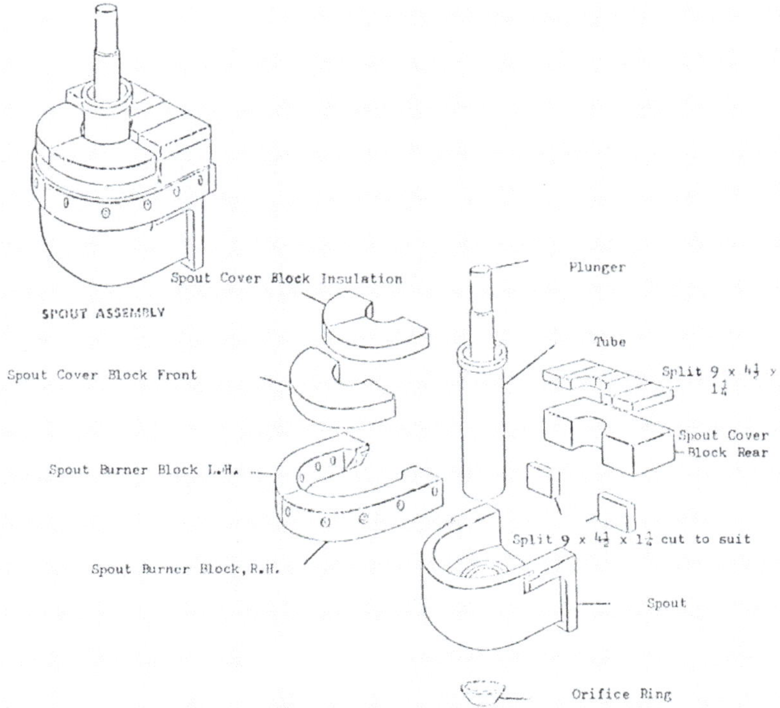

Figure 6,42. Spout assembly of a HF 144 feeder

of plungers that are being used. The following is a practical guide for single and double gob operation.

Single Gob	**Weight in ounces**	**Plunger**
	6-10	2$\frac{3}{4}$" Standard Point #1
	11-19	2$\frac{3}{4}$" Standard Point #1
	20-39	2$\frac{3}{4}$" or 3" Standard Point #1
	40-59	3" or 3$\frac{1}{4}$" Ball Point #2
	60-100	3$\frac{1}{2}$" Ball Point #2
Double Gob	**Weight in Ounces**	**Plunger**
	4-10	2$\frac{3}{4}$" Standard Point #1 or 3" Ball Point #2
	11-16	3$\frac{1}{4}$" or 3$\frac{1}{2}$" Ball Point #2
	17-21	3$\frac{1}{4}$" or 3$\frac{1}{2}$" Ball Point #2

Table 6,4. Plunger type selection

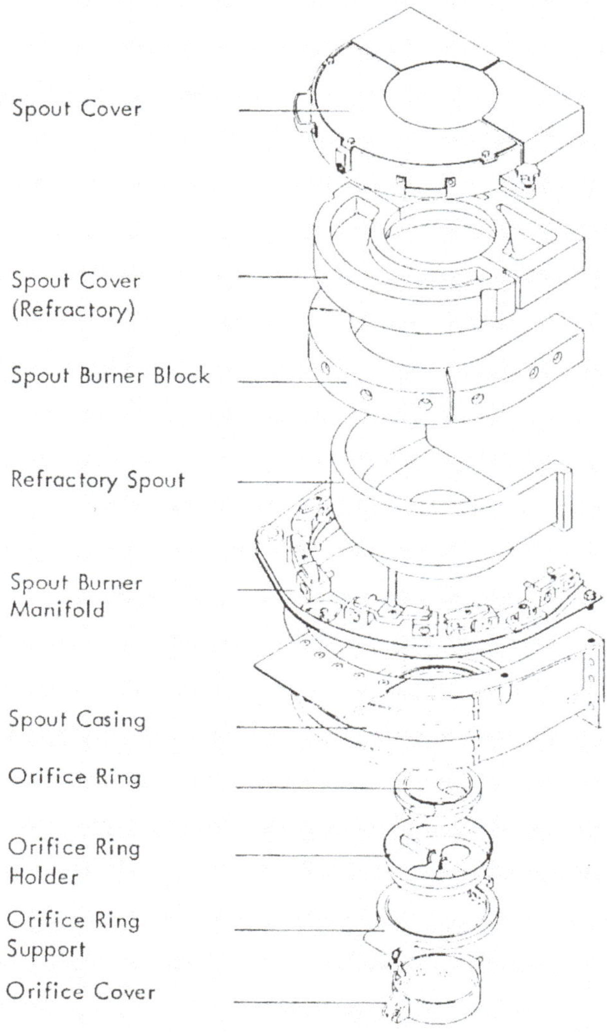

Figure 6,43. Spout Assembly 81 Feeder

These tables are intended as guides only as the actual size and the shape of the plunger will vary with gob length, speed, glass temperature, etc.The following table is for sizing twin plungers, after each twin plunger size is the diameter of a single plunger with the same cross-sectional area as the two twin plungers.

Twin Plungers	Twin Plunger Diameter	Diameter of Equivalent Single Plunger
	$2\frac{1}{8}''$ diameter	$3''$ diameter
	$2\frac{3}{8}''$ diameter	$3\frac{1}{4}''$ diameter
	$2\frac{1}{2}''$ diameter	$3\frac{1}{2}''$ diameter

Table 6,5. Equivalent twin plunger sizes

The action of twin plungers is more like a single-gob operation, and a plunger cam which would normally be used if the job were being run single gob, often works better than a high speed double-gob plunger cam.

Sometimes the glass thermal homogeneity leaves something to be desired - usually when the forehearth is been pulled outside of its design parameters. Here are some instances of methods that have been used to improve the thermal homogeneity of the glass. Figure 6,44 shows some of the modern methods of stirring (four views).

The stirrers are used for other things for example in the 'Colourant Forehearth' where the base glass colour, usually flint, is changed by adding another colour in the form of a 'frit'. Frit is an intensely coloured low melting temperature glass. For a 26 inch wide forehearth the colourant section, which is in addition to the normal forehearth length, would comprise of four sets

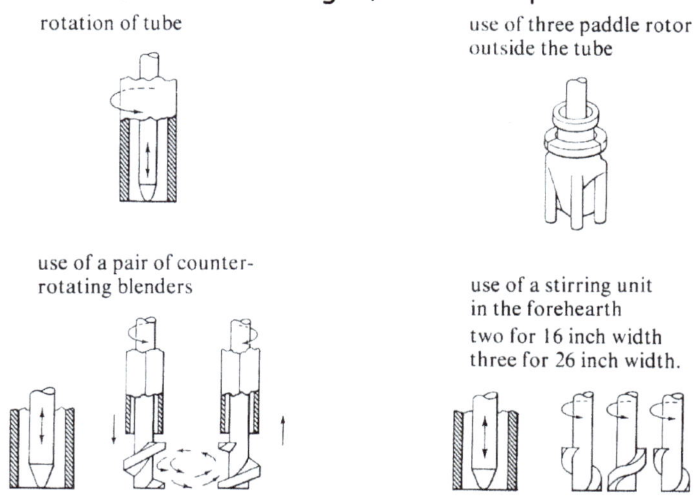

Figure 6,44. Schematic showing various item used to stir the glass

(or rows) of three stirrers. A 36 inch wide forehearth would have either four or five rows of four stirrers, while a 48 inch wide forehearth would have at least five rows of six stirrers. The stirrers in a colourant forehearth are always just after the forehearth entrance, i.e. the back of the forehearth.

Stirrers are used at the other end of the forehearth, they are placed three feet from the spout entrance. Here there are two types, the 'paddle type' used to disperse 'cat scratches' that originate in the glass melter or the 'screw type' used to raise the glass level in the spout or to finally mix the glass to reduce the thermal inhomogeneity.

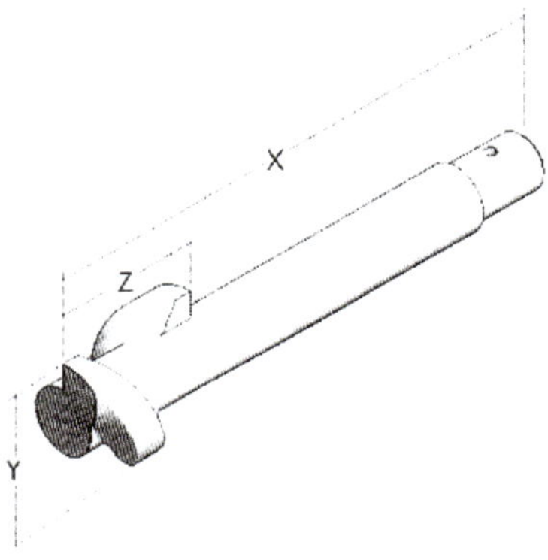

Figure 6,45. Forehearth screw stirrer with right hand helix

Forehearth stirrers have a right hand and a left hand helix. Typical dimensions are; 'X' = 686, 762, 838, 860, and 914 mm. Dimension 'Y' = 146, 152, and 165 mm. Dimension 'Z' = 115 and 190 mm.

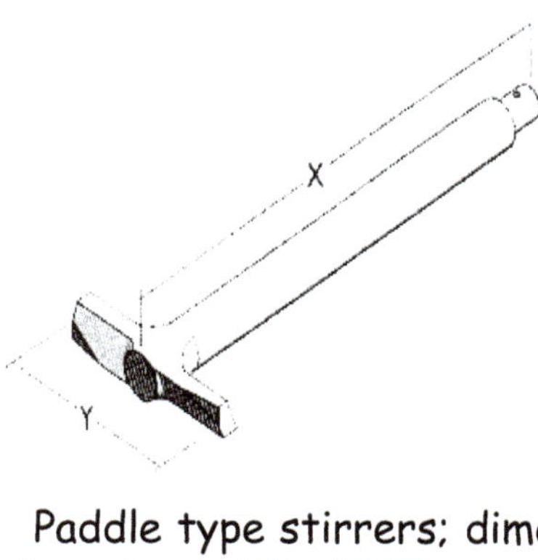

Figure 6,46. Right hand paddle stirrer

Paddle type stirrers; dimensions are maximum 'Y' = 457 mm and maximum 'Z' = 1220 mm.

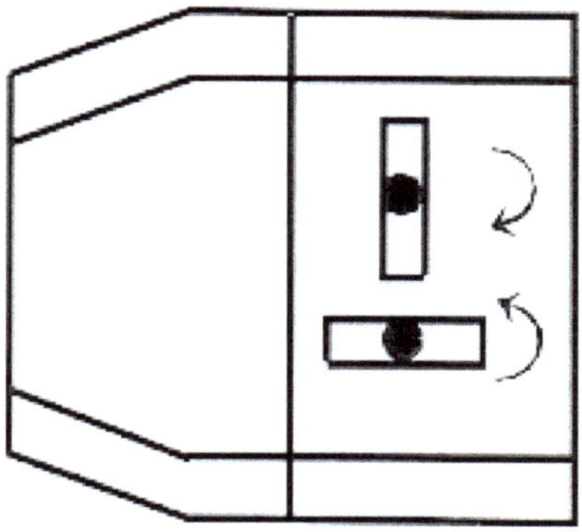
Figure 6,47. Plan of paddle stirrers

For 'cat scratch' dispersion the paddles should be located at the front of the forehearth rather than the rear. If located at the rear then the denser glass has chance to re-form back into heavier cords. The stirrers should be configured to lift the denser glass up from the bottom of the forehearth and place it in the centre of the gob. Right hand stirrers rotate clockwise and left hand one rotate anti-clockwise.

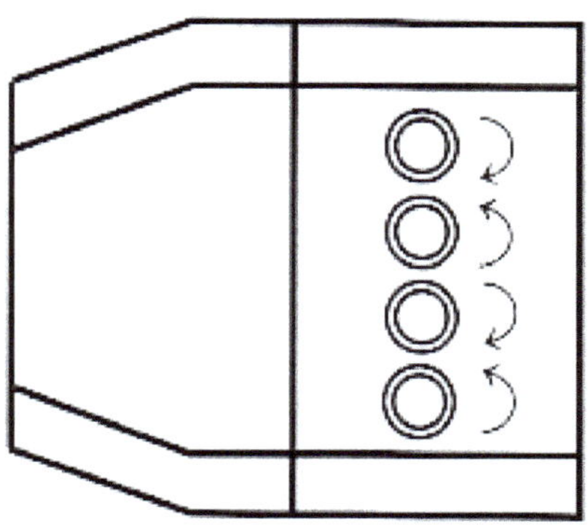
Figure 6,48. Stirrers in a 36 inch wide equalising section

In the equalising section of a forehearth the stirrers should work in pairs, right hand rotating clockwise and left hand anti-clockwise and have the same centre distance i.e. spaced out equally. This lifts the glass and mixes it at the same time. The speed of the stirrers and the glass viscosity is important, if it is too fast for the glass viscosity then blister can be formed or it can cause refractory breakage.

In a colourant forehearth the stirrers should always be to the rear of the forehearth, and unlike stirring of the glass in the equalising section each bank of stirrers (row across the forehearth) should all rotate in the same direction. Alternate rows rotate in opposite directions; this allows the glass to move from side to side which increases the residence time for the mixing of the frit, see Figure 6,49.

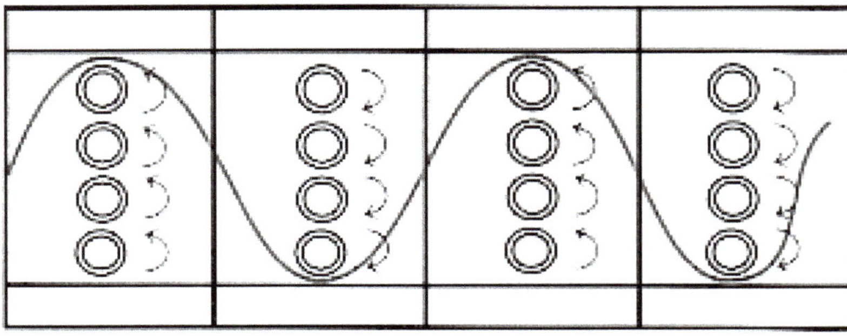

Figure 6,49. Stirrer arrangement for a 36 inch wide colourant forehearth

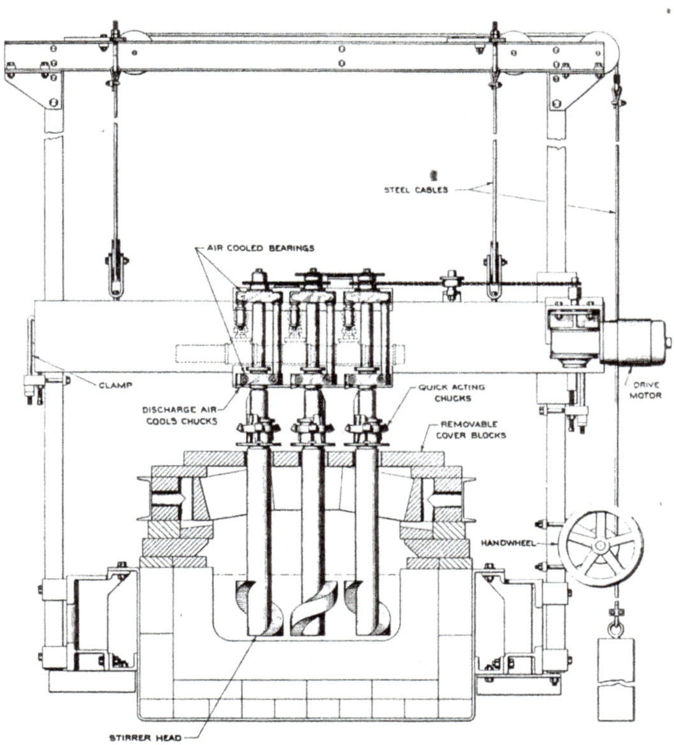

Figure 6,50. Section showing a set of 26 inch wide stirrers. They can be used in a K-26 cooling section or in the transition section of a K-36 conditioning section

A different stirring mechanism is where the tube is fixed and a set of rotors revolves around its outside, see Table 6,6 for types.

Some forehearths that use rotors	
6″ Tube	81 and 501
7″ Tube	81 and 501
8″ Tube	503
10″ Tube	503

Table 6,6. Rotor sizes for a selection of different types of feeder

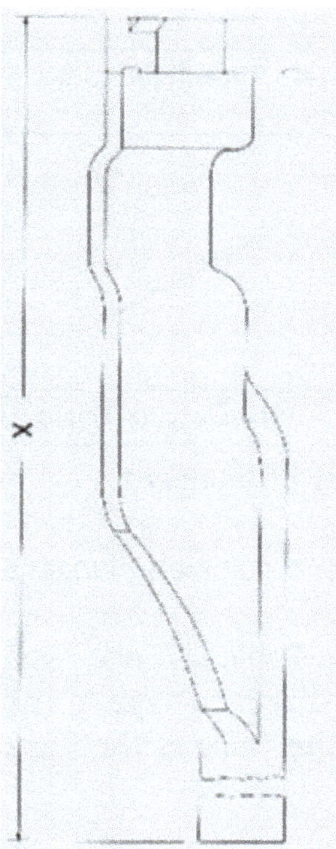

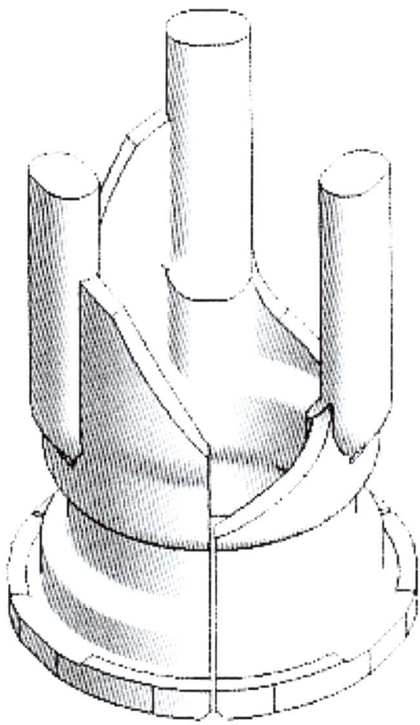

Figures 6,51 (left) showing a single rotor and Figure 6,52 (right) showing a set or rotors

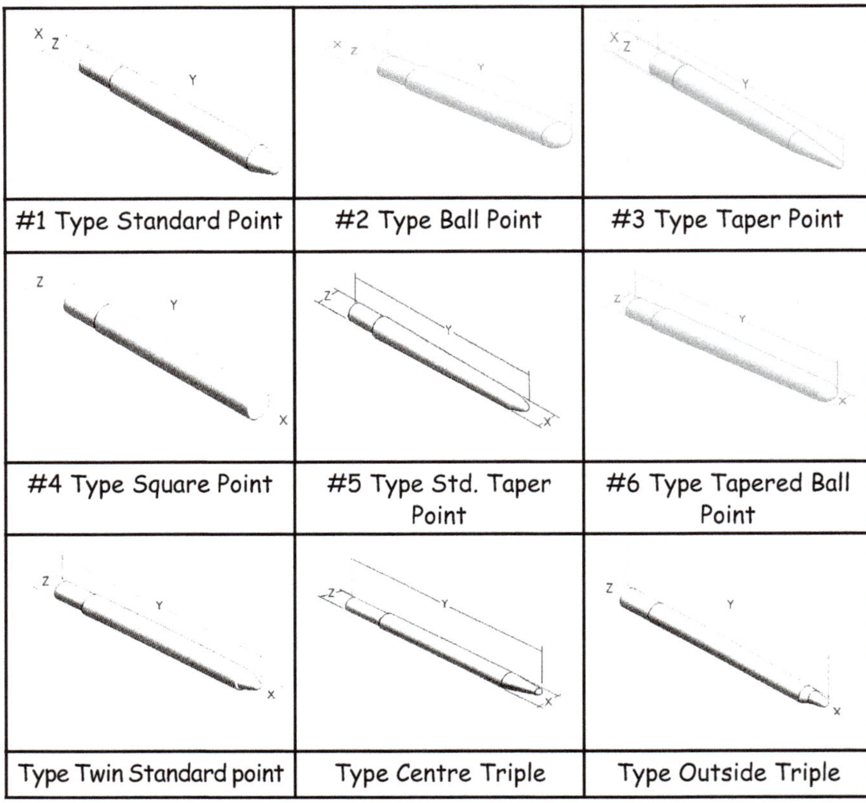

Figure 6,53. The more common types of plungers

The above nine sketches are of the more common plungers that are in use today.

The most common type in this listing, see Table 6,7, are Type Number 2, the Ball Point, the second most common type is the Type Num ber 5 the Standard Taper and the third is the Type Number 1 the Standard Point.

Selection of different sizes in plunger diameters in mm

Type	No. 1 Standard Point	No. 2 Ball Point	No. 3 Taper Point	No.4 Square Point	No. 5 Std Point Tapered	No. 6 Ball Point Tapered
29		89			63, 70, 76	76
58	83, 89, 95, 102, 108, 114, 121, 127	83, 89, 95, 102, 108, 114, 121, 127				
144	63, 70, 76, 83, 89	63, 70, 76, 83, 89, 95, 102, 108, 114, 121	63, 70	76, 89, 105	70, 76	70
81	70, 76, 83, 86	83, 89, 95, 102, 108, 114, 121, 127				
113		140, 152, 165, 178				
501		45, 51, 54, 57, 60, 63			45, 48, 51, 54, 57, 60	45, 51, 57, 60, 63
503		76, 83, 89, 95, 102, 108, 114, 127			45, 48, 51, 54, 57, 60, 63, 70, 76, 83	
515		45, 51, 57, 60, 63			45, 48, 51, 54, 57, 60	

Table 6,7. Some of the more common plungers available on the market. Listed by plunger diameter and by feeder type

Twin Plungers Offset Point - D.G. - C.D. $4\frac{3}{8}$" dia. in mm	
81	45, 48, 51, 54, 60, 63
503	54, 60

Table 6,8. Commonly occurring twin plungers - offset gobs using Emhart model numbers

Triple Plungers

	Plunger diameter	Centre distance
144	$1\frac{3}{4}$"	C.D. 3"
503	$2\frac{3}{8}$"	C.D. $4\frac{3}{8}$"
515	$2\frac{1}{2}$"	C.D. $4\frac{3}{8}$"

Table 6,9. Commonly occurring triple plungers using Emhart model numbers. Triple plungers are different - the centre one is a standard point and each of the outside ones are offset points.

The last of the rotating parts in this discussion is the tube; these come in two types standard and self centering; and each type has two versions plain and vented. Table 6,10 lists a selection of tube sizes that are available on the market the lengths vary according to the feeder mechanism and whether they are for standard or deep spouts.

Selection of the more common tube sizes

Type	Internal diameter of tube: Standard tube Plain	Standard tube Vented	Self centring tube Plain	Self centring tube Vented
144	5", 6" & $7\frac{3}{4}$"	5", 6" & 7"		
81	6", 7", 8" & 10"	6", 7", 8" & 10"		
113	7" & $9\frac{1}{2}$			
115	7", 8" & 10"	7" & 8"		
194	7" & 8"	7" & 8"		
503	7", 8", 10" & 11"	7" & 10"		
515	11"	11" & 12"		
Tube sizes			5", 6", 7", 8" 10" & 12"	5", 6", 7", 8", 10", 11" & 12"
Number of holes	Mostly 4 @ $\frac{7}{8}$" Specials 503 and 515 with 3 @ 2"			Either 3 or 4 @ $\frac{7}{8}$". Specials 4 @ $1\frac{1}{2}$" for 8" & 10". 3 @ 2" for 10" & 11"

Table 6,10. Selection of internal tube sizes using Emhart terminology

Finally we come to the orifice rings, there are too many for us to list but from one reputable catalogue a selection is as follows;

Single Gob

- 5" diameter spout throat, both new and old types. They range from 5/16th of an inch orifice diameter through to $4\frac{1}{2}$ inches diameter. 79 different sizes.
- 7" diameter spout throat, sizes range from $\frac{1}{2}$ inches diameter to $6\frac{1}{2}$ inches. 71 different sizes.
- 10" diameter spout throat, sizes range from $1\frac{3}{8}$ inches diameter to $3\frac{7}{8}$ inches. 31 different sizes.

Double Gob

- 5" diameter spout throat, both old and new designs, sizes range from 2 at $\frac{3}{8}$ths. of an inch in diameter through to 2 at 2 and 3/16th of an inch. The orifice centres are at 3 inches. 2 x 39 different sizes.
- 7" diameter spout throat, 3" centre distance, sizes range from 2 at $\frac{3}{8}$" diameter to 2 at $2\frac{1}{8}$" in diameter. 29 different sizes.
- 7" diameter spout throat, $4\frac{3}{8}$" centre distance, both old (three different types) and new designs, sizes range from 2 at $\frac{1}{2}$" diameter to 2 at 2 and 15/16ths, 2 x 40 different sizes; and from 2 x $\frac{3}{4}$" up to 2 x 2 and 9/16ths of an inch in diameter. 25 different sizes; and from 2 x $\frac{3}{4}$" up to 2 x $2\frac{1}{4}$", 31 different sizes.
- 10" diameter spout throat, $4\frac{3}{8}$" centre distance, both new and old designs and a solid bridge type, sizes range from 2 at $1\frac{1}{4}$" diameter to 2 at $3\frac{3}{8}$" in diameter, 33 different sizes; from 2 x 1 and 7/16ths of an inch to 3", 26 different sizes; and from 2 at $\frac{7}{8}$" in diameter to 2 at $3\frac{3}{8}$" in diameter for the solid bridge type, 33 different sizes.
- 10" diameter spout throat at 5" centre distances, sizes range from 2 at $1\frac{1}{4}$" diameter to $3\frac{3}{4}$" in diameter. 41 different sizes.
- 12" diameter spout throat, $4\frac{3}{8}$" centre distances, sizes range from 2 at 2 and 1/16th inches in diameter to 2 at $2\frac{3}{4}$" in diameter. 12 different sizes.

Triple Gob

- 8″ diameter spout throat, 3″ centre distances, sizes from 3 at $\frac{3}{8}$″ in diameter to 3 at $1\frac{1}{2}$″ in diameter. 19 different sizes.
- 10″ diameter spout throat, 3″ centre distances new and obsolete designs and solid bridge type, sizes from 3 at $1\frac{1}{4}$″ in diameter to up to 3 at $2\frac{1}{4}$″ in diameter. 17 different sizes.
- 12″ diameter spout throat, $4\frac{3}{8}$″ centre distances, from 3 at 1 and 11/16ths of an inch diameter up to 3 at $2\frac{1}{2}$″ in diameter. 14 different sizes.

Spouts

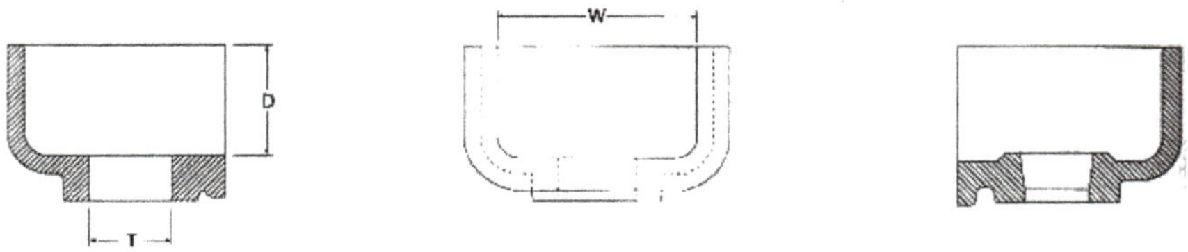

Figure 6,54. Spouts, left hand one is flat bottomed, right hand one is raised seat and tapered throat, centre is an end view

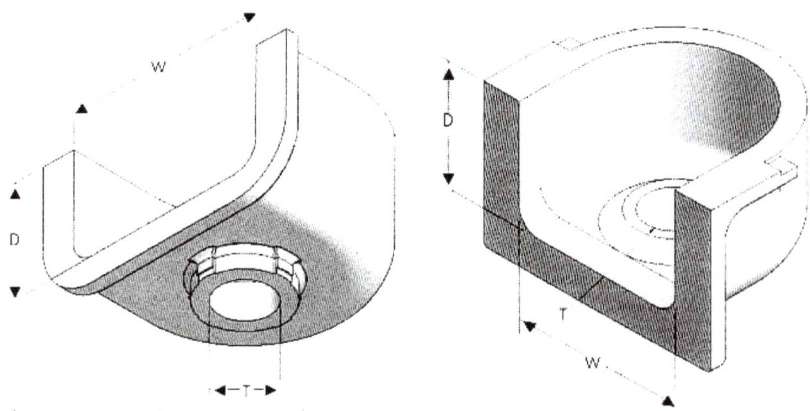

Figure 6,55. View of a spout from underneath and one from above showing a raised seat

Various Spout sizes - Emhart terminology

Type	Throat	Depth	Width	Tube seat	Remarks
144	5"	6"	14"	Raised & flat	Angular shearing available
	$7\frac{3}{4}$"	6"	14"	Flat	8" TG
81	7", 6/5" & 8/7"	$7\frac{1}{4}$" & $9\frac{1}{2}$"	16"	Raised & flat	16" offset
	7", 8" & 8/7"	$7\frac{1}{4}$" & 10"	18"	Flat	18" Centre line
194	7", 8" & 8/7"	$7\frac{1}{4}$" & 10"	18"	Flat	
115	7"	$11\frac{5}{16}$"	22"	Flat	Deep spout only
503	10"	$7\frac{1}{4}$"	22"	Flat	
	7", 8/7", 10" & 11/10"	$11\frac{5}{16}$"	22"	Flat	
515	10" & 12"	$11\frac{5}{16}$"	24"	Flat	Shallow entry
	8/7", 11/10", 10" & 12"	13"	24"	Flat	Narrow entry
81/555	5" & 6/5"	$7\frac{1}{4}$" & 10"	14"	Flat	14" wide
	5", 6/5", 7", 8" & 8/7"	$7\frac{1}{4}$" & 10"	18"	Flat	18" wide

Table 6,11. Using Emhart terminology some of the various spouts in the marketplace

The only items which we have not discussed yet is the forehearth channels these are not an expendable they last the whole of the forehearths life so they are often in 'electrocast' high alumina material.

Channel Blocks

These come in two shapes, i.e. the internal contours are different, the normal or classic contour or the profiled contour, see Figure 6,56 to see the difference in shape and Table 6,12 for the standard sizes. All 'L' dimensions are 24" except ** which can also be bought at 18".

There are two other cooling channels available, one that changes a profiled channel into a normal channel Figure 6,57. The sizes that are available to do this are; 16", 26", 36" and 43", all are 24" long and 7" deep, and a transition channel, see Figure 6,58, which come in the following sizes; 26"/16", 26"/22", 36"/26"**,

36"/26", 43"/26", 43"/36" and 48"/36". All are 24" long and 7" deep except for ** which is 7"/10".

Channel Blocks - Cooling Section

Normal Profile		Profiled Profile	
W	D	W	D
16"	7" & 10"	16"	7"
26"	7", 7/10" & 10"	26"	7"
36"	7"	36"	7"
43"	7"	43"	7"
48"	7" **	48"	7"
54"		54"	7"

Table 6,12. Some of the various channel sizes that are obtainable

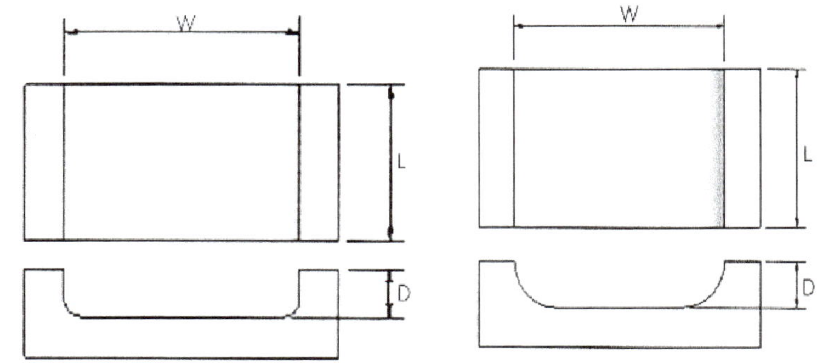

Figure 6,56. Normal and Profiled Channels

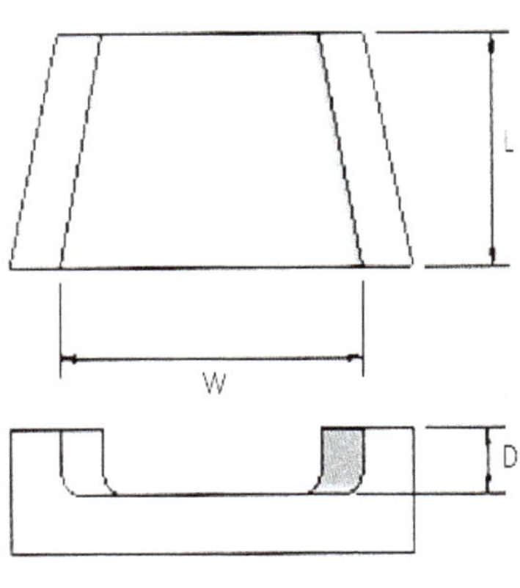

Figure 6,57. Profiled to Normal Channel

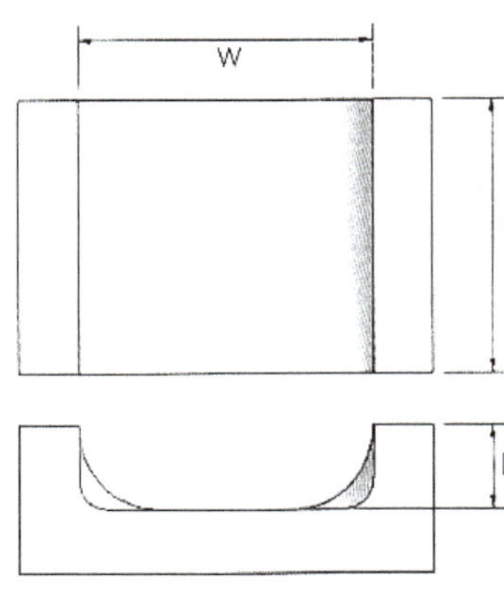

Figure 6,58. Transition Channels

Working Chambers or Ends, Alcoves and Distributors

The discovery or development of the regenerative and recuperative furnaces by Siemens around 1870 led to the development of the working or the refining end of the furnace, i.e. a place from which the glass was gathered and withdrawn from the furnace and then fed into the blank mould.

At first they were part of the furnace proper, then with the advance of throated furnaces first a shadow wall was placed above the throat to help cool the glass and finally the melting chamber and the working chamber where separated. Then with the development of the automatic machine were each machine required its own forehearth working ends became just a distributor of the glass.

At first they stayed the same size but eventually companies reduced them until they became about 20% of the size of the melting end, this was an area relationship but the glass depth was usually less than that of the melting end due to devitrification problems, especially when melting the old style emerald greens. This situation continued up to around the 1970s; due to their popularity there had been a gradual change in the machine line up to IS machines and also, the size of the machine was increasing, which meant that the forehearth pulls were increasing. From now on the 'rate of change' was been driven by the manufacturing companies, the extra glass meant initially longer forehearths followed by wider ones.

The design of the working end started to become more rectangular in shape, with the smallest being square in plan; with either one, two or three forehearths coming from it. A four machine line up used a rectangular working end, all machines coming out from the working end wall at right angles this led to the machines in straight lines down the shop but not across. This then led to larger machines being installed on the outside with longer cranked forehearths to bring all the orifices in line.

The next step was the forehearth and distributor design - now being carried out by specialist companies who supplied everything from the throat to the orifice, including in some cases the shearing mechanism. The glass manufacturer supplied the machine positions and their type, and the job portfolio that they expect them to work, and then the specialist company would do the design.

Much investigation has gone into the removal of heat from glass in channels and shows that the deeper layers of glass can only be affected in shallow depths, especially in coloured glasses. Hence modern designs of forehearths and distributors dictate that the glass depths are much shallower than in the 1960s, 400 to 500 mm (16 to 20 inches). would be the deepest at the centre of the distributor as against the 750 to 900 mm (30 to 36 inches) all over the working end that was the then norm, the width would be up to 220 mm (84 inches). The normal depth of a forehearth would be 6" deep, the 10" deep has gone out of fashion, sometimes the glass depth is 4 or 5" in coloured glasses, when this is the case the forehearth is made wider.

Modern distributor / forehearths systems are now designed with the aid of computer models which uses the glass viscosity, gob temperature, to calculate the 'head loss' for a given or calculated forehearth configuration, (there will be more about this in Appendix 2).

As a résumé regarding the history of forehearth widths; in 1942 Hartford Empire published information on the typical 'K' type forehearth, including how the temperature varied along the length for different weight articles. This basic design lasted and by the 1960s the width was 26 inches, by 1966 the main forehearth in use was the 26 inches wide type 'KWS'. In 1967 the 36 inch wide K36 was introduced followed by the K48 with 48 inch channels in 1978. Each of the width increases being made to accommodate higher pulls.

British Hartford-Fairmont (BHF)

Now we will investigate the early history of British Hartford Fairmont. Prior to WWI (1914-1918), W. A. Bailey imported high quality colourless glass from the Leerdam Glassworks in the Netherlands, in 1917 he took the risk of purchasing the patent rights of the paddle feeder for the UK and the Netherlands and formed British Hartford Fairmont Syndicate Ltd. on the 27th of December that same year. This gave Hartford Empire a sales outlet in Europe.

The first BH-F feeder, a paddle feeder, was installed at the New Cross glassworks of Key Glass. I have two dates for this, the first was April 1918 (from B. Moody writing an article about BH-F Ltd.'s Queens Award for Industry published in Glass Technology, Vol. 27 No. 5, 1986) and the second of June 1918 (published by J Edgington in his book Emhart Glass), see Figure 6,60.

In 1919 Paddle - Needle feeders in conjunction with two O'Neill machines were introduced to the East Coast of Scotland (possibly in Fifeshire at the Kinghorn Bottle Company at Pettycur, Kinghorn) to make beer bottles, these two machines were producing between 1100 and 1500 gross of quart bottles per week in the early 1920s. Some 20 of these feeders were built in Britain one which was still working in 1949. Others were installed at

Figure 6,59. W. A. Bailey

Figure 6,60. Paddle feeder and Hartford milk bottle machine installed at Key Glassworks, New Cross, England

Alloa and Greenford - see photographs printed earlier in this chapter.

In December 1919 Bailey formed the Rockware Company and in 1920 at its Greenford factory P-N feeders were installed with a Hartford milk-bottle machine and Lynch L-type machines.

With the 'paddle and needle' feeder complete with a two feet extension was the first commercially available feeder and forehearth, the choice of manual, semi-automatic and automatic methods of production was now available. The following gives the relative outputs of the methods;

	Bottle output (man/minute)
Manual	One
Semi-automatic	One to three
Automatic	Four to 16

The following advertisment is from the Directory of the British Glass Industry, 1923 First Edition, which shows that they were just an agency.

BRITISH HARTFORD-FAIRMONT SYNDICATE, LTD., Works: **Greenford, Middlesex. Offices: 148 Audrey House, Ely Place, London, E.C. 1.** T.N., Holborn 1953. T.A., "Rockware, Smith, London."
Automatic glass making machinery; Hartford-Fairmont feeders; Lynch bottle making machines.
Also accessories of all kinds for the above machines, and moulds for automatic bottle machines.
Directors: W. A. Bailey (Governing Director); Lieut-Col. C. Hope-Murray: S. N. Jenkinson: C. H. Plant: J. H. Stockton.

In 1927 BHF built two 144C type feeders complete with 12 ft 7½in forehearths for Moss Glassworks in Norway. These were the prototypes of nearly 500 that were sold.

In 1929 a delegation of BHF and Rockware personnel visited the Carr Lowery plant in Baltimore to see the 4 section IS machine in operation and then went on to Hartford to negotiate the purchase of two 4 section IS machines and to obtain a licence, under patent, to manufacture IS machines in Europe. Only two 2 section machines were ever built. The first was for St. Prex in Switzerland and the second for St. Gobain in France.

The following is from Directory for the British Glass Industry,

1934 Third Edition showing their growth.

BRITISH HARTFORD-FAIRMONT SYNDICATE, LTD. Offices: **142-143 Audrey House, Ely Place, London, E.C.1.** Works: **Greenford, Middlesex.** T.N. Holborn 1953; Southall 2572. T.A., "Rockware, Telex, London."
Equipment for the automatic production of glassware; batch mixing and handling plants; batch feeders; gas and air valves; furnaces; Hartford feeders; forming machines; cooling systems; bottle conveyors; lehrs; lehr stackers; surface treatment apparatus; Corhart refractories.
Technical and engineering service.
Sole European Agents for Hartford Empire Co.
Sole European Agents for Miller Machine and Mould Works.
Agents for Simplex Engineering Co.
Agents for Lynch Corporation.
Sole Selling Agents in United Kingdom and Northern Ireland for Corhart refractories.
Directors: W. A. Bailey (Chairman and Managing Director); R. G. Bailey; I. M. Bailey, B.Sc., A.R.C.S.; Rear-Admiral W". M. Moir, C.B.E.; Lieut-Col. C. Hope-Murray; C. H. Plant; J. H. Stockton.
Manager: Edward Meigh, M.B.E., M.Sc.
(See advertisements Pages 195 B.H.F; 199 Corhart Co.; 340 Simplex Engineering and 345 Lynch Corporation.)

These separate advertisements are as follows;

British Glass Industry Directory. 195

BRITISH

HARTFORD BHF EQUIPMENT

B.H.F. BATCH FEEDERS:	IMPROVE melting capacity—production—efficiency. ELIMINATE lost batch—irregular charging—fluctuating metal level.
HARTFORD FEEDERS:	NO ROYALTIES—SOLD OUTRIGHT—EXPERT SUPERVISION of INSTALLATION FREE—Wider operating range than any other feeder—All operations mechanically controlled.
FORMING MACHINES:	In addition to the various types of HARTFORD forming machines, THE BRITISH HARTFORD FAIRMONT SYNDICATE being the agents of the LYNCH CORPORATION, which now incorporates the MILLER MACHINE AND MOLD WORKS, are in a position to supply the latest and best machines for the manufacture of narrow neck—pressed and blown—and pressed ware.
HARTFORD LEHRS:	THE HARTFORD LEHR revolutionized annealing practice—A consistently high standard of annealing is maintained with minimum attention—Greater capacity than any other lehr of its type—Running and maintenance costs negligible.
STACKERS:	Save ware losses and eliminate carry-in labour—Improve annealing—Increase pack—Savings soon pay for the investment.
CONVEYORS:	The B.H.F. ware conveyor is designed from experience of requirements in this important link in automatic handling of ware—A self-contained unit—Belt speed and height easily and quickly adjusted to suit conditions.
SERVICE:	Expert technical and engineering advice on all problems relating to glass and its manipulation, to users of HARTFORD EQUIPMENT.
TERMS:	EASY TERMS OF PAYMENT CAN BE ARRANGED

For further particulars apply to:

The British Hartford-Fairmont Syndicate, Ltd.

142-143, AUDREY HOUSE, ELY PLACE, LONDON, E.C.1.

Phone: HOLBORN 1953. Tel. Address: "ROCKWARE TELEX, LONDON."

Figure 6,61. Page 195 of the 1934 3rd edition

CORHART

Creates Confidence

The side walls, tuckstones, throat and doghouse of this white flint tank are built with CORHART ELECTROCAST.

MAJOR ADVANTAGES: Corhart has several major advantages not to be found in any other type of refractory, namely

a. Almost zero porosity, less than [illegible] of 1%.
b. High fusion point (cone 38).
c. Resistance to corrosion.
d. Low co-efficient of expansion.
e. Unique hardness.

Enquiries should be sent to the Sole Selling Agents for the United Kingdom and Northern Ireland :—

The British Hartford-Fairmont Syndicate, Ltd.,
142-3 Audrey House, Ely Place, London, E.C.1.

Telegrams—Rockware, Telex, London. *Telephone*—Holborn 1953.

Figure 6,62. Page 199 of the 1934 3rd edition

340 British Glass Industry Directory.

Simplex Engineering Co.

WASHINGTON TRUST BUILDING

WASHINGTON, PENNA., U.S.A.

EXPERT ENGINEERS IN THE DESIGNING AND BUILDING OF ALL TYPES OF EQUIPMENT FOR THE MANUFACTURE OF GLASS.

SIMPLEX

ELECTRIC LEHRS
COMBUSTION TYPE LEHRS
IMPROVED POT FURNACES
CONTINUOUS REGENERATIVE TANKS
DAY TANKS
LINK BELT COAL STOKERS—for Plants and Homes
POWER PLANTS
BATCH SYSTEMS—Vacuum and Conveyor Types
PRODUCER GAS PLANTS
PLATE GLASS LEHRS AND PLANTS
POLARISCOPES—Dry and Solution Types
OIL SYSTEMS—High Pressure
CONVEYING EQUIPMENT—All Types
CULLET WASHING SYSTEMS
INTERLOCKING ARCHES—Suspended
SINGLE FOURCAULT TANKS
COMPLETE PLANT DESIGNS
APPRAISALS, ETC.

Figure 6,63. Page 340 of the 1934 3rd Edition

MANUFACTURERS OF

AUTOMATIC GLASS FORMING MACHINES

FOR ALL CLASSES AND SIZES OF WARE,

NARROW MOUTH MACHINES, WIDE MOUTH PRESS & BLOW MACHINES, PRESS MACHINES, TUMBLER GRINDING AND FINISHING MACHINES.

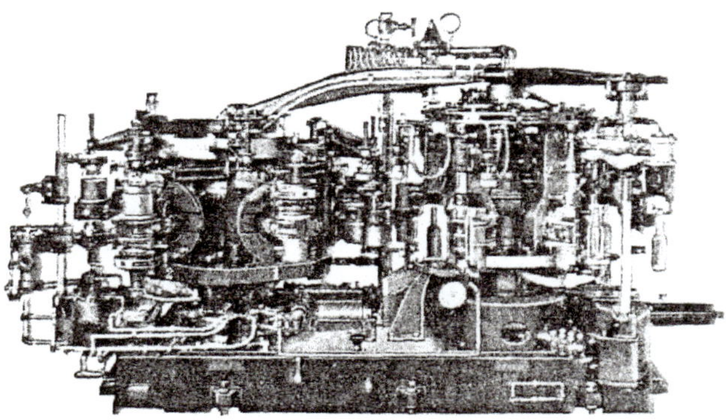

MODEL-10 MACHINE

LYNCH CORPORATION

ANDERSON, INDIANA, U.S.A.

European Selling Agents:

THE BRITISH HARTFORD-FAIRMONT SYNDICATE, LTD.

142-143, AUDREY HOUSE, ELY PLACE, LONDON, E.C.1

'Phone: Holborn 1953. Tel. Address: "Rockware Smith, London."

Figure 6,64. Page 345 of the 1934 3rd Edition

In the abstracts in Volume XXIII (1939) and Volume XXIV (1940) of the Society's Journal there are three entries regarding what was to be called later the 'KU' and 'KW' forehearths. The two in the 1939 edition were for the British Patents for these forehearths, the first was applied for by British-Hartford Syndicate, Ltd. and the second by the same company but with Hartford-Empire Co. added, i.e. British Hartford Syndicate Ltd. (Hartford-Empire Co.). The second was an amendment to the first. The abstract titles were (1) Glass Furnace Forehearth, abstract No. 325 and (2) Delivering and Gathering Molten Glass, abstract No. 556, and one the one the following year by K.E. Peiler and W.T. Barker Jr. of West Hartford, Connecticut, USA, abstract No. 82.

The abstract entries were;

1. Glass Furnace Forehearth. British Hartford-Fairmont Syndicate, Ltd. (Brit. Pat. 498 204, Oct 20th, 1937, No. 28 616. Convn. Date, April 1st [Class 56] Group 23). Also refers to a specification 28 617/37 which is about the air/gas mixture firing system. (Figure 6,65).
2. Delivering and Gathering Molten Glass. British Hartford-FairmontSyndicate, Ltd. (Hartford-Empire Co.) (Brit. Pat. 503 035, Oct. 20th, 1937, No. 28 617. [A specification was laid open to inspection under Sect. 91 of the Acts, Oct. 3rd, 1938]. [Class 56.] Group 23) (Figure 6,66).
3. Temperature Control of Forehearth for Molten Glass. K.E. Peiler and W.T. Barker, Jr., West Hartford, Conn., Assignors to Hartford-Empire Co., USA (USA Pat. 2 139 911, Dec. 13th, 1938. Appln. April 1st, 1937. Serial No. 134 356) (Figure 6,67).

The first patent, Number 498 204, (Figure 6,65), refers to - a forehearth which comprises a rear cooling section, 10, Fig. 93, and a forward heating or homogenising section, 11, and blocks, 26, 27, may be used to separate the space above the glass in section 10 from that in section 11 and in the tank furnace, 18, respectively. The sections 10, 11 are suitably heated, for example, by burners disposed in tunnels, 30, Figs 93 and 94, and are supplied with a predetermined mixture of gas and air as described in Specification 28 617/37.

The roof or cover over section 10 comprises blocks arranged to constitute a flat crown which is provided with vertical vents or exhaust ports, 46.These vents extend into two exhaust manifolds 47, 48, which are provided with discharge outlets 52, 53, respectively, through which the products of combustion escape. These outlets are controlled by dampers, 54 and 55, which can be opened to any desired degree by means of a lever 57 raised or lowered by a screwed rod 60.

Fig. 93

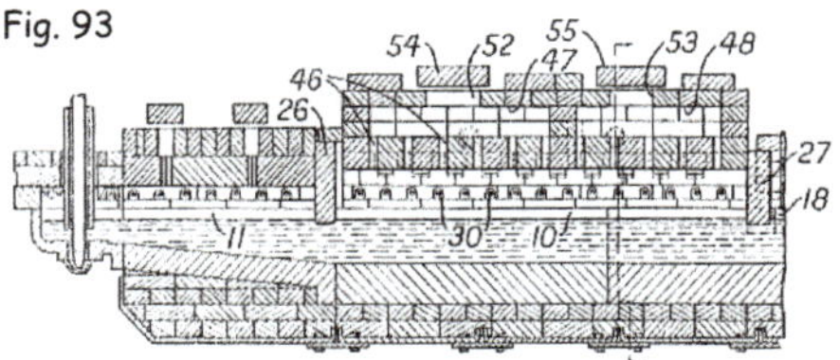

Fig. 94

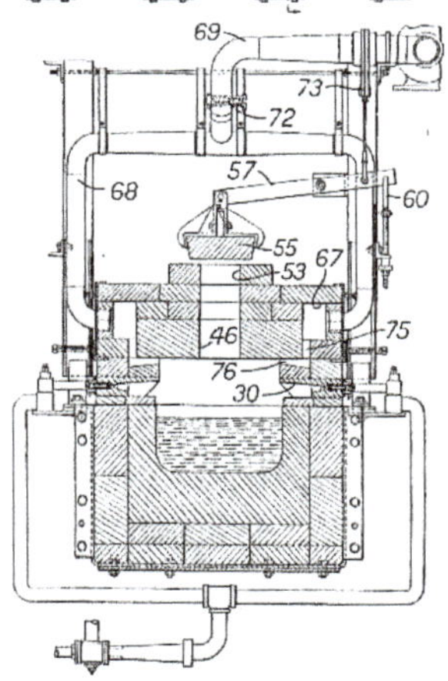

Figure 6,65. Brit. Pat. No. 498 204

The cover is provided with longitudinal intake manifolds 67, Fig. 94, each of which is supplied with air under pressure from a branch pipe 68 attached to a supply pipe 69. The volume of air blown through the pipe 69 is controlled by an adjustable damper 72 and also by a slide valve 73 which is connected to the lever 57, so that opening the discharge outlets 52, 53 is accompanied by an increase flow of air into the manifolds 67, and vice versa. Each manifold 67 communicates through passages 75 with shallow horizontal cooling ports 76 situated close to the underside of the crown of section 10. Air from the ports 76 is blown across the to the vents 46 and cools the crown without changing combustion conditions in the tunnels 30 and without impinging on to the glass. Adjustment of the valves 73 and of the levers 57, together with adjustment of the heating devices in the tunnels 30, enables any desired temperature gradient to be obtained in the length of the glass flowing through section 10. In a modification, additional cooling air from the pipes 68 is blown into annular slots at the upper ends of the flues 52, 53 to assist the escape of the cooling air from the ports 76 through the flues.

The second patent, an amendment to the first, is Patent Number 503 035 and is about - A forehearth for conveying molten glass to a feeder or gathering device is provided with heating devices which enable a suitable gradient to be obtained in the glass longitudinally of the forehearth, and substantial uniformity of temperature across the width of the glass.

The forehearth may be of the type comprising a rear cooling section 1, Fig. 207, a forward heating or homogenising section 2, and a delivery chamber 3. A block 13 separates the space above the glass from the body of the tank 10, and a similar block 14 separates the spaces in sections 1 and 2. The roof above both sections is fitted with flues 29 and 33 provided with dampers 32 which can be adjusted by a suitable device such as lever 38, Fig. 208, operated by a screwed rod 41.

FIG. 207.

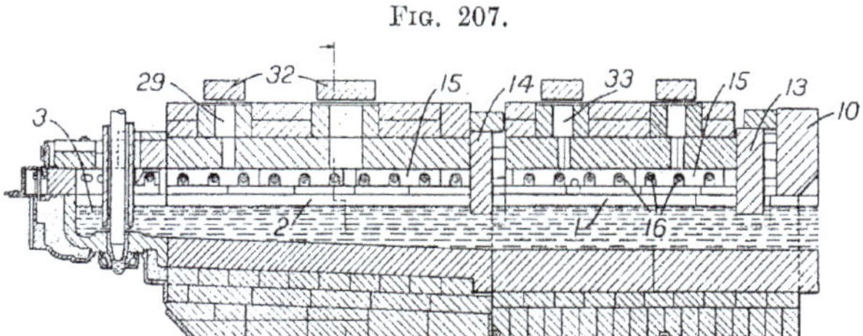

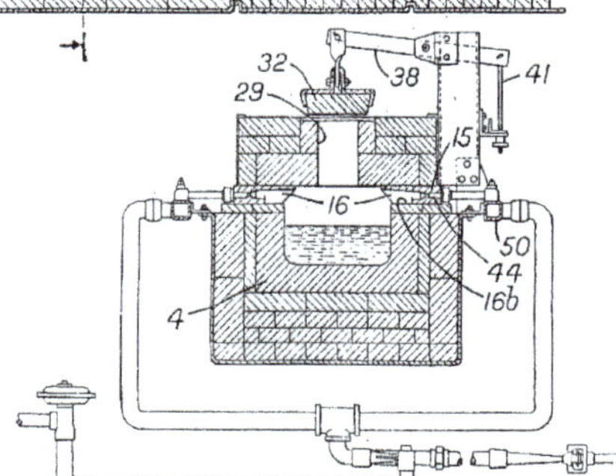

FIG. 208.

Figure 6,66. Brit. Pat. No. 503 035

The walls of the chambers 1, 2 include refractory blocks 15 which are placed above the level of the glass and have conical combustion tunnels 16 therein. The larger end of each tunnel is bevelled off at an angle and its bottom wall is cut away at 16b. The inner end of the tunnel is fitted tightly with a nozzle 44 so that the air cannot enter around the nozzle. The nozzles are connected to manifolds 50 which are supplied with a predetermined mixture of gas and air so arranged that combustion in the tunnels is complete.

The combustion causes the inner surfaces of the tunnels to become incandescent, and the shape of the bevelled ends causes heat to be radiated downwards into the glass at the sides of the flow channel 4 and thus ensure uniformity of temperature across the width of the glass. Suitable devices are provided for controlling the supply of gas to each of the chambers 1, 2 in order to obtain a temperature gradient along the forehearth to suit the rate at which the glass is withdrawn. The delivery end of the foreheath is semi-circular in shape and is provided with combustion tunnels. Specification 498 204 is referred to.

The third abstract was taken out by the Hartford-Empire Co., in the USA Pat. No. 2 139 911 - A forehearth consisting of a cooling section 1, (Fig. 9), a heating or homogenising section 2, and a delivery chamber 3, was provided with a series of ports 16, 17, each of which could be controlled independently. The glass flow-channel 4, was a U-shape in cross section.

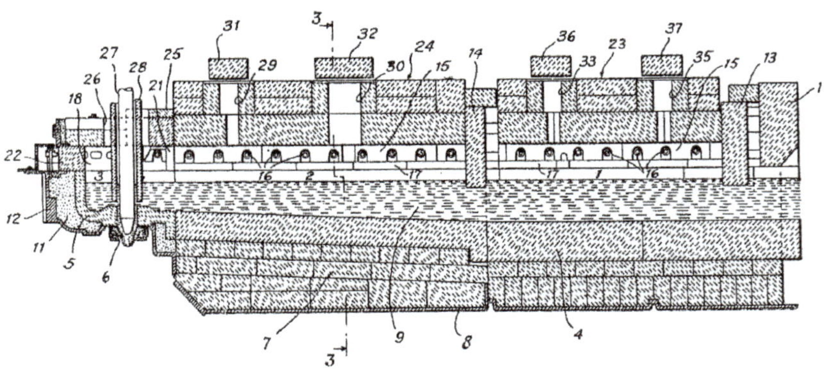

FIG. 9.

Figure 6,67. US patent number 2 139 911 December 13th, 1938

Exhaust outlets 29, 30, 33, 35, which were controlled by movable dampers 31, 32, 36, 37, permitted the loss of heat from the glass by radiation if so required. Heating section 2 and delivery chamber 3 were provided with special layers of insulating blocks, 7 and 11, respectively. The space above the glass in the cooling section 2 could be separated from the corresponding space in the heating section by a block 14.

The manifold surrounding the nose of the feeder feeding the burners with premixed combustion mixture was so arranged that replacement of the "bowl" 5 of the feeder was possible without dismantling of the air and gas system. It was claimed that the temperature gradient of the glass in the forehearth could be controlled satisfactory both longitudinally and transversely.

Additional information from the USA patent, although not included in the abstract, is that if the exhaust covers were lifted high enough then the rate of heat exchange from the glass could be increased by direct radiation to the atmosphere. It mentions that the underside of the glass channels and round the feeder bowl where supplied with special layers of insulation blocks. As already mentioned that, the 'horseshoe' of burners and its manifold around the spout bowl could be removed easily for replacement of the 'bowl' without dismantling the air and gas system.

Abstract number 83 Temperature Control of Forehearth for Molten Glass. K.E. Peiler and W.T. Barker, Jr. West Hartford , Conn., Assignors to the Hartford-Empire Co. Hartford, Conn., USA (USA Pat. No. 2 139 770, Dec. 13th, 1938. Appln. Sept. 24th, 1935. Serial No. 41 881). This was a precursor for abstract number 82 which was made two years previous.

Further information from the British Glass Industry Directory of 1956 Edition states that BH-F were;

Agents for The Lynch Corporation, Lynch International Ltd., and L'Electro-Refractaire. And that they were suppliers of;

Batch chargers and batch handling equipment; automatic feeders; forming machines; machine conveyors; stackers; lehrs; decorating machines; engineering services; equipment and ancillaries of all kinds for automatic glass container works; electrocast refractory blocks.

Similarly from an advert in the 1959 edition of the directory states that they were;
Engineers to the Glass Container Industry; and they were manufacturers or agents for;

Corhart Electrocast Blocks
Lynch Forming Machines
B.H.F. Feeders
Automatic Stackers
Machine Conveyors
Batch Chargers
Decorating Machines

In 1950 Lynch International Ltd. was formed which manufactured under licence Lynch glass forming machines and spares and in 1952 the first British made Lynch machines was installed. In 1954 Lynch International Ltd. was sold to the Lynch Corporation of America.

In either 1952 or 1953 with the opening of the Emhart manufacturing facility BH-F lost their agreement to act as agents, a situation that had been in existence since 1918.

In the 1961 BH-F developed a successor to the 144 feeder, in the form of the 50 series. The range included an epicyclic drive gearbox. The first was installed in the Greenford works of Rockware Glass Ltd. in February 1961, and soon after, in 1964, the first of the type 60 series was installed at Albion Bottle Company in Birmingham. Today over 200 of these 50/60 feeders have been installed worldwide. The 60 type feeder was designed to suit the Roirant machine.

Other information from the same advertisement is:

1. Epicyclic mechanism box giving full 360° differential timing adjustment between plunger and shear cuts can be carried out during feeder operation.
2. BHF 50/60 feeders are fitted with a revolving tube with

drive as standard. The tube drive is reversible up to 13 rpm. Orifice sizes are 5 inches and 6 inches.

3. The shear mechanism is of conventional design driven from a one piece cam by the epicyclic mechanism box. The main features are:
   - angular shearing available from 30° right to 90° left.
   - single of double operation
   - pneumatic or mechanical spring
   - automatic shear retraction in the event of a power failure
4. The plunger mechanism and linkage are of conventional design and are driven from a one piece cam by the epicyclic mechanism box.

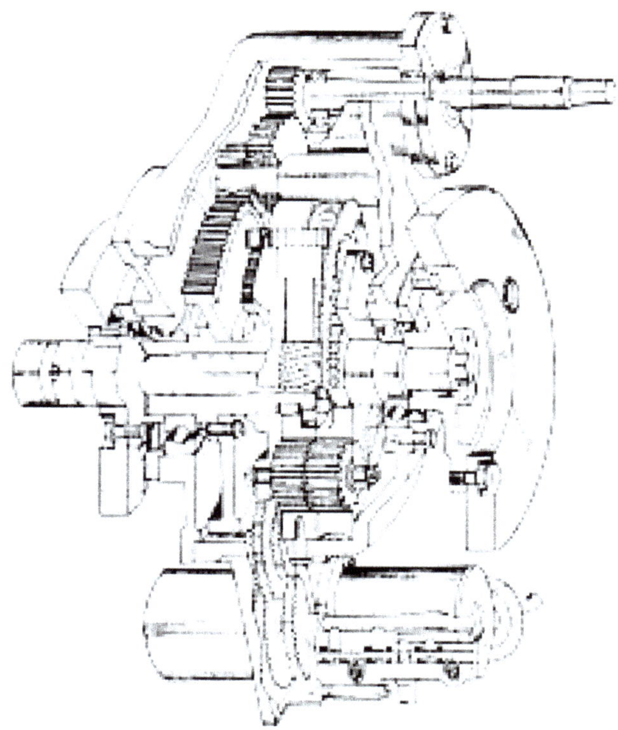

Figure 6,68. BH-F's epicyclic feeder drive

Production capacity of the BH-F type 50 and 60 feeders

	50 Std	50 Medium	50 Deep	60 Medium	60 Deep
Spout bore	5 in	5 in	5 in	6 in	6 in
Depth of glass in spout	4 in	6 in	8.5 in	6 in	8.5 in
Max orifice diameter	S 4.5 in D 2.125 in	S 4.5 in D 2.125 in	S 4.5 in D 2.125 in	S 4.5 in D 2.5 in	S 4.5 in D 2.5 in
Max Weight	S 65 oz D 12 oz	S 75 oz D 13.5 oz	S 100 oz D 15 oz	S 75 oz D 18 oz	S 100 oz D 21 oz
Min weight	0.5 oz	0.5 oz	1.5 oz	1.75 oz	1.75 oz
Orifice centres	2.875 & 3 in	2.875 & 3 in	2.875 & 3 in	3.5 in	3.5 in
Max glass flow /min	750 oz	900 oz	1200 oz	900 oz	1200 oz
Max glass flow /day	30 t	36 t	48 t	36 t	48 t

Table 6,13. Production capacity of the 50 and 60 type BH-F feeders

The following photograph is of a feeder mechanism and the other is of a complete Type 60 forehearth and feeder. The photograph also show that its angular shearing.

Figure 6,69. A feeder mechanism from the same advertisement

Figure 6,70. Type 60 feeder with angular shearing

In 1963 Electrocast Ltd. was formed to sell refractories produced by Electro Refactaire and in 1971 this interest was sold to raise capital.

In the 1969 the type 70 was put onto the market, the largest of its kind, which could handle up to 100 tons per day throughput. Between say 1970 and 1980 the new 900 series were being sold, an advertisement in 1981 stated that the 900 series smooth operation with up to 175 c.p.m. and up to 130 t.p.d. are still working below design potential.

In 1977 the company stopped the manufacture of annealing and decorating lehrs and other ancillary equipment and started to develop new forehearths,

In 1980 in the New Years Honours list the Chairman Eric Lyall, a solicitor, was awarded the C.B.E., he was appointed five years earlier when the company was in dire straights, the technical expertise and flair was not in doubt but good people were doing the wrong jobs and a great deal of its resources was being wasted on fruitless diversifications. Very quickly he worked out a strategy and returned the company back to its former glory.

In Feburary 1980 the company was called BH-F Ltd. of Rockware Avenue, Greenford , Middlesex, England. By September of the same year it had been renamed as BH-F (Engineering) Ltd. at the same address, this continued up to 1987.when they moved to 174 Milton Park, Abingdon, Oxfordshire. The new site was officially opened by the Duke of Kent on the 19th February 1988 during a conference on forehearths, with input being given by attendees for customers. Later they moved to 4a Churchward, Southmead Park, Didcot, Oxfordshire, possibly after BH-F went into receivership in 1993.

Figure 6,71. Eric Lyall C.B.E. Chairman of British Hartford-Fairmont

In August 1981 first new design of the 400 series of forehearths was installed at CWS Worksop, it had been unveiled at a seminar in London in 1979. Early in 1982 BH-F signed an agreement with St. Gobain with gave BH-F worldwide rights for the manufacture and sales of their SG2 forehearth as well as full patent protection for it's 400 series. The 400 series is based on this design.

Saint-Gobain had patented a central, longitudinal cooling system that was licensed, developed and widely marketed by BH-F Ltd. With this new competition Emhart then developed their Type 540 cooling section.

Other advertisements show the companies evolution; Advertisements in 1981 said that the Types 50, 60 and 70 were earlier triumphs, now a new peak has been reached with the 900 Series. A continuation of the theme later in the year said that there are more than 200 of Types 50/60 in operation with IS and Roirant machines and similarly with Type 70 for higher throughputs. The 900 series can work at 175 c.p.m. with a throughput in excess of 130 t.p.d.

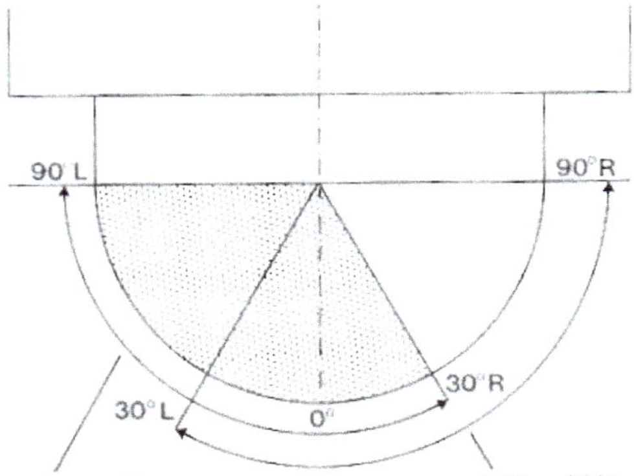

Figure 6,72. New angle of shearing

In 1982 a small article appeared in the press detailing a new development of the 900 series.

A customer Westerwald AG, of The Federal Republic of Germany a producer of flint, amber, and light green bottles and glass bricks requested a left handed feeder to work with a Pöting press instead of the usual IS machine. Up to this point BH-F had only supplied right handed feeders for IS machines.

To meet this requirement a left handed feeder (907/22M) was designed and produced, after successfully developing the Westerwald feeder BH-F realised that the rest of the industry could benefit from this development as it could eliminate many of the machine layout problems. From this point BH-F now offered both R.H. shearing and L.H. shearing, operating between 90°L and 30°R or 90°R and 30°L for the 900 series of feeders.

Since 1985 developed the straight line shears which overcomes the problems encountered with angular shearing at high cutting speeds, i.e. up to 200 cuts per minute having been achieved on a double gob machine. Also in 1985 BH-F announced that due to the vast improvements in the design, supplying and commissioning of their forehearths and to ensure that they keep up their claims of fuel efficiency, as a service to their customers they are to set up a new specialised training department.

Through their US representative they had secured an order for a Type 436 to feed a plate glass rolling machine at AFG of Tennessee. This was a new venture but their customer is satisfied that the control of temperature across the forehearth can be maintained to their standards and still give the energy saving. The forehearth can supply glass to 2°C across its width and it is possible to achieve up to 50% saving in energy. The order comprised a 44 ft Type 436 forehearth including burners, pipework and an alcove in a position normally occupied by a feeder unit.

In 1986 they stated that the fuel cost have been halved at a BSN plant in France. The company had achieved a record daily production rate of 720 000 25 cl beer bottles from one line using a Type 900 feeder mechanism with an adapted Owens-Illinois parallel shearing system. The company had had favourable experience with BH-F's Type 900 triple gob shearing system and required its vibration free, high speed action to be placed on a quadruple shearing system to nullify the possible effects of the problems that could occur when using arcular motion shearing. See Figure 6,73. This arrangement placed great demands on the

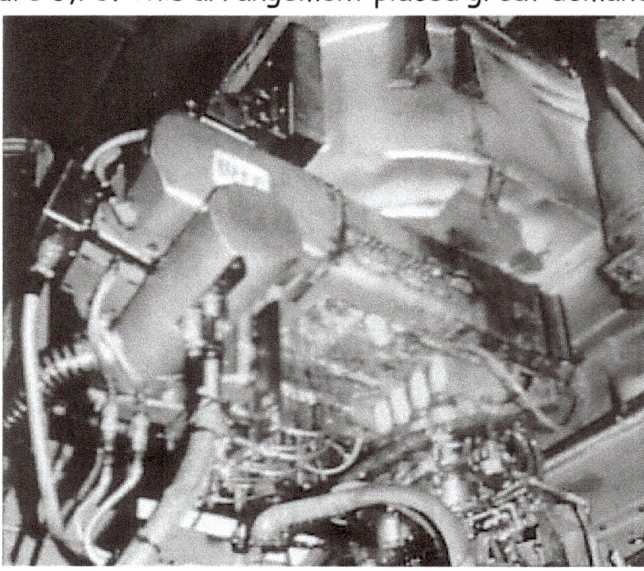

Figure 6,73 BH-F's Quad gob straight line shearing mechanism

forehearth and the homogeneity of glass temperature was vital, the longitudinal cooling employed by the BH-F 400 series promised the characteristics needed. This was confirmed by the company and they ordered their first 400 Series, type 436 in 1984.

1986 and 1988 shows some of the many countries that BH-F are now supplying, 1986 records show that Vetropack in Switzerland after consultations with St. Gobain had adopted the concept of longitudinal cooling placing an order for a 27ft 443 forehearth complete with casings and two 20ft 436 conversions. All will feature the BH-F high pressure systems and unique longitudinal cooling system. Coutinho Glass have ordered for National Glass and Crystal SAE of Ramadan City, Egypt's new factory two 907-22D double gob feeder mechanisms complete with rotors, and two 22ft 436 forehearths, each including three zones of 132 type temperature control equipment.

While in 1988, BH-F records business in the Philippines, Malaysia and Zimbabwe. In the Philippines BH-F are supplying San Miguel for their Farola plant in Manila an 18ft 426 forehearth conversion to serve a triple gob machine. In Malaysia, Malaya Glass have chosen BH-F to supply equipment for the modernisation programme at their Johore Bahu factory, they have installed two 20ft foreheaths types 426 and 436 each equipped with their high pressure gas firing, three zone temperature control system and their longitudinal cooling system.

While in Zimbabwe in early 1987 at Zim Glass in Gwera BH-F supplied a 900/400 feeder/forehearth combination. The line is for small size cosmetic and pharmaceutical production so high temperature and fast cutting speeds were essential. The line is a conversion to a 426 19ft 6in. forehearth complete with 130 series digital temperature control an insulated to a higher standard than normal. The 907-18D feeder mechanism can be worked single or dual.

In early 1987 the press announced two new innovations, the first was a range of spout cover blocks suitable for all types of feeder mechanisms, they have a weight reduction of 85% compared to conventional ones and a 80% reduction of heat loss gives improved glass homogenisation in the spout. An estimated reduction of at least 10% energy by the reduction of spout firing. The new materials give longer life due to their improved thermal shock resistance. These items can also be bought from BH-F Sales Inc. of Lilburn, Georgia, USA.

The second was that BH-F has developed a Microtherm feeder bowl insulation kit for their 900 series. This method of insulation is already being used but it can be purchased in kit form.

In April of that year in a move to expand its glass business BH-F purchased DSF Refractories Ltd. from The Tilbury Group plc. DSF manufactures refractory materials for the glass, steel, cement, aluminium and petrochemical industries. The previous owners stated that it made more sense for DSF to be owned by a company that was refractory oriented rather than one more inclined with contracting in the civil engineering business.

In 1988 the company held a seminar to open their new headquarters, over 120 delegates from 25 countries attended. Professor Jottrand of Brussels University delivered a paper on 'Mathematical modelling of glass conditions in forehearths', and other speakers including representatives from St. Gobain, Holmegaards Glassvaerker, Oberland Glas, American Glass Research and Yamamura.

P. Vilk and A. Stephens presented BH-F's ideas on total glass conditioning with respect to working ends and alcoves, and forehearths including their temperature control. P. Spruenken delivered one on the 'Performance of the Straight Line Shear' after G. Wright of BH-F had had outlined its history and its future development, including a one to be given field trials in conjunction with an electronic servo feeder.

Delegates later had the opportunity to see a mechanism in action and to see a complete working end with two forehearths and feeder mechanisms with refractories supplied by DSF who also exhibited a regenerator to show the different materials used.

On the final day, Friday 19th February 1988 the new premises were officially opened by HRH the Duke of Kent.

Also in 1988 the acquisition of Warren Technical Associates Inc. (WTA) of Toledo, Ohio, in the United States was finalised.

They were producers of computer based process control equipment and the company was re-named BH-F Systems Inc.

In 1989 Robert Bosch announced their new Servodyn AC brushless servo drives which had been developed for BH-F Ltd's. new 1000 Series feeder system which is designed to give greater cutting speeds. The drives have succeeded the costly and complicated cam arrangement used before.

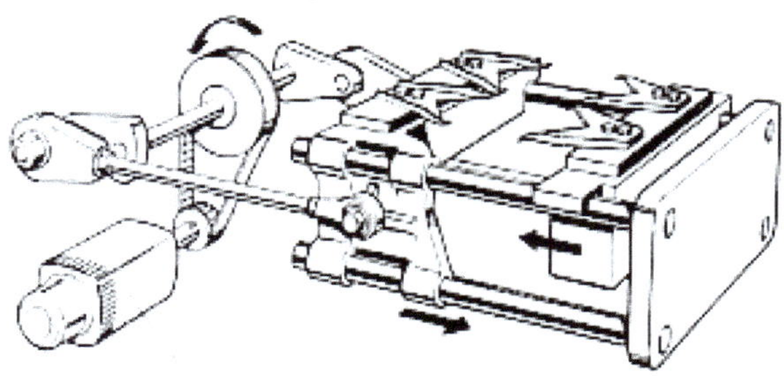

Figure 6,74. The Bosch shear system

In 1989 BH-F devoted much of their bulletin to their Warren Technical Associates mentioning in the Spring Bulletin the services and products that they provide It also reported that the straight line shears undergoing trials in Italy are going well and is so satisfactory in France that the company have ordered more. While in the Autumn issue it contains details of forehearth and control equipment that has been supplied to three companies in the Philippines and Warren Technical Associates supply of a 'Pacesetter' to a firm in Italy.

The preceding entries give some examples of how much of the firm's sales were for the export market. This was acknowledged by the company receiving the Queens Award for Export Achievement three times, in 1981, 1986 and in 1991.

In the early part of 1993 BH-F (Engineering) Ltd. along with the rest of the group filed for bankruptcy. In the October issue of Glass Technology it was reported that the Stein Heurtey Group has acquired the business and industrial activities of BH-F Engineering from the receivers of the BH-F Group. The acquisition took place at the Group's UK subsidiary, Stein Atkinson Stordy of

Wombourne, West Midlands, England. The Stein Heurtey Group were themselves owned by the Compagnie Groupe Fives Lille which included BGE and Penelectro.

Also announced in the same issue, DSF Refractories an operating subsidiary of the BH-F Group has been bought out of receivership by the management with institutional support. In future the companies name will be DSF Refractories and Minerals to reflect the wider sales range of its products.

In the February 1994(?) issue of Glass Technology it was reported that Airco Gases the US gases business of BOC had purchased BH-F (Systems) of Maumee, Ohio a leading supplier of process control systems for the glass industry. The new company will trade under the name of BH-F Systems Inc. and will continue their relationship with BH-F (Engineering) and its new parent company to supply and service working ends, forehearths and feeder mechanisms in North and South America.

Other advertisements appearing since 1996 include; a layout of a distributor with two forehearths; Figure 6,75.

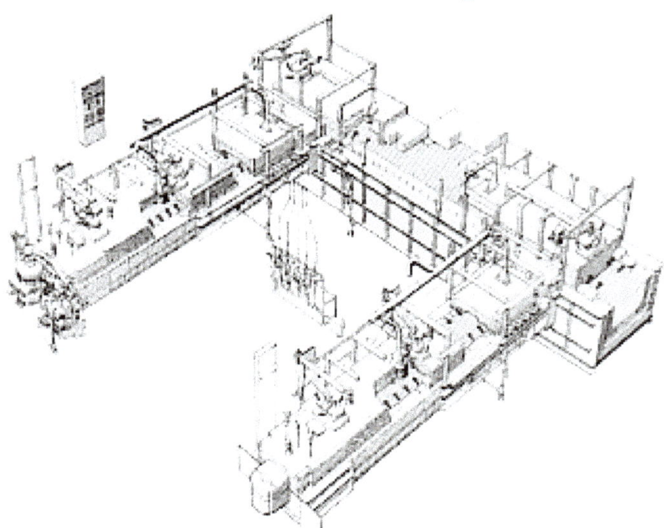

Figure 6,75. BH-F's distributor and two forehearths. The two possibly marries up to a working end situated at the right hand side of distributor channel

By February 1996 they had moved to 4A Churchward, Southmead Park, Didcot, Oxfordshire.

In an advertisement dated Feb 2003 BH-F (Engineering) Ltd of 4A Churchward, Southmead Park, Didcot, Oxon had subsidiary offices in Liberec, situated in the Czech Republic (opened 1991), in Tianjin, in the P.R. China, (opened 1992) and in the USA with an office in Los Angeles, California (opened 1999). An office was opened in 1980 in Rochdale, New York State and one in Lilburn, Georgia in 1987 whether or when these had been closed I don't know. From February 2008 they are now called Fives-Stein Ltd.

Other achievements were; in 1993 they released the Cascade Bowl and the first Mass Heat Extractor was installed at Gallo Glass in California. In 1994 they released the Pneumatic Feeder Mechanism and in 1996 they installed the first BH-F Oxy Gas Combustion system in Jeanette, Pennsylvania and released the Stem Press Feeder (Double Gob Platinum Tube Feeder replacement).

In 1997 they installed the first Electronic Slow Gob Feeder at Bohemia Crystal in the Czech Republic, followed in 1998 by the first colourant rolled glass forehearth from the side of a float glass furnace. In 1999 they installed the first BH-F Archial servo shears mechanism in Bohemia Crystal in the Czech Republic and in the same year.

Like many forehearth suppliers BH-F markets a colourant package, which in their case, they do in conjunction with the Ferro Corporation of the United States of America. It uses the 400 series forehearth technology and the proven Ferro forehearth colouring system. BH-F in their advertising literature gives the following technical data for a typical forehearth, while Figure 6,76 shows a typical colourant forehearth.

The following extract is based on a speech given at the Society of Glass Technology in May 1999 by P. J. Maitland of Ferro.

The colouring section increases the forehearth length by 12 to 20 feet, and has three sections or zones;

1. Preheat zone - refiner to feed tube; where the colourant is added.
2. The melt zone - feed tube to first bank of stirrers.

3. Stirring zone – from the first bank of stirrers to the last bank.

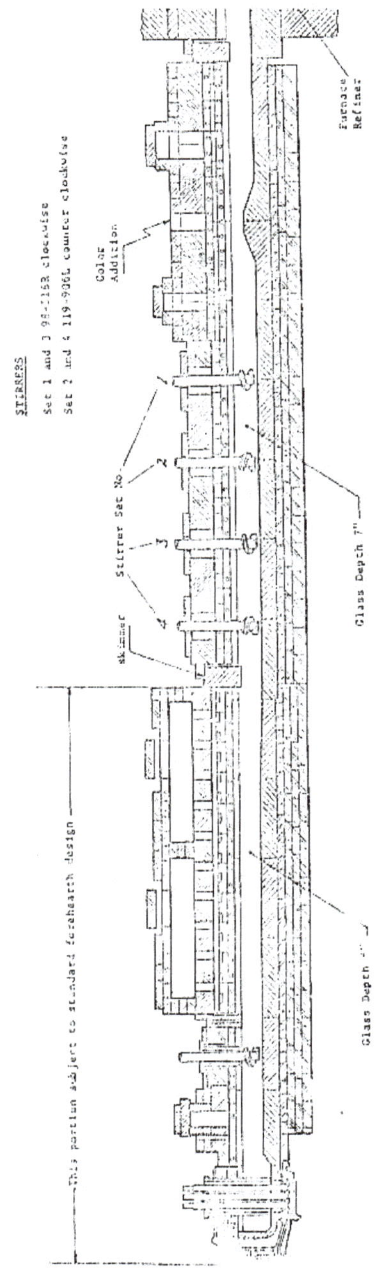

Figure 6.76. A typical colourant layout

The requirements for each zone are;

Preheat zone

This zone must have sufficient firing capacity and control to achieve and maintain the required glass temperatures to allow the colour concentrate (the frit) to melt. This often requires the burners to be double stacked.

The melt zone

This zone needs to be long enough for the colourant to melt and de-gas on the glass surface prior to stirring. This part of the forehearth must have sufficient firing capacity with the capability of controlling the combustion ratio to maintain a precise oxidising atmosphere over the glass surface.

Stirring zone

The stirring zone must be long enough to ensure that the melted colour is properly mixed into the molten glass without creating seeds, blisters or colour streaks. This 'stirring capacity' as it is referred to depends upon the glasses depth, the total number of stirrer banks and the number of stirrers in each bank. There should be adequate inspection holes to observe and evaluate the melting of the frit and for the insertion of temperature sensors.

The exact layout of a colourant forehearth depends on;

1. Forehearth width
2. Glass pull
3. Desired colours and the number of them
4. Limitations in glass depth
5. Any limitations to the space and/or layout
6. Base glass analysis, batch composition and redox state

Pull rates can be as low as four tons per day and as high as 120 tons per day. The amount of colourant added depends upon the desired shade and its colour density they can vary from 0.2% up

Colour range from clear oxidised soda-lime-silica glass is Blues, Greens, Browns, Purples, Pinks, Greys and Blacks

Pull Range	416 Forehearth (16″ wide)	2.5 to 15.0 t/d (2.7 to 16.5 US t/d)
	426 Forehearth (26″ wide)	10.0 to 50.0 t/d (11.0 to 55.0 US t/d)
	436 Forehearth (36″ wide)	40.0 to 120.0 t/d (44.0 to 132 US p/d)
	443 Forehearth (42″ wide)	80.0 to 200.0 t/d (88.0 to 220.0 US t/d)
Typical Zoning	Warm-up zone	Increases the inlet glass temperature to that required to melt the colour concentrate
	Melt zone	Allows the concentrate to melt and diffuse into the glass
	Stirrer zone	Uses three or more banks of stirrers to ensure an even colour distribution in the glass
	Cooling zone	One or more 400 series longitudinal cooling zones to condition the glass to the right temperature and homogeneity
	Conditioning zone	Finishes the glass conditioning and delivers the glass to the forming process

Table 6,14. BH-F's technical data regarding the whole forehearth layout to 3.0%. Glass temperature in the feed area typically range from 1300°C down to 1220°C at the exit from the stirring section.

In the container industry typical end products are say mineral waters, wines, spirits, alcopops and soft drinks with colours ranging from Bordeaux green, Georgia green, and dead leaf green through to blacks, blues and smoke greys. In the tableware market colouring can be used for table tops, ovenware, ornamental and for promotional items while for the architectural glass market the colouring of glass building blocks and rolled patterned glass.

Generally speaking the colourants used are one or more of the following cobalt, chromium, copper, nickel and manganese along with some other transition elements. Table 6,15 shows some of the C.I.E. values and the proposed federates for a number of both oxidised and reduced glasses in the container industry. Other colours not already mentioned are for oxidised glasses; cobalt blue, Perrier green, half white, Bacardi green, emerald green, pink and black. For reduced glasses it is possible to change the base glass colour to another one, for example; reuced amber to black, light amber to dark amber, emerald green to Bordeaux (Champagne) green and dead lead green to French antique green.

Similar colours have been developed by the Ferro Corporation for the tableware and the flat glass industry. The feed systems can be volumetric or gravimetric due to control brought about by the use of screw feeds in the volumetric system gravimetric system are always recommended.

For oxidised glass from White Flint			**Typical C.I.E. values**		
	Colourant	Feed %	D W L	%B	%P
Smoke grey	Ni	0.40%	574	67.52	12.86
Ramlosa Blue	Co & Cu	0.80%	486	78.66	9.95
Clearly Canadian	Co & Cu	0.50%	480	65.52	17.18
Bacardi Limon	Ni, Co & Cr	0.25%	566	77.81	4.00
Dead-leaf Green	Mn & Cr	1.00%	562	90.30	5.09
For reduced glass from Dead Leaf					
Antique Green	Ni & Cr	0.55%	576	10.00	95.00

Table 6,15. Typical colours showing feed rate and base colour

The next three photographs are of a mass heat extractor Figure 6,77 shows one in close-up, and then two views of a bank of them Figures 6,78 and 6,79, Mass heat extractors are situated either in the distributor roof or the first section of the

Figure 6,77. Mass Heat Extractor

Figure 6,78. BH-F's Mass Heat Extractor, side view

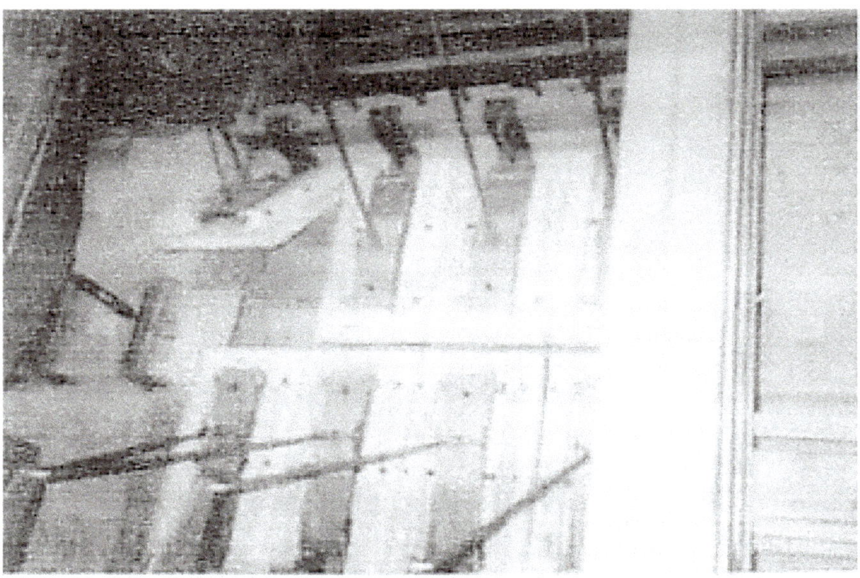

Figure 6,79. BH-F's Mass Heat Extractor, end view

forehearth, they have a high extraction heat and are used to reduce the 'footprint' of the distributor/forehearth arrangement where the installation of wider channels or a working end are not an option. The control of them is usually manual.

After these three photographs follows the inside of a BH-F type 484 working end Figure 6,80 is shown a long with an arrangement of an oxy-gas fired forehearth Figure 6,81 and Figure 6,82 shows a P90-18D feeder mechanism.

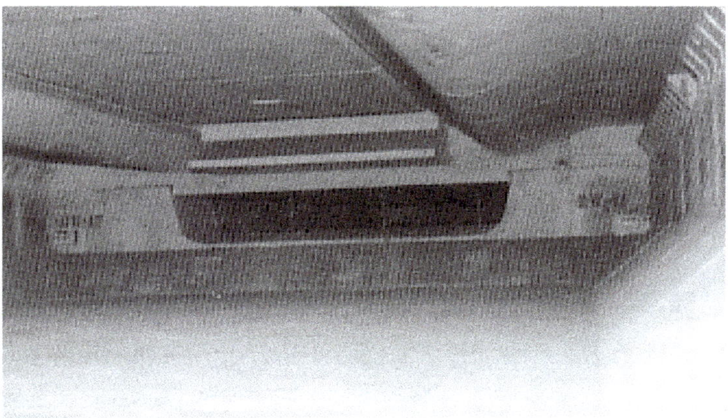

Figure 6,80. 84" wide working end

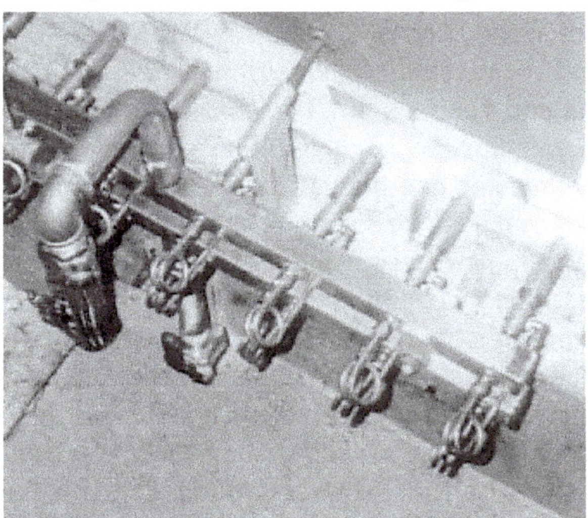

Figure 6,81. Oxy-gas firing system

Figure 6,82. The BH-F's P90-18D feeder mechanism

Other people that went on to design and sell forehearths and feeders were, Frank O'Neill, J. W. Miller, Carl Rankine, H.H. Pitt and W. Mitchell. Owens-Illinois also designed and built forehearths and feeders but they only sold them to their own company, their subsidiaries and affiliates, they never sold them on the open market and there is very little information about any of them. Sorg, Parkinson and Spencer Ltd, and Zedmark of the Dyson Group now design and make working end, alcoves, and forehearths along with their firing systems and control gear. Two of them also manufacture all or some of the refractories.

The following three photographs show the various parts of the Zedmark 'Universal' forehearth which in 1993 was the next step in their forehearth design package. They show the 'Universal Arch Block', and views of the 'Advanced Mode' and the 'Rapid Cooling Modes'. They are both based around the 'universal arch block' which forms part of the roof section along the entire length of the forehearth.

In the conditioning section, called here the 'advanced mode' the holes to the outside flues are sealed with refractory bricks leaving only the centres one open, via the exhaust system, to the atmosphere.

Figure 6,83. Zedmark's Universal arch block

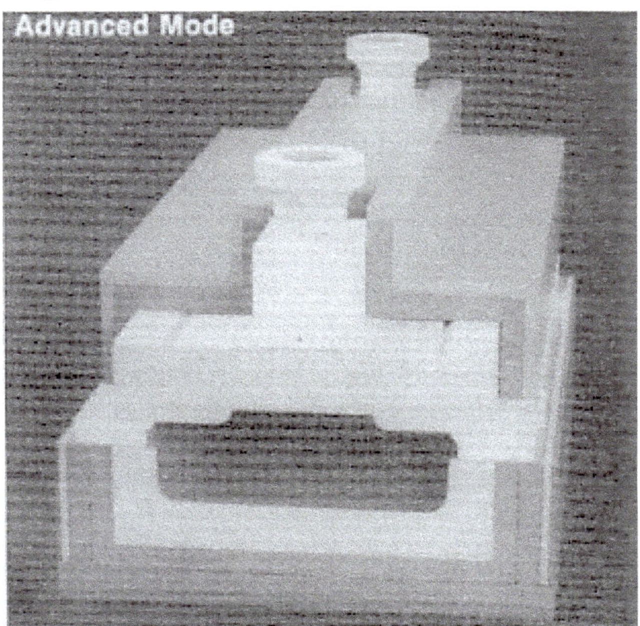

Figure 6,84. The advanced mode

Figure 6,85. The rapid-cool mode

In the 'rapid cooling zone' (the cooling section) the outside holes on the arch block are left open and exhaust flues are built over them including the required controlled venting system, while the centre flue is converted to an air channel with the necessary controlled equipment. By having two independently controlled outer channels it means that different amounts of heat can be extracted 'side to side', which makes for better control when going round a cranked section.

In the rapid cool mode the air is introduced through the centre channel and enters the inside chamber through widening slots which reduces the airs velocity causing to enter the forehearth and fall towards the glass without chilling it. The hot air is then vented to atmosphere through the outside exhaust channels.

There is built into the design the means for later expansion, i.e. higher pulls, if the cooling is insufficient. If the top of the existing conditioning section exposing the universal arch block then if the covers to the outside channels are removed then the top can be rebuilt as a rapid-cool zone.

There are brief biographies of four of these people elsewhere, the work by O'Neill and Mitchell is included in other sections, Pitt has very little information other than a patent application Number 195 448 entitled 'A Device for Delivering Charges of Molten Glass'. Apparently he lived or worked at Christon Bank, Lesbury, Northumberland, England, see abstract number 288 in the 1923 year book of the Society of Glass Technology. In October of that year Professor W.E.S. Turner's note book records two of his feeders working at Wood's Bottle Works (1920) Ltd., Portobello, Edinburgh, Scotland.

Carl H. Rankin and the Rankin Automatic Feeder Co.

There is very little recorded history of this man and much of this information comes from the Presidential address given by W.U.H. (Bill) Gregg in February 1983. Rankin was originally a bottle maker, who came from America to sell his glass feeder. He developed strong associations with the city of York, where he became a Director of the York Glass Company, later renamed as the National Glass Works and he also had a strong working relationship with Sheppee Engineering who built his feeders and forehearths. He was the inventor of the 'tube' or 'sleeve'.

He made a great impression on Bill Gregg when he was a child when he came to visit the factory driving his Arol-Aster motor car which he had shipped from American, the main impression was that it had a bonnet which seemed to be half a mile long. It is a little known fact Rankin along with Alex Lawson his draughtsman designed a six head suction machine along with its own pot. Gregg's bought the first machine which was built at the Sheppee Motor and Engineering Company works in York.

Carl Rankin used to visit the factory every day and run the machine for a few hours, the first job to be produced was the 8 oz HP sauce, see Figure 6,87. The machine was expensive to run with everything except for the electric machine drive being actuated or driven by compressed air, it also had one major drawback in that the blank was transferred to the finished mould by lowering the neck-ring mechanism by means of a long cylinder and piston

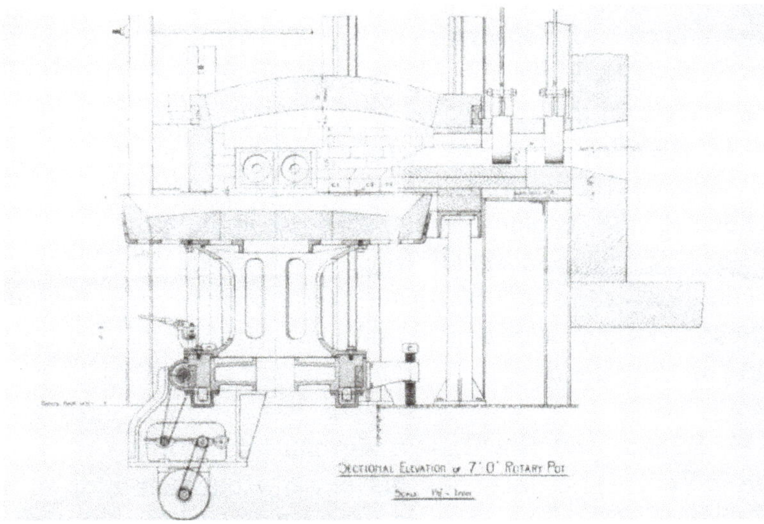

Figure 6,86. Schematic showing Rankin's six head suction machine

mechanism, every bottle produced had an 'off-set' finish. Gregg's pulled the plug on the venture and returned to Rankin feeders and Lynch machines leaving Carl Rankin a very disappointed man.

Rankin set about designing a second machine incorporating the lessons learnt with the first machine and one head was built. A diagram is shown in Figure 6,86 but this machine never worked.

One of Rankins important customers was Forsters Glass of St. Helens in Lancashire, the following is an excerpt from their book 'The History of Forsters Glass' which is entitled 'The Rankin Feeder'

> "This feeder has been entirely developed and constructed in England, and is recognised as being extremely successful in operation. The feeder

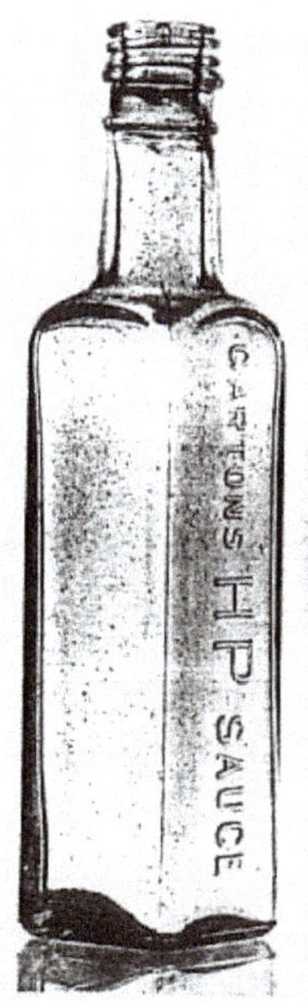

Figure 6,87. HP sauce

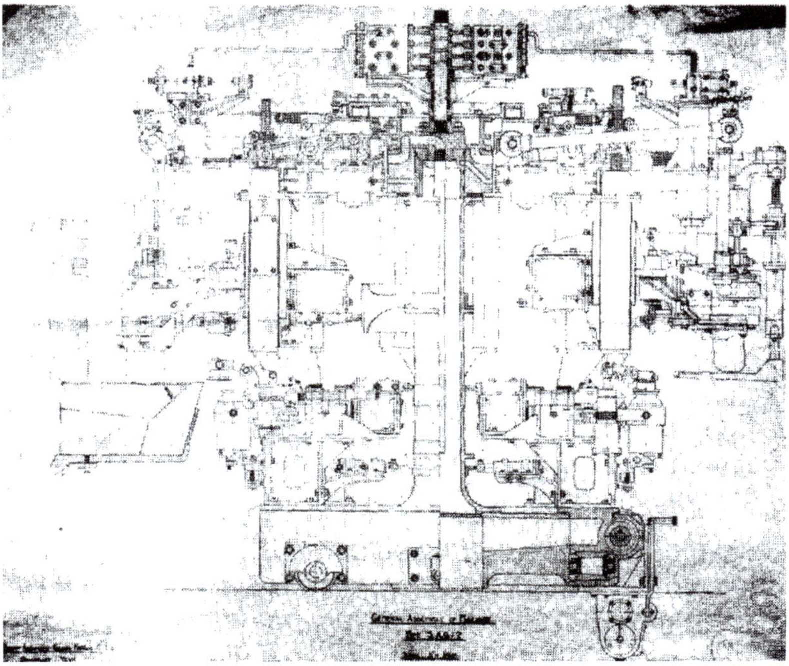

Figure 6,88. Schematic of Rankin's second machine

proper consists of an extension or forehearth to the furnace built up of insulated refractory blocks, carried in a cast iron casing. The molten glass passes from the furnace to the forehearth or boot through a throat block of a special shape into a circular 'nose'. In the bottom of this 'nose' is a small circular hole or orifice through which the glass is extruded by the movement of a vertically reciprocating plunger.

Surrounding the plunger is a refractory sleeve capable of being raised or lowered at will by a hand wheel. Owing to the plunger working inside this sleeve its action is localised, whilst the height at which the sleeve is placed enables the weight of glass ejected by each stroke to be controlled within very narrow limits. After being forced through the hole in the bottom of the feeder, the sausage-shaped piece of glass is severed by means of pneumatically operated shears, the piece of glass cut off is known as a 'gob', and it falls into the waiting parison mould of the forming machine. The 'brain' of the equipment is to be found in the cam-box, which is always placed away from the direct heat of the feeder. By means of this cam- box the timing of the various movements is adjusted both as to the actual duration of the movement and also their time relative to one another.

The 'Rankin' feeder is sometimes fitted with a gear which enables the sleeve to be rotated when this is considered necessary".

Further information from the same book informs us that in 1920 a Miller feeder was installed to feed a Miller automatic machine. About the same time, 1921, several Rankin drop feeders were attached to the Pocket Nook plant and used to feed O'Neill and Lynch machines. These were successful and soon replaced the Miller feeder. In 1922/23 a third type of feeder underwent trials at the works, the Hartford Fairmont's 'paddle' feeder of which there were three, one in each type of glass colour, these were soon removed and replaced by Rankin feeders. By 1925 the Rankin feeder had become the standard feeder with 27 in total across seven tanks. In 1926 they were all converted to have revolving sleeves.

Advertisement from the British Glass Industry Directory shows that in the 1923 directory:

RANKIN AUTOMATIC GLASS FEEDER CO. Works **Fishergate York**. Offices: **1 Charterhouse Street, London, E.C.1**.
T.N. London Holborn 4591; York 0495. T.A., "Natiglaso, London"; "Glasscoy, York"
Automatic glass feeding machines for blown and pressed glass of all descriptions.
Directors: Charles J. Pratt; C. H. Bluemenauer; C. H. Rankin.
(See advertisement Page 201)

There is very little recorded material left to highlight Rankin's career but his main success was his forehearth and feeder. The only records that we have are to be found in the abstracts in

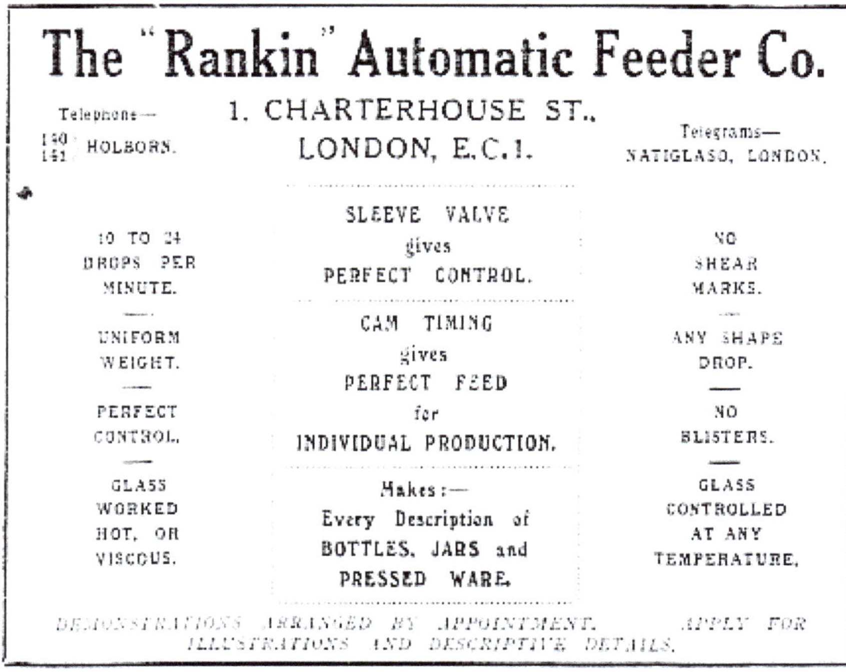

Figure 6,89. The Rankin advertisement from the 1923 edition of the British Glass Industry Directory

the Journal of the Society of Glass Technology, we have included a few but whether they came to fruition or not we just do not know , but some are obvious. The illustration is sparse so to give it some body we have included the write up.

#69. **Glass Delivery from a Tank Furnace**. W.L. Pratt, London (C.H. Rankin, USA) British Patent, December 30th 1919, No. 143 159. Provisional number, 32 634. Abstract No. 69, Volume V, 1921. This abstract relates to modifications to Patent Number 144 552 - see next abstract.

According to the invention, the upper edge of the discharge hole, C, is formed as a seating, D, for a tubular valve, E, which is raised and lowered pneumatically. The flow of metal is further controlled by the rise and fall of a plunger, F, which is moved independently of the tube, E. The strokes of the plunger, F, and the tube, E, may be varied by adjustable discs, S, L, in the pneumatic cylinders, R, K. The compressed air may be controlled by the valve mechanism described in specification 143 794. The tube ,E, is provided with flues, G.

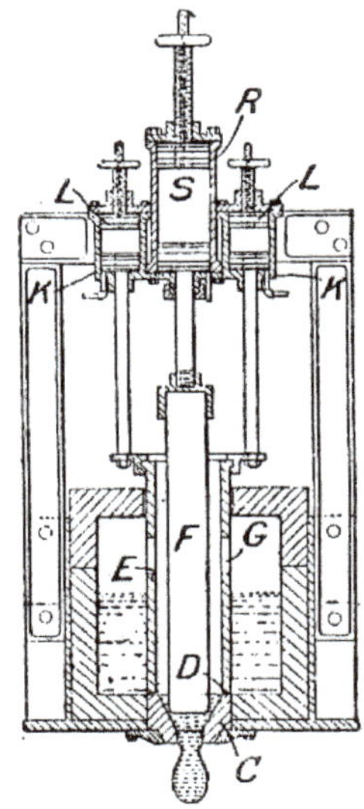

Figure 6,90. Glass Delivery from a Tank Furnace

Figure 6,90 and Figure 6,91 refers to **Glass Delivery from a Tank Furnace**. British Patent No. 144 552, November 6th, 1919. Provisional No. 32 634, and relates to apparatus for delivering uniform charges of glass or other molten material from a tank furnace. Both can be found in Abstract No. 70, Volume V, 1921. This patent application refers to another Patent, No. 143 794.

According to the invention, the charge during extrusion is surrounded by a tubular member or box to prevent chilling, the box being movable downwards to permit shears to

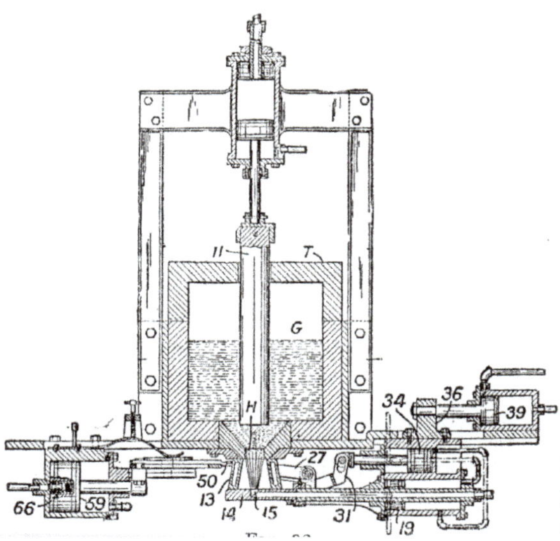

Figure 6,91. Glass Delivery

operate close to the orifice. The apparatus is operated by pneumatic power, the timing of the various movements being controlled by a valve of the kind described in specification 143 794. The molten metal, G, (see Figure 6,91), flows from a hole, H, in the bottom of a bay, T, in the melting tank, the rate of flow being made to fluctuate by the rise and fall of a plunger, 11. As the plunger, 11, rises after the delivery of a charge, a water-cooled conical box, 13, is raised against the tank bottom by the action of the piston, 34, and levers, 27, 31.

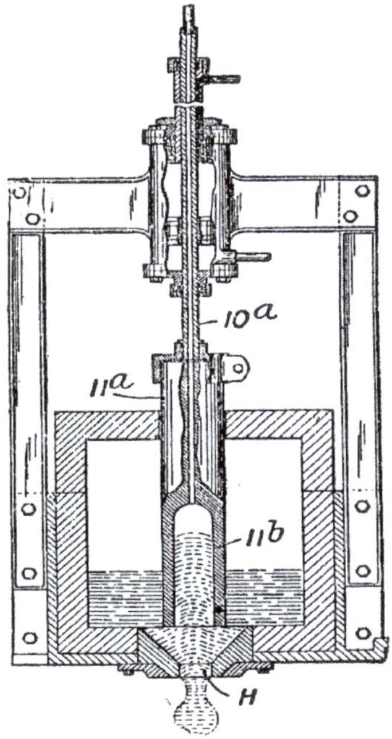

Figure 6,92. Modified plunger

A slide, 14. is moved to the left by a piston, 19, so as to close the bottom of the box, 13, and a blast of air and gas issues from a hole, 15. This drives the molten metal back into the hole, H, and also re-heats it. The slide, 14, is then drawn aside, the box, 13, is lowered and the plunger, 11, descends, causing he delivery of a chargewhich is cut of by shears, 50. The shears are operated by a piston, 59, having a cushioning valve, 66, to reduce shock when the shears, 50, open. The box, 13. and the gear connected with it are mounted on a slide, 36, which can be moved to the right by a piston, 39, so as to move the box, 13, clear of the hole, H.

An oil spray is provided to cool the shears, 50.

In a modification, plunger, 11^a (Figure 6,92), is formed with a chamber, 11^b , in which the air pressure is varied by means of a o pipe, 10^a, so as to assist the action of the plunger in expelling the charges from the hole, H.

Figure 6,93 is from Abstract 394, Volume VII, 1923. the abstract reads, **An Automatic Feeding Device**. C.H. Rankin,

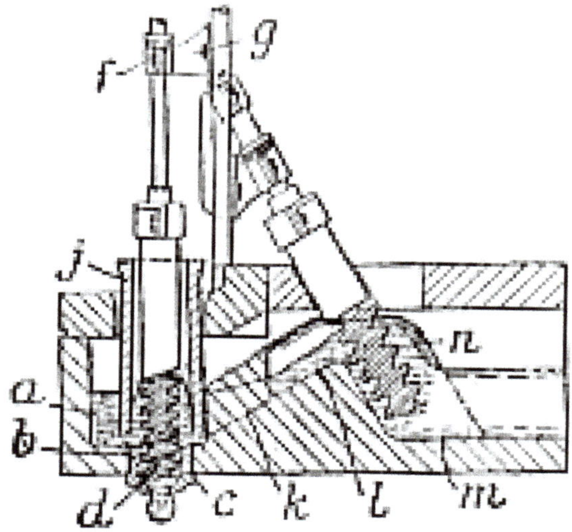

Figure 6,93. Schematic of Rankin's screw feeder

Wellsburg, West Virginia USA (British Patent, No. 198 952, September 28th, 1922). In a device for feeding glass towards and through a discharge orifice, the flow of glass is effected and controlled by a rotary Archimedean screw dipping into or immersed in the molten glass. The device comprises a forehearth, a, having in the lower part a cup, b, fitted with a bush, c, forming the discharge orifice. An Archimedean screw, d, extends into the cup, b, and by its rotation assists or retards the extrusion of the metal through the bush, c. The screw is supported in bearings, f, carried by a bracket, g, which can be moved to raise or lower the screw to regulate the discharge.

An adjustable sleeve, j, may surround the screw to co-operate in controlling the discharge and to serve, as a dross trap or stop valve. Metal is led to the forehearth, a, from a pocket, l, by way of gutters, k, which feed the metal to diametrically opposite parts of the forehearth. The metal is raised into the pocket from a trough, m, by a second screw, n, which has its axis inclined at a convenient angle to the end wall of the trough.

A further patent, by the same inventor (British Patent No. 202 496, July 26th, 1922), relates to a feeding device of the same

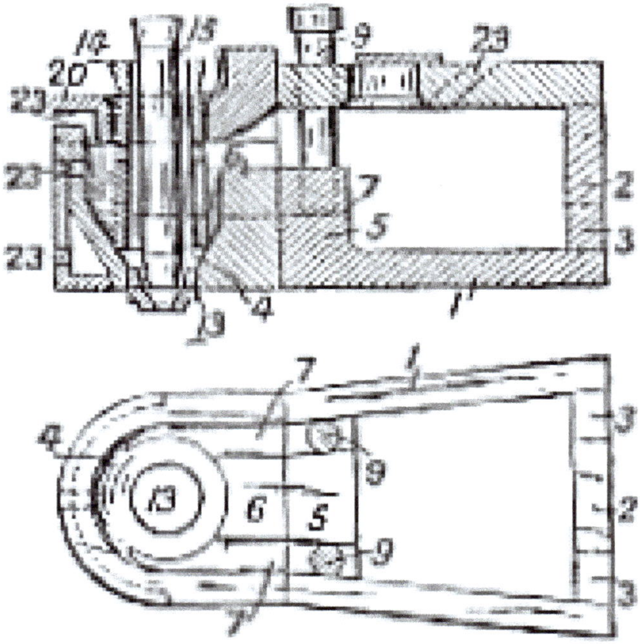

Figure 6,94. Glass feeding device

kind comprising a throat block, 1, and a nose cup, 4, having ducts or gutters, 7, arranged therein to deliver the glass in separate streams directed to opposite sides of the delivery aperture, 13. According to the invention, adjustable gates or plugs, 9, are provided in the ducts, 7, to regulate the flow of glass. The throat block, 1, receives glass from a tank furnace through submerged apertures, 3, in one of its end walls, 2, which acts therefore, as a trap to prevent entrance of scum from the metal in the tank. The nose cup is fitted to the other end wall of the throat block, the abutting 4 walls, 5, 6, forming a dam in which two ducts, 7, are formed. The ducts lead the molten glass, as shown, around opposite sides of the sleeve valve, 14. and feeding plunger, 15, so that the temperature is kept more uniform. The adjustment plugs, 9, enable the ducts to be closed or restricted to assist in regulating the temperature of the glass. The lower part of the nose cup is made conical and a removable corner, 20, is fitted.

Orifices, 23, for the admission of gas flames are provided in various parts of the apparatus.

Figure 6,95 is the illustration for **An Automatic Glass Feeder**, Abstract No. 127 Volume IX, 1925 (British Patent No. 218 888) submitted by W.L. Pratt, London and C.H. Rankin, Wellsburg, West Virginia, USA on October 1st, 1923.

It relates to a feeder of the kind comprising a receptacle having an orifice in its lower part through which the glass is discharged, and a plunger moveable in alignment with the orifice within a vertically moveable controlling-tube which dips into the glass in the receptacle.

According to the invention, the receptacle, plunger, and controlling tube are rotated at the same angular velocity, so that the molten glass as it runs into the receptacle is distributed around

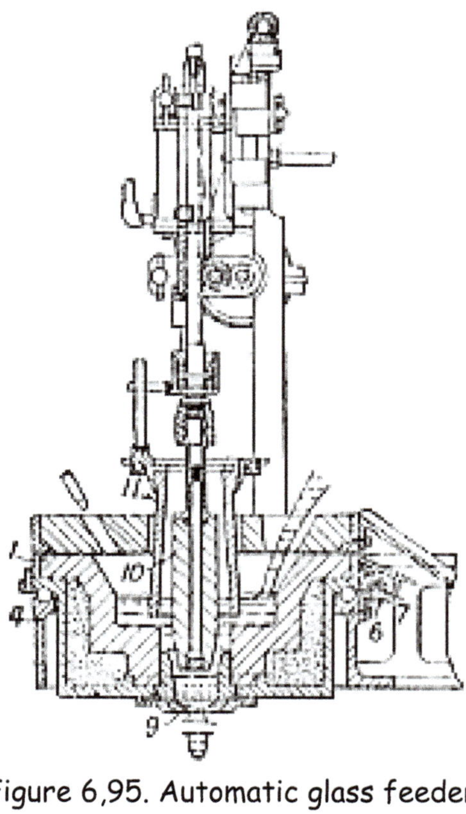

Figure 6,95. Automatic glass feeder

the plunger and ensures an even distribution of temperature in the column of glass discharged from the orifice.

The receptacle, 1, is rotatably supported by a ball race on a fixed part, 4, of the framing, and is rotated by a worm, 6, on a driving shaft, 7.The bottom of the receptacle is provided with an orifice, 9, above which a vertically reciprocating plunger, 10, surrounded by a flow-controlling tube, 11, is arranged. The receptacle receives a stream of molten glass from a tank and this glass is expelled in the gobs by the reciprocation of the plunger, the tube, 11, being adjusted vertically to control the size of the gobs. Both the tube and the plunger are rotated at the same angular speed as is the receptacle, so that the inflowing glass is spread evenly on the surface of the glass already in the cup, and the flow to the orifice takes place evenly from all parts of the receptacle.

In Volume X, (1926), of the Journal two abstract are attributed to Rankin, they are numbered 391 and 586. Number 391 (See Figure 6,96) is for **Modification of the Rankin Feeder** submitted by C.H. Rankin, West Virginia, USA (British Patent No. 243 459, September 3rd, 1924; from Ill. OfficialJ. Pat. No. 1931, January 20th, 1926). Where it relates to a glass feeding apparatus of the kind comprising a discharge aperture, 5, in the bottom of a cup, 1, of conical or other shape, and a sleeve or tube, 7, open at both ends, which dips into the cup and is adjustable vertically to or from the aperture. According to the invention the sleeve is rotated or oscillated about its axis to stir the glass in the cup to secure uniformity of temperature in the glass surrounding the aperture. A plunger, 8, reciprocates within the sleeve to control the formation of the charges expelled from the aperture. In addition to the sleeve, either the plunger or the cup, but not both, may be rotated.

Abstract No. 586 is for an **Automatic Feeding Device.** Submitted by C.H. Rankin, West Virginia, USA (British Patent No. 250 664, January 15th, 1925; from Ill. Official J. Patents, No. 1953, June 23rd, 1926). It relates to a feeder of the kind in which a plunger or sleeve is moveable to and from a discharge

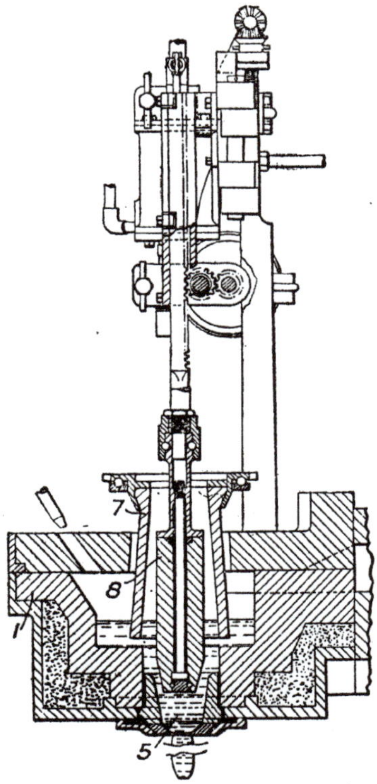

Figure 6,96. Rankins modified feeder

orifice and an adjustable stop is provided to limit the extent of the movement. According to the invention, means are provided for periodically removing from and returning to the operative position one or more stops, so that it is possible to deliver a succession of 'gobs' or mould charges of different weights. Means are also provided for varying the cross-sectional area of the discharge orifice, and for controlling the level of glass around the reciprocating plunger.

The feeder comprises a forehearth having at its front end a well, 1, Figs. 156 and 157, in the floor of which is the discharge orifice, 2. A reciprocating plunger, 3, controls the formation of the gobs of glass discharged through the outlet, and a sleeve, 4, concentric with the plunger and capable of vertical adjustment, controls the flow of glass to the orifice. The plunger, 3, is carried

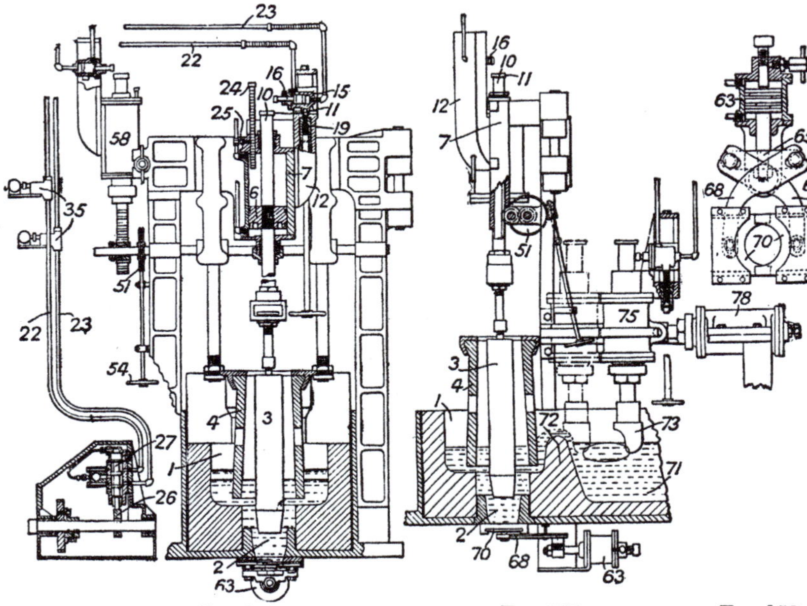

Figure 6,97. Rankin's Automatic Feeding Device. Abstract 586, Vol. X, 1926

by the piston rod of a cylinder, 7, and is reciprocated by the admission of compressed air to the ends of the cylinder. The piston, 6, has a tail rod, 10, which extends beyond the cylinder and terminates in a collar, 11. The device for limiting the stroke of the piston consists of a small air cylinder, 15, Fig. 156, arranged with its axis at right angles to the cylinder, 7, and carried by a bracket, 12, secured to the cylinder. The position of the small cylinder, 15, is adjustable vertically by means of a screw, 19, and the piston rod, 16, of this cylinder can be moved into or out of the path of the collar, 11.

The rod, 16, thus acts as a stop for limiting the movement of the plunger, 3, and by mounting a number of these small or stop cylinders at different heights, it is possible to control the stroke of the plunger, 3, the rods, 16, projecting either above or below the collar, 11. Another form of stop consists of a rod, 24, mounted in the end of the cylinder, 7, and secured in its adjusted position by a screw, 25. By mounting such a stop at each end of the cylinder, 7, the maximum stroke of the plunger can be limited, lesser variations in the stroke being controlled by the cylinders, 15.

Compressed air for operating the pistons of the cylinder, 7, is controlled by means of a timing-mechanism comprising cams, 26, which operate distributing-valves, 27, and the timing of the cams can be varied by mechanism such as that described in Specification 213 351. The speed at which the plunger, 3, moves is controlled by cushion valves, 35, in the pipelines, 22, 23, between the valve, 27, and the ends of the cylinder, 7. Each valve, 35, consists of a spring-operated ball which admits air to, but prevents it escape from, one end of the cylinder. The regulation of the escape of the air controls the speed of the piston, 6, and therefore of the plunger, 3.

The sleeve, 4, may be adjusted vertically, in order to vary its distance above the floor of the wheel, 1, by means of a worm wheel, 51, turned by a handwheel, 54. Alternatively, the sleeve may be moved by an air cylinder, 58, the movements of its piston being controlled in the same way as that of the piston, 6. The outlet, 2, may be expanded or contracted by means of two plates, 70, Figures 157 and 158 having opposed semicircular edges, which slide over one another beneath the orifice. The separation of the plates is controlled by an air cylinder, 63, that operates levers, 68 pivoted to the plates, 70. The movement of the plates may be co-ordinated with that of the plunger, 3.

For controlling the level of the glass in the well, 1, a paddle-shaped impeller, 73, Fig. 157 is used to propel the glass along the forehearth, 71, over a wall, 72, into the well, 1, in the manner described in Specification 105 672. The paddle is reciprocated horizontally by an air cylinder, 78, and vertically by another cylinder, 75. The stroke of the pistons of these cylinders is varied in the same way as that of the piston, 6, so that the path of the paddle can be varied in order to regulate the amount of glass fed to the plunger.

The foregoing abstracts from the Journal gives a flavour of

Rankin's inventiveness and expertise, there are other entries in the abstracts but his main contribution to the glass industry in this country was his forehearth and feeder which were used until the 1960s and 1970s.

Wm. J. Miller

Wm. J. Miller was another pioneer who gave much to the glass industry and there are even fewer records of him than they are of Rankin. Two such records are to be found in the British Glass Industry Directory of 1923 (First Edition) and the one in 1928, the second edition which gives a feeling of what he was marketing;

MILLER, W. J., 12/13 Henrietta Street, Strand, London, W.C.2.

T.N. Regent 1639. T.A., "Autoglasma, Rand, London." Specialities: Presses; wide mouth bottle machines; paste mould machines; automatic glass feeders (six models).

Also suppliers of needle etching machines, disc grinders, mould cleaning lathes, cracking off machines, sand blast machines, tumbler grinding machines.

All machines made in Swissvale, Pa. USA They range from toy marble machines to presses exerting 400 tons pressure to make pressed articles 75 in. x 26 in. x 18 in. Fifty eight distinct models are made in 174 sizes, and agency held for 28 models in 44 sizes. Proprietor: Wm. J. Miller. (See advertisement Page 281)

And the third edition of the 1934 British Glass Industry Directory says;

MILLER (WM. J.) INC., 105-107, Cecil Chambers, 96, Strand, London, W.C.2. T.N., Temple Bar 5647, T.A., "Autoglasma, Rand, London." Head Office and Works: **Swissvale, Pa., USA**

Hand-operated, semi-automatic and fully automatic glass house machinery, either air or motor driven, for the production of bottles, jars, pressed tableware and turned ware.

The 1923 advertisement shows that they made 58 distinct types of machine, presses, press and blow machines, narrow

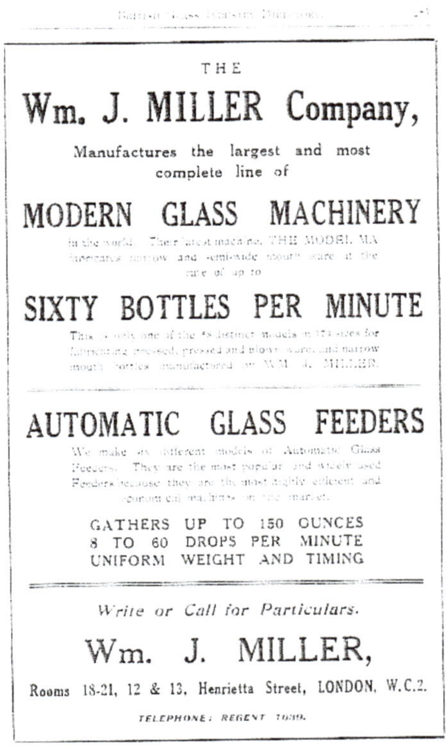

THE

Wm. J. MILLER Company,

Manufactures the largest and most complete line of

MODERN GLASS MACHINERY

SIXTY BOTTLES PER MINUTE

AUTOMATIC GLASS FEEDERS

GATHERS UP TO 150 OUNCES
8 TO 60 DROPS PER MINUTE
UNIFORM WEIGHT AND TIMING

Write or Call for Particulars.

Wm. J. MILLER,

Rooms 18-21, 12 & 13, Henrietta Street, LONDON, W.C.2.

Figure 6,98. Miller's advertisement on Page 281 in the 1923 Edition of the British Glass Industry Directory

mouth machines and paste mould machines, which came in 174 different sizes. Along side these he made six different types of automatic glass feeders, capable of 8 to 60 drops per minute and could make gathers up to 160 ounces. He was also agent for another 28 different machines in 44 sizes.

He also supplied; needle etching machines, disc grinders, mould cleaning lathes, cracking off machines, sand blast machines and tumbler grinding machines. So seemingly he had quite a substantial business.

In 1952 he sold out to the General Glass Equipment Co. of Atlantic City, New Jersey, USA The source of this was the 1952 Journal of the Society of Glass Technology -Abstract #944 stated that - General Glass Equipment Company. Anon. (Glass,

Figure 6,99. A 1928 advertisement by Wm. J. Miller

1952, **29**, (2). 58) - The General Glass Equipment Company have taken over the manufacture of the 'Miller' feeder on which many improvements are being incorporated. It may be of considerable interest to the present users of the 'Miller' feeder that some of the new changes will be adaptable to their existing plant.

This is their entry in the sixth edition of the Directory of the British Glass Industry which was published in 1956;

GENERAL GLASS EQUIPMENT CO. LTD. Head Office: Atlantic County Trust Building, Atlantic City, New Jersey, USA European Office: 50 Argyle Crescent, Portobello, Scotland. T.A., "Glasequip, Edinburgh". PORtobello 2045.

Specialists in design and construction of glass plant. Batch chargers; melting tank-furnaces; glass-forming machinery; glass-level recorder and controller.

Managing Director: A.C. Langham (European Office); Louis Plumbo (Head Office, USA)

Secretary: H.M. Langham.

While the seventh edition published in 1959 had the advertisement shown in Figure 6,100.

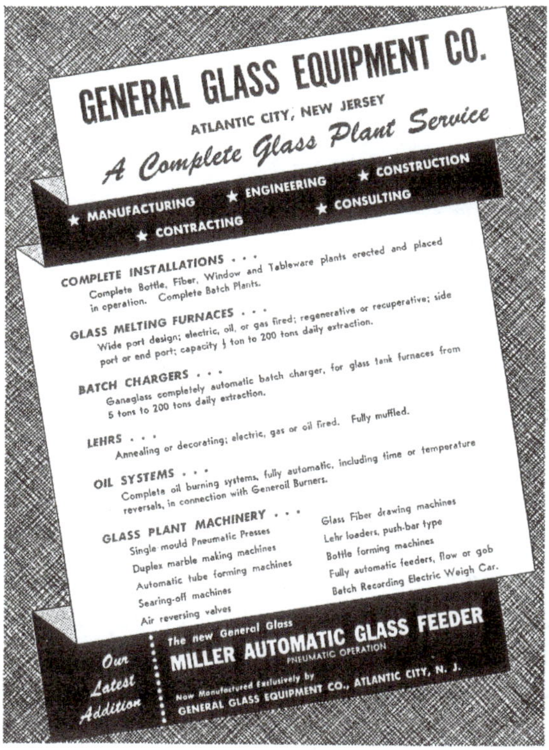

European Representative:

A. C. LANGHAM, A.M.Inst.B.E., 50 Argyle Crescent, Portobello, Scotland

Figure 6,100. General Glass Equipment's advertisement in the 1959 Edition of the Directory of the British Glass Industry

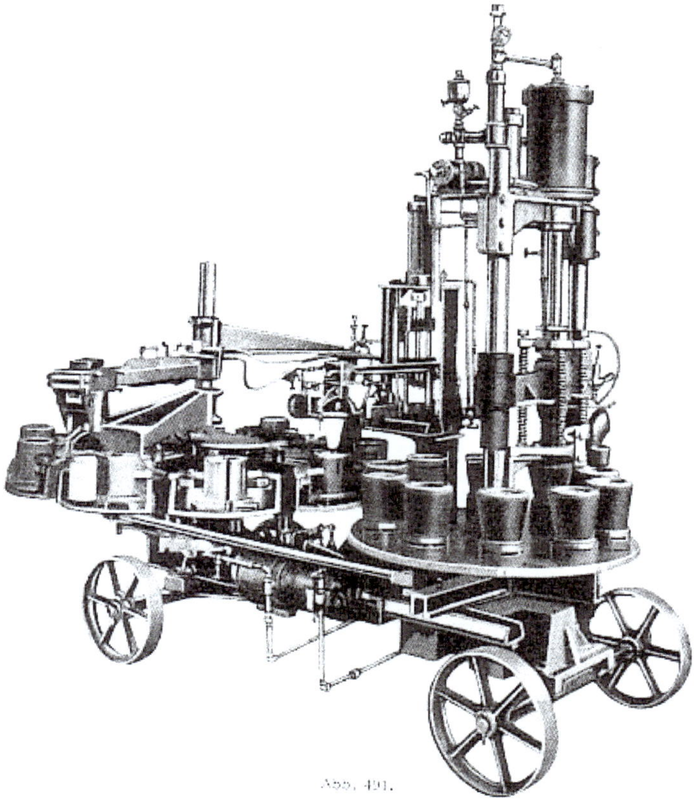

Figure 6,101. Miller's Press and Blow machine. British Patent No. 152 567 and later No. 174 645

Other than the photograph in the 1928 advertisement Figure 6,101 is the only other photograph that I have of a Miller machine.

The abstract for the above machine is from the 1921 Journal of The Society of Glass Technology, Volume V, and is as follows; **Abstract # 146. Bottle-blowing Machine.** W.J. Miller, Swissvale, Pa. USA (Brit. Pat., September 24th, 1920. No. 152 597. Provisional No. 27 263. convention date, October 16th, 1919. Not yet accepted).

It relates to a double rotating table, semi-automatic, pneumatically operated bottle-blowing machine. The parison moulds, H, and the blowing moulds, I, are mounted on tables, E_1,

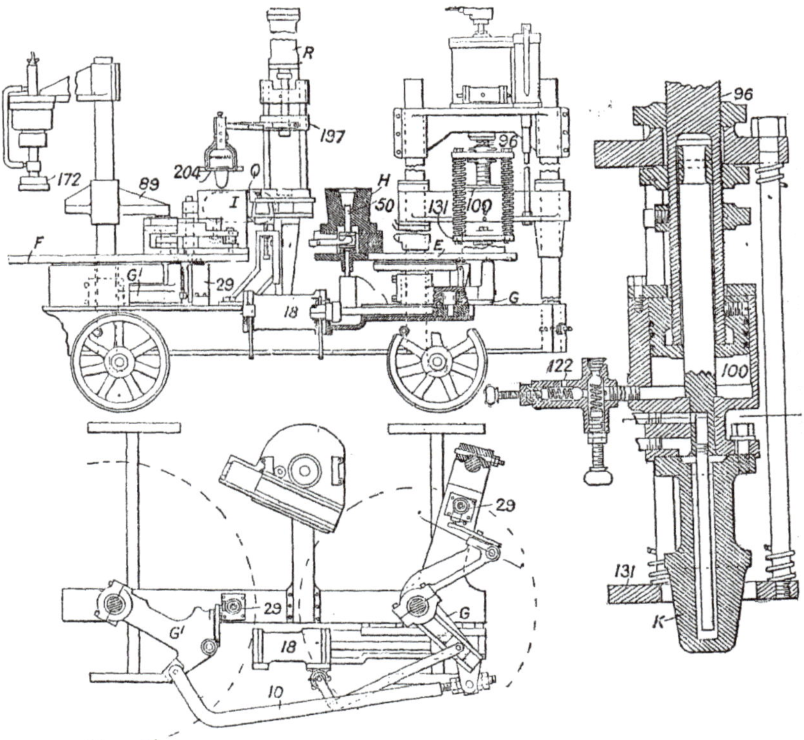

Figure 6,102. Wm. J. Miller's Press and Blow machine

F, respectively which are rotated intermittently and in unison by swinging arms, G, G_1, connected by a rod, 10. Power for this operation is supplied by the cylinder, 18.

The tables are locked by automatic bolts, 29. The parison is pressed by a water-cooled plunger, K, the rod, 96, of which is fitted with a pneumatic cushioning cylinder, 100, provided with an adjustable release valve, 122, by which the pressure of the plunger, K, may be varied. The usual spring cover plate, 131, is employed. The parison is raised at the transfer station by a cam or hand-operated push-rod, 50, and then is seized by transfer jaws 204, lifted clear of the parision mould, swung round, and lowered into the blowing mould, I. The transfer mechanism, 204, is mounted on a slide, 197, which is raised and lowered by a pneumatic cylinder, R. and swung round by a cylinder, Q. The blowing moulds, I, are operated by a fixed cam, 89. A blowing head, 172, of the usual kind is provided.

The next four abstracts are the first entries for each type of machine whether they worked successfully or not. In total Wm. J. Miller had nearly 60 patent applications recorded in the first twenty years of the Journal of the Society of Glass Technology, obviously too many to record here. The first one is for an automatic glass-blowing machine.

#334 Automatic Glass-blowing Machine. Wm. J. Miller, Pa., USA (Brit. Pat., October 30th, 1918, No. 142 156).
In a machine for blowing glass articles, and of the particular kind comprising a series of blank and blowing moulds movable step by step to different stations for successive operations, means for blowing the glass in the moulds, for transferring the blank, by turning it over independently of the blank mould, to the blowing mould, and for sucking down the gather by vacuum into the blank and neck moulds, are provided.

A series of blank moulds, J, (Fig 187), adapted to receive gatherings of glass in succession, and a series of blow moulds, K, each in vertical alignment with the corresponding blank mould, are mounted on supports, H, D, which are intermittently rotated round a hollow pillar, B, and a suction and blowing mechanism, M, mounted between each blank mould and the corresponding blowing mould to rotate therewith, includes neck forming elements, 99, and is reversible independently of the blank mould, so that glass is subjected to suction and blowing actions in the blank mould, and then to a final blowing action in the blowing mould, from which the finished article is removed by grippers.

The supports, H, D, are aligned by vertical bars, 15, to which they are adjustably connected, and screw mechanism, 7, 8, is connected by rods, 9, to the support, H, for adjusting this support to permit moulds of different heights to be used. A sleeve, E, on the pillar, B, is oscillated by an attached arm, in which slides a member, 33, engaging a crank-pin, 32, on a wheel rotated by worm gearing, 27, 28, and the sliding member, 33, has a cam

Fig. 187

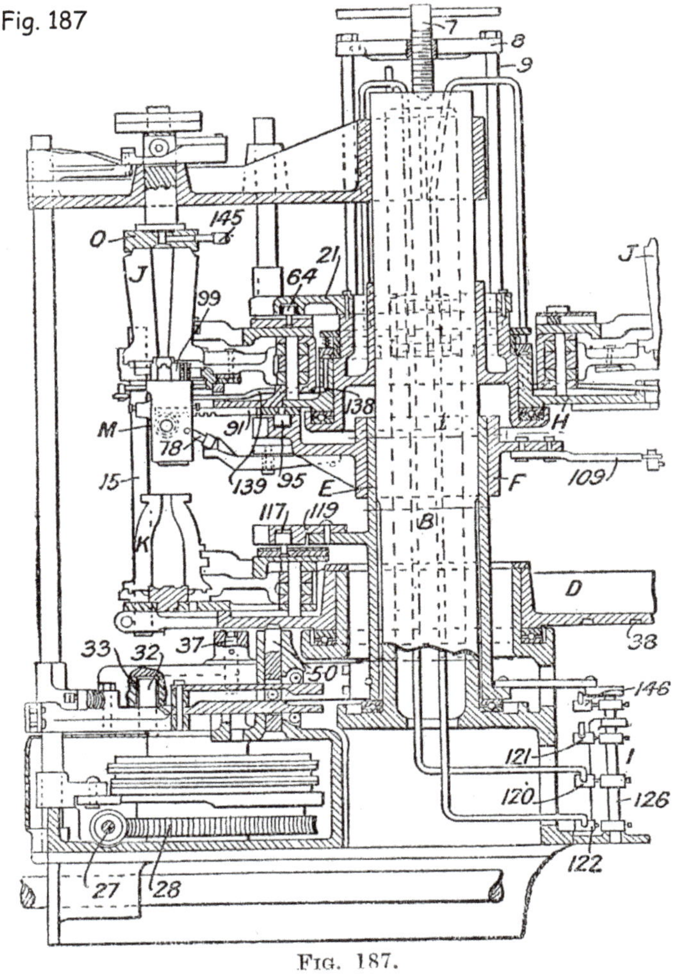

FIG. 187.

Figure 6,103. Automatic Glass-blowing Machine. J.W. Miller's application for a British Patent. October 30th, 1918, No. 142 156. Abstract No. 334, Volume IV, 1920

groove adapted to project and retract a pin, 37, into and from engagement with recesses, 38, in the support, D, so that the support is moved forwards and then freed, the arm at the end of its forward movement actuating a pin, 50, to lock the support during the reverse movement of the arm.

FIG. 188.

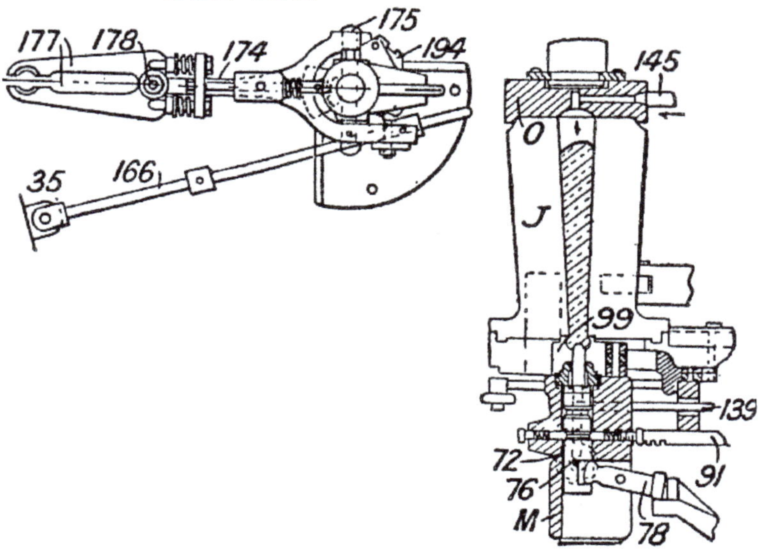

FIG. 189.

Figure 6,104. Part of Miller's application No. 142 156

The blowing mechanism includes a cylinder, 72, (Fig. 189), containing a plunger, 76, the outer end of which is initially extended to form a depression in the neck end of the gather in the blank mould, and is subsequently retracted by an arm, 78, and a cam surface to permit the entrance of pressure fluid to the cylinder by a pipe, 139, to act on the glass. The outer end of the plunger is screwed thereto, so as to be readily changeable. The pipe, 139, is connected to a passage, 138, which is brought successively into communication with a suction chamber and three pressure chambers connected by pipes passing down to valves, 120, 121, 122, respectively, the valves being controlled by a tappet shaft, 126, oscillated by connection with the sleeve, E.

The arrangement is such that the glass is forced down to the neck of the blank mould by pressure fluid entering a blow-head, O, through a pipe, 145, connected to a valve, 146, controlled by the tappet shaft, 126, is sucked into the neck mould when the suction chamber becomes operative, and has two blowings in the blank mould. This mould is then opened, and the device, M, is reversed to carry the partly formed article into the open blowing mould, which is then closed, and the final blowing takes place.

Each blank mould comprises two sections pivoted on a vertical pin and connected by links to a yoke yieldingly connected to a slide carrying a roller, 64, engaging a cam grove in a member, 21, whereby the sections are opened and closed as the support, H, is rotated. The blowing mould is similarly formed and operated, but its actuating roller, 117, engages a cam grove in a member, 119, on the oscillating sleeve, E. The neck mould also comprises two pivoted sections, which are closed by a toggle mechanism operated by the blank mould closing means, and are opened by a cam striker, 109, on a member, F, on the sleeve, E, to release the neck of the article after it has been positioned in the blowing mould.

The blowing mechanism, M, is mounted on the support, H, and is reversed by a pinion and a rack, 91, operated through a yielding connection with a slide carrying a roller, 95, engaging a cam groove in the member, F. The articles are taken fro the opened blowing moulds by gripping members, 177 (Fig. 188) pivoted on a vertical axis, 178, on an arm, 174, movable about horizontal trunnions, 175, on a rotary supporting member.

The members, 177, are adapted to be depressed and closed to grip an article in the opened mould, then elevated to raise the article clear of the mould bottom and swung to remove it from the mould, then lowered and opened to release the article, and finally raised and returned to their initial position, the opening and closing and raising and lowering movements being effected by cam and roller devices mounted on arm, 174, and on a relatively moveable operating member connected by a rod, 166, to the arm, 35, that operates the sleeve, E, and adjustable in height to suit the size of the mould.

At the termination of each swinging movement of the arm, 174, the supporting member of the trunnions is locked by catches, 194, which are released after the parts have moved relatively to cause the cam and roller devices to operate at the next movement of the rod, 166.

This second abstract one is for a press, No. 56, Volume IV, 1920 entitled Glass pressing machines; in reality the abstract is for four patents a press, and three refinements.

#56 Glass Pressing Machines. W. J. Miller USA, (Brit. Pat., August 8th, 1918, No. 19 289/19). It relates to a press for moulding glass articles. The plunger, P, (Fig. 17), is mounted on a moveable cross-head, N, connected with a fixed cross-head, 126, by toggles, 174,176, which are operated through links, 175, by a cross-head, 172, moved up and down by fluid pressure acting in the cylinder, J.

The capacity of the cylinder, J, can be varied by moving a false head or piston, 133 (Fig. 17), which is fixed to a screwed sleeve, 132, passing through the true cylinder head, 130. An extension, 134, of the piston rod, 129, is grooved at 136 to allow

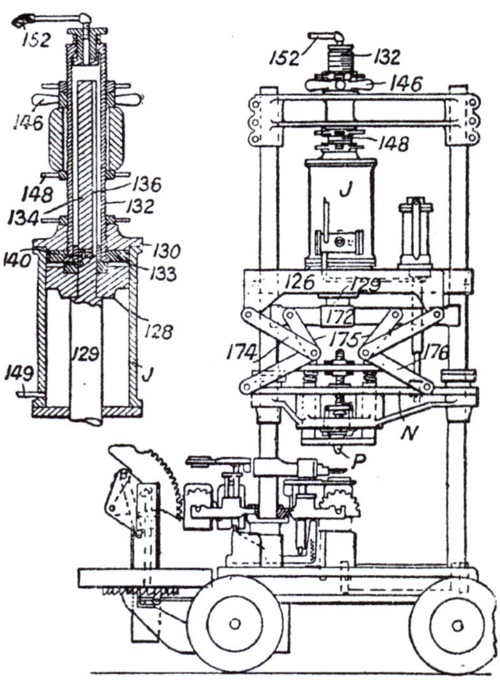

Fig. 17.

Figure 6,105. Glass Pressing Machines. WJ Miller's press application for a British Patent Aug 8 1918, No 19 289/19

fluid pressure to pass from the supply pipe, 152, through the sleeve, 132, into the space between the false head, 133, and the piston, 128, and so cause this to descend. The piston is raised by air admitted through a pipe, 149. The last part of the upward movement of the piston, 128, closes the entrance to the groove, 136, and the imprisoned air is allowed to escape slowly through a small port, 140, thus producing a cushioning effect. The height of the cylinder, J, and of the cross-head, 136, attached to it, is adjusted by screw collars, 146, 148.

The second part of this abstract is as follows; further devices by the same inventor are as follows:- (Brit. Pat., August 8th, 1918, No. 198 290/19.) - This relates to apparatus for discharging the

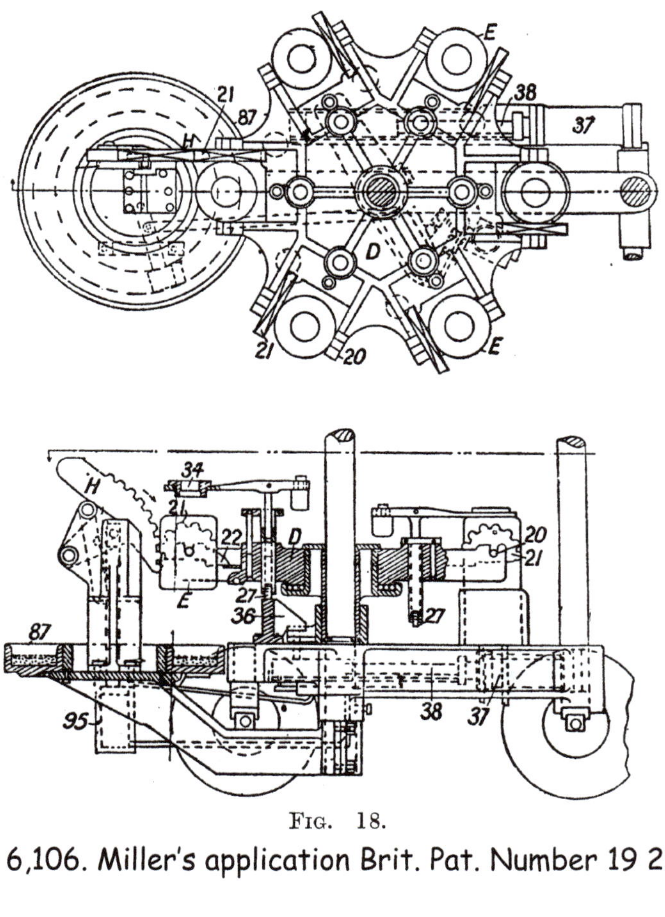

FIG. 18.

Figure 6,106. Miller's application Brit. Pat. Number 19 290/19 finished glassware from the moulds of a press.

The moulds, E, (Fig. 18), are carried on trunnions, 20, on a frame, D, which is rotated step by step by a fluid-pressure cylinder, 37. The gatherer places the charge in the mould and presses the punty, upon a valve, which starts the machine, after which the cycle of operations is completed automatically by fluid-pressure mechanism. Shears cut the depending charge from the punty, and the charged mould is carried round to the pressing station, where the plunger descends and compresses the charge.

The further rotation of the frame, D, causes a rod, 27, to strike a cam, 36, and so raise the mould top, 34, which it supports. A toothed arc, 21, on the mould, E, engages another toothed arc, H, which is swung downwards by a fluid-pressure cylinder, etc., 95, and so inverts the mould and discharges the finished article into an angular trough, 87, containing sand, etc., to cushion the fall.

The tappets of the slide valve controlling the fluid-pressure supply to cylinder 95 are operated by a block fixed to the piston rod, 38, of cylinder, 37. The annular trough, 87, is rotated intermittingly in unison with the frame, D, so that successive articles do not fall upon one another. The trunnions, 20, are off-set, so that the mould, E, turns back to its normal position by its own weight when disengaged from the arc, H, and a stop, 22, keeps it upright.

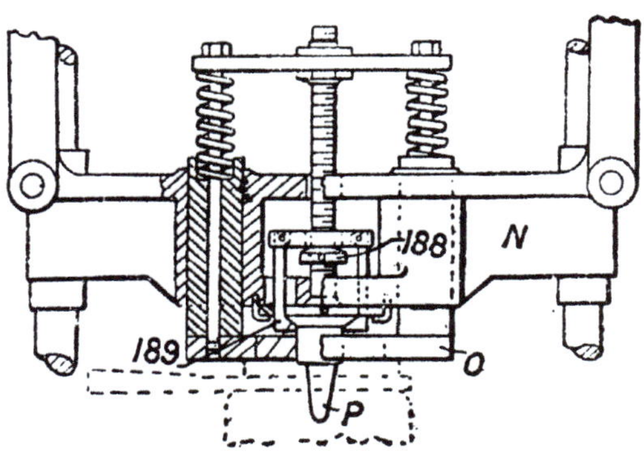

FIG. 19.

Figure 6,107. Millers application regarding a pressing head

The third part, Brit. Pat., dated August 8th, 1918, No. 19 291/19, refers to a pressing head.

In a machine for moulding glass articles, the plunger, P, is centred above the mould, and then clamped to the cross-head, N, by hooks, 189, which are drawn upwards by a nut, 188. The hooks, 189, are also moved radially by a cam ring. The plunger, P, passes through a spring-pressed plate, O, which rests upon the top of the mould during the pressing operation (Figure 6,107).

Finally the fourth part, British Patent, 8th August, 1918 No. 19 292/19 states that in a press of the type in which the moulds, E, are carried on rotating arms, 16, a cylindrical block, 113, is mounted directly below the pressing plunger and is surrounded by a sleeve, 114, having a bevelled upper edge, T, and is thus raised slightly, being supported during the pressing operation by a block, 113, the arms, 16, thus being relieved of the strain. The sleeve, 114, is raised to the position shown by fluid pressure acting on a piston, 119, as soon as the mould, E, is I position above the block, 113. Accurate centring is thus ensured (Figure 6,108).

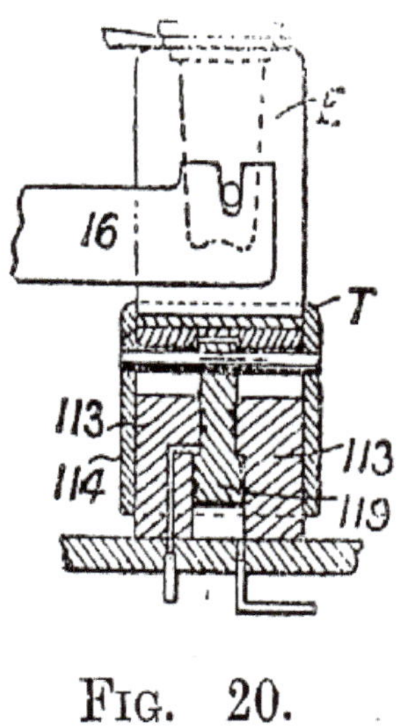

FIG. 20.

Figure 6,108.
Centring device

Two abstracts later Number 58 there is another patent application by Miller, this time with a British Patent number, and the drawing with it seems to be of the same press.

Abstract # 58, Machinery for Moulding Glass. W. J. Miller, Swissvale, Pennsylvania, USA (Brit. Pat. No. 118 282.) (August 8th, 1918, No. 12 883). - Relates to a rotary-table moulding machine actuated by fluid pressure. A series of moulds, E, are suspended on trunnions at the ends of radial arms on the table, D, connected to a boss turning on a pillar, C, and driven intermittently by a oscillating arm actuated by a piston in a cylinder, 37. An automatically actuated pin engaging holes in the underside of the table serves to lock the table in successive positions. The glass is fed into a mould when in the position shown at the right of Figure 22, and the workman then actuates a valve by means of his punty, causing shears, 83, to be pneumatically operated to sever the charge.

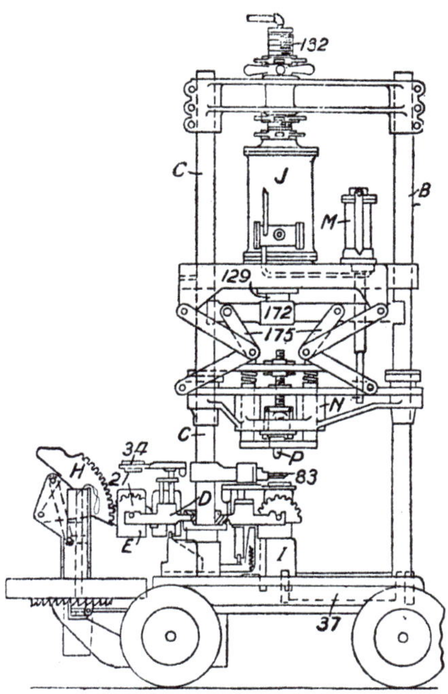

FIG. 22.

Figure 6,109. Miller's press Abstract # 58, 192

The table then moves automatically, and the cover, 34, is brought down upon the mould, which rides up on a block, I, when it reaches the moulding station. The block consists of an outer sleeve, which is forced upwards by pneumatic pressure to support the mould. The die, P, (Fig. 22), is adjustably mounted on a sliding frame, N, which is connected by adjustable toggle levers, 175, with a cross-head, 172, carried by the rod, 129, of a piston working in a vertical cylinder, J, supported by the collars, B, C. The cylinder is fitted with an adjustable false end to regulate the movement of the piston, and air is first admitted slowly to give a slow initial movement, and them more rapidly, the final pressing being effected by the toggle levers.

The time that the die remains in the glass is determined by timing mechanism, M, of the kind described in the US Specification 941 935. After the moulding has been completed, the table moves forward, and when it reaches the discharging position the mould is turned over by a tooth wheeled segment, H, engaging a segment, 21, and the finished article is delivered into an annular rotating sand box. The cylinder, J, is supported by a screwed stem, 132, which can be adjusted vertically for moulding articles of different lengths. It is stated that the apparatus is applicable to blowing machines.

This next abstract No. 156 is from the 1921 Volume of the Journal, and it is for a Paste Mould machine.

 # 156. Paste Mould Machine. W. J. Miller (Brit. Pat. No. 151 606, September 13th, 1920, Provisional No. 26 248). This patent application relates to rotary paste mould machines in which the blowpipe with a gather on the end is placed in the machine by hand.

The upper part of the machine is shown in Fig. 66 and the lower part in Fig. 67. The blowpipes, I, are gripped and rotated by gear-driven rubber rollers, 58, and an idler roller, 59. The rollers, 58, are automatically separated at the required times for the release of the blowpipe bearing the blown ware by a cam-operated slide 53. The blowhead, 67, is lowered to the blowpipe, I, by a fixed cam, 75. Air is supplied through a rotary valve, 83. The moulds,

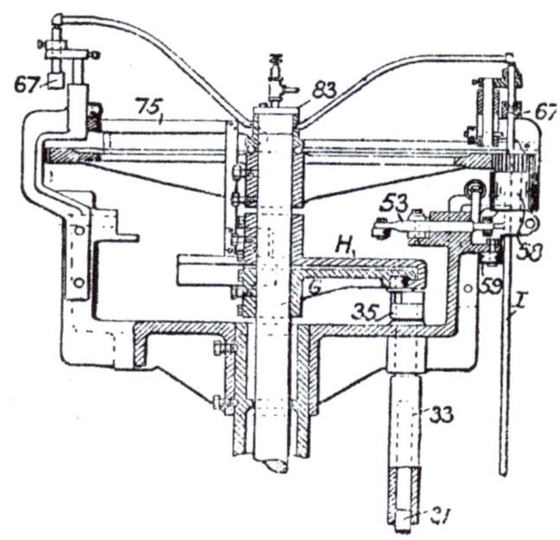

Fig. 66.

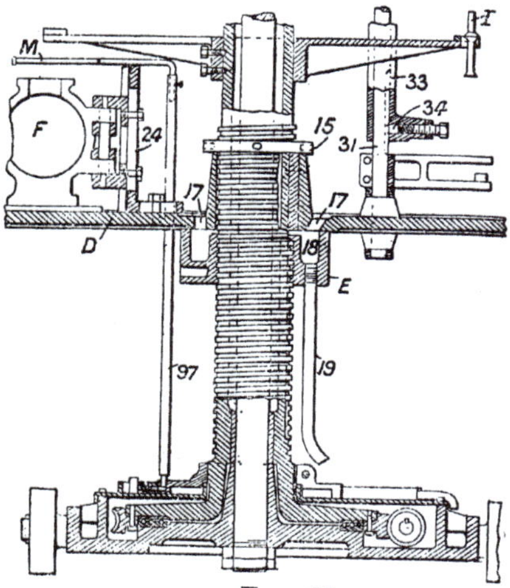

Fig. 67.

Figure 6,110. Miller's paste machine, abstract No. 156, Vol. V, 1921

F, are mounted on brackets, 24, so that they can be adjusted vertically and radially. They are opened and closed by adjustable stationary cams, G, H, through cranks, 35, and telescopic shafts, 31, 33. A spring release, 34, comes into action if a mould jams. Cooling water is discharged on the moulds, F, from pipes, M, 97, the excess passing away through passages, 17, 18, and a pipe, 19. The height of the mould table, D, is adjusted by nuts, E, 15.

Finally the last abstract which is for a glass feeding device can be found in Vol. V, 1921.

#163, Glass Feeding Device. W. J. Miller, Swissvale, Pa., USA (Brit. Pat. No. 151 605, September 13th, 1920, Provisional No. 26 247. Convention date, September 23rd 1919. Not yet accepted)

It relates to automatic pneumatically operated apparatus for delivering uniform charges of glass from a tank furnace. The molten metal is impelled through a hole, 13, in the bottom or side of the tank, 4, by a water-cooled reciprocating plunger, 24. The charge is cut of by water-cooled shears, J, and the metal above the shears is blown back into the hole, 13, by a blast of hot air from an annular orifice, 51. The air for this blast is heated by

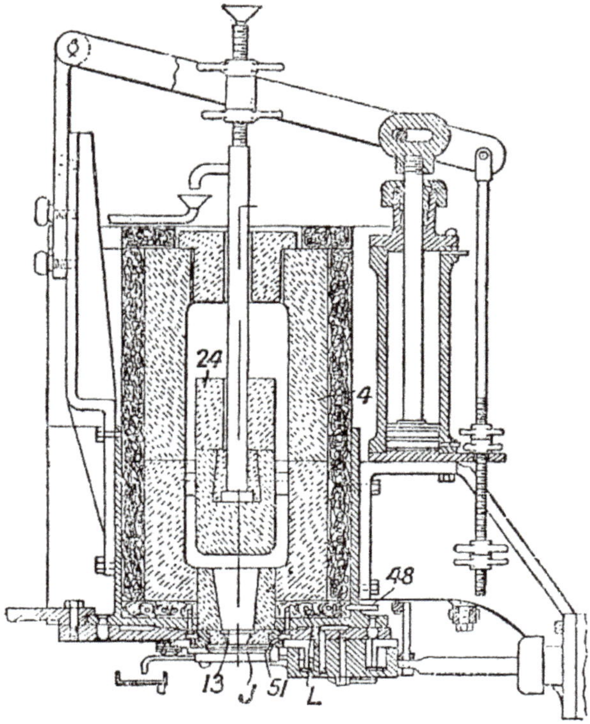

Figure 6,111. Glass feeding device

passage through a pipe, 48, coiled in asbestos lagging, L. The blast pressure is arranged to diminish gradually during each application. An oil spray is provided to lubricate the shears.

Frank O'Neill

Finally we come to Frank O'Neill who, just like W.J. Miller, had inventions covering the whole spectrum of bottle making. Here we are just going to deal with some of his contribution to feeders and forehearths, his main biopic can be found in Chapter 8, The Rise and Fall of the Twin Table Machines.

In the abstracts in 1922 issue of the Journal of the Society of Glass Technology, Volume VI, the first instances of his work, there are three patent applications by Frank O'Neill. Obviously they are from a much later date than 1896. Two of the abstracts are for a Glass-feeding Device while the third is for a Glass Flow Controller.

#76. Glass-feeding Device. F. O'Neil, Toledo, Ohio, USA (Brit. Pat., No. 166 605, January 28th, 1920).
According to the invention, molten glass flows continually in a stream, 28, from a furnace extension, 12, and is cut off into charges by shears, 27. The charges drop into the mouth of a fixed funnel, 34, which makes close contact with the parison moulds, 53, and are drawn down into the neck moulds by suction applied through a hollow piston-rod, 72, and passages, 75, 83, in the plungers, 73, and parison moulds, 53, respectively. The funnel, 34, is provided with gas burners, 37, to heat the incoming charge, with oil-holes, 44, to lubricate it and with water or steam cooling channels, 47, to prevent excessive heating of the lower part.

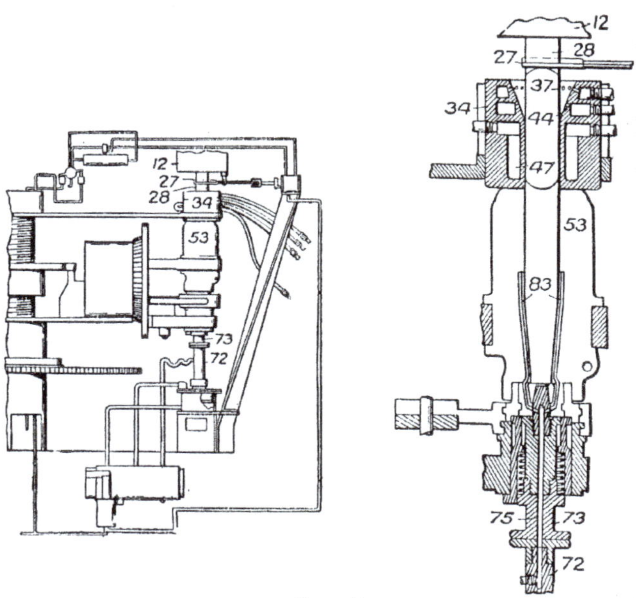

FIG. 44.

Figure 6,112. Glass-feeding device, Volume VI, 1922

#411. Glass Feeding Device. F. O'Neill, Toledo, Ohio, USA (Brit. Pat., No. 179 977, January 12th, 1921). - Relates to apparatus for delivering uniform charges of molten glass from a tank furnace.

According to the invention, a plunger, d, is reciprocated in a cylindrical opening, c^2, which ends in a conical outlet, c^1, formed

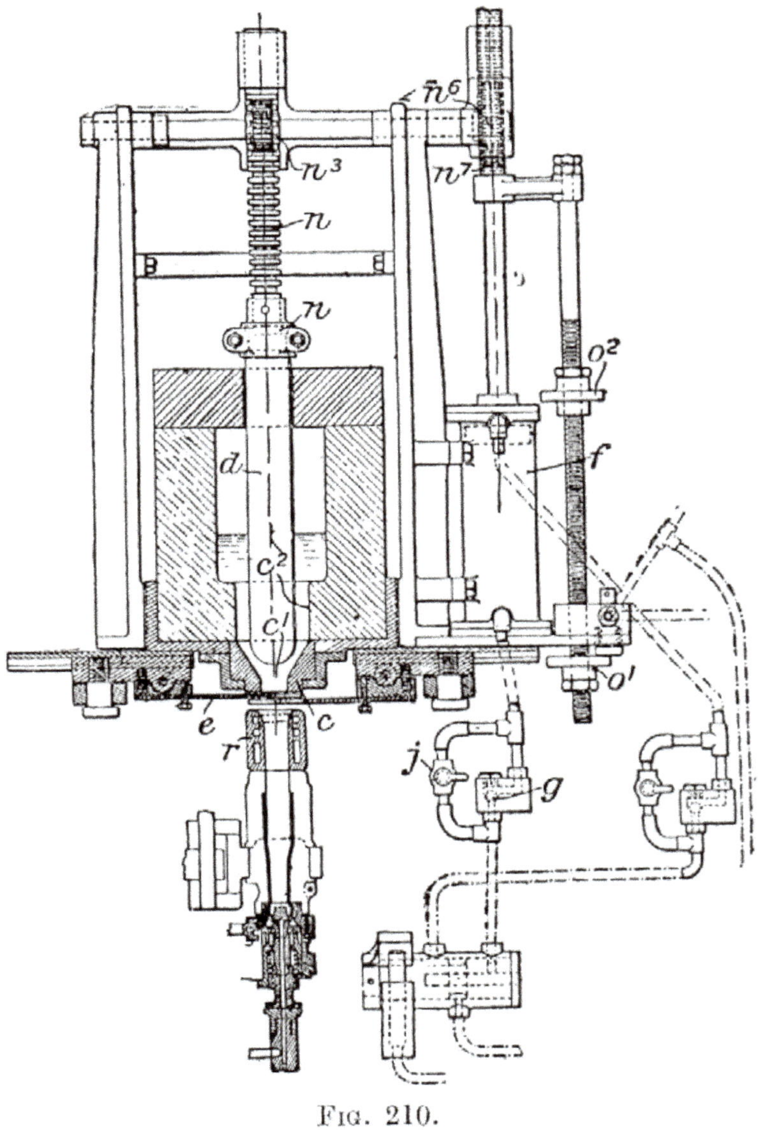

FIG. 210.

Figure 6,113. Glass feeding device

in a removable block, c, the plunger, d, being given a quick down-stroke and a slow up-stroke by a pneumatic cylinder, f, working through rack-and-pinion gear, n, n^3, n^6, n^7. The speed of movement of the plunger on the up- and down-strokes is controlled by valves comprising a ball valve, g, and an adjustable by-pass, j. The extent of movement is limited by adjustable stops, O^1, O^2. Shears, e, are operated by pneumatic gear, and the charge falls into a funnel, r, as described in Specification 166 605. Specification 27 227/05 is referred to.

#286. Glass Flow Controller. F. O'Neill, Toledo, Ohio, USA (USA Pat., No. 1 413 183, April 18th, 1922. Filed October 17th, 1921, No. $508321\frac{1}{2}$). - This patent relates to the securing of a clay portion of a glass feeder plunger to the operating rod.

The operating rod with its rack, 9, is securely fastened to the metal clamp, 10, into which the refractory stirrer, 11, fits. The operating rod and the upper portion of the refractory stirrer are hollow, and through them passes a metal rod, 14, with a head, 18, engaged by a threaded connection with the interior of the lower portion of the clay rod.

Yielding contact between the end of the operating rod, 9, and the inner rod, 14, is provided by a spring, 16, which allows for any differential expansion of the iron and refractory material and avoids any loosening of the plunger.

By holding the refractory material at these two points any tendency to break is reduced, and it is possible to continue using the intermediate portion of the plunger even when the latter breaks.

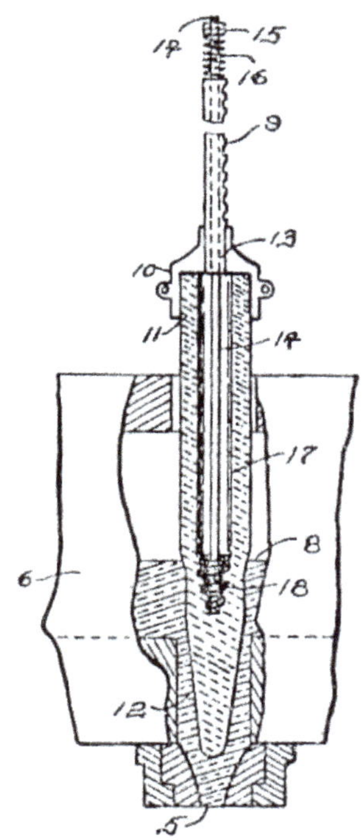

FIG. 165.

Figure 6,114. O'Neill's Patent application for a glass flow controller

Three of W.E.S. Turner's tables from his 1937 Presidential Address regarding the installation of forehearths are shown as Tables 6,16, 6,17 and 6,18;

Machine or process	Author or source	Country	First Patent	First commercial use
Tucker-Reeves feeder		USA	Brit. Pat. No. 105,564 Conv. Date 10/4/1916	
W. J. Miller Feeder		USA	USA. No. 1,260,428, March 26, 1918, (Appin Jan. 1917)	1918
Hartford-Empire Paddle needle feeder	K. Peiler	USA	USA No. 1,199,108 26/9/1916	USA, March 1915
Rankin feeder	C.H. Rankin	England	Brit. Pat. No. 202,496 26/7/1922	England, Aug. 1921
Hartford-Empire Single Feeder	K. Peiler	USA	USA No. 1,655,391	USA, Nov. 1922 3/2/1928

Table 6,16. Introduction of the forehearth and feeder mechanisms

Type	Date	Works
Severin	1914	Queenborough
Pyle	1915	Shaw-Brunner
Hartford-Fairmont (double)	1918	Key
McCauley	1919	Queenborough
Daubenspeck	1919	Canning Town
Miller	1920	Forster
Rankin	1920	Gregg & Co.
O'Neill *	1921	
Hartford (single)	1923	Rockware
Miller (rotating)	1924	
J.T. Wood	1927	Branscombe
Pearson	1928	Jackson

* There was at least one working at Forster's in November 1922 - W. E. S. Turners diary

Table 6,17. Introduction of gravity glass feeders into England

First non-USA sales of Hartford-Empire machinery

	Date	Country
Paddle Feeder	27/8/1917	Canada
	29/4/1918	England
	30/4/1918	Scotland
Paddle Needle Feeder	30/10/1918	Australia
	14/3/1919	Holland
	22/1/1919	Cuba
	2/12/1920	Argentine
	6/12/1920	Brazil
	17/2/1925	Canada
Single Feeder	22/3/1923	Japan
	19/9/1923	Argentine
	8/10/1923	England
	21/12/1923	Australia
	26/8/1926	Canada
	21/11/1927	Chile
	4/10/1929	Brazil
	9/12/35	China
	6/10/1936	USSR
IS Machine	23/3/1928	England
	12/10/1929	Brazil
	26/6/1936	Australia
	11/9/1936	USSR

Table 6,18. The first non-USA sales of Hartford-Empire or Hartford-Fairmont machinery by country

Finally we come to the various methods of gob distribution which all came from the abstracts in the Journal of the Society of Glass Technology, we will cover just four of these earlier ones and they all are for delivery systems that feed more than one machine. These are the first four illustrated journal abstracts; number one is from Journal Number III (1919), the next two are from Journal IV (1920) and the fourth one is from Journal VIII (1924). The first, from the 1919 Journal is;

#355. Apparatus for Feeding Molten Glass. W. A. Lorenz, Hartford Fairmont Co. (USA Pat., April 8th, 1919, No. 1 300 180. Filed August 7th, 1917, No. 184 874). - The object of this invention

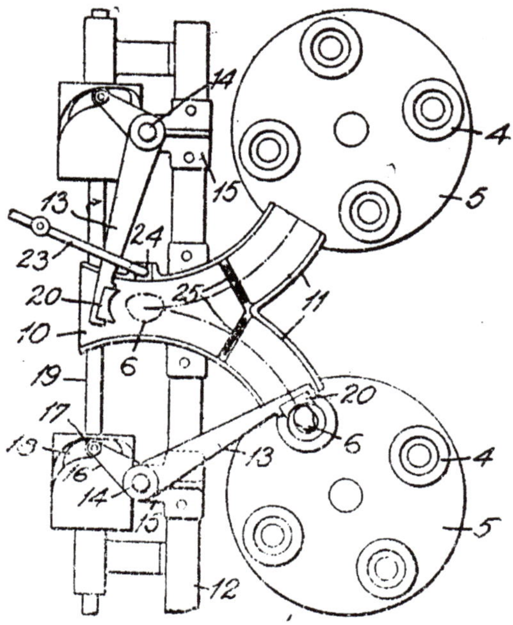

Figure 6,115. Lorenz's glass delivery system

is to provide means for transporting glass from a gathering apparatus to the moulds of various machines in a substantially horizontal plane, thus avoiding the necessity of bringing the moulds directly beneath the glass outlet.

The molten glass on being separated by the shears in the usual manner falls onto a trough, 25, with a suitable refractory lining. The trough has two or three curved branches, 11, according to the number of machines to be fed, each trough ending vertically above the machine moulds.

Horizontal arms, 13, reciprocating on vertical shafts, 14, move alternately along these troughs, and have fitted at their free ends blocks of carbon or other refractory material, 20.

Thus the gobs of glass as they fall are swept along troughs and in to the moulds.

The troughs may have flowing along them small steams of water, which is allowed to escape through small holes in the bottom just before the end of the trough is reached. The pusher arms are operated by rolls, 17, running in cam grooves, 18, in drums rotating on a horizontal axis, 19, and mounted underneath the rocker shafts.

The next two are from the 1920 Journal.

#60. Feeding Glass-making Machines. R. E. McCauley, London. (Brit. Pat. No. 118 585.) (May 31st, 1918, No. 9 030). - Relates to means of conveying molten metal from the melting tank to machines, and consists of an arrangement by which separate machines may be supplied in succession. In the arrangement shown, shoots, c, c^1, mounted on pivots, are actuated by compressed-air cylinders, e, so that they alternately brought under the outlet, d^1, of the melting tank, d.

The shoots respectively deliver the charge to the moulds, a^1, b^1, on the rotary tables, a, b. In another arrangement, one of the shoots is stationary and the other is moved so as to intercept the discharging glass at intervals. The shoots may have a sliding instead of a swinging movement, and may be driven by any suitable mechanism. To prevent overheating, the shoots are formed hollow, and cooling fluid is circulated through them. The metal is severed by the usual shears, j.

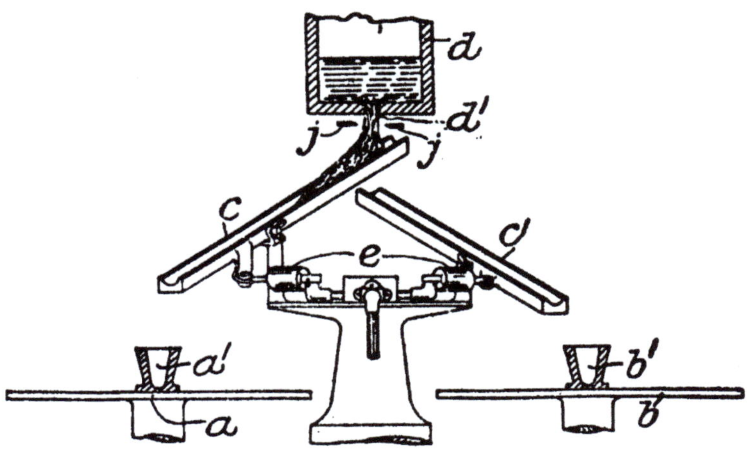

Figure 6,116. Schematic for R. E. McCauleys patent application

#365. Feeding Device for Glass-making Machinery. Hartford-Fairmont Co., New York, USA (Brit. Pat., November 12th, 1919, No. 141 320). - Relates to apparatus for feeding to moulding, etc., machines the charges delivered regularly from a tank, 2.

The charges fall upon a curved distributor, 14, pivoted at 17 to a ring, 10, which is mounted in a bearing, 11, so that it can be turned automatically by a cam mechanism to deliver the charges down a short trough, 15, successively to a number of troughs, 25, (Fig. 238), each leading to a glass-working machine.

When it is desired to stop feeding all of the machines without stopping the delivery from the tank, 2, the distributor, 14, is tilted downwards, as shown in Figure 238, by a handle, 24, at the distributor, or an extension, 63, at the machine. This allows the charges to fall through a ring, 10, into the cullet pit.

If a particular machine is to be put out of action, its trough, 25, is raised, as shown in Fig. 238, so that the charges leave the short trough, 15, and are deflected by a plate, 50, down a shoot, 65, into the cullet pit. The trough, 15, may be arranged to discharge directly to the machines, and in this case the machine

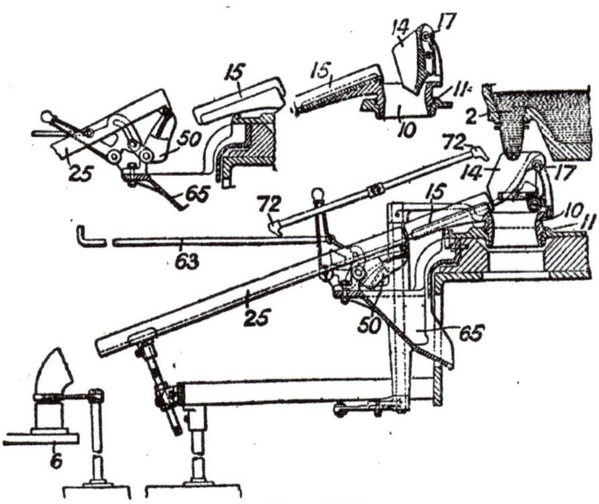

Fig. 238.

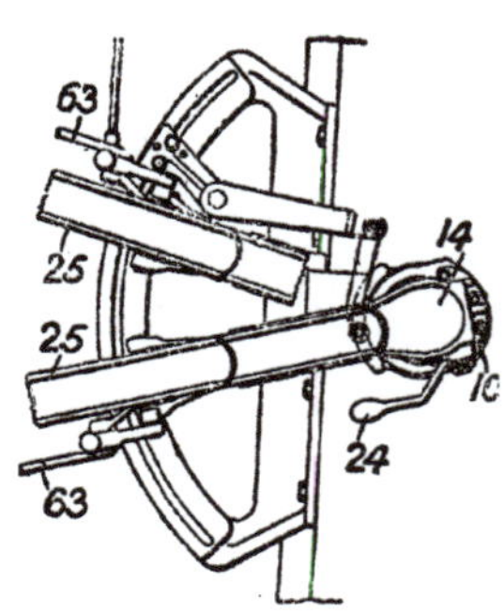

Fig. 239.

Figure 6,117. Schematic of Hartford-Fairmont Co. patent application tables, 6, may rotate continuously. The troughs, 25, etc., are sprayed with liquid from nozzles.

The final abstract is from the 1924 Journal.

#126. An Automatic Feeding Device. C. W. Pleukharp and H. Raynes, San Francisco, USA (Brit. Pat., No. 203 858, July 28th, 1922). - Delivering molten metal - an automatic feeding device has means by which the molten metal can be supplied from a single delivery aperture to any one of a number of stations, and comprises a number of reciprocating troughs each of which receives a charge and delivers it to an inclined shoot, the mechanism permitting the delivery of charges of different weights to the different shoots.

The metal is contained in a forehearth, 11, and is fed through a delivery aperture, 12, beneath which is a shearing device, 13, worked by a piston, 14, in a cylinder, 15. A number of inclined shoots, 18, extend radially outwards beneath the forehearth, and a reciprocating feed trough, 19, co-operates with the upper

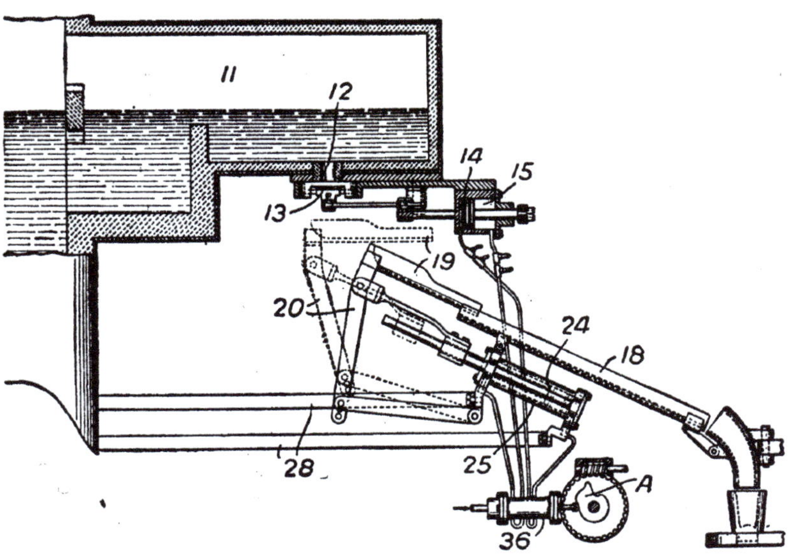

Figure 6,118. Schematic of Pleukharp & Raynes distribution device

end of each shoot. Each trough is fixed on an arm, 20, which is pivoted at its lower end to the framework, 28, and is rocked between the positions shown in full and dotted lines by means of a piston, 24, in a cylinder, 25. The pivot point of the arm, 20, can be adjusted to vary the inclination of the trough to suit different conditions of work.

Compressed air is admitted to the cylinders, 15 and 25, by a valve, 36, operated by a cam, A, the cylinder, 15, being connected to each of the valves, 36, so that the shears are operated in conjunction with each of the reciprocation troughs. The cams, A, are so set on their shafts that the troughs are operated successively, and these cams may be of varying shapes in order that the time during which the trough is beneath the outlet, 12, and therefore the weight of charge it receives maybe varied.

Chapter 7

The rise and fall of the single table or column machine

'his chapter has a peculiar title, it covers presses - single tables ıachines and the single column machines such as the press and blow, low and blow, and suck and suck processes. For bottle making in most f the world the use of presses for making bottles is now non-existent. 'he same applies to press and blow machines as the containers made owadays are usually made on in-line machines due to the requirements f speed of production - but there still a remote chance that in some hird world countries single tables machines are still being used. Blow nd blow machines such as the Lynch 10 or suck and suck machines such s the Roirant R7 or S10 can still be found but again, but world wide it ; only a remote chance.

In 1993 according to the American Glass Industry Directory, a Vorld Wide listing of Glass Factories, entries were factories listed heir bottle making machines there were ten factories that used Lynch 0 machines, one factory using a Lynch 'LA' machine and one factory sing a Lynch 'B' machine. Roirant machine were listed at nine factories or the R7 machine, S10 machines were listed in four factories and one sted a S10 dual machine.

In 1937 when Professor W.E.S. Turner gave his second Presidential peech "Twenty-one Years" he gave some interesting facts applicable o the growth of automatic machines, there is no suitable place where hey can be included in this book so they have been included here;

Year	Number of machines Wide mouth	Narrow mouth	Number of operators
1906	168	----	630
1907	191	----	710
1908	223	----	830
1909	205	19	944
1910	216	----	----
1911	173	52	1004
1912	170	96 †	----
1913	201	96 †	----
1914	210	102 †	----
1915	193	265	1700 ‡
1916	167	292	2000 ‡
1917	Total =	428	2000 ‡

* Glass Bottle Blowers Association Reports, USA. † Three man machines only.
‡ Partially estimated.

Table 7,1. Number of semi-automatic glass bottle-making machines in the USA, 1906–17*

Spread of Mechanisation in the European Glass Industry (Owens machines excluded)

	Gravity fed machines W. J. Miller	Ed. Miller	O'Neill	Lynch	Suction fed machines Monish	Roirant
Great Britain	1917	1919	1917	1919	1929	1932
Italy		1920	1920			1926
Germany		1924		1924	1932	1923
France		1926		1919-20	1931	
Sweden		1929				
Norway		1932		1927		
Russia			1920			1923
Belgium			1919		1932	
Spain			1920			
Switzerland						
Holland					1932	
Roumania					1930	

Table 7,2. Spread of mechanisation in Europe (Owens machines excluded)

Machine or process	Country	First patent	First commercial use
Hartford-Empire Milk Bottle M/c	USA	USA No. 1 316 550 16/9/1919	USA May 1916
Hartford-Empire IS M/c	USA	USA No. 1 654 731 3/1/1928	USA 19/10/1925
O'Neill Bottle M/c No 21	USA	Brit. Pat. No. 109 314 13/9/1916	
O'Neill Bottle M/c No 25	USA	Brit. Pat. No. 144 344 6/2/1919	
O'Neill Bottle M/c No 28	USA		
O'Neill Bottle M/c No 42	USA		
O'Neill Bottle Suction Machine	USA		1928
Lynch No-Boy Bottle M/c	USA		1917
Lynch L.A. Bottle M/c	USA		1921
Lynch R Bottle M/c	USA		1925
Lynch R.A. Bottle M/c	USA		1927
Lynch 10 Bottle M/c	USA		1930
W.J. Miller (many forming machines)			
Ed. Miller (many forming machines of Press, and P & B types			
McNish Auto Suction M/c	England	Brit. Pat. No. 257 641 6/9/1926	1925
Monish Suction M/c	Scotland	Brit. Pat. No. 320 034 28/11/1928	1928
Monish Twin Table suction M/c	Scotland	Brit. Pat. No. 426 275 27/10/1934	1936
Roirant Semi-Auto M/c	France	France No. 556 378 30/12/21	1923
Roirant Single Mld Suction M/c	France	France No. 575 544 30/3/23	1924
Roirant 'F' M/c	France	B.P. Nos. 213 576 & 378,768	
Roirant 6 Mld M/c	France	France No. 727,065 31/1/31	1934
Redfern Bottle M/c	England	England 10/7/1917	1920

Table 7,3. Important developments to the container industry 1916–1937

Now we will go back to this chapter's topic where we will first look briefly at the different machines types that will be discussed here.

Press machines.

The automatic press as against the semi-automatic ones did not come until after the invention of the forehearth, up to this time they would have been fed by hand gathered glass. The definition of a press that we are using here is a machine that has more than two moulds and the table rotates, we are not interested in fixed table presses, which are only used by the hand gathering industry.

One of the first machines to be patented was by Frank O'Neill in 1893 see Figure 7,1 below. Frank O'Neill also invented a feeder in 1897. There will be more about Frank O'Neill's history, as most of his machines were twin table machines the bio-pic will be found in that chapter.

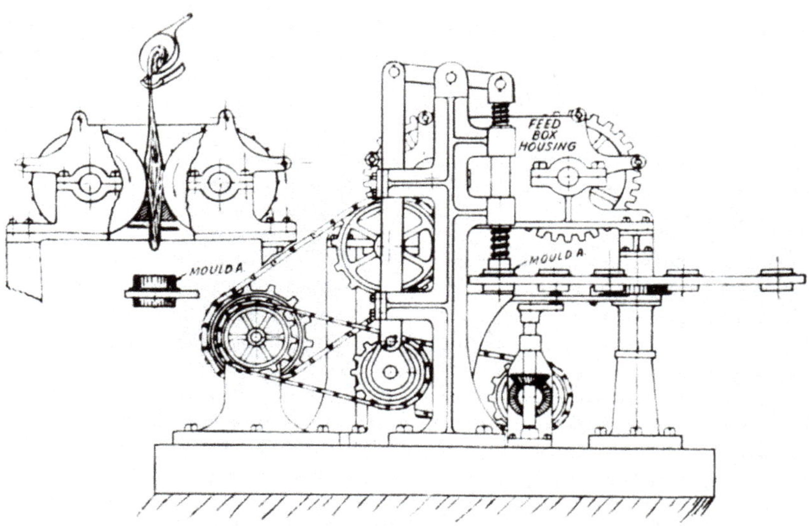

Figure 7,1. The first O'Neill machine

Press and Blow machines.

One of the first to be patented (March 1888) was that by Rylands and Stoner, an 8 mould press, based on Windmill's patent again a single table machine. Many manufacturers made both single and twin table machines, when we are discussing an individual manufacturer, his work will be listed in which ever section he sold most of his machines. Typical instances of this is Edward Miller who made single and twin type machines, both presses and press and blow types, then there is William Mitchell who made presses, blow and blow machines, press and blow machines, in-line machines and suction machines, which were mostly single table types.

Blow and Blow Machines.

The Graham machine and later the Owens Flow machine are examples of a single column machine.

Suck and (blow) and Suck Machines.

Examples of machines in this category are the machines made by Roirant.

The following is a biographical picture of Edward Miller:

Edward Miller, The Miller Machine and Mould Company and The Lynch Corporation.

At the end of the 19th century and the beginning of the 20th century Edward Miller was working at the Atlas Glass Company. He had served a full apprenticeship in an engineering shop and he was employed there as the Machine Shop Supervisor and worked on moulds. One of his first jobs was to convert a hand machine into a semi-automatic one, which he did by applying compress air to rotate the original fixed table into one that was now movable. As early as 1900 he designed and put on the market his first wide mouth jar machine, it was of the telescopic type and it obviously had been modelled on the Charles Blue machine. The first Blue machine that was ever installed was installed in 1896 at the Atlas Glass Company's factory.

Miller set up his company the Miller Machine and Mould Company in 1902 based at 705-719 Ann Street, in Columbus, Ohio. Confirmation of this date is the brochures used for the next two machines which are the original Miller ones and on the back both give the start of the company as 1902. In 1933 he sold his company to the Lynch Corporation and then retired.

Probably one of the first series of machines that Edward Miller made was his 'M' series. Two of which are the air driven **'M' machine** a single table press and blow machine using the 'lift over' or 'transfer' type and the motor driven **'M-T' machine** a twin table machine that could be utilised as a press by using one table. The 'M-T' machine used 10 moulds and blanks while the 'M' machine used 8.

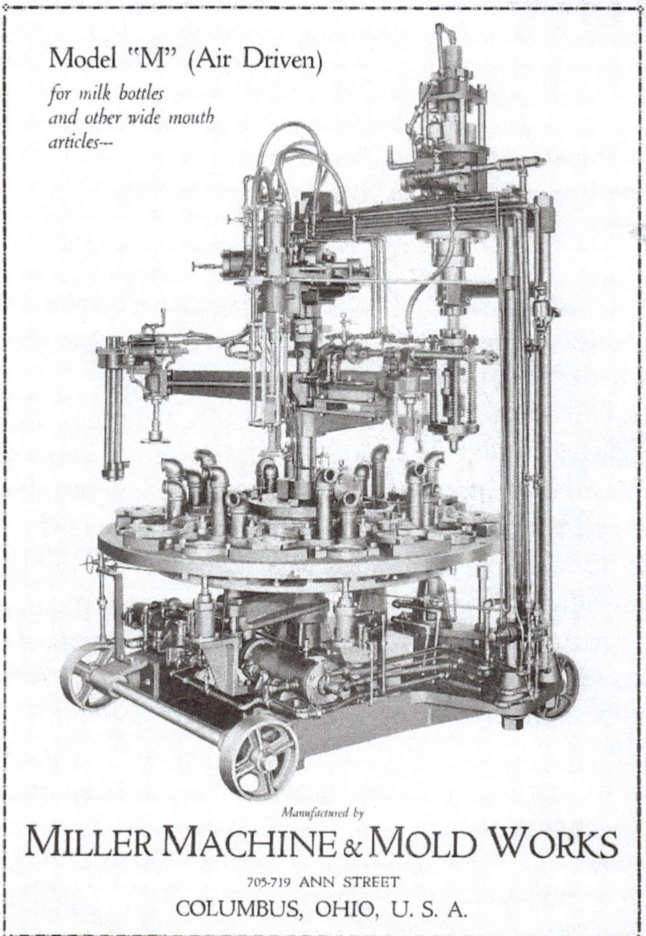

Figure 7,2. The air driven Type 'M' press and blow machine

The types of ware that these machines were designed for was;

For the 'M' machine designed for the manufacture of milk bottles, globes, shades, packer jars, etc.

For the 'M-T' machine designed for either the manufacture of pressed ware such as tumblers, jellies, nappies, beer mugs, sherberts, lemon reamers, salad plates, etc. or for press and blown ware such as milk bottles, globes, shades, etc.

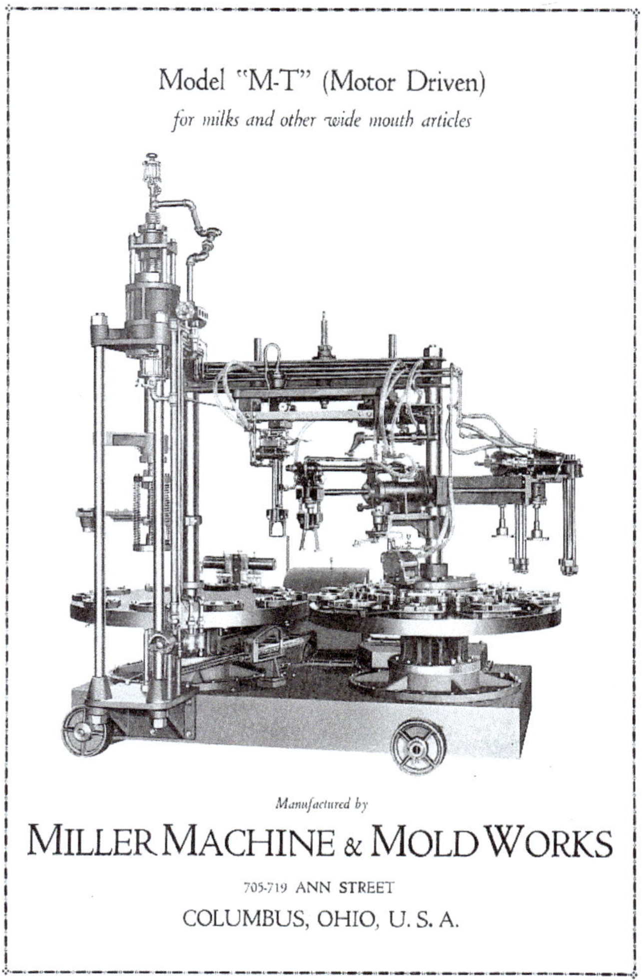

Figure 7,3. Millers 'M-T' type press and blow machine

The specifications for the two machines are shown in Table 7,4.

		Specification Type 'M' machine	Type 'M-T' machine
Capacity		4 to 64 ounces	4 to 64 ounces
Max. dia. of ware	- Mould cages	$4\frac{1}{2}$ inches	$4\frac{1}{2}$ inches
	- Hinged moulds	7 inches	7 inches
Height		10 inches max.	10 inches max.
Mould Centres	- Blank to blank	$19\frac{7}{8}$ inches	13.597 inches
	- Blank to mould	$10\frac{1}{8}$ inches	
Radius ctr. columb to ctr. of mould		26 inches	22 inches
Table diameter		64 inches	52 inches
Speed, depends on article		up to 25 p.p.m.	up to 30 p.p.m.
Air requirement 30 psi		Free air 60 c.f.m.	Free air 60 c.f.m.
Power requirement			3 HP
Floor space	Length	6.5 feet	8.0 feet
	Width	5.5 feet	5.0 feet
	Height	8.5 feet	9.0 feet
Mould equipment.			
Blow mould with bottoms		10	12
Blanks with valves		16	20
Plungers		2	2
Plunger rings		2	2
Or			
Press moulds			12
Plungers			2
Plunger rings			2

Table 7,4. Specification of the Type 'M' and 'M-T' press and blow machines

The later versions were equipped with two moveable blow heads. Obviously for machines which are produced over many years they are improved upon so that they change from the original model the next illustration shows what say 30 years of improvement has taken place.

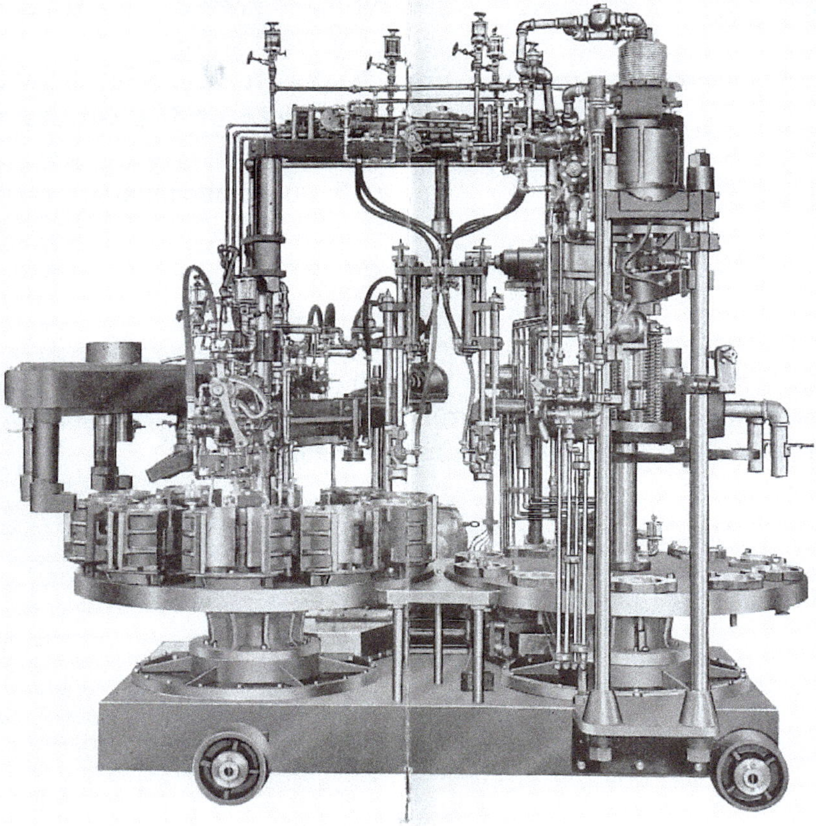

Figure 7,4. The 1946 version of the 'M-T' machine

It goes without saying that the machine specification has also improved;

Ware range -	Milk bottles, gas and lantern globes, shades, vases, fish bowls, thin blown (hot mould) tumblers, etc.
Height of Article -	Up to 12 inches
Maximum finish -	Mason, or G-finish ($2\frac{5}{8}$inches)
Maximum diameter -	5 inches

Compress air requirements -	200 c.f.m. of free air at 40 to 50 psi at the inlet valve
Cooling wind requirements -	From 10 000 to 12 000 c.f.m. with static pressure at the outlets of 10 to 12 inches
Speed. -	Up to 35 pieces per minute

The speed depends on the article being made.
For example - Milks
Half gallon milk bottles 16 to 18 per minute
Quart milk bottles 22 to 26 per minute.
Pint milks 28 to 32 per minute.
Smaller sizes 32 to as high as 35 per minute.

Milk Bottles of the Common Sense Finish . . .

Half Gallon, Quart, Pint, ⅓ Pint and ¼ Pint

Figure 7,5. Range of milks that the M-T machine will make

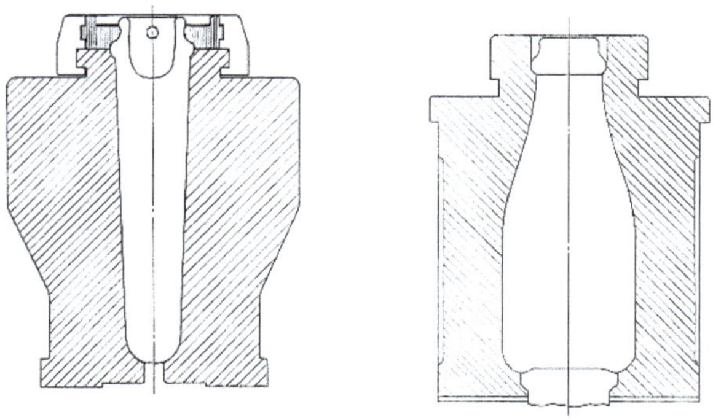

Blank-Mold Construction *Blow-Mold Construction*

Figure 7,6. Mould equipment when making the Crown finish

Mould equipment for the 'Dacro' or 'Crown' finish.

Millers original naming for some of his machines was the 'Universal', the '2-A Universal', and the 'Telescopic' jar machines. The 'Telescopic' was built in 1913, while the 'Universal' and the '2-A Universal' were started in 1918.

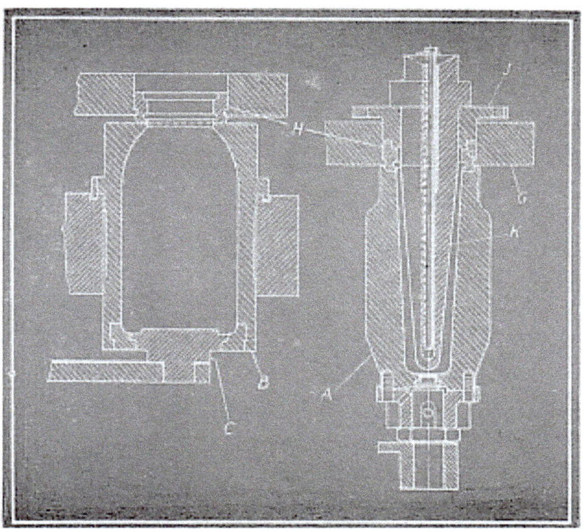

Figure 7,7. Millers interpretation of the Blue principle

A - the blank mould
B - the blow mould
C - the bottom plate of the blow mould
D - the blow mould cage
E - the blank mould adapter
F - the bottom plate slide
G - the neck ring holder
H - the neck, or threaded ring
J - the plunger ring
K - the plunger

Ed Miller used the knowledge that he had gained at the Atlas Glass Company regarding glass and its properties and his knowledge of his existing machines which allowed the Miller Machine & Mould Company in 1925 to design its next series of machines the 'J- P.' series which used the 'Blue' principle of the telescopic blank. The machine was a 'press and blow' machine using ten blanks and ten blow moulds.

Quoting from a Lynch catalogue - The Miller J-P machine is suitable for producing all types of glass containers of the wide and semi-wide mouth varieties, from 4 ounce to 64 ounce maximum capacity.

The MILLER Model "J-P" Machine is designed for wide mouth and semi-wide mouth glass containers from 4 ounce to 64 ounce capacity inclusive...

Code Name "UKER"

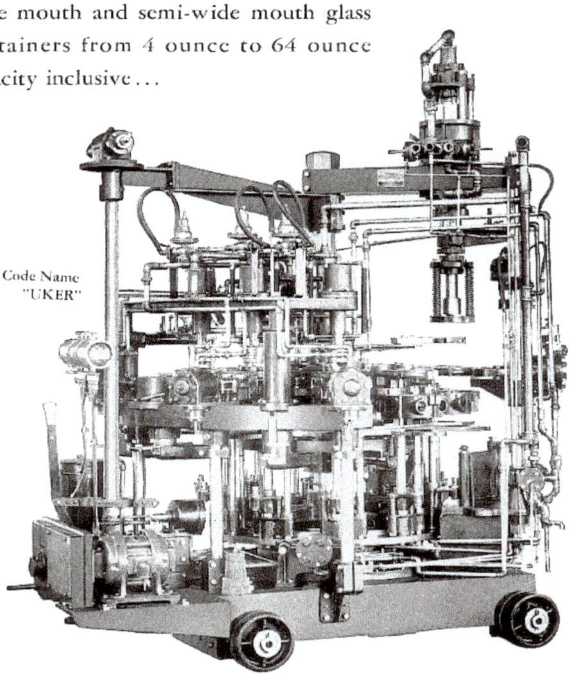

The "J-P" carries ten blank-molds and ten blow-molds, and one plunger. The article is blanked out roughly in the blank-mold by the plunger, but in the blow-mold the article acquires its final shape with the aid of compressed air.

The machine-made glass container is more uniform in weight, capacity and finish than the hand-made article. And it is ideal for screw-on lids and other metal closures.

Figure 7,8. The Miller 'J.P.' machine

The operation of the Miller machine is as follows;
"When the neck ring and the blank mould are in position and the parison is pressed, the blank mould descends on a cam slide to clear the end of the parision which is held by the neck-ring holder. The blow mould bottom moves into position, the blow mould closes around the parison while it is held by, and suspended from, the neck-ring holder. The blowing takes place at the following station.

The blank mould is mounted on the same vertical axis as the blow mould, the latter being split to permit blank mould to move upwards within the blow mould to interlock with the neck ring holder.

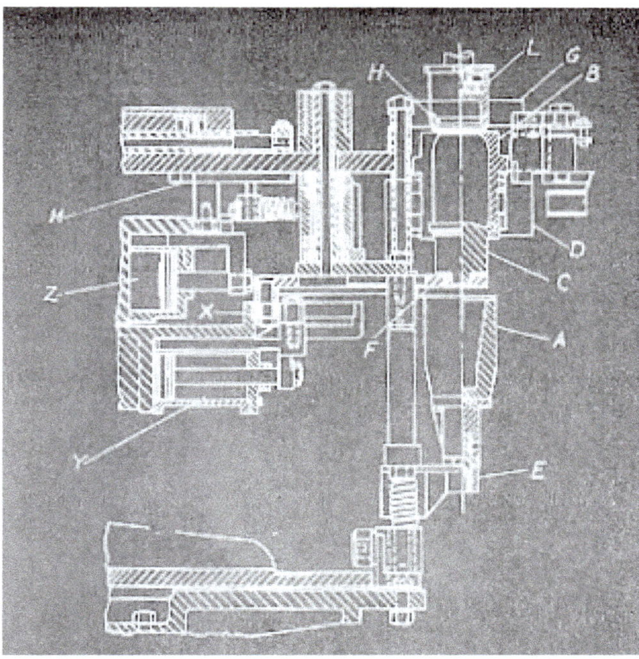

Figure 7,9. Schematic of the mechanism

The machine specifications are;

Class of Ware -	Wide-mouth and semi-wide mouth glass containers.
Capacity -	From 4 ounce minimum to 64 ounce maximum capacity.

Finishes -	Suitable for all kinds, regardless of style or shape.
Diameter of Finish -	From $1\frac{1}{4}$ inches to $4\frac{3}{4}$inches maximum.
Maximum Height of Article -	$9\frac{1}{2}$inches overall, $8\frac{5}{8}$ under the finish ring holder.
Finish Ring Holder Sizes -	From 3 inch minimum to 6 inch maximum.
Speed -	Up to 42 pieces per minute. Depends upon weight and size of article.

The machine needs an automatic feeding device, the machine is not recommended for hand gathering. Figure 7,10 shows a speed curve for one of the earlier J-P machine while Figure 7,11 show the same for a later version around about 1946

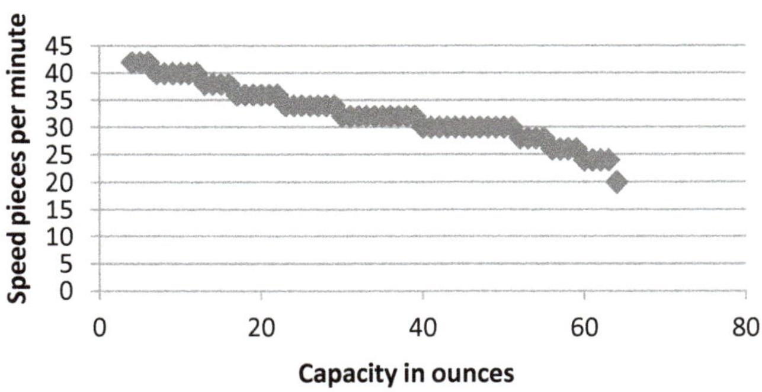

Figure 7,10. Speed curve for an early Miller JP machine

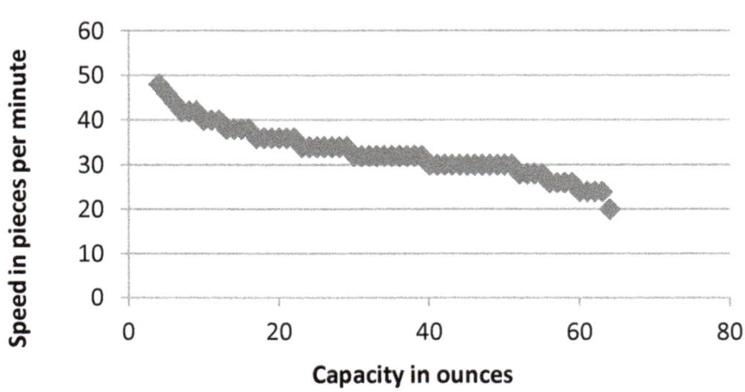

Figure 7,11. Speed curve for a later version of Miller's machine made by Lynch after the buy out

The machine incorporates three blow heads that can be worked individually or in a combination.

Figure 7,12 shows a selection of jars made on the J-P press and blow machine. The machine was no doubt a big improvement in productivity terms but today these types of container are made on the IS machine.

The improved speed curve shown above would only make the smallest - bottom jar on the right hand side at just under 50 pieces per minute. Whereas a modern machine would make them at, at least 200 per minute, the ultimate speed would be decided by the size of the order - a large one would be made on at least a 8 section machine working either dual or triple moulds.

Other J-P machines that were designed and marketed were the J-P-M machine, and the J-P-S machine. Both of these machines were single table press and blow machines the J-P-S had ten blank and ten blow moulds while the J-P-M machine used twelve of each. There was a fourth machine the J-P-L but other than its name I do not have any further information.

The machines were based on the Blue method, i.e. the Windmill

Figure 7,12 shows a selection of jars made on the J-P press and blow machine

method, and the operation of these machines is identical to that of the original J-P machine. The machines were designed to manufacture wide-mouth and semi wide-mouth jars which would give a complete range of products when using two machines. The J-P-S weight range was from a minimum capacity of half ounce up to a maximum capacity of 32 ounces, typical products were; ink bottles, mayonnaise jars, ointment jars, fruit jars, packer's jars, coffee jars, etc. While the J-P-M machine covered the range of 4 ounces minimum capacity to 105 ounces maximum capacity. Typical products can be seen in Figure 7,13.

Figure 7,13. Typical products that have been manufactured on the Miller J-P-M machine. All of the above can be classed as containers

The machine specifications for both of the machines can be seen in Tables 7,5 and 7,6.

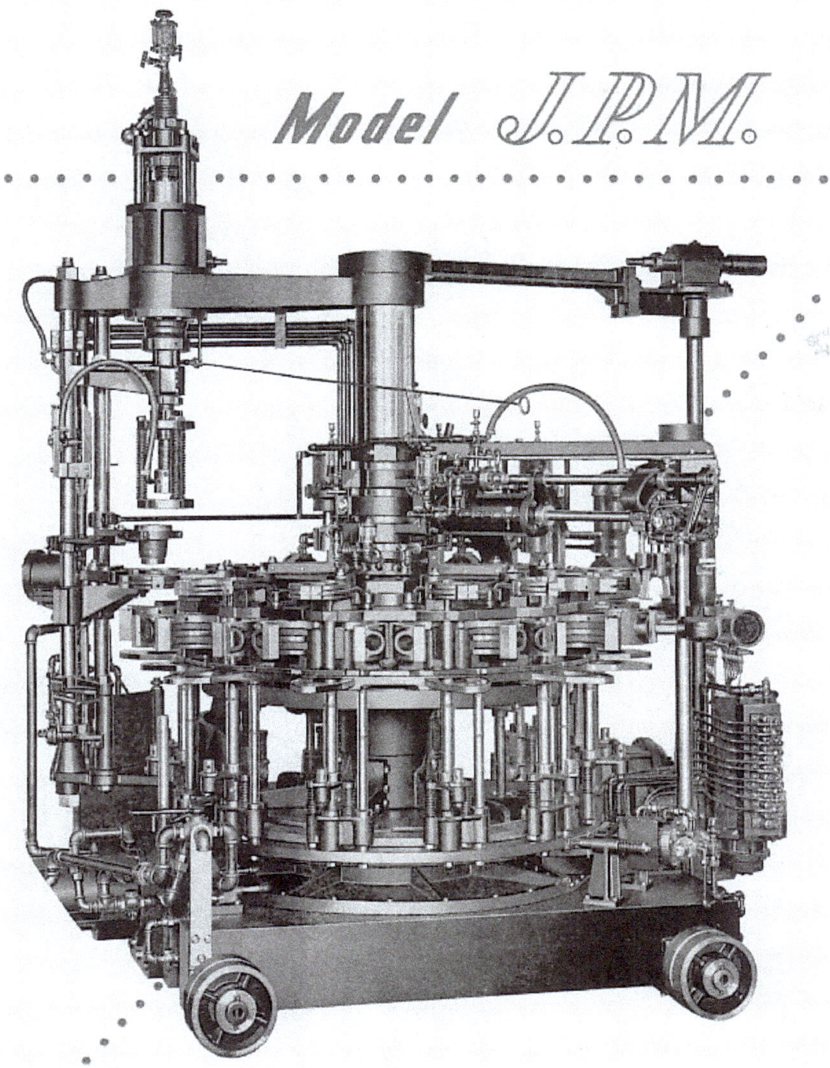

Figure 7,14. The Miller J-P-M machine. The machine has three blow heads which can be worked individually or in any combination.

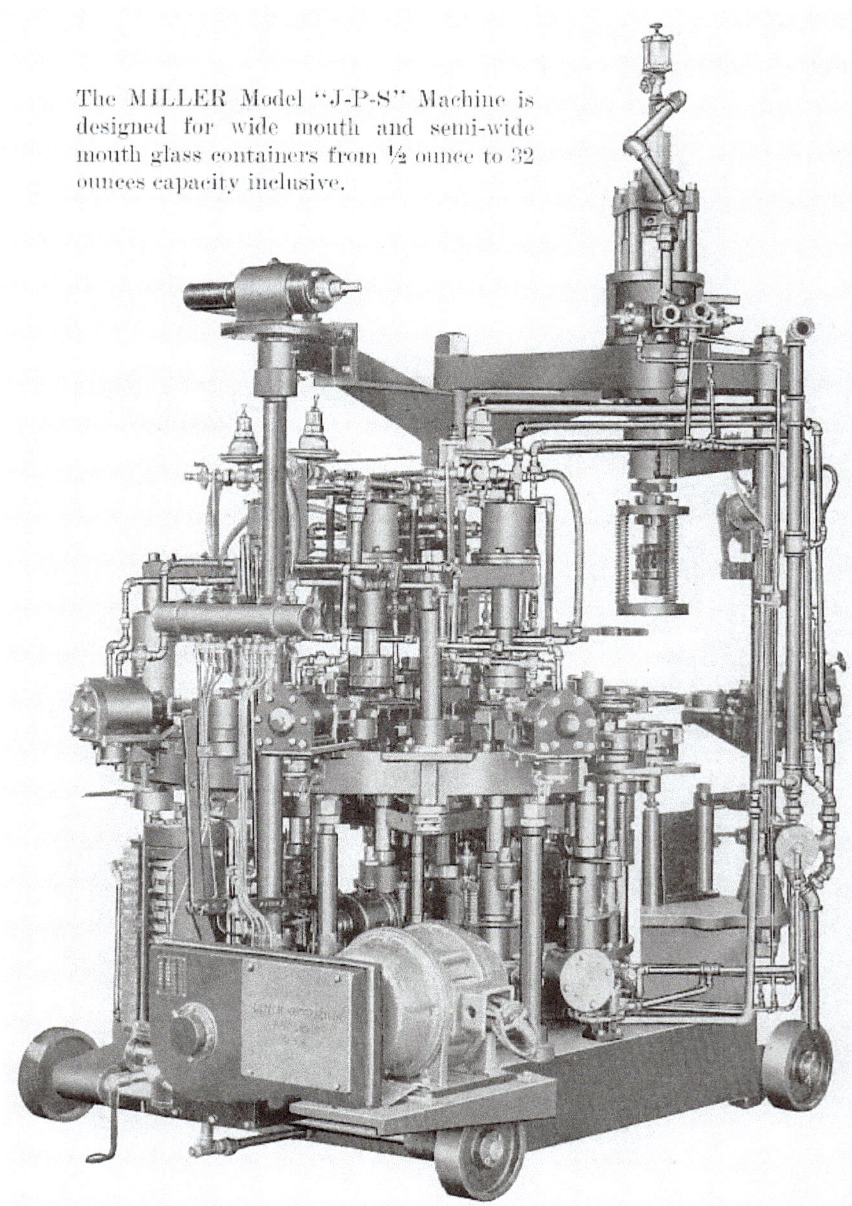

Figure 7,15. The Miller J-P-S machine. The machine has three blow heads which can be worked individually or as a combination.

Specification

	Type J-P-M machine	Type J-P-S machine
Class of ware	Wide and semi wide mouth glass containers by press and blow methods	Wide and semi wide mouth glass containers by press and blow methods
Capacity of articles	From 4 to 105 ounces inclusive	From $\frac{1}{2}$ ounce minimum to 16 ounce capacity with regular $4\frac{1}{2}$ mould cages. $5\frac{1}{2}$ inch mould cages required up to 32 ounce maximum
Finishes	Suitable for all types	Suitable for all types
Diameter of finish	From $1\frac{1}{4}$ inch to $4\frac{3}{4}$ inches maximum	From 1 inch to $3\frac{1}{2}$ inches maximum
Height under finish		Minimum of 1 inch to maximum of $5\frac{5}{8}$ inches
Overall height		$6\frac{1}{4}$ inches includes $\frac{7}{8}$ inch for the finish ring
Max. Diameter of article	Up to $4\frac{3}{4}$ inches with $6\frac{1}{8}$ inch bore mould cages Maximum $5\frac{3}{4}$ with $7\frac{1}{2}$ bore mould cages	Up to $3\frac{1}{2}$ inches in standard or $4\frac{1}{2}$ when using special mould cages
Finish-ring holder sizes	From 3 inch to 6 inch maximum bore	$2\frac{1}{2}$ inches minimum bore to $4\frac{1}{2}$ inch maximum bore
Blow mould holders	Standard $6\frac{1}{8}$ bore. Specials required for above 64 ounces $7\frac{1}{2}$ bore	$4\frac{1}{2}$ bore standard to $5\frac{1}{2}$ inch specials
Speed	Up to 58 per minute, see curve	Up to 55 per minute, see curve

Table 7,5. Machine and ware specifications

Minimum Specification for machine services

	Type J-P-M machine	Type J-P-S machine
Machine services		
Motor requirements	5 HP 1800 rpm	3 HP 1800 rpm
Compressed air	300 c.f.m. of free air at the machine inlet valve	300 c.f.m. of free air at 40 pounds gauge pressure at the machine inlet valve
Cooling wind requirements		6000 to 8000 c.f.m. at a static pressure of 10 to 12 inches at the outlets
Water pressure	Minimum of 25 p.s.i. of treated water	Minimum of 25 psi of treated water
Mould equipment		
Blow moulds with bottoms	12	10
Blank moulds with adapters and valves	12	10
Pairs of finished or threaded rings	12	10
Plunger	1	1
Plunger (or follower) ring	1	1
Guide ring or funnel	1	
Pair of take-out jaws	1	1

Table 7,6. Minimum services that are required by the machines

The following two figures indicate what type of speed that the machine should run at versus the containers capacity. Please note that its not the containers weight.

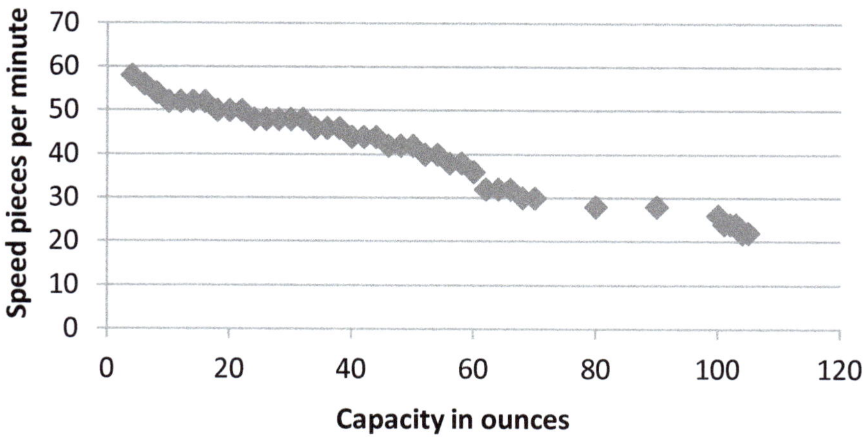

Figure 7,16. Speed curve for the Miller J-P-M machine

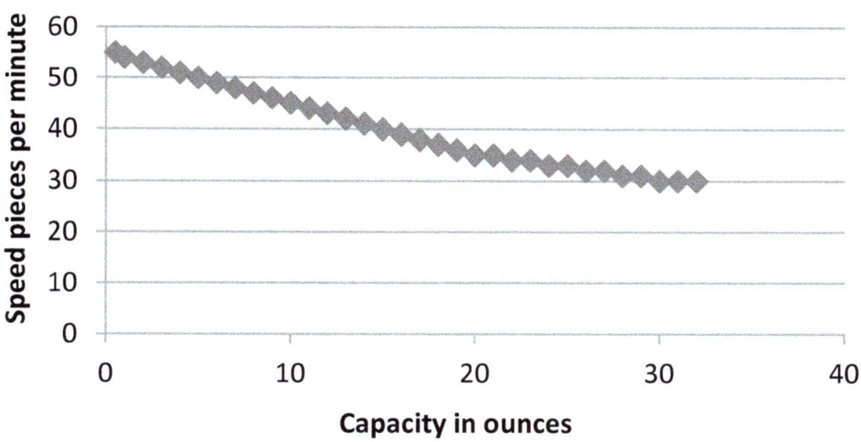

Figure 7,17. Speed curve for the Miller J-P-S machine

Jars and bottles from 0.5 to 16 ounce capacity are made with the regular $4\frac{1}{2}$ inch bore mould cage. From 16 to 32 ounce capacity they are made in the special $5\frac{1}{2}$ inch bore cage.

The Miller Model **'T-T'** makes all of the conventional ware that the press and blow machine makes but at the upper end of the range, from 8 ounces to 256 ounces.

Example of a barrel shaped jar; suitable for pigs feet, pickles, relishes, olives, jellies, etc.

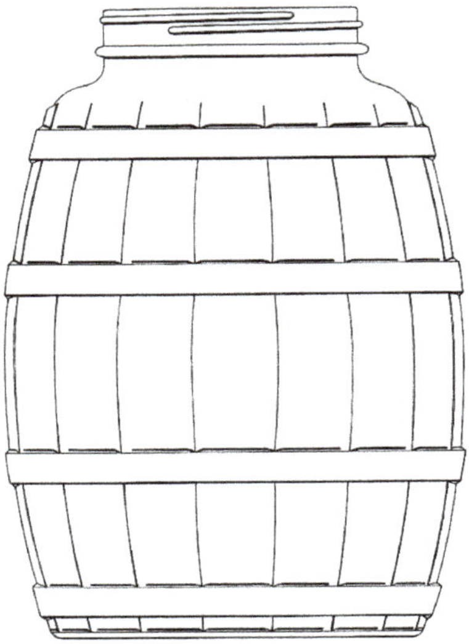

The barrel-shaped jar shown here is a product of the "T-T" Machine. Suitable for pigs feet, pickles, relish, olives, jelly, etc.

Figure 7,18 Barrel shaped pickle jar

Example of a wide mouth jar suitable for crackers, cookies, rolls, etc.

This jar was also made on the "T-T" Machine . . . suitable for crackers, cookies, rolls, etc.

Figure 7,19. Wide mouth cookie jar

The next two figures numbered 7,20 and 7,21 show the T-T machine in both the side view and the end view.

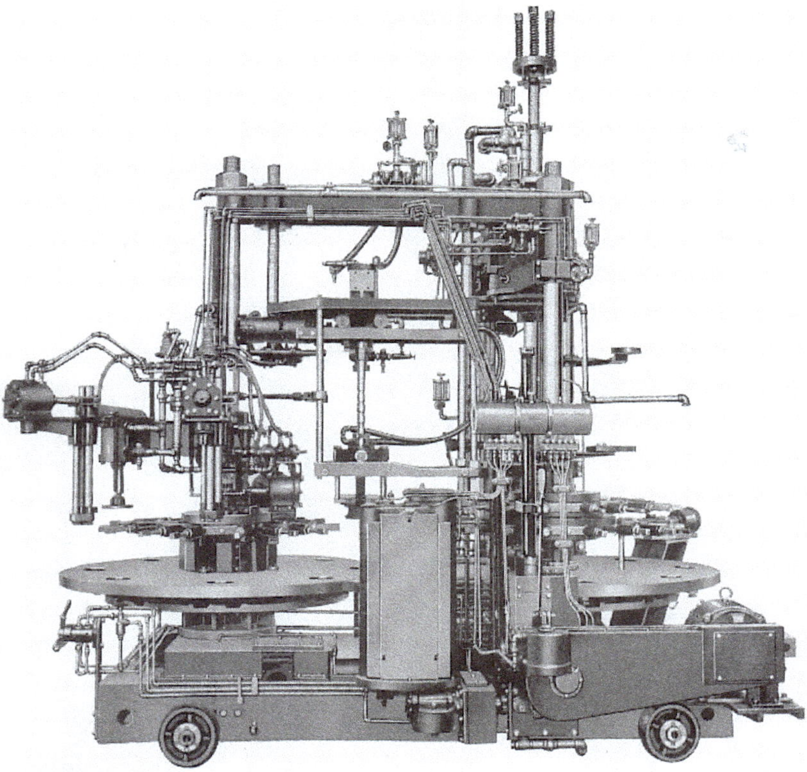

Side elevation of the "T-T" Machine showing the tables, timer, mold opening and closing device with linkage, blow-heads, transfer, etc.

Figure 7,20. Side view of the Miller T-T machine

The machines specifications are;
Capacity range - 8 to 256 ounces,
Diameter of press and blow articles - 9 inches maximum
Maximum height under finish - 9 inches
Finishes made all types regardless of style or shape.
Diameter of tables - 64 inches
Mould radius - 26 inches
Mould centres - 19.899 inches
Speed up to 25 pieces per minute depending on weight and size of article being made.
Motor requirements – 5 HP, 1200 rpm

Compressed air - 10 000 to 12 00 c.f.m. at 10 to 12 inches static pressure at outlets.

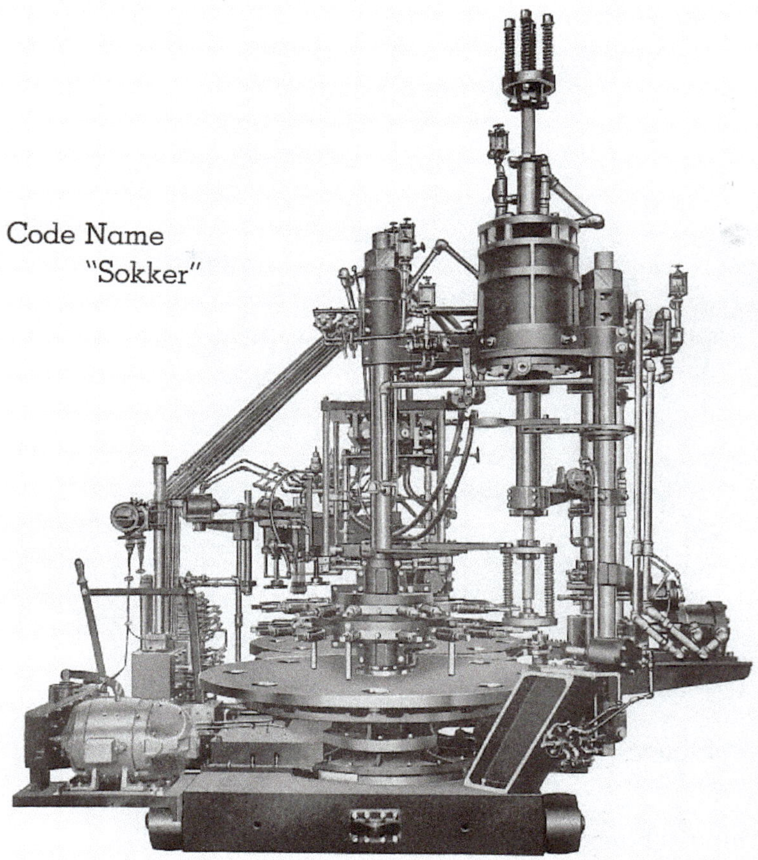

End view of the "T-T" Machine showing pressing cylinder, spring plate, grip at charge station, grip at the press station, motor, etc.

Figure 7,21. End view of the Miller T-T machine

The machine also will work as a press using just one table, according to the Lynch literature it can make pressed tumblers, jellies, nappies, etc. in both block and open-and-shut moulds. Ware specifications are that it will make up to a maximum diameter of 12 inches and to a maximum height of 8 inches.

Minimum mould equipment required is;

	Press and Blow	Pressed ware
Blank moulds complete	8	-
Neck ring holders	8	-
Plunger	1	1
Plunger ring	1	1
Blow moulds complete	8	-
Mould bottoms	8	-
Pair of Transfer jaws	1	-
Press moulds complete	-	8

Table 7,7. Minimum mould equipment for the Miller T-T machine

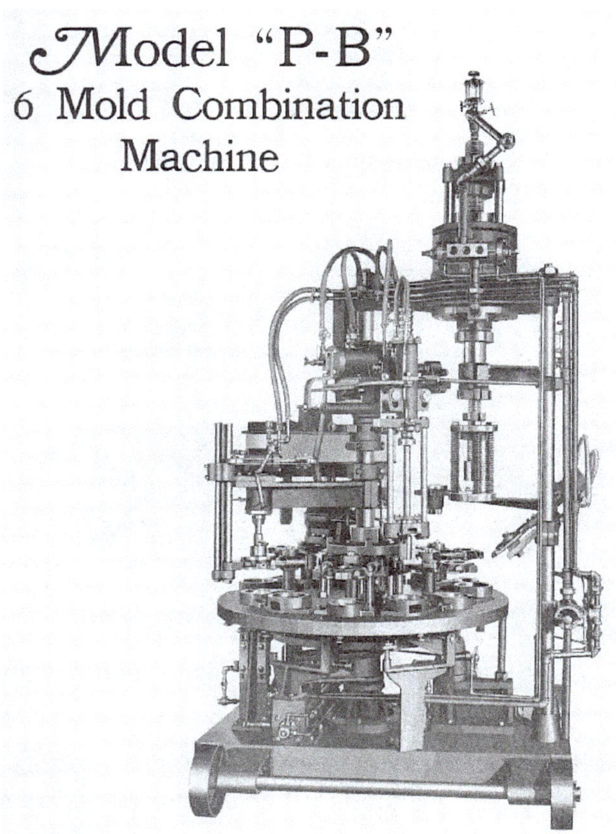

Figure 7,22. The Lynch P-B combination machine

In the 1940s the Lynch Corporation were marketing two combination machines models type P-B and P-B-L both had six moulds, the P-B-L was the larger of the two its maximum ware capacity was 128 ounces against the P-B's 64 ounce.

Medium Size Combination Machine

Figure 7,23. The Lynch P-B-L Medium Size Combination Machine

Both of the machines could work either the press and blow method or the press method. When in the pressing mode the machine could be used as 12 station machines. Figures 7,22 and 7,23 show the two machines.

The specifications for the machines and their services and mould equipment are as follows in Table 7,8.

The operating speeds of the machines depends upon the style of the glassware, the conditions of the glass, mould and plunger radiation capacity, cooling wind efficiency, etc. Typical speeds are given in Table 7,9.

	Specifications Model P-B	Model P-B-L
Class of glassware	Press and blow of the wide mouth variety; also pressed glassware in block, or open and shut moulds	Press and blow of the wide mouth variety; also pressed glassware in block, or open and shut moulds
Capacity of P & B articles	From 1 ounce to 64 ounces inclusive	From 6 ounces to 128 ounces inclusive
Max. height P & B	Up to 9 inches	Up to $11\frac{1}{2}$ inches
Max. diameter P & B	Up to 5 inches	Up to $9\frac{1}{2}$ inches
Max. diameter pressware	Up to 5 inches with 12 moulds, or 9 inches when using 6 moulds	Up to 10 inches with 12 moulds, or 15 inches with 6 moulds
Max. height pressware	Up to and including 4 inches	Up to and including 8 inches
Speed	Up to 40 pieces per min. See amendment	Up to 35 pieces per min. See amendment
Mould centres	$9\frac{3}{8}$ inches	13.4375 inches
Mould radius	18 inches	26 inches
Table diameter	48 inches	64 inches
Floor space requirements	$4\frac{1}{2}$ feet wide x 5 feet long x $8\frac{1}{2}$ high	5 feet wide x 6 feet long x $8\frac{1}{2}$ feet high

Table 7,8. Specifications for the P-B and the P-B-L machines

The service requirements are shown in Table 7,10, as with all of the other machines the mould equipment numbers are for the minimum requirements, i.e. one of each for each position and one for common pieces.

	Indication of Speed Model P-B	Model P-B-L
Press and Blow		
Thin blown tumblers - hot moulds	24 to 28 pieces per minute	22 to 26 pieces per minute
Quart milk bottles	9 to 12 pieces per minute	10 to 12 pieces per minute for Qt. and Half gallons
Gas and lantern globes	14 to 18 pieces per minute	14 to 18 pieces per minute
Pressed tumblers, jellies, nappies, etc.	Approximately 18 to 20 pieces per minute	Approximately 16 to 24 pieces per minute
Press glassware with 12 moulds	Can be made at between 30 and 40 pieces per minute	Can be made at between 30 and 35 pieces per minute

Table 7,9. Speed indication for selected articles

	Services and minimum Mould requirements Model P-B	Model P-B-L
Air pressure requirement	200 c.f. of air per minute at 40 pound gauge pressure	160 c.f. of air per minute at 40 to 45 psi
Cooling wind requirement	3 to 5 ounces is ample or 5 inches	From 3 to 5 ounces is ample
Min. Mould equipment		
Press & Blow		
Blow moulds with valves	6	6
Blank moulds with valves	6	6
Plunger	1	1
Plunger ring	1	1
Pair transfer jaws	1	1
Funnel	1	1
Pressed glassware		
Press moulds		
(Open and shut type)	6	6
(Block type)	6 or 12	6 or 12
Plunger	1	1
Plunger ring	1	1
Funnel or guide ring	1	1

Table 7,10. Machine and mould service requirements

Finally the last of the press and blow machines, although strictly it does not fit in with our requirements, the Miller H-M-T machine was used for making lightweight tumblers and not bottles, it used solid moulds the open and shut of moulds would not fit onto the machine.

The machine worked remarkably like a 'J-P' type using the dropping down of the blank mould. Due to it not meeting or requirements the actual statistic are a minimum. The machine uses 10 moulds and blanks. Capacity of articles from 1 ounce to 16 ounce maximum capacity. Maximum diameter of article $3\frac{1}{2}$ inches, maximum height of finished article 6 inches. Speed up to 40 pieces per minute.

Figure 7,24. The Miller Model 'H-M-T' machine

As stated earlier Edward Miller sold his business to the Lynch Corporation in 1933 who have manufactured both his machines and their own since, although this biopic is about Miller for completeness we are going to include some of the other Lynch presses. These are all pressed ware machines besides the name we will include a photograph of the machine along with the sizes of ware that they can make.

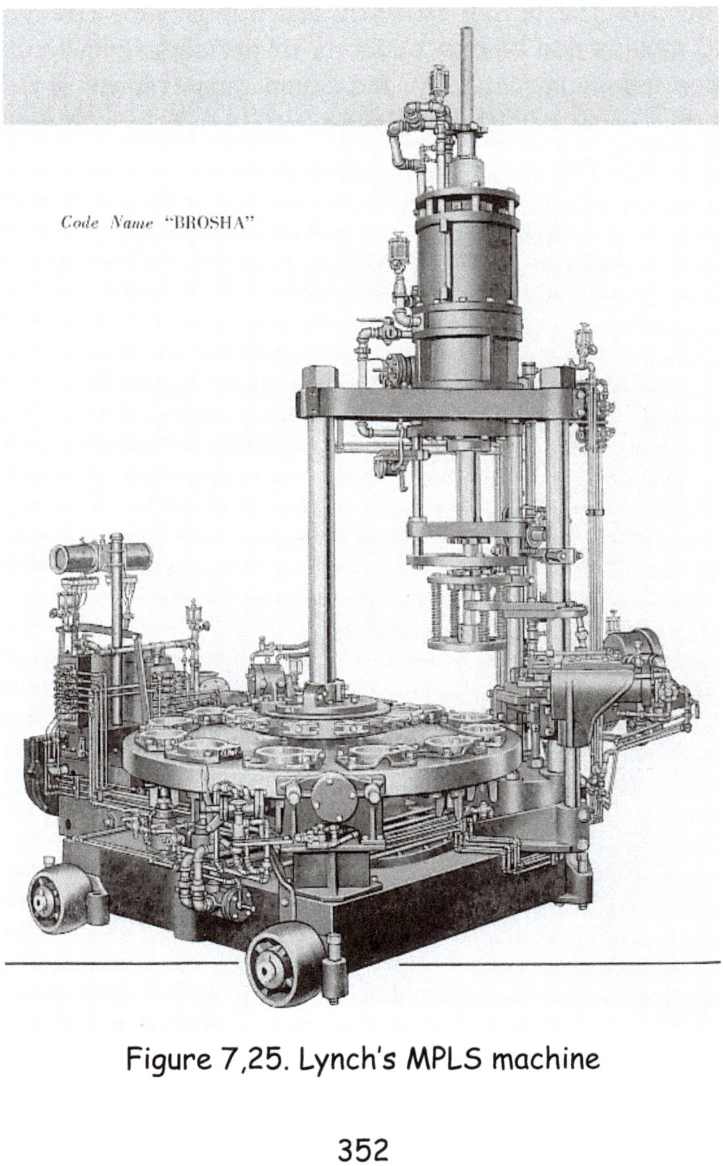

Figure 7,25. Lynch's MPLS machine

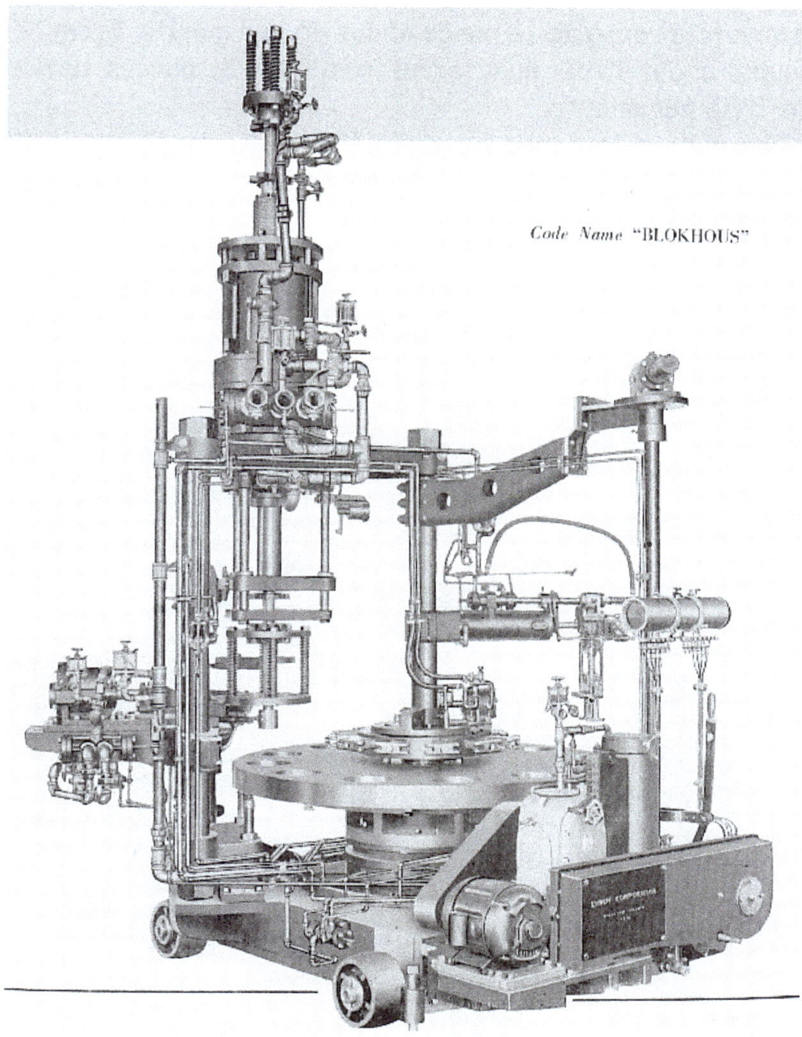

Figure 7,26. Lynch's MPLC machine

For both machines MPLS and MPLC; 12 pressing moulds with 12 mould baskets. Ware range up to 12 inches maximum diameter, height flat to 8 inches. Capacity from 1 ounce minimum to 30 ounces maximum inclusive. Table diameter 64 inches. The maximum speed is 50 pieces per minute. Indication of the machine speed throughout the range is; 1 to 6 ounce tumblers about 50 per minute. From 6 to 12 ounce about 45 per minute. From 12 to 20 ounce about 40 per minute and from 20 to 30 ounces, between 30 and 35 per minute.

The main difference between these two machines is that the table height of the MPLC is approximately 3¾ inches higher than the MPLS.

Figure 7,27. Lynch's MPS machine

A 12 mould press. Capacity of article from 1 up to and including 12 ounce. Maximum diameter of article up to 10 inch, maximum height from flat to 6 inches. Table diameter 56 inches. Speed up to 55 pieces per minute. From 1 ounce to 6 ounce tumblers, about 55 pieces per minute. From 6 ounce to 9 ounce tumblers approximately 50 pieces per minute. From 9 ounce to 12 ounce tumblers, from 40 to as high as 45 pieces per minute.

MODEL "ADP" PRESSING MACHINE

SUITABLE FOR GLASS BAKING DISHES, NAPPIES, MIXING BOWLS, GLASS BUILDING BLOCKS, AND OTHER LARGE HEAVY PIECES.

Figure 7,28. The Lynch ADP machine

A 6 or 12 mould press. Diameter from 4 to 16 inches when using 12 moulds, and an article up to 30 when using 6 moulds. When using open and shut mould maximum diameter up to 10 inches when using 12 moulds and up to 20 inches with six moulds. Height, from flat up to 10 inches. Table diameter 89 inches. Speed up to 25 pieces per minute.

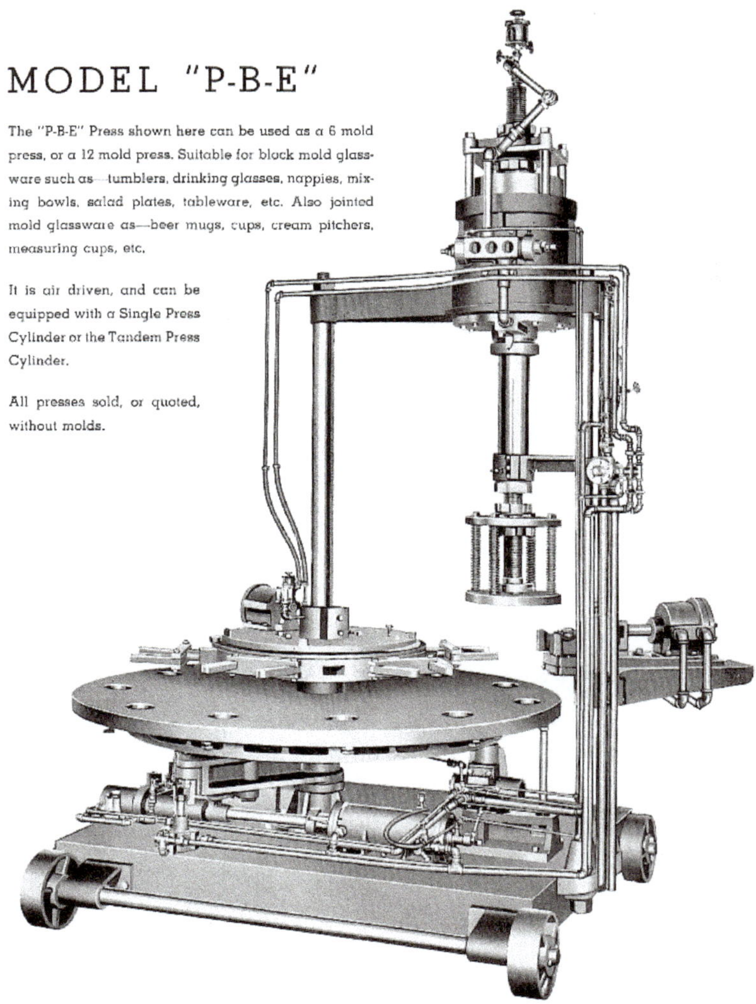

Figure 7,29. The Lynch P-B-E machine

Can be used as either a 12 mould or a six mould machine. Diameter from 1 inch up to 12 inches. Height from flat up to 8 inches. Table diameter 72 inches. Speed up to 40 pieces per minute.

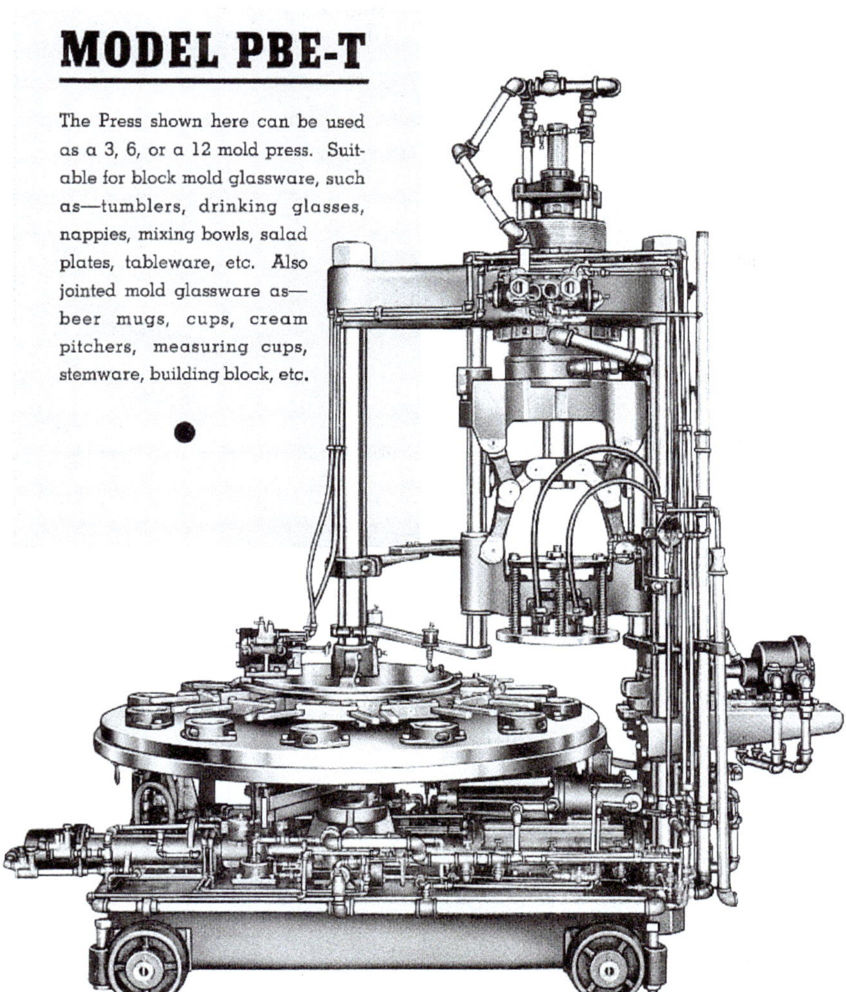

Figure 7,30. The Lynch PBE-T machine

Can work with 3, 6 or 12 moulds. Diameter from 1 inch minimum to 12 inches maximum. Height from flat to 8 inches maximum. Table diameter 72 inches. Speed up to 30 pieces per minute.

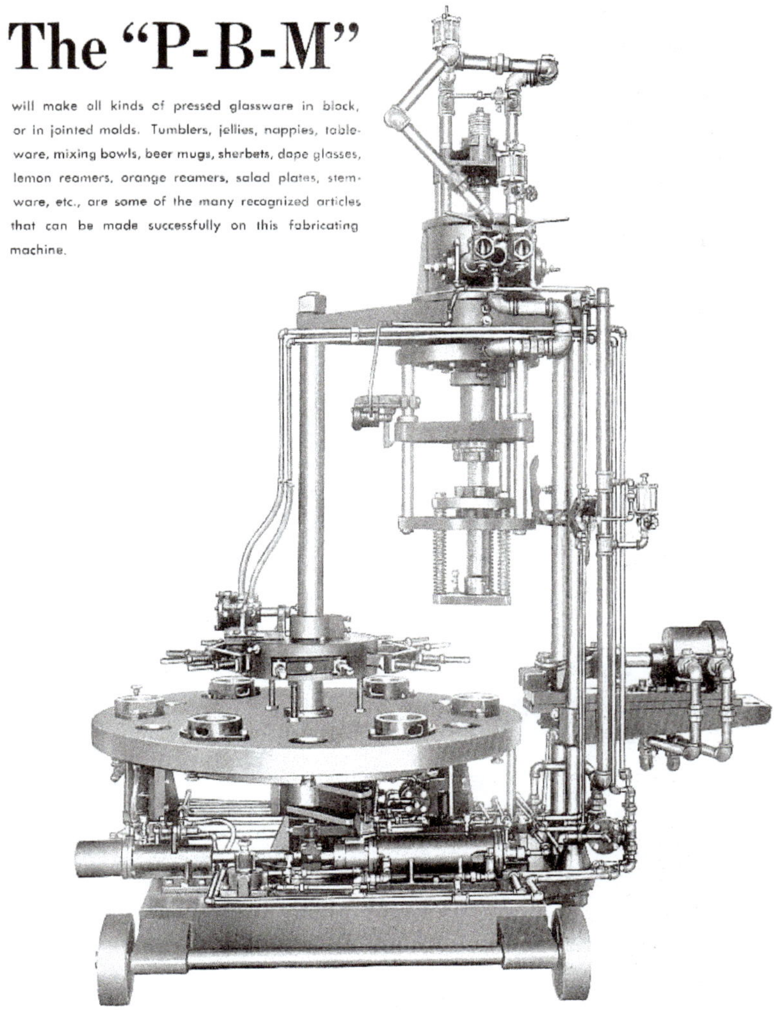

Figure 7,31. The Lynch P-B-M machine

Diameter; from 2 minimum to 10 maximum with 12 moulds, and from 2 inch minimum to 15 inch maximum when using six moulds. Height from flat up to 8 inches. Table diameter 64 inches. Speed up to 40 pieces per minute, when using six moulds between 18 and 22 pieces can be made, when using 12 moulds from 30 to as high as 40 pieces per minute can be made.

The next press is the Model 'P-B-S'.

Figure 7,32 shows it as a press machine and Figure 7,33 shows it equipped with dumping baskets.

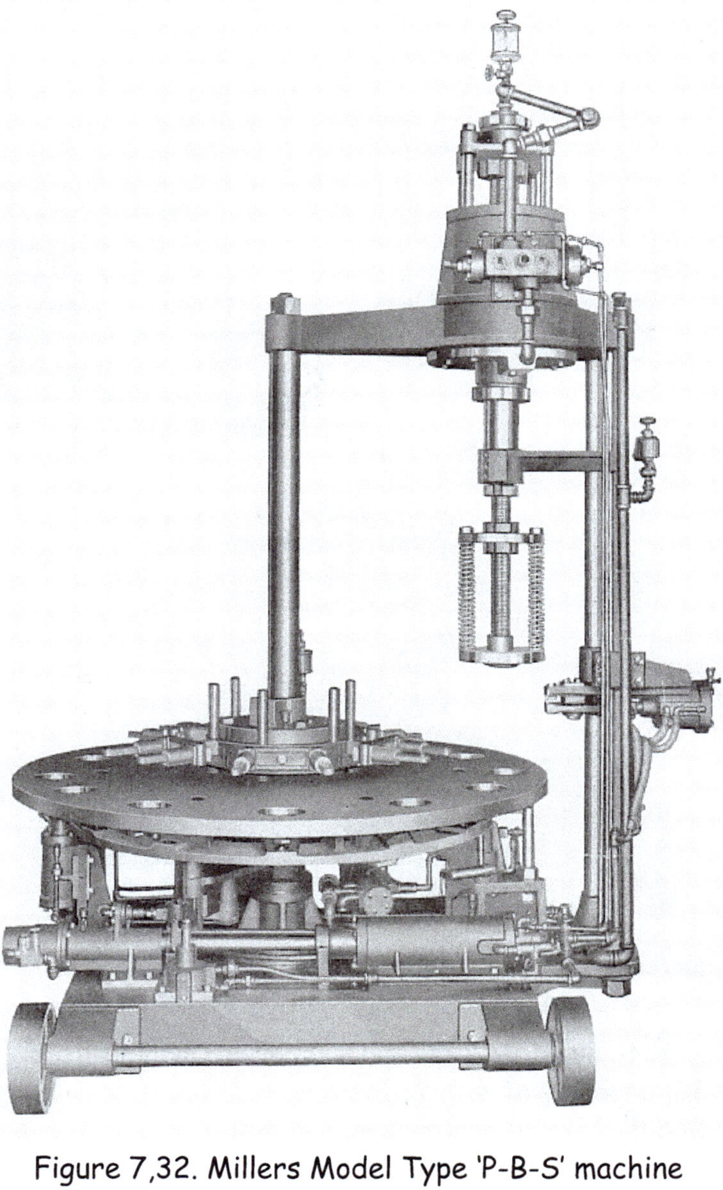

Figure 7,32. Millers Model Type 'P-B-S' machine

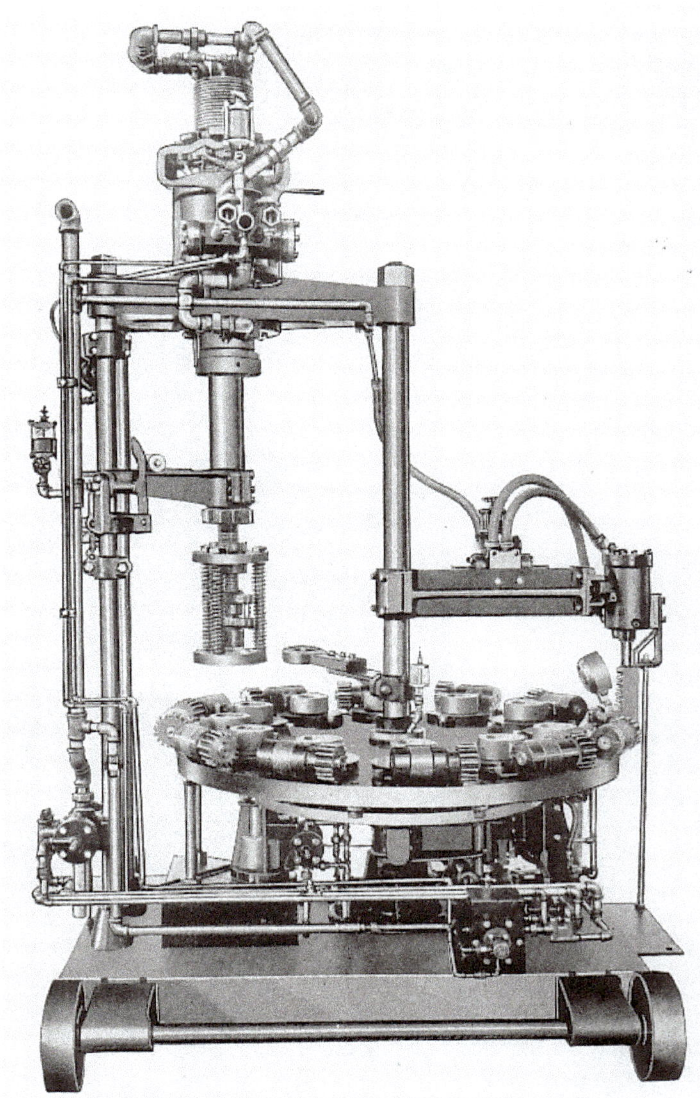

Figure 7,33. Millers Model 'P-B-S' with dumping baskets

The machine statistics are number of moulds used either six or 12. Capacity of article from a fraction of an ounce up to and including 12 ounce maximum. Diameter of article, up to 10 inches when using six moulds and up to 5 inches when using 12 moulds. Height from flat up to 5 inches maximum. Speed up to 50 pieces per minute. When using six moulds from 18 to 20 pieces per minute, with 12 moulds 30 to 40 pieces per minute can be achieved. With the machine equipped with dumping baskets for lens manufacture speed as high as 50 pieces per minute can be achieved. See Figure 7,34 for rates.

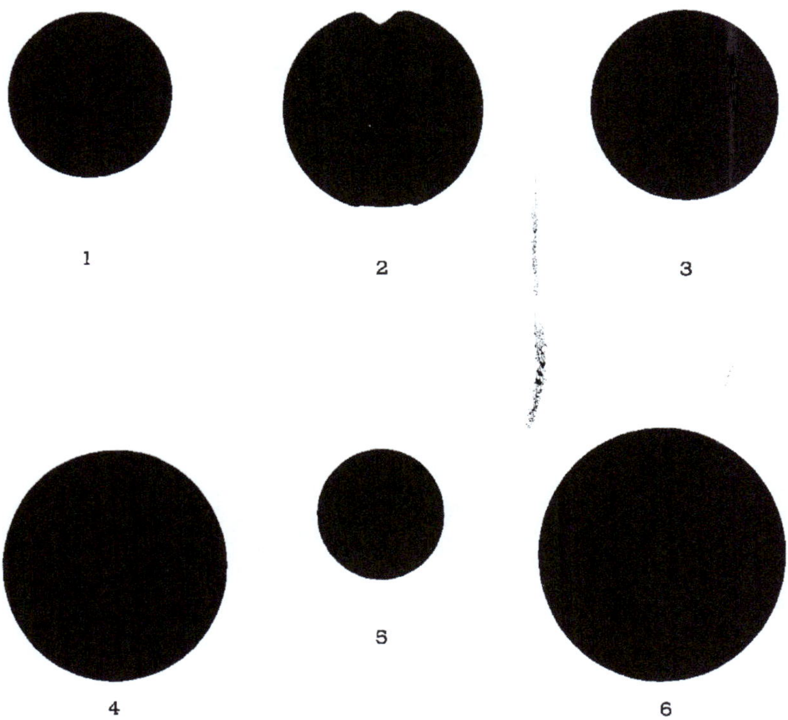

ITEM	NAME	SPEED PER MINUTE
1	*$2\frac{3}{4}$ inch Round Top Lens*	*38 to 42*
2	*Tail Light Lens*	*36 to 40*
3	*Beehive Type Lens*	*40 to 46*
4	*Flat Type Lens*	*34 to 38*
5	*$2\frac{1}{2}$ inch and 3 inch Flat Lenses*	*46 to 50*
6	*$3\frac{1}{2}$ inch Flat Type Lens*	*32 to 34*

Figure 7,34. Speed of manufacture when using dumping baskets to make lenses

The following photograph shows some of the articles that have been made on this press, the ink bottle on the top row, the handled beer mug on the second row and most of the ware on row four could be classed as container ware.

Figure 7,35. Ware made on a P-B-S press

If a press can make this type of ware they could be classed as a container making machine but nowadays most of this type are made on an IS machine, the output from press is just too low for the ware to be commercially competitive.

Finally the last series in the Lynch - Miller family of presses all these are termed heavy duty presses. The first is the model 'P-G-A', the second is the Model 'P-G-B', the third is the Model 'P-G-C' and finally the Model 'P-G-T'.

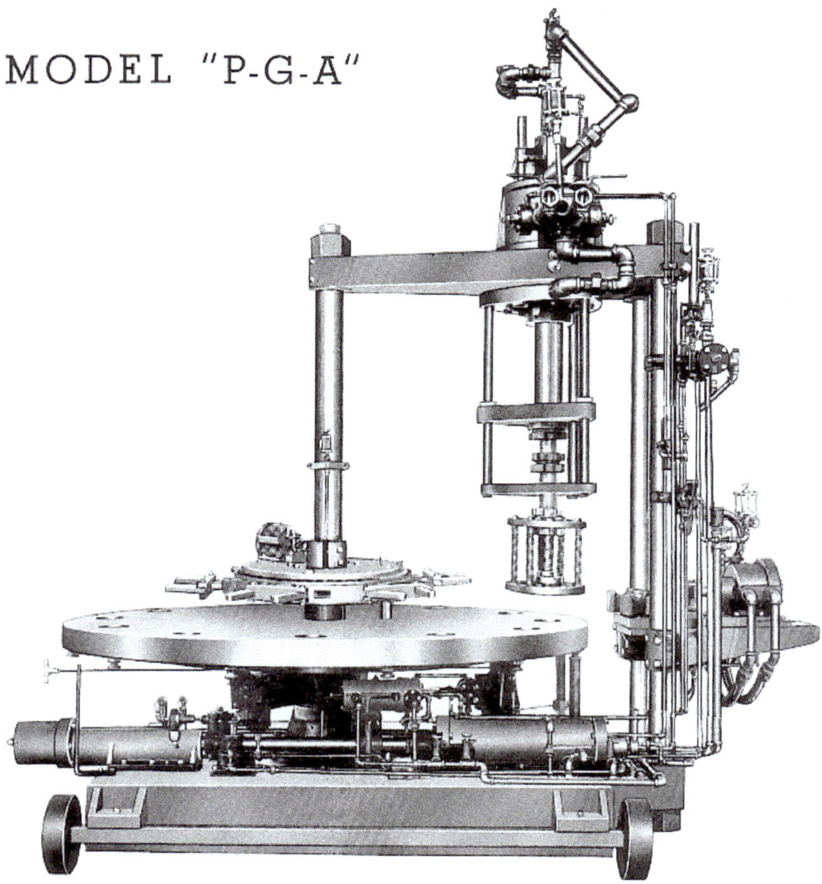

Figure 7,36. The Lynch Model P-G-A

The machine can be worked with either six or 12 moulds. Diameter from 4 inches minimum to 12 inches maximum, height from flat up to a maximum of 10 inches. Table diameter 80 inches.

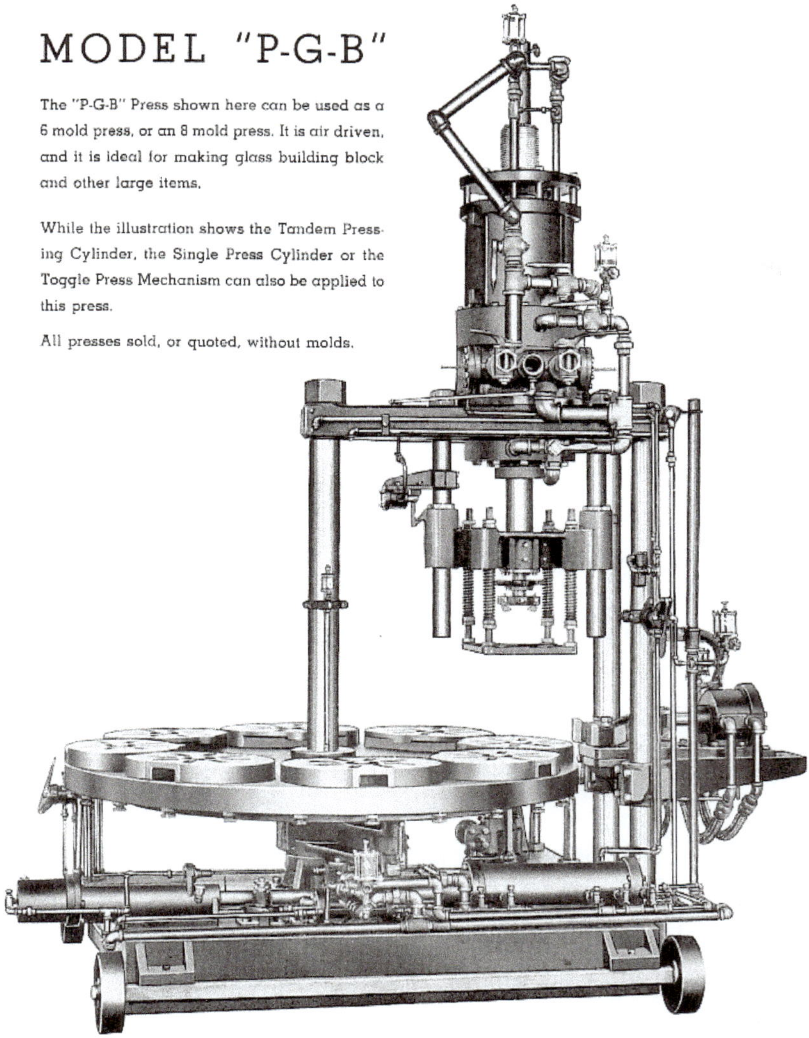

Figure 7,37. The Lynch type P-G-B machine

The press can be worked with either six or eight moulds. Ware diameter from 4 inches minimum to 16 inches maximum, ware height from flat up to 10 inches maximum. Table diameter 72 inches. Speed up to 25 pieces per minute.

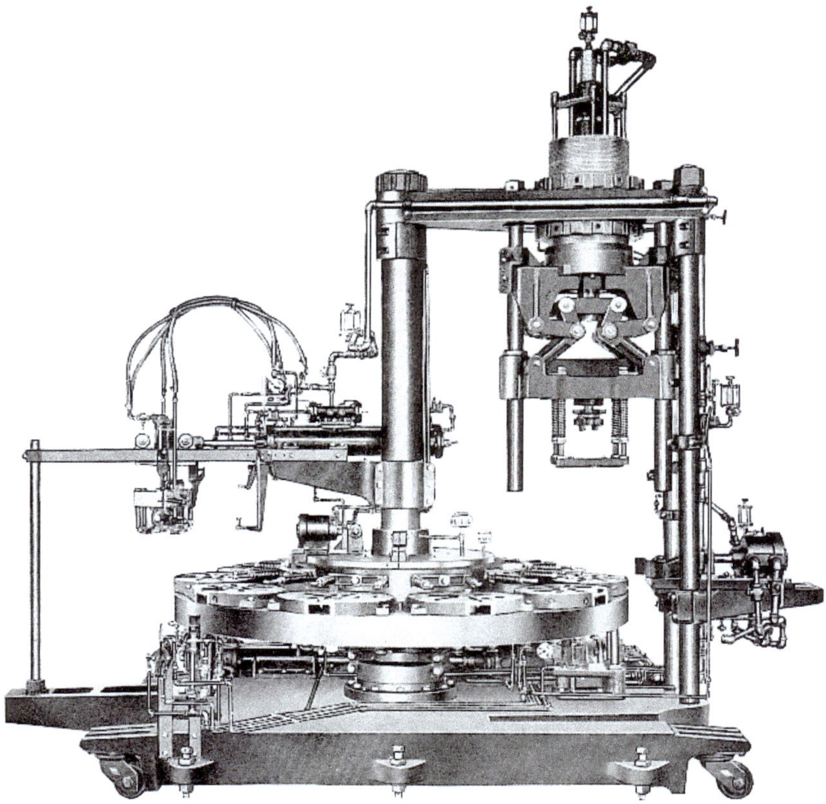

Figure 7,38. The Lynch Model P-G-C press

Number of moulds that can be used is either 8 or 10. Diameter of ware from 4 inches minimum up to 16 inches maximum. Height of ware from flat up to 10 inches maximum. Diameter of table 89 inches. Speed up to 25 pieces per minute.

And finally the last of the Lynch's presses is the Model 'P-G-T'.

The number of moulds that can be used at anyone time either eight or 16. Diameter of ware from 3 inches up to 12 inches maximum, height from flat up to 10 inches maximum. Diameter of table 88 inches. Speed up to 30 pieces per minute.

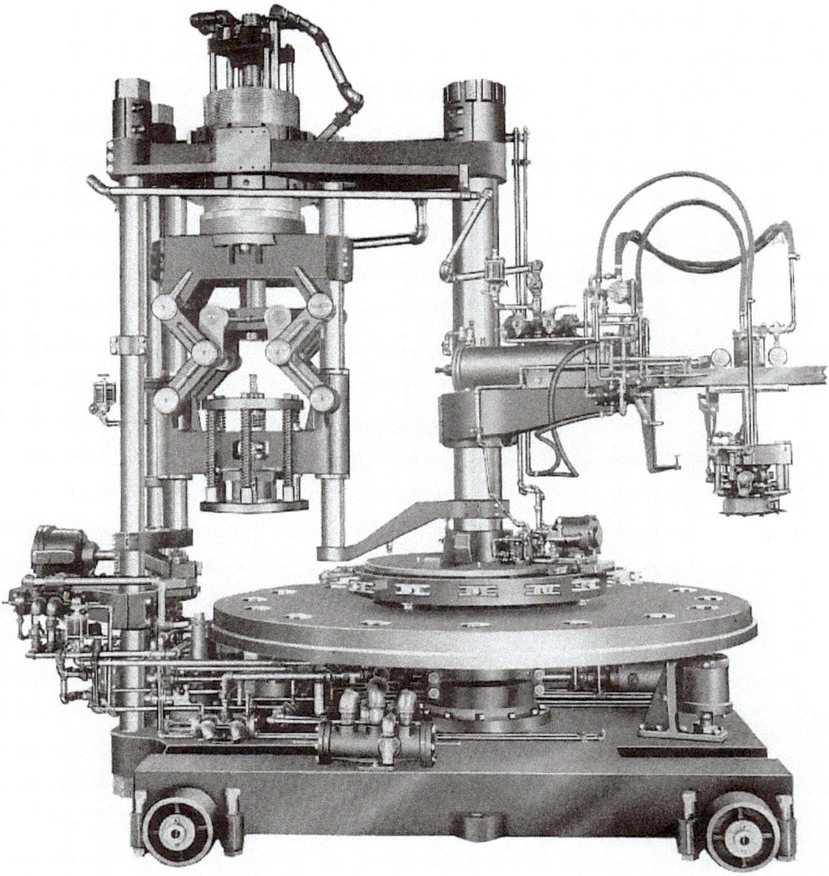

Figure 7,39. The Lynch press, Model P-G-T

Now we will move on to Blow and Blow machines starting with a biopic of J. B. Graham.

J. B. Graham, The Graham Glass Company, The Owens Bottle Company, Owens-Illinois.

J. B. Graham owned or was associated with the Graham Glass Company of Evansville, Indiana; along with glass factories at Loogootee in Indiana, Okmulgee in Oklahoma, and in Chekotah,

Oklahoma which they owned from 1913 to 1916. Besides owning these plants Graham was a good engineer who had developed his own bottle making machine which looked somewhat similar to an Owens machine but it was gravity fed not by suction like the Owens.

By the mid 1910s Owens were starting to get perturbed by the growth of the newer flow machines that were coming onto the market which they considered a threat to their market domination. These machines were now becoming 'automatic machines' with the emergence of the forehearth and 'flow' feeder. One of their first moves came in 1916 when they were bought out the Graham Glass Company, the company was then owned by the Owens Bottle Company up to 1929, and when the Owens Bottle Company merged with the Illinios Glass Company they were absorbed into the new company under the O-I banner.

When the merger between the Owens and Illinois companies took place all of the plants were given a number, the Evansville factory became Plant Number 11 and the Okmulgee factory Plant Number 15, both of these original Graham Glass Company plants were closed in 1939. The Chekotah plant was closed in 1923 by Owens and the Loogootee plant was closed on or before 1929 again by the Owens Bottle Company.

The Grahams machine was an indexing machine, i.e. it stopped while the gob was being loaded, the Owens engineering staff in Toledo redesigned the machine and put it on the open market. From their 1925 catalogue there were three different machines for sale, the six arm AY6; the 10 arm AY, and the 15 arm AZ machines.

The Owens Flow Type machine has two mould tables - one above the other, which revolve about a stationary central column that is fastened to the base. The upper table carries the blank moulds which are used to form the parison, while the lower table carries the blow moulds with their bottom plates which forms the finished bottle. The blank mould is adjustable vertically in relation to the blow mould table for the manufacture of different types of bottle.

Interposed between the two tables and in registering position

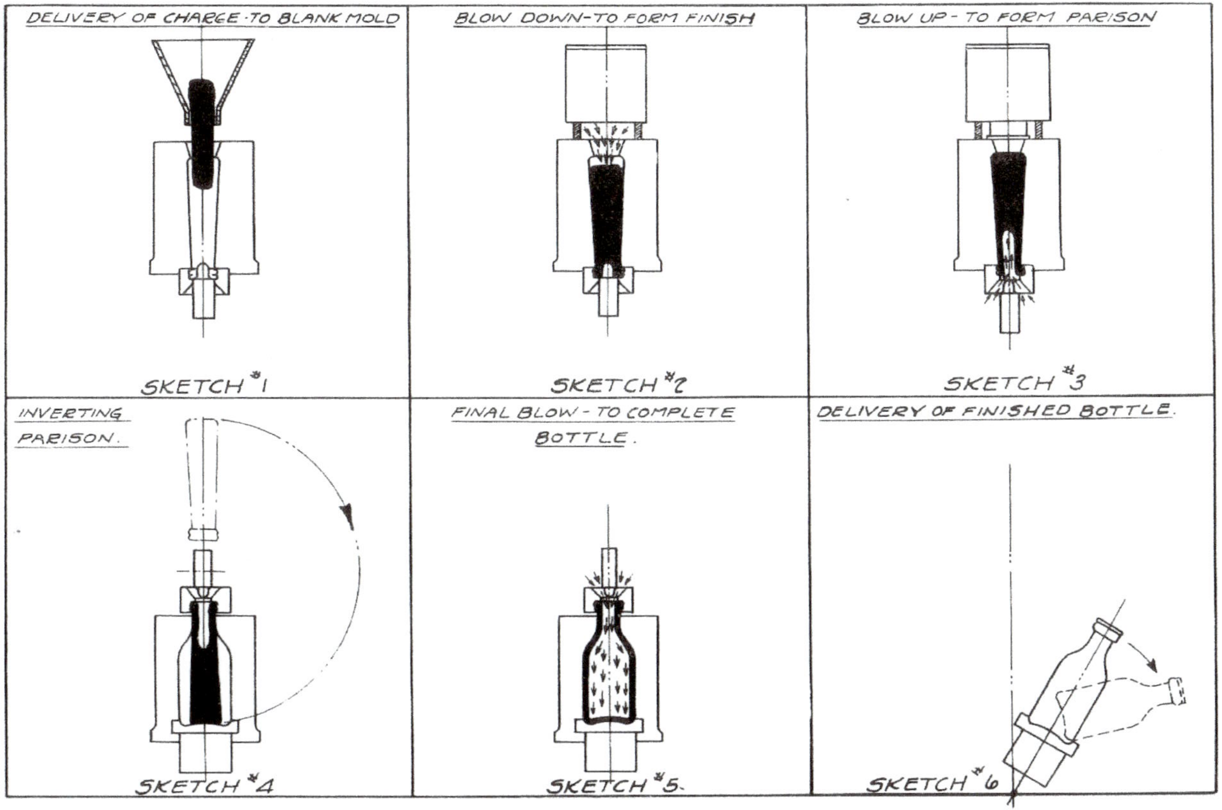

Figure 7,40. Sketches of the Owens Flow Process

with each are neck mould turnovers which carry the plunger tip and neck mould for making the lip or finish of the bottle. This neck mould turnover unit when in registration with the blank mould is inverted by cam motion to transfer the bare parison from the blank mould to the blow mould. The neck mould turnover unit is later brought back to its original position so that it can repeat the operation on the machines next cycle.

In conjunction with Figure 7,40 the machines operation is as follows; the Owens Flow Type Machine receives its glass from an Owens Flow Feeder attached to the side of the furnace. The charge of glass drops from the flow feeder into the passing blank moulds, the sketches in the following figure will help to explain the rest of the process.

Sketch 1 shows a charge of glass which has dropped from the flow feeder, passing down through the funnel into a passing blank mould which is closed at its bottom by the plunger tip and the neck mould. The plunger tip is a pointed piece of iron the size of the opening in the bottle. The plunger tip and the neck mould form the finish of the bottle.

Sketch 2 shows the funnel moved out of position from the blank mould and it is replaced by the blow head which now covers the top of the blank mould. This blow head blows down compressed air on the top of the charge of glass, and forces it into the neck mould to form the "finish." The plunger tip is still in position during this operation.

Sketch 3 shows the blow-up plug lowered into the top of the blank mould to close up this opening. The plunger tip is removed from the neck mould which is still in position at the bottom of the blank mould. Compressed air is admitted through the opening formed by the plunger tip in order to expand the glass to the shape of the blank mould and to set the parison. This operation completes the forming of the glass parison.

Sketch 4 shows the glass parison swung downward into the blow moulds by inverting the neck mould turnover. The neck mould still securely grips the "finish" of the parison.

Sketch 5 shows the blow mould closed about the parison

and the bottom plate locked into position at the bottom of the blow mould. Compressed air is admitted through the neck mould into the finish and the parison is expanded to the walls of the blow mould. After this the neck mould opens and the neck mould turnover returns to the blank mould, while the bottle is cooling in the still closed blow mould.

Sketch 6 shows the blow mould opened up. The bottom plate is tilted by means of a smooth cam movement, causing each bottle to all out in a uniform manner so a "setting-up device" may set it up on a conveyor which will carry it away.

An unusual and important feature of the above operation is that the bottle is allowed to remain in the blow mould while the blank mould above it is gathering glass from the Owens Flow

Figure 7,41. The ten arm Owens Flow machine

Feeder. This permits the machine to manufacture the bottle in $1\frac{1}{4}$ revolutions which allows greater cooling of the glass and enables the machine to operate at faster speed.

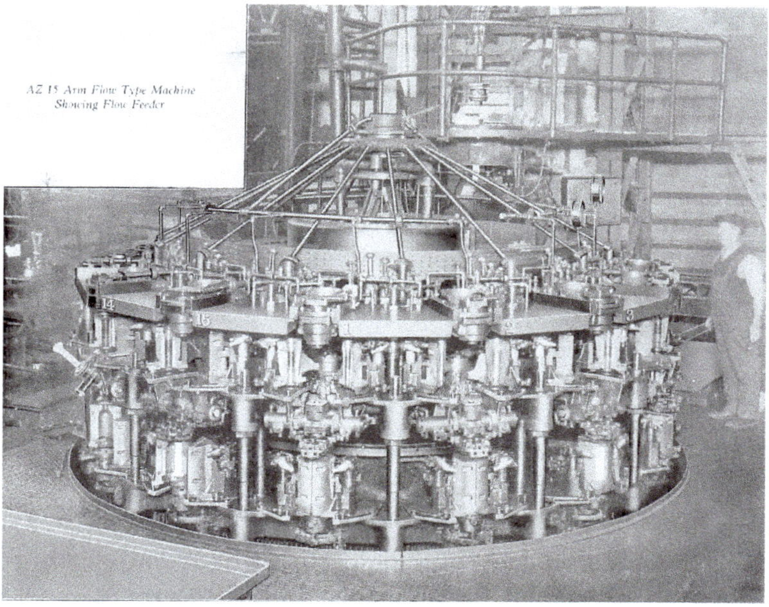

Figure 7,42. The 15 arm Owens Flow machine type AZ

Machine type	**Range of Speeds** Number of moulds	Lowest speed per minute	Highest speed per minute
AZ	15	24	72
AY	10	16	48
AY6	6	10	30

Table 7,11. Speed ranges of the various types of Owens Flow machines

Machine Type	**Typical Speed** AZ	AY	AY6
1.00 L capacity	36	24	14.5
0.75 L capacity	42	28	17
0.50 L capacity	51	34	20.5
0.25 L capacity	60	40	24

Table 7,12. Typical speed for various capacity bottles

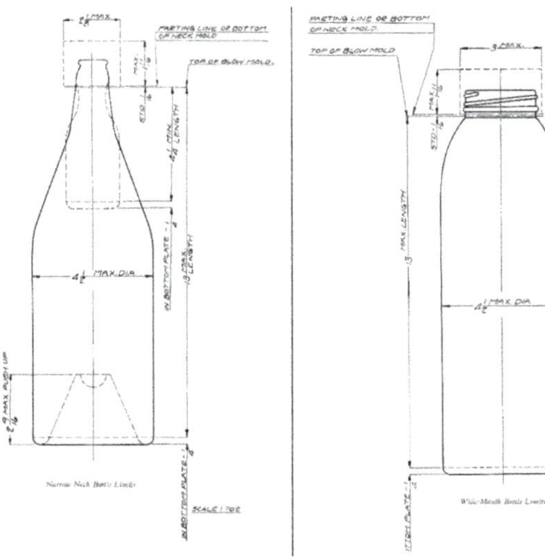

Figure 7,43. Bottle limits

Figure 7,43 shows the limits for narrow neck and wide mouth and semi wide mouth bottles.

The machine can make bottles within the range of minimum length $4\frac{1}{2}$ inches within the mould to a maximum length of 13 inches within the mould. The maximum body diameter is $4\frac{1}{2}$ inches.

The machine can also make 'push-up' bottoms for example for wines up to a maximum of 2.5625 inches.

Obviously the machine requires one piece of mould equipment for each head, orders when placed with Owens usually comprise of;

	AZ	AY	AY6
Plunger tips	50	35	20
N'k m'ld Thimbles	75	50	30
Blank moulds	22	15	9
Blow moulds	20	13	8
Bottom plates	20	13	8
Blow up plugs	30	20	12
Push up bottoms (if required)	22	15	9

These quantities allow for the cleaning and repair of parts.

Figure 7,44 and Figure 7,45 shows a typical layout for an Owens Flow machine shop, both for furnace and machine shop floor.

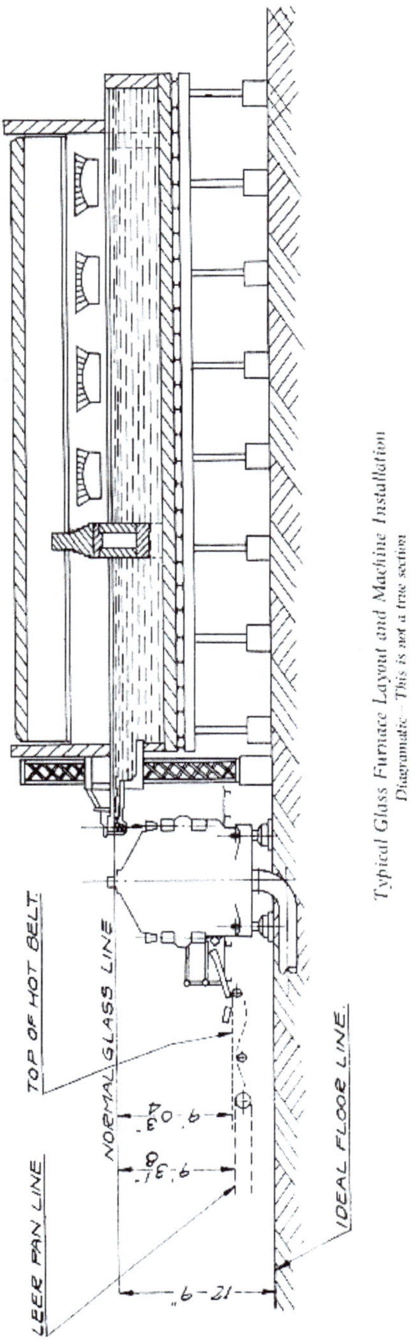

Figure 7,44. Longitudinal section of a furnace and machine shop

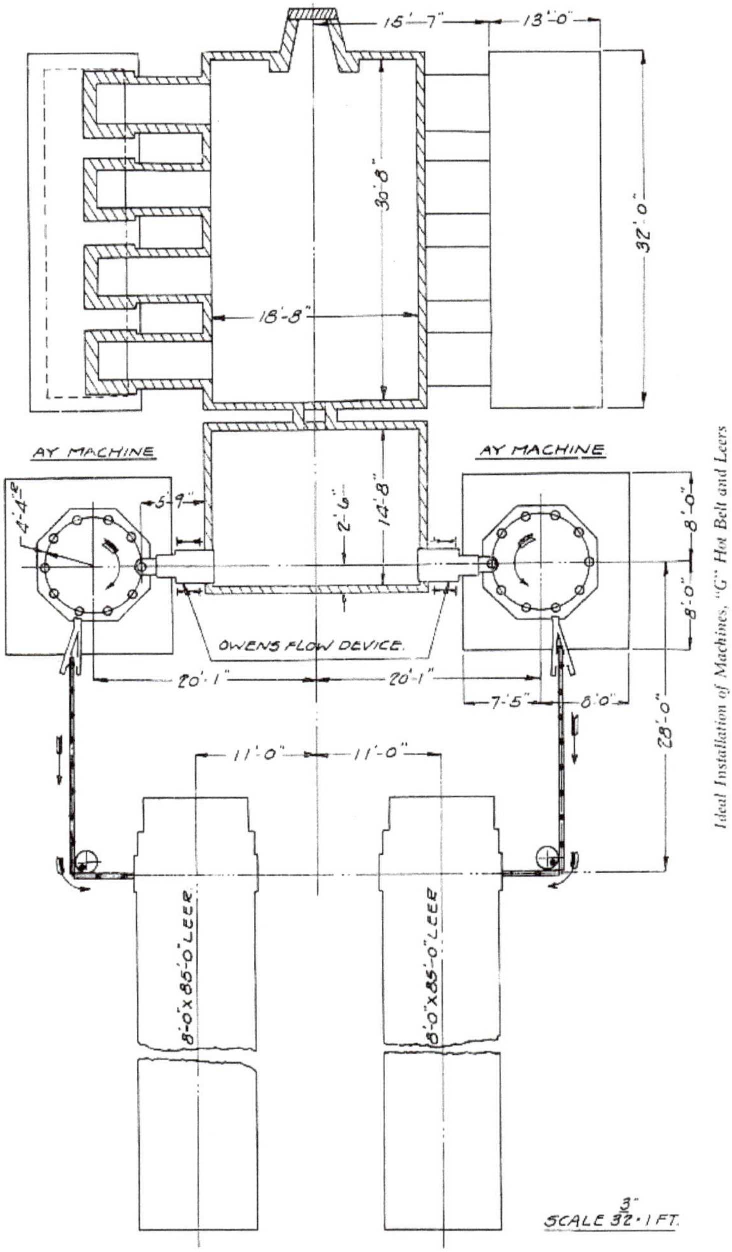

Figure 7,45. Plan view of a typical installation for a Owens Flow machine shop

Figure 7,46. Shows a close-up of a setting up machine attached to a 10 arm machine

The operation of the flow feeder was dealt with in the previous chapter, the machine services required are:

Compressed air.

Compressed air is required for the flow feeder burners in addition to the air used to operate the machine i.e. the mechanical movements and the blowing of the bottles. Based on tests carried out under actual operating conditions, at 35 pounds gauge pressure at the machine, the requirements would be;

AZ machine and Flow Feeder	450 cu. ft. per minute
AY machine and Flow Feeder	300 cu. ft. per minute
AY6 machine and Flow Feeder	180 cu. ft. per minute

There must be a surplus of supply, and the amount of surplus depends upon the total factory demands, for example one 'AY' machine should have an available supply of 500 cu. ft. per minute, where as two 'AY' machines would be amply protected by a supply of 800 cu. ft. of capacity provided.

For sizing the compressor a 'rule of thumb' requirement would be 11 H.P. per 100 feet of free air.

Blower air.

The amount of air required for the cooling of the Flow Feeder and machine are;

AZ Machine and Flow Feeder	10 500 cu. ft. per minute
AY Machine and Flow Feeder	7000 cu. ft. per minute
AY6 Machine and Flow Feeder	4200 cu. ft. per minute

The air pressure is 3 ounces or about 6 inch water column. To achieve this a blower motor of the following sizes should suffice;

AZ Installation	30.0 H.P.
AY Installation	20.0 H.P.
AY6 Installation	12.0 H.P.

Motor requirements for the Bottle Machine and the 'G' Hot Belt (conveyor) are;

AZ Machine Motor and Controller	7.5 H.P.
AY Machine Motor and Controller	5.0 H.P.
AY6 Machine Motor and Controller	3.0 H.P.

Water requirements for the cooling of the glass while being rejected it to the water tray are;

AZ Machine Flow Feeder	25 gallons per minute
AY Machine Flow Feeder	25 gallons per minute
AY6 Machine Flow Feeder	15 gallons per minute

Suck and Suck machines or Suck/Blow and Suck machines.
Here we are basically talking about the Roirant 'flow' machines which used vacuum to make the finish, the blank was then blown and the moulding of the bottle was carried out using vacuum.

Emile Roirant.

Earlier we dealt with Emille Roirant's suction machines now we will look at his 'flow' machines. The first was the 'M8' marketed in 1928, followed by the 'R7' which originated between 1949 and 1954 and the 'S10' which arrived in the 1960s. All of them rotated around a central column, the blank or parison moulds were positioned above the finishing moulds, the transfer was done by a rotatable neck ring mechanism.

The machine is provided with eight parison and eight finishing moulds which revolve round a central column. The process is by vacuum for both the parison and the finishing moulds. The machine can operated with four moulds if the furnace pull will not allow for eight mould.

Figure 7,47. Roirant's 'M8' machine

If all of the articles are of the same height and type them they can be removed automatically from the machine, they do not have to be the same shape but they all must have the same weight, a limitation already imposed by the feeder process.

Ware limitations;

Total height	maximum -	325 mm (12.75 inches)
	minimum -	180 mm (7.125 inches)
Maximum diameter of body -		95 mm (3.75 inches)
Ring	maximum diameter -	36 mm (1.4375 inches)
	maximum height -	36 mm (1.4375 inches)

Auxiliary services;

1. Driving electric motor.

At maximum speed 2 HP is necessary for the driving of the machine

2. Vacuum pump.

A minimum of 600 mm of mercury (23.5 inch) corresponding to an output of about 1,1 c.m. per minute (39 c.f.) is required by the 'M8' machine - measurements taken on the machine. The pump, however, must be able to work under a vacuum of 700 mm (27.5 inches) of mercury - power absorbed under 650 mm (25.5 inches) of vacuum about 11 HP. (working vacuum).

3. Compressor.

Quantity of air required under atmospheric pressure is 360 litres (13 c.f.) per minute. Average working pressure 1 kg/cm^2 (14 pounds/sq. inch).

4. Cooling fans.

Three identical fans are necessary for the cooling of one machine. The characteristics of each fan are;

Air volume per sec. - 1,2 cu. m (45 cu ft)

Air pressure - 230 mm (9 inches) minimum

Power absorbed approximately 8 HP.

The machine is 8 foot 3 inches (2.5 m) long, 6 foot 7 inches (2 m) wide, and 4 foot 8 inches (1.4 m) high. Its weight is about 1 ton 1000 lb (4500 kg).

The 'M8' made way for Roirant's 'R7' the second machine in this series. After design and development work lasting from 1949 to 1954, the machine entered the market in the mid to latter half of the 1950s.

Figure 7,48. The Roirant 'R7' machine

It was a seven arm continuously rotating single table machine driven and controlled mechanically. There are three levels, the top one is were the blanks and funnels fit, the second is were the neck ring mechanisms lie and the third one, the bottom, is were the finishing moulds with their bottom plates are located.

The machine operates along these lines;
First two high pressure spray nozzles lubricate the blank mould. As soon as the gob enters the blank mould and funnel the vacuum comes on. and continues until the plunger rises.

The funnel in the meantime has been set upon the blank mould. After the retraction of the plunger, the counter blowing starts.

Next step is that the blank is opened and the formed parison is swung from the upper level down to the lower one. The stretching of the parison takes place during this manoeuvre and is finally finished as it hangs in the blow mould.

Finally the parison is 'blown' up to the shape of the mould by vacuum.

The machine uses old techniques engineered to the latest engineering standards and production techniques, it performs loading by vacuum, it blows the parision by compressed air, and the finished blowing again by vacuum. Similar methods used earlier were those used by Graham, Severin, Afterguth and Roirant himself with his 'M8' machine.

The limiting dimension for the bottles are;

Height including finish	115–375 mm for flat bottoms 115–341 mm for hollow (push up) bottoms
Maximum outside diameter	125 mm
Maximum finish diameter	58 mm
Puncture maximum	40 mm
Plunger diameter maximum	60 mm
Weight	150–1200 gram

The power requirements were;

Compressed air at 2.2 at.	100–144 cu. m/h	7.0 kW
2.1 at.	30-84 cu. m/h	2.5 kW
Vacuum 82 % on machine	960 cu. m/h	18.5 kW
Cooling air 279 mm WC	21 600 N. cu. m/h	32.0 kW
(in reality 265 mm WC	25 370 N. cu. m/h)	
Drive		2.0 kW
Drive for connecting device		2.0 kW
Feeder drive		0.5 kW

According to the Emhart Machine Production Rates the R7

was reported under the groupings of R7, R7 - double gob - 82 mm centres, R7 - double gob - 92 mm centres, and R7 - single gob - 14 moulds. For the statistics that I have (1983) only the R7 single and the R7 single - 14 moulds were reported, obviously the speed curves are different indicating a higher output machine but its reporting description is not for a dual cavity machine.

The following speed curves are from around 1965, the double gob curve was when it was in its formative period.

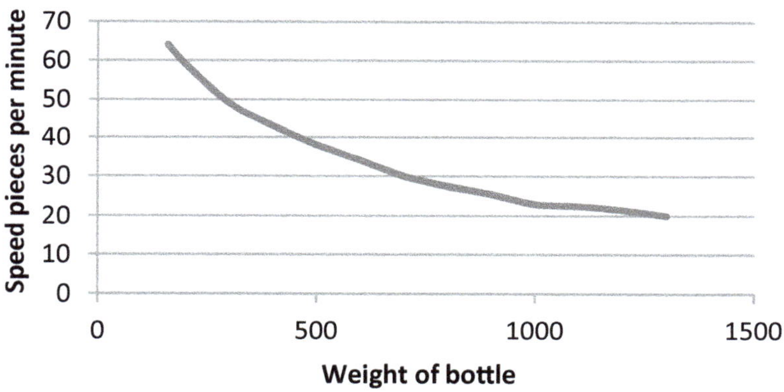

Figure 7,49. Speed curve for the Roirant R7 machine, circa 1965 - single gob

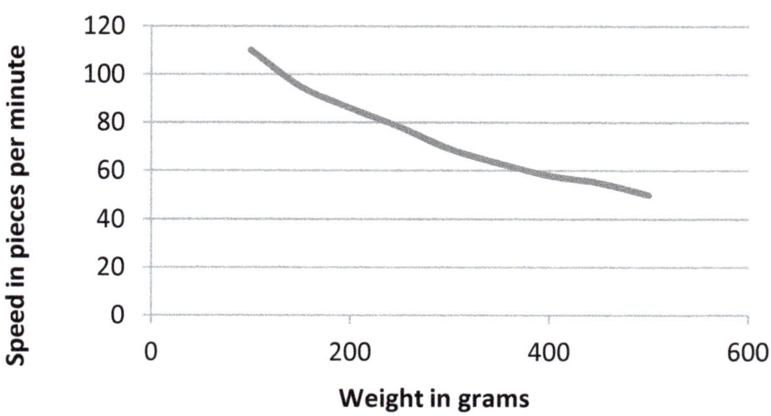

Figure 7,50. Speed curve for the Roirant R7 machine, circa 1965 - double gob

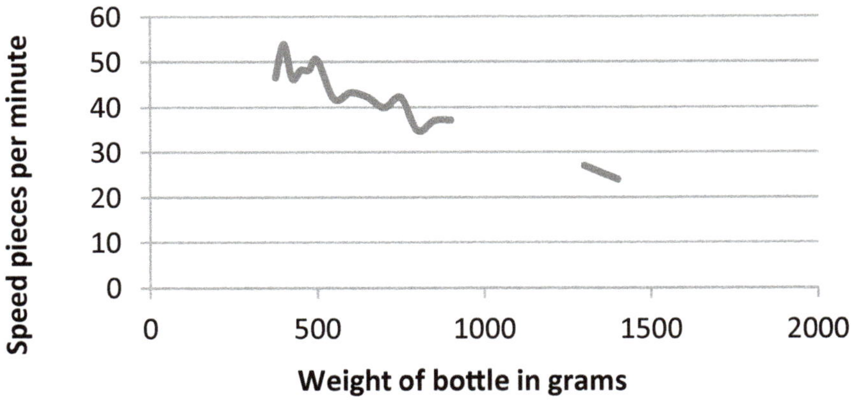

Figure 7,51. Actual speed curve for the Roirant R7 - single mould 1983

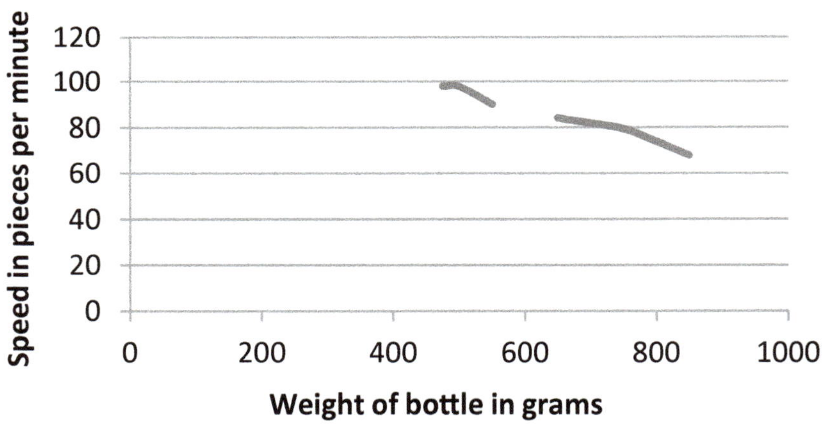

Figure 7,52. Actual speed curve for the Roirant R7 -single gob-14 moulds, 1983

The 'S10' came onto the market in the late 1960s and eventually the family consisted of S10 single gob, S10 double gob 82 mm centres, S10 double gob 92 mm centres; and the S10 double gob 111 mm. centres. Although this is the family only the S10 single and the S10 111 mm. centres were being reported. The speed curves for the can be seen in Figures 7,55 and 7,56.

Figure 7,53. Cooling table of a 'S10' machine

Figure 7,53 is the only photograph of the S10 that I have, it is of the cooling table, I don't dispute the origin but it looks like a R7 cooling table to me. The only S10s that I have seen operating were at the Gironcourt plant of BSN where the company had manufactured their own cooling tables that were nearly as big as the machine itself.

Figure 7,54. The S10 Machine

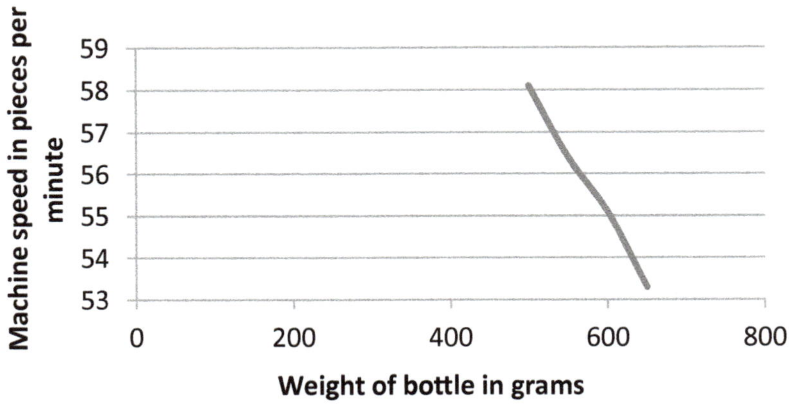

Figure 7,55. Speed curve for the Roirant S10 machine circa 1983

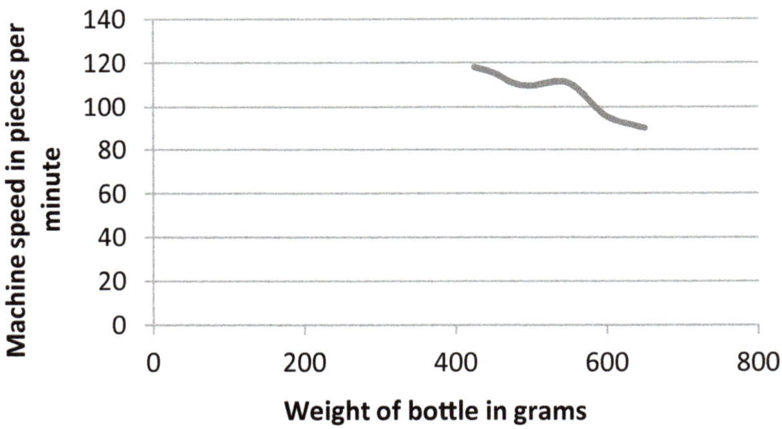

Figure 7,56. Speed curve for Roirant's S10 machine
- double gob - 111 mm. centres

The next biopic is of the Knox partnership and S. E. Winder and H. C. Daubenspick.

Knox, The IK machine and the W-D Machine.

The Press and Blow 'IK' machine is a Knox development of the Lynch JP machine, but it deviates somewhat from the Lynch design. The machine is a single table one with ten stations which can be equipped with either single or double moulds, the drive is mechanical through Geneva gears. All other functions are carried out by cams and air cylinders.

Parison and blow moulds are arranged next to each other in contrast to how they are positioned on the Lynch JP machine, and in contrast to all other press and blow machines the gob is loaded into the running machine just prior to the pressing station. This method of loading gives for a quicker insertion with the corresponding reduction in overall loading and pressing time. This gives rise to improved glass uniformity and gives none or fewer press wrinkles.

The IK machine has only been used in the glasshouses of the Knox Company where their production is said to be greater than the Lynch JP machine.

The 'W-D' machine was developed in the early to mid 1940s, (as reported in the Glass Industry in 1946), and is a Blow and Blow machine of the gob fed continuous motion type. The Knox W-D machine is a single column machine with the blank moulds underneath the blowing moulds (opposite way to most of the other machines of this type). The blank mould gives a full length parison equal to that of the bottle which is required by the means of transfer). The diameter of the blank table is greater than that

Figure 7,57. The Knox W-D machine

of the mould table to permit the gobs to fall in easier.

After the parison is formed, a special vacuum type bottom plate holds the parison by a number of small vacuum holes and transfers the parison through 180 degrees, from being vertically blown to vertically up. The machine carries ten sets of moulds and is fed by an oscillating gob chute. Speeds of up to 100 pieces per minute are claimed with a range from 2 ounces to 1 gallon containers.

The machine is driven by electrically activating cams, and rollers and guiding notches are attached to the centre column and to the stationary framework, both the cams and guides can be adjusted in operation. The activation processes are pneumatic for each of the stations individually. Moulds are opened, closed and locked pneumatically, the parison is transferred from parison the final blow mould pneumatically. Plunger and blowheads are dropped and lifted by compressed air.

The parison is blown by vacuum and the transfer to the blow mould is also by vacuum. The production range is between 50 cm^3 and 3.8 l. Minimum cooling air requirements are 283.2 m^3/minute at 354 mm water column. For production rates see Figure 7,59.

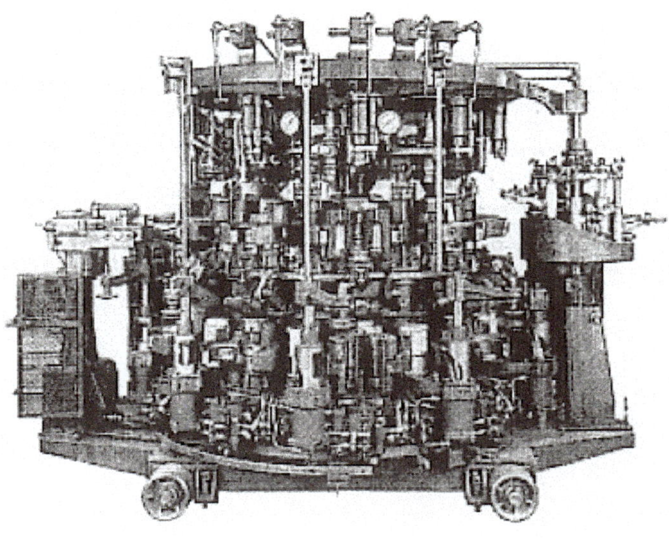

Figure 7,58. A non-working view of the Knox machine

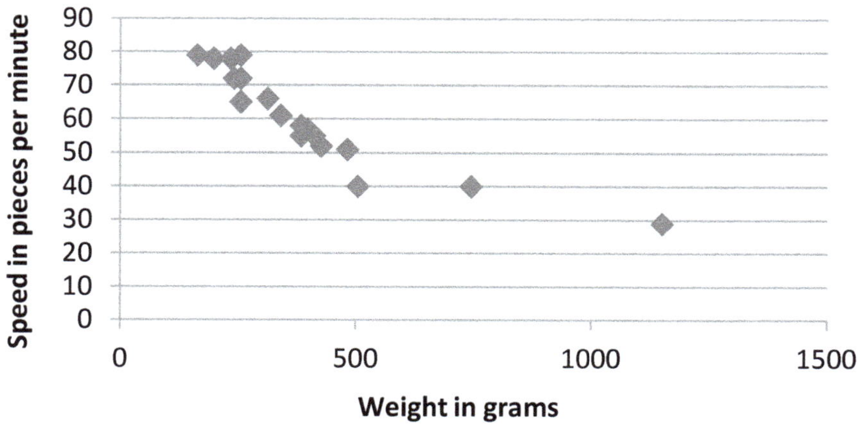

Figure 7,59. Speed curve for the Knox W-D machine

Space requirements in metres; length 3.43 × width 3.56 × height 3.05, weight 15 000 kg. (Packaged for export 16 700 kg)

Services required;
Compressed air 14 m^3/minute at 3.16 atmospheres, - 25% blow, 74% control, 1% others (plunger cooling, etc.) air.
Vacuum 13 m^3/minute at 645 mm of mercury.

The Knox W-D machine I believe was designed by S. E. Winder and H. C. Daubenspeck US patent Number 2 366 669. The machine was offered to the open market when the patents ran out but very little interest was shown. There are a number of abstracts in the Society Journals many of which show that Winder lived in Chicago, Illinois, USA while Daubenspeck lived in Blackheath, London, England which seems somewhat odd for a design partnership.

From the abstracts other Winder and Daubenspeck patents taken from Volume V, 1921 are;

#147. Bottle-blowing Machine. S. E. Winder, USA, and H. C. Daubenspeck (Brit. Pat. No. 152 414, July 12th, 1919. No. 17 493) Figure 7,60. - Relates to bottle and like blowing-machines in which the parison moulds and the blowing moulds are mounted on separate rotating tables, a, b.

According to the invention, a gripper transfer device, d, of the kind described in Specification No. 149 402, is used to transfer the blank from a solid undivided parison mould, a^1, to the blowing mould, b^1, along the path, e. A similar device, j, may be used to transfer the finished article to a delivery table, k. The finishing moulds, b^1, are yieldingly operated through springs contained in casings, b^4. Two blowing heads i, i^1, are provided, the ware being partially blown at the first station, i, and finished by the second blowing head, i^1.

The blank moulds are fitted with ejectors as described in Specification No. 146 602. The plunger is pneumatically operated and is provided with a cushioning device also as described in Specification No. 146 602.

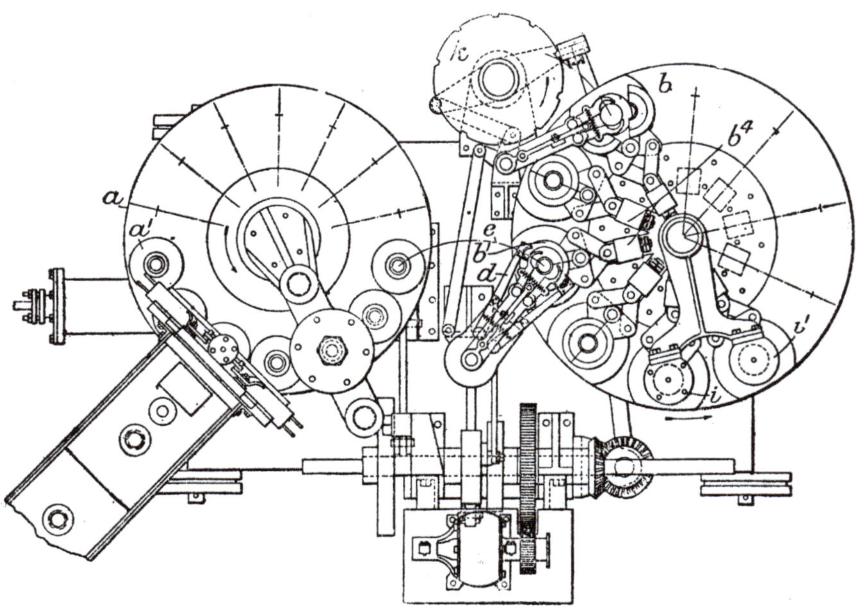

Figure 7,60. W & D's bottle blowing machine

The second one being;

#155 Glass Pressing Machine. H.C. Daubenspeck, London, and S.E. Winder, Chicago, USA (Brit. Pat. No. 146 602, April 8th, 1919, No. 8 841) Figure 7,61 - Relates to a machine for making pressed glassware of the type in which the moulds are carried on a rotating table, 59. According to the invention, the plunger is operated through an adjustable pneumatic cushioning device. The molten metal flows continuously from a tank, 15, and is separated into charges by pneumatically-operated water-cooled shears, 33.

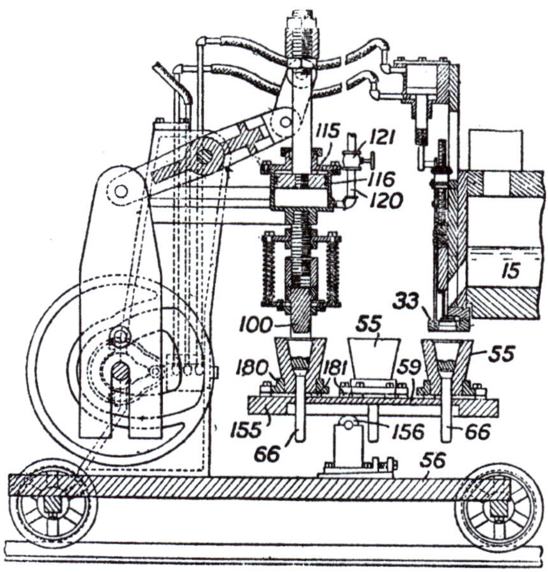

Figure 7,61. D & S's Glass pressing machine

Each charge drops into a mould, 55, and is carried round beneath the plunger, 100, which is carried by a short cylinder, 116, mounted in a vertical slide. A piston, 115, is reciprocated by the cam mechanism shown and, in rising, strikes the top of the cylinder, 116, and lifts it. The interior of the cylinder, 116, is connected by a pipe, 120, with a chamber containing air at a constant pressure, and communication is controlled by a valve, 121.

The descent of the piston, 115, forces down the cylinder, 116, and with it the plunger, 100, thus compressing the charge in the mould, 55. By regulating the pressure in the cylinder, 116, the pressure of the plunger can be controlled. The finished article is subsequently ejected by a cam-operated ejector, 66. The table, 59, is formed with a thick rim, 155, and is supported in the neighbourhood of the plunger, 100, by adjustable rollers, 156. The moulds, 55, are mounted on plates, 180, supported above the table, 59, by distance-pieces, 181, so as to reduce conduction of heat from the moulds, 55, to the table, 59. The machine is moved on its carriage, 56, towards or away from the tanks, 15, by pneumatic power when desired. Please note the different method of gob formation.

There is a third abstract (same issue) which deals with a modification to the parent specification, it is as follows;

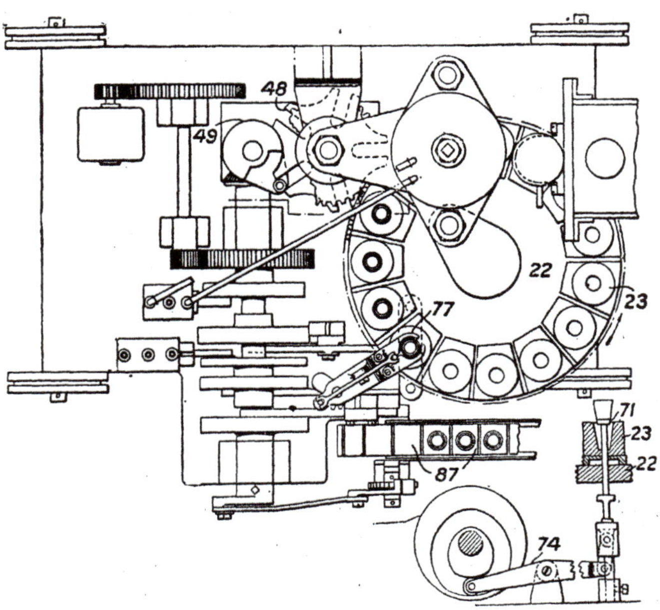

Figure 7,62. Daubenspeck and Winder's Glass Press modifications

#360 Glass-pressing Machine. H.C. Daubenspeck, Blackheath, London, and S,E, Winder, Chicago, USA (Brit. Pat., No. 158 711, November 12th, 1919). - It relates to a modification of the press-moulding machine described in the parent specification.

According to the invention, the plunger, 25, is operated by a pneumatic cylinder, 51, and piston, 111, instead of by a cam-and-lever device. The stroke of the piston, 111, is adjusted by a screw, S. The plunger, 25, is water-cooled.

The mould table, 22, is rotated by a Geneva-stop mechanism, 48, 49. The ejectors, 71, Fig. 168, are raised by a cam-operated lever, 74, instead of by a fixed inclined cam-surface.

The finished article is lifted clear of the mould by grippers, 77, and transferred to the conveyor 87.

The molten metal, 38, flows through a regulating-gate, 20,

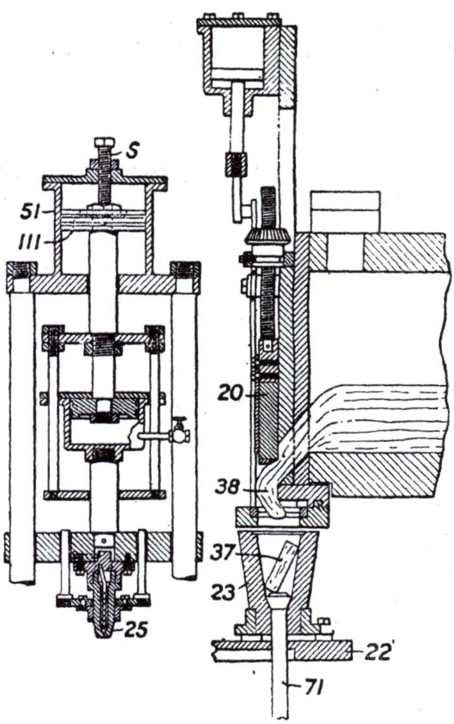

Figure 7,63. D & S's press modification

and is cut off into charges, 37, which drop directly into the moulds, 23.

Daubenspeck and Winder used the Colonial & Foreign Glass Industries, Ltd. as their agents. The following is an advertisement found in the 1923 issue of the Directory for the British Glass Industry.

178 British Glass Industry Directory.

The FOURCAULT Automatic SHEET-GLASS Manufacturing PROCESS.

Now operating in Belgium, Holland, Czecho-Slovakia, Germany, Greece. Factories will also shortly be operating in Great Britain, France, Italy, China, Japan, U.S.A.

The DAUBENSPECK Automatic WIDE-MOUTH CONTAINER and TUMBLER Manufacturing PROCESS.

Now operating in Great Britain.

MOHN'S OIL-BURNER.

The Patent Interchangeable

STOPPER and BOTTLE-GRINDING PROCESS

For information about these Processes, including conditions of licence, apply to

COLONIAL & FOREIGN GLASS INDUSTRIES LTD.,

40-43, Norfolk Street, Strand,
LONDON, W.C. 2, England.

Telephone—Central 808C. Telegrams—GLASINDUS, ESTRAND, LONDON.

When writing, kindly mention this Advertisement.

Figure 7,64. 1923 advert for the Colonial & Foreign Glass Industries Ltd

COLONIAL & FOREIGN GLASS INDUSTRIES, LTD., 40/43, Norfolk Street, Strand, London, W.C.2. T.N., Central 8090-8089 (10 lines). T.A., "Glassindus, Estrand, London."

The Fourcault automatic sheet-glass manufacturing process. The Daubenspeck automatic wide-mouth container and tumbler manufacturing process. Mohn's oil-burner. The patent interchangeable stopper and bottle-grinding process.

The Company is a patent-holding and licensing Company, licensing other Companies to use the processes specified.

Directors: Sir Walrond Sinclair (Chairman); M. D. Booth; Andre Gobbe; G. I. C. Marchand.

(See advertisement Page 178)

Will now go on to discuss William Mitchell and the Mitchell Glass Company.

William (Bill) Mitchell, The Mitchell Glass Company.

The first recorded reference that we have of Bill Mitchell is in the Presidential address by Bill Gregg when he was reminiscing about events that took place during his earlier working life.

When the Society made a visit to the United States of America in 1920 the present was one Mr. Foster who was the Secretary and a Director of the Fletcher's Sauce Company of Selby who were listed in the 1919 edition of the Ministry of Munitions as a bottle makers - they manufactured their own bottles which were 'burst offs'. This was when sauce bottles had a glass stopper with a cork ring.

Mr. Foster realised that he would have to invest a lot of money to purchase the 'know-how' and the equipment, and have all the problems that came with 24 hours production. His answer to this conundrum was to close the glassworks down. The glasshouse manager, who was also on the trip, was a certain Mr. W. B. Mitchell from Shipley near to Bradford, he obviously lost his job and gained employment in a cement factory and by his own admission said that he knew nothing about cement but "he had a very good foreman".

In his heart Bill Mitchell was a glass man and in 1928 he set up his own company to manufacture machines. He built a small oil fired furnace in Shipley that melted cullet and designed and built a small automatic stopper press. The press was erected in a garage owned by a Mr. Jeffery who sold motorbikes, Bill was married to his daughter, Bill worked his press but soon came to realise that it was the making of machines that interested him not the running of them.

So he contacted his old friend Henry Gregg who he had met in America to come along to see his machine, with the idea of selling it to him. Henry said, remembering his American trip where he had seen the Owens Machine in operation there, "said that if Mitchell would design a 10 head press making 3 marble per head that he Henry Gregg would buy it and we'll pay for it as you build it". The machine was duly installed at Knottingley achieving fantastic production results for that time, manufacturing 4000-5000 gross of stoppers per week as a norm which were sold at 1s 6d (7½p) per gross and sometimes as low as 1s 3d per gross.

Bill spent over a year at Gregg's (on their payroll), working on one of the 10 heads utilised as a straight press where he produced the 'one penny jams' that where used by the railways and he pressed glass marbles but couldn't compete with the competition to supply marbles for 'Codd Bottle' manufacture.

This experimental work led to the design of the ten head 'J' type press, see illustrations for the types B (Figure 7,65) and C (Figure 7,66) models.

The first machine was installed in the Queenborough works of Canning Town Glass Ltd. on the Isle of Sheppey where it primarily made drinking glasses in the Jacobean range. Later this machine was redesigned and marketed as the eight head 'J Type model C'.

By 1938 other presses had been bought and installed, or were under construction namely at;

United Glass Bottle Manufacturers Ltd. at their London and St. Helens factories.

The Rockware Glass Syndicate Ltd. at their Greenford factory.

The Co-operative Wholesale Society Ltd. at their Manchester

and Worksop factories.
The Canning Town Glass Works, Ltd. at their Queenborough factory.
Gregg & Co. Ltd. of Knottingley.

Figure 7,65. Mitchell ten head press machine. Type J, Model B

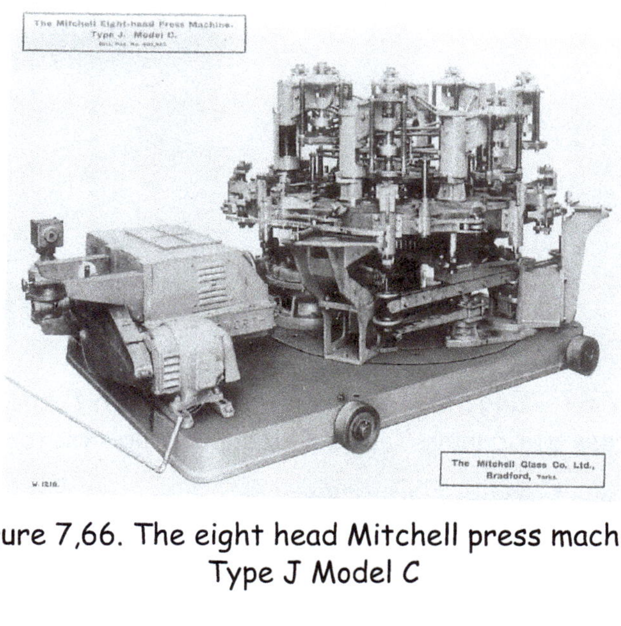

Figure 7,66. The eight head Mitchell press machine, Type J Model C

Figure 7,67. Range of Jacobean ware made on the Mitchell press at Canning Town's Queensborough factory

Other ware made by these machines can be seen in Figure 7,68 - decorative ware.

Figure 7,69. Some examples of the more conventional types of pressed glass articles. Some of these bottles are container types

The next machine to be designed was the four head suction machine. The glass manufacturers were interested in vacuum forming after seeing or having heard of the Owens machines. The first four head suction machine was installed at one of the Beatson Clark factories, the interesting feature of this machine was the method of opening the blank moulds, which opened outwards on an axis adjacent to the neck ring mechanism. The machine was not very successful and Bill Gregg thought that Mitchell bought it back from Beatson Clark and sold it on to a man called Greenwood who had a factory in Hunslet, where he made ink bottles on it for quite a number of years.

Just before the outbreak of World War II Gregg's ordered one but it never made a bottle, Bill Gregg stated that when he returned from the war it was his job to break it up.

Company placing orders for the machine in 1938 were; Beatson Clark Ltd. of Barnsley and Rotherham; Gregg & Co. Ltd. of Knottingley

Figure 7,70. The Mitchell four head suction machine

Mitchell also produced two press and blow machines, a six head and an eight, see Figures 7,71 and 7,72, and two blow and blow machines one for small ware and the other for larger sizes see Figure 7,73. These machines are feeder fed just like the presses.

Companies who had on order or installed by 1938 were; Forster's Glass Company Ltd. of St. Helens; Lax and Shaw Ltd. of Leeds.

Figure 7,71. Six head Mitchell press and blow machine

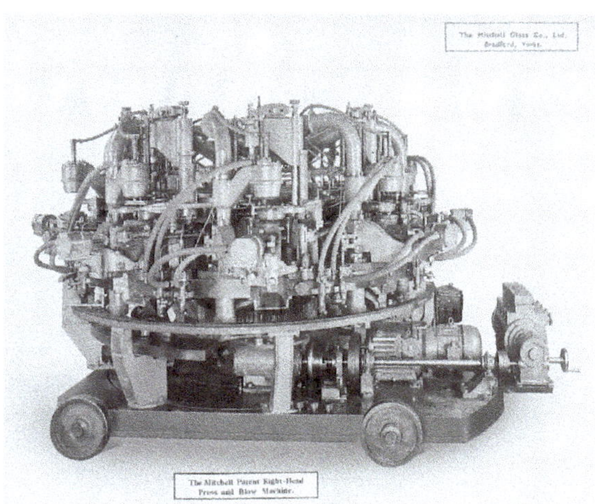

Figure 7,72. Eight head Mitchell press and blow machine

Figure 7,73. Examples of the range of ware made on the Mitchell press and blow machine

Figure 7,74. Mitchell's blow and blow machine

The machine specifications of the various types are;

For pressed ware Type J	Model C
Machine speed	Between 8 to 48 pieces per minute
Ware dimensions.	
Inside diameter	Up to 8 inches wide
Depth	Up to 5 inches deep

From a later pressed ware machine catalogue

	8 Head machine	10 Head machine
Machine speed	Up to 64 pieces per minute	Up to 80 pieces per minute
Inside diameter	Up to 6" wide	Up to 4" wide
Inside depth	Up to $4\frac{1}{2}$" deep	Up to $4\frac{1}{2}$" deep

For press and blow machines

Machine speed	6 Head	8 Head
	Between 6 and 48 per minute	Between 6 and 60 per minute

Ware dimension for standard machine

Neck diameter	4"
Body diameter	4"
Height under ring	8"
Minimum height under ring	1"

For the blow and blow machines,

Machine speed	Between 6 to 54 pieces per minute

Ware dimensions	Small ware Type BD	Large ware Type BE
Maximum outside dia. of neck	3"	$2\frac{1}{2}$"
Maximum height under ring	$7\frac{1}{2}$"	$11\frac{1}{2}$"
Maximum outside dia. of body	4"	4"
Minium height under ring	1"	$3\frac{1}{2}$"

The last of the machines designed by the Mitchell Glass Company were the AU and the BU, these were independent section machines, see Figures 7,75 and 7,76. Only two were ever made, both were hand gathered and hand fed. They were sold

Figure 7,75. The Mitchell 'AU' machine

Figure 7,76. The Mitchell 'BU' machine

to Fountain Glass at Liversedge, Huddersfield who worked them until the factory was closed down.

Bill Mitchell also made feeders and forehearths but there is very little information about them. The following three advertisements from the British Glass Industry Trade Directory gives some indication of the growth of the company.

From the 1934 Edition (Third).
Mitchell Glass Co., LTD. 40, Peckover Street, Bradford.
'Manufacturers of the Mitchell patent automatic press machine'
Directors: C. Mitchell; W. B. Mitchell

From the 1956 Edition (Sixth).
Mitchell Glass Company LTD., 40 Peckover Street, Bradford, Yorkshire. T.A., "Automitch". T.N., Bradford 20463/4 Established 1928.
Automatic glass-forming and fabricating machinery.

Managing Director W.B. Mitchell,
Secretary Miss K.A. Lennon

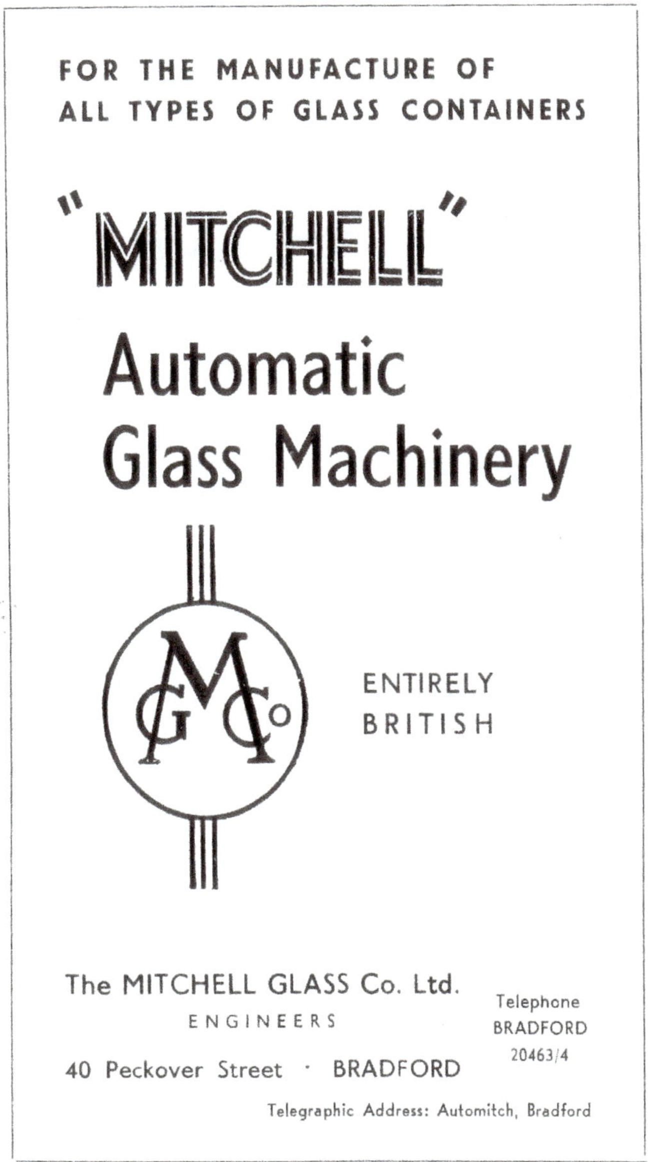

Figure 7,77. The 1956 Mitchell advertisement

From the 1959 Edition (Seventh).
An advertisement giving details of their products.

A-13

Automatic machines for the Glass Industry

Press and blow machinery

Press machinery

Blow & blow machinery—straight line, multi-head

Blow & blow machinery—¾ automatic

Feeder forehearths

Mechanical feeders

Stackers

Conveyors

Timers

Phase changing units

Electrical synchronised drives

Variable speed motors

RADIO VALVE MACHINES

Automatic sealing machines

Automatic base making machines

Bench machines

May we send you further particulars?
Write to

The Mitchell Glass Co. Ltd

Engineers & Consultants

VICTORIA WORKS, RAGLAN STREET, THORNBURY, BRADFORD 3, YORKS

Telephone: 64339 Telegrams: 'Automitch'

Figure 7,78. Mitchell's 1959 advertisement

Chapter 8

The rise and fall of the twin table machine

This chapter is basically two biopic's, the first is Frank O'Neill the company that he set up and what happened to them when he severed his connection with them and the second is about James Lynch, his company and what happened to it after he resigned.

Frank O'Neill. The O'Neill Machine Company

Frank O'Neill was born in 1862 and little did the public know how much influence he would have on the development of the glass bottle-making machine. At the age of 15 he was apprenticed to the Buckeye Glass Works in Ohio in America to learn mould making and the repair of moulds. When he had served his time he worked as a journeyman mould-maker in some half a dozen different glassworks.

As early as 1893 when even hand-operated machines were just starting to be used O'Neill set out to automate the pressing process of a side lever press of which there were many in use. His endeavours resulted in him developing a power driven press that incorporated its own feeding device. The machine can be seen in Figure 7,1 in the previous chapter on single table machines. Further information regarding his work, first started in 1896-1897, on feeders can be found in Chapter 6, The Development of the Feeder and the Forehearth.

In 1902 he developed a press which was possibly installed at the Hemmingray Glass Works in Muncie, Indiana. This was followed in 1908 by a twin table, four head, press and blow machine designed to make milks. Installed at the Travis Glass Company, of Clarksburg, West Virginia. See Figure 8,1.

Figure 8,1. O'Neill's press and blow milk bottle making machine, circa 1904

In this formative period he developed a machine for the manufacture of tumblers, and a press and blow machine which was capable of pressing 27 lids per minute before settling down to inaugurate a long line of blow and blow and suck and blow machines with which his name is permanently associated. From the beginning and right up to 1916 all of his machines were fed by hand, they never became automatic machines until the advent of the forehearth.

From now we will be confining ourselves to **Blow & Blow machines.**

His earlier models

Possibly the first O'Neill blow and blow machine to appear in any quantity was the No. 20, see Figure 8,2. By present standards this was a peculiar machine, it was equipped with two moulding tables each of which were equipped with two finishing moulds. A

Figure 8,2. One of Frank O'Neill's earlier machines

central column, located between the two tables, carried a blank head for each table. More moulds were considered as a means of increasing output.

After considerable research they launched their No. 21 machine which was the forerunner of more or less all of his subsequent machines. This machine was a semi-automatic machine, it incorporated six parison moulds and six blowing moulds. The machine was hand fed, the glass being packed into the ring and then blown back to form the parison. This was done automatically. The transfer from the parison to the blowing mould was done manually by a boy, the process being that he would remove the parison or blank, invert it, and placed it into the blow mould. The blow mould them advanced to the blowing station and finally to the take-out position, where the same boy removed the finished bottle and replaced it with another blank. The same cycle was then repeated.

The next machine to appear was the No. 25 which incorporated many improvements. This was also a six parison-mould, six blow-mould, two table, indexing machine, similar to the No. 21 but using a turnover drum, an automatic transfer from the blank-mould to the blow-mould, and an automatic take-out. This was a considerable improvement on the No. 21, in addition to all of the automatic features it was considerable more robust.

His intermediate period

The No. 25 machine was later re-designed by replacing the gear and rack method of turnover with rollers mounted on the drum face which meshed with a cam having gaps similar in size.

Figure 8,3. O'Neill's No. 21 machine

The transfer mechanism was speeded up as well. Speed was the objective of all of these modifications and additions.

About this time it was decided that if time was made available to finish the bottle than was needed to form the blank, production increases would be obtained. For this reason the No. 28 was born, it had six parison moulds and eight blow moulds mounted on a larger table. This allowed the parison table to complete one revolution more quickly than the finishing table, which due to its larger circumference required more time to make its cycle. This machine was recommended for large ware.

At this time the counterpart of this machine was the No. 30. As this unit was recommended for smaller ware it was equipped with six parison moulds and six blow moulds. Both of these machines were fitted with the double-pincer revolving type take-outs, which were found to be much more positive.

The next step was to upgrade and improve the twenties series of machines, these were named the No. 40, the No. 42, and the No. 44 machines which were to succeed the No. 25 and the No. 28 models. The smaller No. 48 machine replacing the No. 30 machine. Although there is evidence of a No. 36 machine I do not know if there was a Series 30 of indexing machines.

All of the improvements listed above were incorporated in the new machines along with improvements brought about by 'in the field' observations and the experienced gained from them operating in working conditions. The general construction of the units was considerably strengthened, and ball and roller bearings incorporated where none had been before. The tables were constructed to ensure that glass particles could not get at the bearings surfaces causing damage to them. A track arrangement as to a blank mould support was added along with many other improvements that were incorporated. These new improvements made the 40's series the most popular series of its day.

Until now all of the O'Neill machines had been air driven but the cost of compressing sufficient air to drive them led to the development of the No. 50 machine. The change to electric power for the drive meant that the machine would continuously rotate

Figure 8,4. O'Neill's No. 25 machine

the stopping and starting of the indexing machine was machine was now over.

The No. 50 was a two table, continuously rotating machine unit that performed all of its operations while on the move. The No. 50 machine was strongly built, and due to its smooth performance when turning it was exceptionally easy on wear and tear. It was almost completely cam controlled, and is therefore very positive in all operations.

This unit is not well known and has not proved very popular. D. M. Moody in his paper to the Society's Symposium on Machines for the Fabrication of Glass Containers held in Sheffield in February 1952, from which most of this information comes, stated that in his opinion that it was the difficulty in educating the operators and engineers to new ideas. Discerning people have

Above Fig. 8,5. O'Neill's No. 42 machine while below, Fig. 8,6 his No. 48 machine

O'NEILL MACHINE NO. 48

made the statement that this was the best O'Neill machine to be marketed.

Later developments

We now come to the later machines the 30's series, which include the No. 30 built for small ware manufacture which incorporates eight blank (parison) moulds and eight blow (finishing) moulds, the No. 36 for large ware using six blank moulds and six blowing moulds and the No. 38 with eight blank and blow moulds for larger volumes of large ware such as quarts and pints beers and minerals.

An important improvement was the drive, this in the case of the No. 30 was by a 3 H.P. motor, through a P.I.V. gear-box and worm shaft, on which is incorporated a clutch and limit switch which breaks the power when under strain, illustrated in Figure 8,7 to left. This shows the transmission of power from a shaft through to a heart shaped cam.

Figure 8,7. Transmission system

The power from the shafting is transmitted through a worm-and-gear to a heart shaped cam which makes a complete revolution for each station (or each bottle). This cam engages with one of eight rollers which are connected to the mould table gear.

Please do not confuse this No. 30 machine with the earlier No. 30 which was pneumatically driven and it was the last of the 20 series. This machine is the first of the 30 series.

The timing of when these machines were introduced to the market is unknown but reading old year books and factory descriptions we can put forward a tentative date when they were introduced.

Figure 8,8. The O'Neill No. 30 machine with electric drive

Number 21 - early 1920s
Number 25 - working on 3/11/1922
Number 28 - working on 3/11/1922
Number 30 - first No. 30, on 3/11/1922, number of machines working working 3, each with 6 Parison Moulds & 8 Finishing Moulds
Number 36 - working before end of the 1939-1945 war, assume that this took place before the No's 42, 44 and 48 were installed
Number 42 - by 1937
Number 44 - by 1937 (Abstract says working by 1929)
Number 48 - by 1937
Number 50 - probably before 1935
Number 80 - Knox-O'Neill. Working in 1929–1930. Designed in 1929 and 95 of them sold by the end of the first year.
Number 96 - two purchased in 1936-1937, one before end of 1939-1945 war, different user.

In 1936 Frank O'Neill sold his American interests in the company to the Owens-Illinois Company of Toledo, Ohio and the Hartford Empire Company of Hartford, Connecticut after which he retired. The O'Neill European Machine Company was formed to continue the business in European countries, originally it was based in Montreal, Canada, later the business was transferred to the O'Neill Machine Company Ltd, of Dudley, Worcester, in England.

Frank O'Neill was elected a Fellow of the Society of Glass Technology in April 1946, and he passed away on the 30th of November 1954 at the grand age of ninety two.

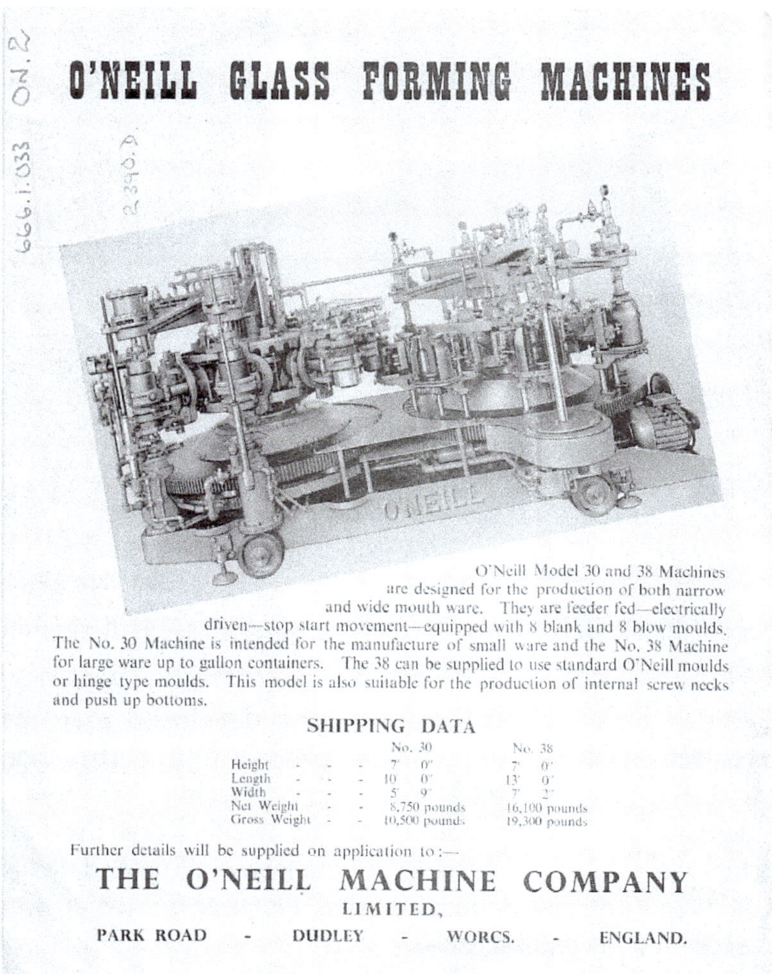

O'NEILL GLASS FORMING MACHINES

O'Neill Model 30 and 38 Machines are designed for the production of both narrow and wide mouth ware. They are feeder fed—electrically driven—stop start movement—equipped with 8 blank and 8 blow moulds. The No. 30 Machine is intended for the manufacture of small ware and the No. 38 Machine for large ware up to gallon containers. The 38 can be supplied to use standard O'Neill moulds or hinge type moulds. This model is also suitable for the production of internal screw necks and push up bottoms.

SHIPPING DATA

	No. 30	No. 38
Height - - -	7' 0"	7' 0"
Length - - -	10' 0"	13' 0"
Width - - -	5' 9"	7' 2"
Net Weight - -	8,750 pounds	16,100 pounds
Gross Weight - -	10,500 pounds	19,300 pounds

Further details will be supplied on application to:—

THE O'NEILL MACHINE COMPANY
LIMITED,
PARK ROAD - DUDLEY - WORCS. - ENGLAND.

Figure 8,9a. Brochure for Models 30 and 38 O'Neill machines

MODEL No. 30

Limits	
Max. Dia.	3"
Max. Neck Dia.	2⅜"
Max. Height	8½"
Min. Height	1½"

Mould Requirements

- 8 Blank Moulds
- 8 Blow Moulds
- 8 Finish Rings
- 8 Plungers
- 2 Baffle Plates
- 1 Funnel
- 8 Bottom Plugs
- 8 Plunger Sleeves

MODEL No. 38

Limits	
Max. Dia.	4¼"
Max. Neck Dia.	3"
Max. Height	13¾"
Min. Height	3¾"

Mould Requirements

- 8 Blank Moulds
- 8 Blow Moulds
- 8 Plungers
- 2 Baffle Plates
- 8 Finish Rings
- 1 Funnel
- 8 Bottom Plugs

ADVANTAGES

★ Double Blow Back—Superior Ware
★ Economical Electrical Drive
★ Adjustable Timer
★ Automatic Lubrication
★ Both Vacuum and Compressed Air to form Finish
★ Internal Cooling

CONSTRUCTION

These are very robust in design being composed of Alloy Iron and Steel castings—where necessary parts are hardened and ground. Drive is powered by a 3 h.p. Motor on the Model 30 and 5 h.p. on the 38. Ball and Roller Bearings ensure smooth operation and economy in maintenance.

Further details will be supplied on application to:—

THE O'NEILL MACHINE COMPANY

LIMITED,

PARK ROAD - DUDLEY - WORCS. - ENGLAND

Figure 8,9b. Brochure for Models 30 and 38 O'Neill machines

Besides blow and blow machines he developed a series of two table suction machines, these you will find in the chapter entitled 'The rise and fall of the Suction machine'. In it you will find descriptions of his main machines which were the No. 80 and the Nos. 93 and 96 series. The company in England designed and developed an in-line machine the Number 55 which will be discussed in that chapter.

Two sales leaflets describing the company machines are shown here, and these are followed by extracts from the British Glass

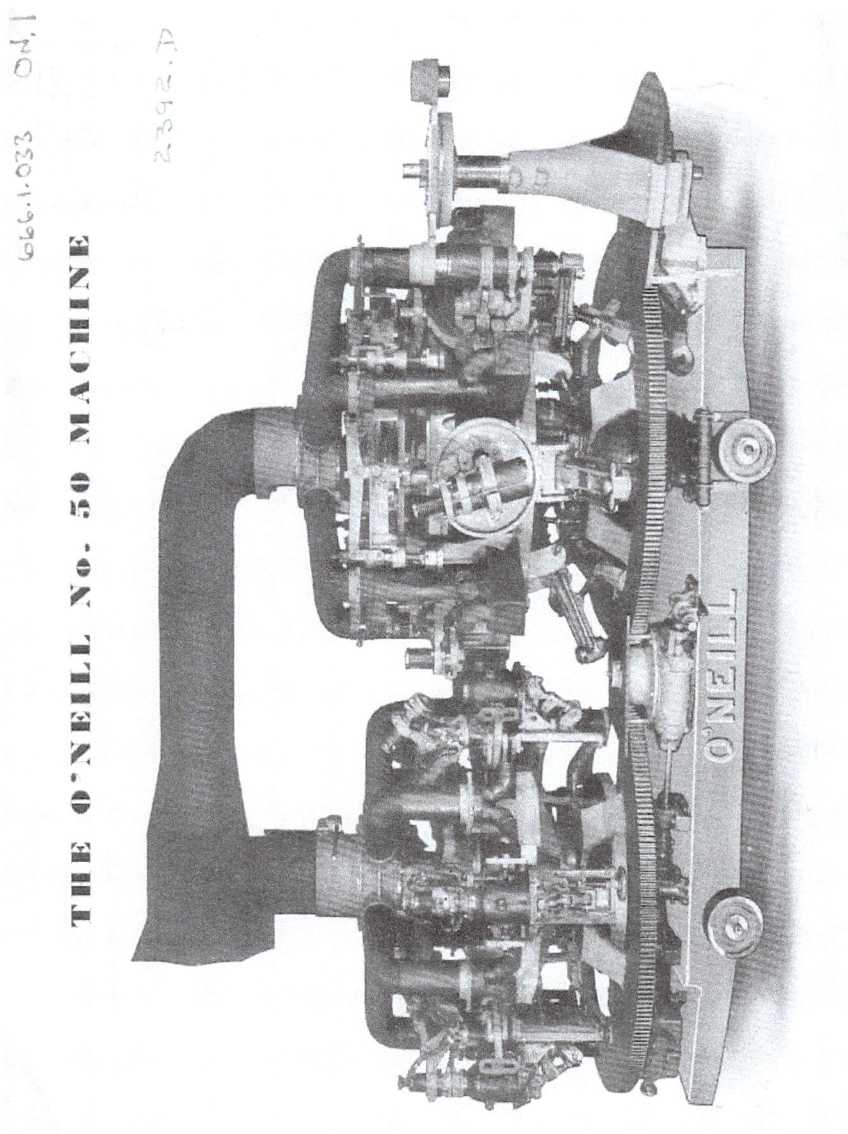

Figure 8,10a. Brochure for the O'Neill No 50 machine

THE O'NEILL No. 50 MACHINE IS A CONTINUOUSLY TURNING—ELECTRICALLY-DRIVEN—TWO-TABLE MACHINE, EQUIPPED WITH 6 BLANK AND 6 BLOW MOULDS. THIS MODEL CAN BE SUPPLIED TO USE EITHER THE STANDARD O'NEILL 40 SERIES OR WAIST BAND HINGE TYPE MOULDS, AS REQUIRED.

THIS MACHINE IS POWERED BY A 3-H.P. MOTOR AND FOR THIS REASON IS CONSIDERABLY MORE ECONOMICAL THAN THE OLDER AIR DRIVEN MACHINES WHICH REQUIRED A LARGE COMPRESSOR FOR DRIVING AIR. FURTHER ECONOMY IS MADE DUE TO THE CONTINUOUSLY TURNING MOVEMENT WHICH ELIMINATES THE STRAIN ON WORKING PARTS, THUS REDUCING COSTS OF MAINTENANCE.

THE MODEL 50 IS COMPLETELY CAM CONTROLLED AND IS THEREFORE VERY POSITIVE IN ALL MOVEMENTS. IT HAS THE ABILITY TO FORM FINISH BY VACUUM—AIR—OR BOTH. COMPLETE ADJUSTMENT IS PROVIDED ON BLOWING. INTERNAL COOLING CAN BE PROVIDED AT HIGHER SPEEDS.

LIMITS OF WARE

MAX. HEIGHT UNDER FINISH	12½"
MAX. BOTTLE DIAMETER	5½"
MAX. NECK DIAMETER	3"

SHIPPING DATA

HEIGHT	8′ 0″
LENGTH	12′ 6″
WIDTH	7′ 0″
NET WEIGHT	16,360 POUNDS

FOR FURTHER DETAILS PLEASE APPLY TO:—

THE O'NEILL MACHINE CO.

LIMITED

PARK ROAD DUDLEY, WORCS. ENGLAND

Cables: ONEILLANY-DUDLEY

Figure 8,10b. Brochure for the O'Neill No. 50

Industries Trade Directory.

Directory of the British Glass Industry - 1923 First Edition. O,NEILL MACHINE CO., Albany Works, Trenmar Gardens, Willesden Junction, London, N.W.10.

T.N. Willesden 2070, T.A., "O'Neillany", Kensal, London.

Moulds of all descriptions and the O'Neill glassworking machines and glass feeders. Also spare parts for O'Neill machines and feeders. (American Works: Toledo, Ohio, USA)

President; Frank O'Neill; First Vice-President Wilson M. O'Neill.

(See advertisement Page 199)

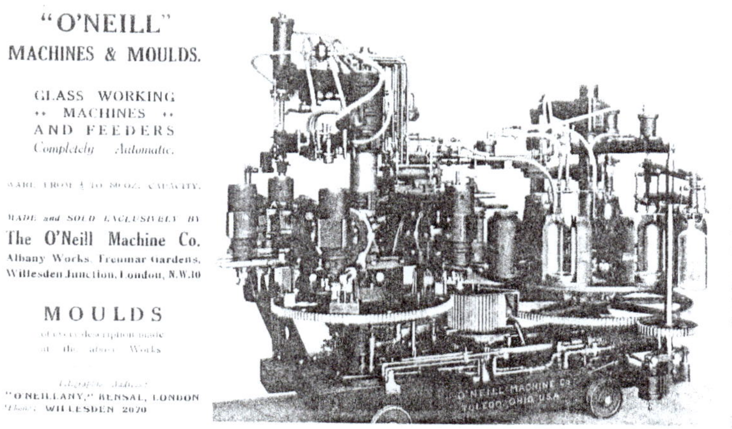

Directory for the British Glass Industry - 1934 Third edition. **O'NEILL MACHINE Company, 10/11 Dentral Buildings, 24, Southwark Street, London, S.E.1**. T.N. Hop 2330, T.A., "ONeillany, Boroh, London". (American Works: **Toledo, Ohio, USA**)

Moulds of all descrition. O'Neill glassworking machines and glass feeders. Also spare parts for O'Neill machines and feeders.

President: Frank O'Neill; First Vice-President: Wilson M. O'Neill.

After the severing with O'Neill's organisation in Toledo in 1936, an entry in the 1956 edition of the Directory of the British

Glass Industry along with an advertisement stated;
O'NEILL MACHINE Co. Ltd., Dormston Trading Estate, Burton Road, Upper Gornal, Nr., Dudley, Worcs. T.A., "Oneillany, Dudley". T. N. Sedgley 2038. Established 1910.

Glass-froming machines and glass feeders; proprietors of John Ferguson Patents for the electric heating of glass furnaces and feeders.

Managing Director: Douglas M. Moody.

Secretary: D. L. Lewis.

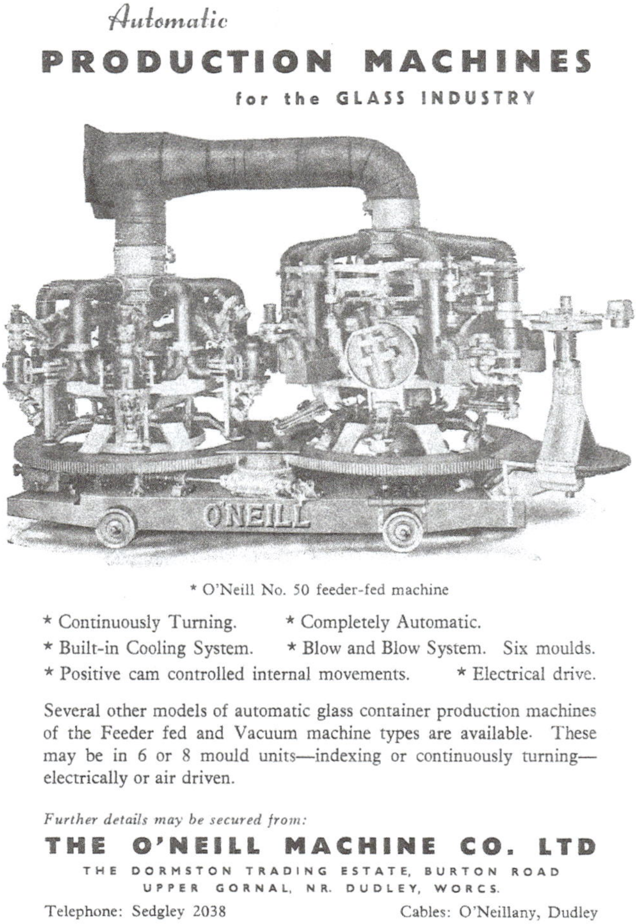

Figure 8,11. O'Neill's 1956 advertisement

An excerpt from The History of Forsters Glass entitled 'The O'Neill machine' describes the machines method of operation. This was written after the company was taken over by Rockware Glass in 1968:

"This machine consists of two circular tables, each carrying six (or eight) moulds and each rotating round a central column carried by a very firm base. The machine is driven by compressed air (true at this plant but there were electrically driven models as well) and controlled automatically in all off its actions by air valves. It is also synchronised to work with the plunger and cutting-off shears of the feeder.

The table immediately beneath the feeder carries what are known as the 'parison' moulds. In these the neck-ring and the neck of the bottle or jar are formed and the article partially shaped in the body. The second table carries the 'blow moulds' in which the article receives its final shape. The two tables pause and rotate together.

The making of a bottle can be described as follows:

1. The 'gob' of glass drops into the first parision mould which is in an inverted position with the neck-ring and the 'plug' at the bottom.
2. As soon as the 'gob' is in the parision mould, a plate slides over the bottom and closes it. A valve is then opened automatically and compressed air entering forces the glass round the plug at the bottom and into the neck-ring. The plug serves to pierce the lump of glass leaving a hole for air to enter subsequently and blow the bottle.
3. The plate next slides back, the plug is withdrawn, and the table rotates, the mould passing to the next position.
4. At this point a solid head or 'baffle' is pressed down upon the parison mould to close the opening and air is admitted fro below, blowing the glass to the shape of the parision mould.
5. The baffle is now removed and the table once more rotates. During this rotation the parison mould revolves so that when it is nearest the blow table it is no longer

inverted but is neck upwards.

6. As the table stops the mould opens, leaving the parision 'blank' as it is called, suspended by the neck-ring. This 'blank' then moves across until it is over the corresponding 'blow mould' on the second table.
7. The blow mould now closes on the glass and simultaneously the neck-ring opens leaving the 'blank' in the blow mould.
8. The table once more rotates and at the next stop a 'blow-head' descends over the glass neck-ring. Compressed air enters the 'blank' and the bottle is blown to fill the blow mould, the inside shape of which is that of the body of the bottle.
9. At the next stopping place of the 'blow' table a further blowing takes place or the mould may move on to the final position when the bottle is removed, either automatically or by hand. It is then placed on the conveyor from whence it is transferred to the annealing lehr."

Written information regarding Frank O'Neill is sparse except for that supplied by D. Moody in 1952 and that from the Forster's history, the following are miscellaneous articles regarding his work the first two are from abstracts in the Journal of the Society of Glass Technology and the others are unidentified photographs of O'Neill machines.

#65. Semi-automatic Glass-blowing Machine. F. O'Neill, USA (Brit. Pat., February 6th, 1919, No. 144 344. No. 2932). [1921 J.S.G.T., Volume V]. - Relates to a semi-automatic pneumatically-operated machine for blowing glass bottles, etc, of the type in which turn-over mechanisms are carried on rotating tables.

According to the invention, the parisons are each transferred mechanically from the parison moulds to the blowing-moulds and the circular paths of the moulds do not overlap. The table, 17, supporting the parison moulds, 95, and the table, 35, supporting the finishing-moulds, 197, are rotated intermittently through an angle of 60 degrees by a rack, 25, and long pinion, 32. At the first station, the inverted parison mould is charged by hand, a valve lever being tripped by the punty so as to start the mechanism.

The excess of metal is cut off automatically by shears, a cover is brought down on to the mould and air admitted so as to consolidate the metal, the neck is formed by a plunger, which rises from below, and a small puff of air is admitted to prevent the soft glass from falling into the initial neck opening.

At the second station the blank is blown in the parison mould. In passing to the third station it is inverted by the action of gearing, 115, 148. At the third, or transfer, station, the parison

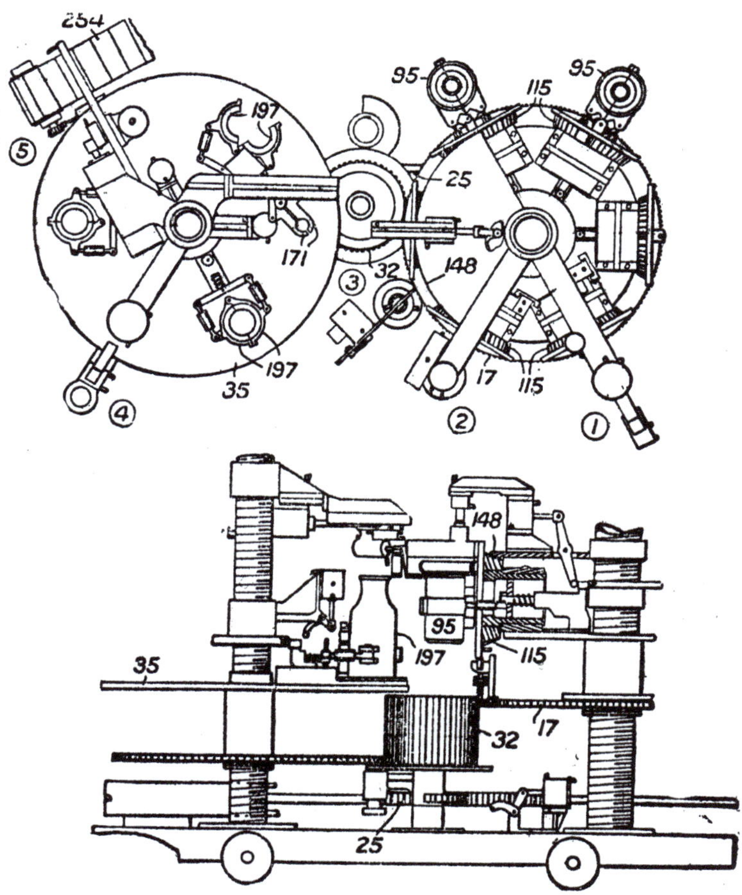

Figure 8,12. F. O'Neil's 1919 Patent application

and the neck moulds are opened and the blank is transferred by tongs, 171, to a finishing mould, 197.

In a modification, each finishing mould moves radially outwards on the table, 35, to receive the parison instead of transfer tongs, 171, being employed. At the fourth station, which is 120 degrees distant from the third, the final blowing takes place. At the fifth station the finished bottle is placed on a conveyor, 254, which removes it to the lehr.

#227. Transfer Device for an Automatic Bottle-blowing Machine. F. O'Neill, Toledo, Ohio, (USA Pat. No. 1 462 652, July 24th, 1923. Filed June 14th, 1920, No. 388 724). [1924 J.S.G.T., Volume VIII]. - The invention, relates to the O'Neill "No-boy" machine, and consists chiefly of a device for transferring the parisons from the blank moulds to the finishing moulds.

The blank moulds, 11, are mounted on vertical circular plates, 5, which, as they approach the transfer station, are inverted so as to bring the parison, 70, mouth upwards. Projecting from the plates are parallel guide rods, 72, on which a slide member, 73, is mounted. The halves of the ring moulds are pivoted on a pin,

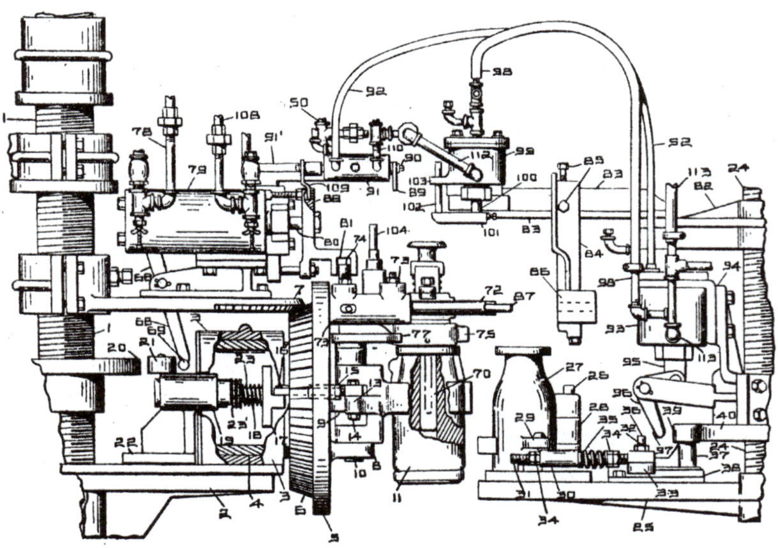

Figure 8,13. O'Neill's Transfer Device

75, carried by the slide, and a U-shaped spring, 77, keeps the ring mould closed.

When the mould comes to rest in the transfer position, the pin, 74, rests in a fork, 81, attached to the piston rod, 80, of an air cylinder, 79. As soon as the parison mould opened, the piston rod, 80, moves outwards carrying the slide, 73, and the parison is transferred to the open finishing mould, 27. This mould then closes on the parison and air is admitted to a cylinder, 99. Its piston rod, 100, carries a foot, 101, which depresses a rod, 104. This brings about the opening of the ring mould, and the reversal of the operations, leaving the parison in the mould, 27.

The following are unidentified photographs of O'Neill machines. The first is from the 1923 edition of the British glass trade journal, so it must have been in operation before that date, the same for the second which was printed in 1924, and finally the last one which worked at Foster's glass in St. Helens.

Figure 8,14. An O'Neill machine prior to 1923. It has two six section tables so it cannot be a No. 28 machine

Above Figure 8,15. An early O'Neill machine. Prior to 1924 while below Figure 8,16 one of Foster's O'Neill machines

James W. Lynch. The Lynch Glass Machinery Company and The Lynch Corporation.

James Lynch's history follows that of Mike Owens in that he served an apprenticeship in glass blowing but unlike two other earlier pioneers, Edward Miller and Frank O'Neill, James Lynch found when he started his business that times had changed quite radically as the industry was now becoming more mechanised.

According to the information kindly given to the writer (Meigh) by Andrew G. Doll (who was a colleague of Ed. Miller for thirteen years before joining the Lynch Corporation in 1933 - this was when Ed. Miller sold out to the Lynch Corporation) James Lynch 'could fashion vases of such intricate beauty that they were an exact image of a collectors rarity, including bottles of vari-coloured sizes and shapes, door stops, and many other kindred items'.

According to Doll, the idea of forming the Lynch Glass Machinery Company was born in the minds of James Lynch and Edward G. Bridges when in 1916 they were employed, the latter as a mould maker, at the Root Glass Company, at Terre Haute, in Indiana, USA. They had witnessed the installation by Ed. Miller of several of his early glass making machines at the Root factory, these were single table machines suitable for press & blow but Lynch and Bridges believed that a two table machine, which would be an improvement on the old Teeple-Johnson machine, would find a good market.

The only other reference that I have found about this machine was in "Glass - The Miracle Maker" by C.J. Philips under 'press and blow' machines in his book he says that "The Teeple-Johnson machine which had two turrets each carrying three moulds. A gatherer and either two or three operators were still required. In later developments the number of moulds was further increased and electric motors or compressed air were used to rotate the moulds and to operate the plunger and blow heads.

Both single and double table machines were used. In the single-table system the blank mould and the blow moulds alternated; in the double-table system the blank moulds were all on one table and the blow moulds on the other, the two tables being synchronised with one another." It has been included here but the second paragraph seems to me to be talking about semi-automatic machines in general, not about one particular machine.

They resolved to go into business together, James Lynch to be responsible for administration and sales and Ed. Bridges for design and engineering. Their first commission was from the Obear-Nester Glass Company of St. Louis, Missouri, in the United States of America, which was a success. It became known as the 'Cock-Roach' machine and its sale provided capital for the Lynch Glass Machinery Company to go ahead with their plans.

Instead of embarking on the erection of a manufacturing plant, at the outset they had the machines that they designed built by the Dice Machine Company. This decision was to have an important effect on the Lynch Company later, because there was at that company one Thomas Chandler Werbe who afterwards took over the business of the Lynch Glass Machinery Company.

Meanwhile the company was progressing well, the 'L' machine named after Lynch followed the 'Cock-Roach' which at once achieved popularity. By the middle of 1921 some 230 L-type machines had been sold. The L was followed by the LA model, which incorporated both automatic transfer and automatic take out. This model eliminated both the 'transfer' boy and the 'take out' boy so it was marketed as the 'No-Boy' machine. Other models soon followed the 'B' in 1921, the 'F' in 1923, the 'R' in 1925 and the 'RS' in 1927.

In 1928 James Lynch sold his interests to T. C. Werbe and others and the name of the company changed to the Lynch Corporation. Lynch's decision to retire from business was largely due to his dislike of litigation suits which at that time there were far too many in the glass industry.

One patent lawsuit in which Lynch was involved was for infringement of the 'Badger' patent which was held to cover both the transfer and take-out mechanisms. Ultimately the Lynch Corporation had to pay $25 000 to use the Badger patent, a large sum of money for a small company to find, equal to the return on eight of their machines.

The work of the Lynch Corporation continued and in 1929 they built the 'LB' machine and in 1930, the most popular of all of their machines, the model '10', so designated because it was the tenth model designed by the Lynch organisation. Type '16' followed and wide-mouth varieties of the model 10. In the late 1940s the dual version of this popular machine made the market and it found popularity due to the increase in production that it gave.

The repeal of the Volstead Act in December 1932 brought prohibition to an end and created a large new market for beer bottles. The demand for glassmaking machinery increased correspondingly and the Lynch Corporation were not slow in taking advantage of it.

In 1933 the Lynch Corporation acquired the business of Ed. Miller, the Miller Machine and Mould Works, which was solely owned by him.

With the extensive range of Miller machines added to their own, Lynch Corporation were in an excellent position to cover the requirements of the glass container industry, for the two companies had to a large extent been non-competitive, Miller specialising in press and blow while Lynch had concentrated on blow and blow.

The eleven years spent by James Lynch in the glass machinery world, although comparatively short, were sufficient to number him among the leaders of the glass machine age, and his name will remain in the annals of the industry. He was of a reserved disposition but a very agreeable personality. To quote Doll again: "Jim Lynch was a quiet man. He usually chose the right words and closely defined them in such a way that he never used the 'you-know-what-I-mean' school of diction. He weighed his words carefully'" He was a pioneer and his initiative has served the industry well.

His first machine was the "Cock-Roach", we do not have a photograph of this machine but in the abstracts in the 1923 Journal of the Society of Glass Technology there are three patent applications by O. Y. Imray of London, England and the

Obear-Nester Glass Company of St. Louis, Missouri, USA. All are applications for British patents, all are dated the same day - September 12th, 1921 - one is for a machine, Abstract Number 284 and two others for handling equipment - one for a take-out mechanism Abstract Number 191 and the other for a glass steadying device Abstract Number 192.

There is no evidence that these are connected to the "Cock-Roach" machine installed at Obear-Nester Glass but as the machine one is for a 'hand fed' machine, the United States patent if there was one, would have been made a number of years before this date. The rational behind this theory is that these patents are possibly for improvements to the original "Cock-Roach" machine.

The first one for the machine is very long and includes three illustrations, the other two are well illustrated but much shorter.

#284 Improvements in Glass Bottle Blowing Machines. O.Y. Imray, London, and the Obear-Nester Glass Co., St. Louis, Missouri, USA (Brit. Pat., No. 193 080, September 12th, 1921) - Relates to machines of the kind comprising two rotatable tables carrying the blank and finished mould respectively and provided with inverting and transferring mechanism between the moulds and with a punty-operated element actuated by the lowering of a gather into a blank mould to control pneumatic means for severing and packing the gather, opening the mould, operating the transferring and inverting mechanism to convey the blank within the finishing mould.

According to the invention, the shearing mechanism is combined with a blow head for packing the gather, the transfer mechanism is adapted to grip the necks of the inverted bottles in the blank moulds and to deposit them in an upright positioning the finishing moulds, and the blank moulds are covered by a vertically movable, but non-rotary top disc carried by the piston rod of a pneumatic cylinder upon which the mould table rotates, the disc being cut away at one sector to enable the bottles to be removed from the blank moulds.

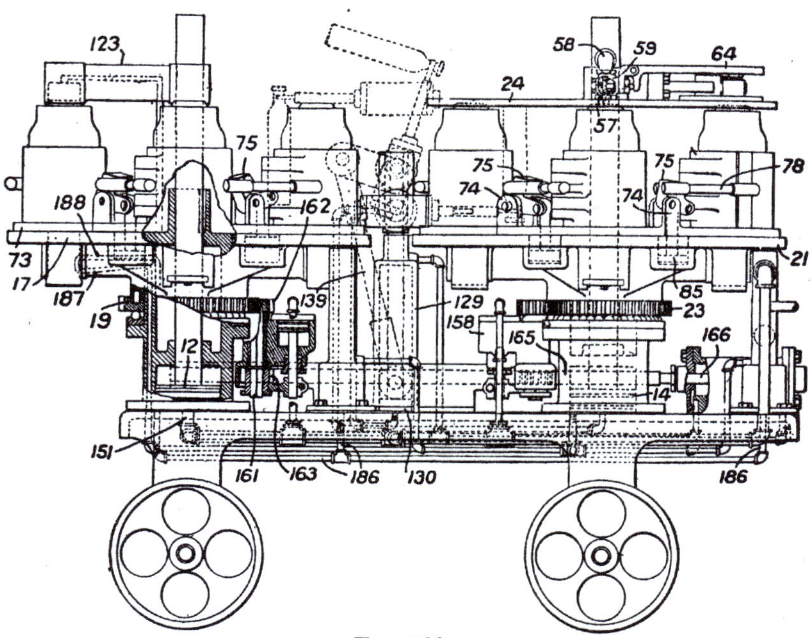

FIG. 130.

Figure 8,17a. Imray's and Obear-Nestor Glass Co. patent application. Part 1

Figs. 130, 131 and 132 illustrate a five-station machine, the operation at the various stations being as follows, namely, for the blank table, at station 1, charging, shearing, packing, and removal of the neck pin; station 2, initial blowing; station 3, transfer to finish mould; stations 4 and 5 idle; for the finishing table, at station 1, blank received and mould closed; station 2, idle, allowing glass to cool; station 3, bottle blown; station 4, idle; station 5, finished bottle removed.

A combined shearing-plate and blow head, 38, is mounted upon a stud, 35, on the top disc, 24, which is provided with an opening, 28, at the first position for the introduction of the charge. As the latter is introduced, the punty depresses a lever, 64, opening a valve, 59, which admits compressed air from pipes 182 and 154, pipe 72, and pipe 57, to a pipe, 58, and a cylinder, 46, to rotate the shearing-plate, 38, and sever the charge. As the movement of the shearing-plate finishes, compressed air is admitted through pipes 66, 67, and 69, to a port, 36, in the stud, 35, and passes through a passage, 43, and cluster of air vents, 45, in the shearing-plate to the mould, thus effecting the packing of the gather about the neck pin at the bottom of the mould.

A portion of the compressed air passes from the pipe 67 to the pipes 68, 189, thus depressing the piston, 172, throwing the pipe 154 open to exhaust and admitting compressed air from the pipe 182 to the pipe 170. While the piston 172 is in its upper position, pistons 12, 14 (Fig. 130), are held in their lower positions by compressed air admitted above them, thus depressing the blow head, 123, and top plate, 24, respectively into contact with the moulds beneath.

As soon as the piston 172 rises, the spaces above the pistons 12, 14 are thrown open to exhaust and air is admitted from pipe 170, beneath these pistons, which rise to their upper positions freeing the mould from engagement with the blow head, 123, and top plate, 24. The upward movement of the piston 172 is affected by compressed air admitted from the pipes 188, 187, 186, after the rotation of the tables 17, 21 is completed. The upper end of the piston 175 engages with holes, 174, in the rotary table, 21, when the upper position so as to lock it and prevent rotation.

At the moment when piston 14 rises, compressed air is admitted from the space below the piston to the space above a piston, 100, Fig. 131, withdrawing the neck pin as shown.

The pipes 154, 170 are connected to the upper and lower ends of two vertical cylinders, 158, adapted to raise and lower gear wheels, 163, slidably mounted on shafts, 161, carrying pinions, 162, in engagement with ring gears, 19, 23, secured to the rotary mould tables. When in the upper positions, the gears, 163, engage with the rack, 165, operated by the piston rod, 166. When compressed air is admitted to the pipe 170, the gears, 163, first engage with the rack and the piston rod 166 is then forced outwards, thus rotating the moulds tables through 73° to bring the moulds to the next station. When the piston 172 rises, the gears, 163, drop out of engagement with the rack, and the piston rod 166 moves

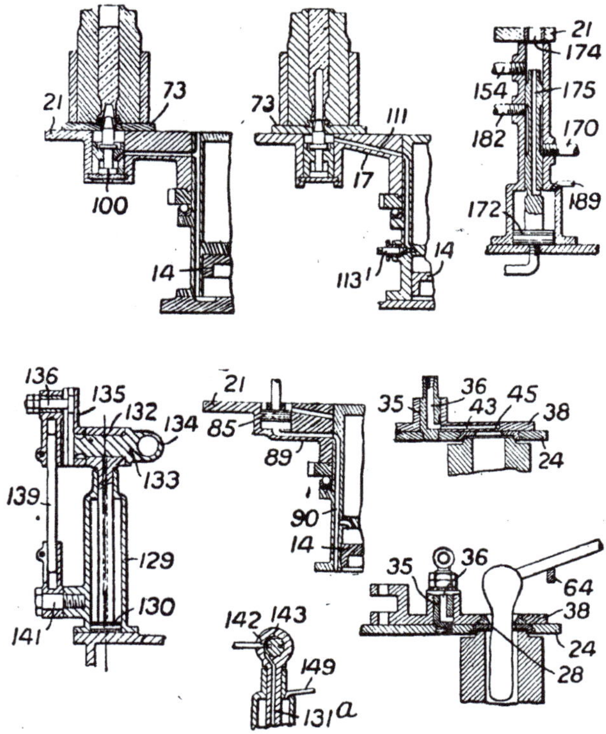

FIG. 131

Figure 8,17b. Illustration Part 2.

inwards ready for the next operative stroke.

Initial blowing of the blank is effected at station 2, by the admission of compressed air through a port, 111, from the space above the piston 14.The extent of blowing is regulated by a throttle valve, 113.

The mould plates, 73 (Fig. 130), carry projections, 74, on which are mounted bell crank levers, 75, connected by links, 78, to the half sections of the moulds. Cylinders are formed in the mould plates, 21, beneath each bell crank lever containing pistons, 85, which are raised just before a mould reaches the third station by the engagement of the ports, 89, 90, as shown.

Simultaneously with the opening of the mould at the third station, air is admitted from the port, 90 (Fig. 131), through a

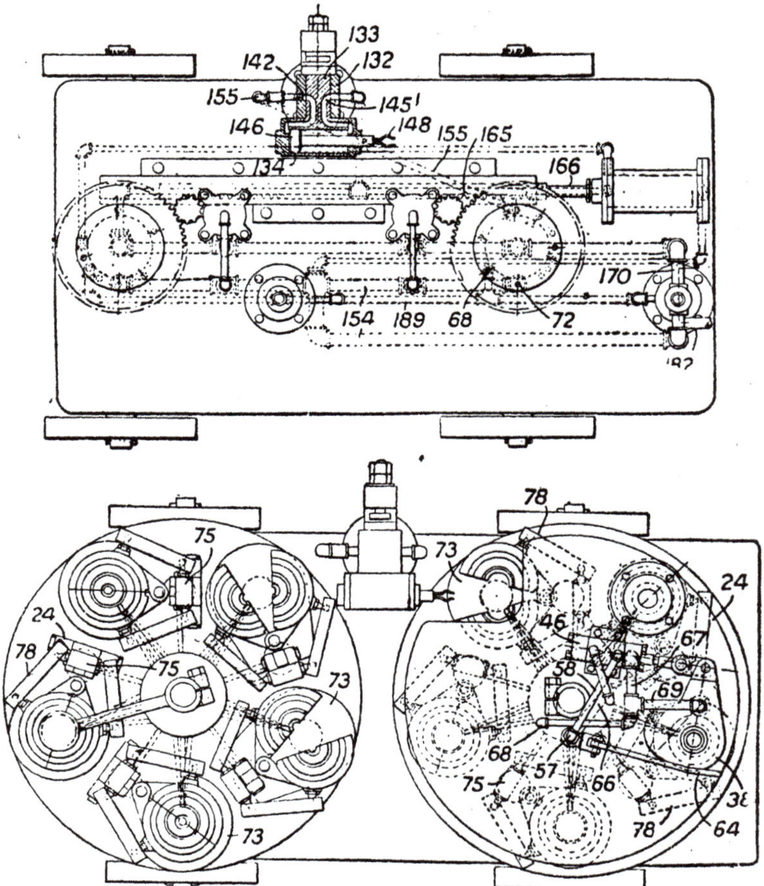

FIG. 132.

Figure 8,17c. Illustration Part 3.

pipe, 155 (Fig. 132), to a port, 142, and cylinder, 134, driving a piston, 146, outwards so that the grippers, 148, carried by carried by the piston rod engage the neck of the bottle in the open mould. The cylinder, 134, is mounted on a stud, 133, rotatably mounted in a bearing, 132, which can be raised and lowered by a piston, 130, working in a cylinder, 129. A crank arm, 135, is carried by the stud, 133, the crank pin, 136, being adjustably coupled by a rod, 139, pivoted at 141 to the cylinder, 129.

Immediately after the grippers, 148, have engaged a bottle, compressed air from the port, 142, passing through passages, 143, 131, to the space below the piston, 130, forces the latter upwards and operates the transfer device as indicated in dotted lines in Fig. 130, to deposit the bottle in an upright position in an open finishing mould.

As the movement is completed, the port, 145[1], comes into engagement with the port, 142,and the gripper is retracted from the bottle. The transfer mechanism is returned to its original position by compressed air passing from the lower side of the piston, 12, through a pipe, 151, and pipe, 149, to the space above the piston, 130.

The mechanism for opening and closing the finishing moulds and for raising and lowering the blow head, 123, are of similar construction to those employed for those employed for operating the blank mould. Specification 193 365 is referred to.

As promised the next two are much shorter, the first is;

#191. A Take-out Mechanism for Glass-blowing Machines. O.Y. Imray, London, and the Obear-Nester Glass Co., St. Louis, USA (Brit. Pat., No. 190 224, September 12th, 1921).

A pneumatically-operated transfer mechanism for glass-

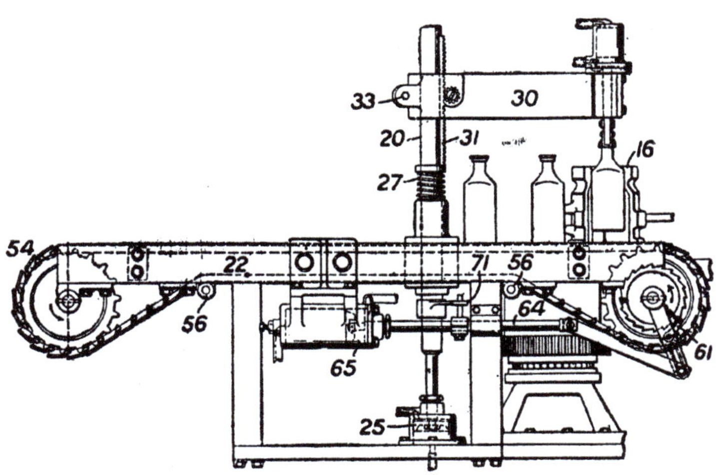

Figure 8,18. The Imray & Obear-Nesters take-out mechanism

blowing machines is mounted upon a cross-arm carried by a vertical shaft which is lifted and rotated pneumatically to raise the article from the mould and deposit it on a conveyor.

The take-out mechanism comprises a pair of grippers, 43, which are raised and lowered by a pneumatically-operated piston, 35, and caused to engage or disengage the neck of the bottle or other article when in their lowermost position through action of levers, 42, engaging with cam slots, 40, of the shape shown. The take-out device is mounted on the end of an arm, 30, adjustably supported by a rack, 31, and clamping device, 33, on an upright shaft, 20, which is raised and lowered by a pneumatically-operated piston, 25. A buffer spring, 27, is provided to cushion the downward movement. The arm, 30, and shaft, 20, are oscillated, to convey the bottles from the moulds, 16, to the conveyor, 54, by a pneumatically-operated piston, 65, and piston-rod, 64.

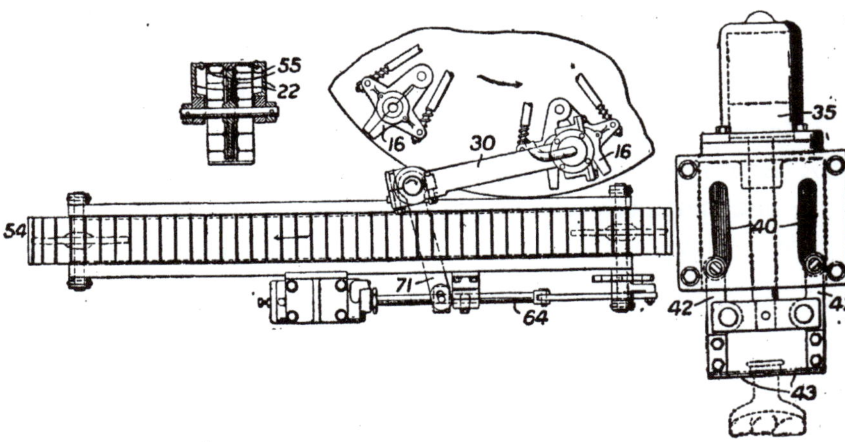

Figure 8,19. Imray and Obear-Nestors take-out device

The reciprocation of the piston-rod, 64, also operates the conveyor band, 54, through the ratchet device, 61, shown in Fig. 84, the arrangement being such that the conveyor moves forward while the arm, 30, is swinging over the moulds, 16, and remains stationary while the arm, 30, swings over the conveyor. Sagging of the conveyor, 54, is prevented by metal strips, 55, riveted to the flanges of the side rails, 22, and by idler rollers, 56, on the underside of the conveyor.

#192. A Steadying Device for Glass-blowing Machines. O.Y. Imray, London, and the Obear-Nester Glass Co., St. Louis, USA (Brit. Pat., No. 190 225, September 12th, 1921).

In glass-blowing machine as described in the preceding abstract, a steadying device is combined with the take-out mechanism to steady the article prior to its engagement by such mechanism and also after the article has been released therefrom. The arm carrying the take-out device may also be pivoted to swing vertically, in which case the vertical shaft carrying the arm is given a horizontal oscillation only and not a vertical movement.

The steadying device comprises a spring-pressed disc, 77, carried by a stern, 78, provided with a cross-pin, 79, which is free to slide in slots, 80, formed in the piston-rod of the piston, 35. The arm, 30, carrying the take-out device is pivoted at 24, and oscillated vertically at the proper intervals by a pneumatically-operated piston, 27. The vertical shaft, 20, is

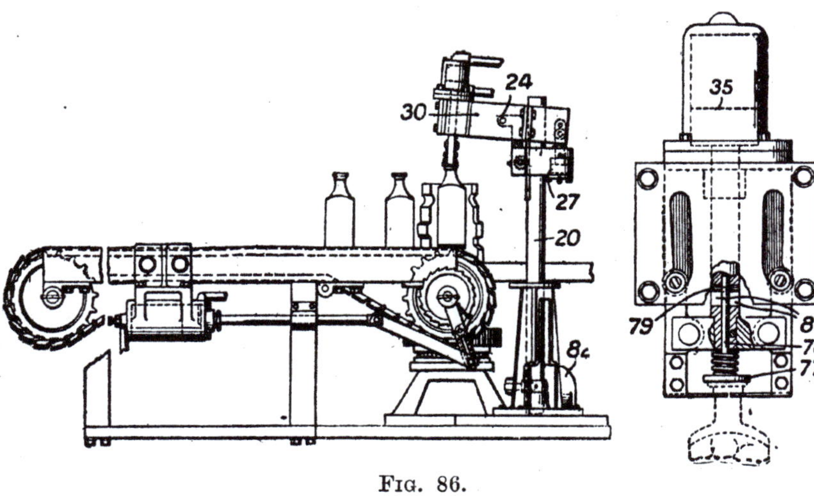

FIG. 86.

Figure 8,20. Imray & Obear-Nester's steadying device patent application oscillated horizontally by a pneumatically-operated piston working in a cylinder 82, mounted at the base of the shaft.

The first machine which we do have photographs for is the Lynch model 'B', this was introduced in 1921 and is illustrated in Figure 8,21 below, the source is from much a later catalogue. The machine is an air driven, intermittently rotated machine.

The brochure says that 'This machine is suitable for all kinds of small bottles. It is ideal for perfume, medicine, toilet, cosmetic, prescription, vaseline, or other bottles from $\frac{1}{4}$ ounce to 8 ounce contents; or for bottles that do not exceed a height of 6 inches under the finish'.

Figure 8,21. The Model Lynch 'B' also known as the 'Baby Lynch'

The machine specifications are;

Class of Glassware	- narrow neck and semi-wide mouth bottles exclusively.
Capacity	- from $\frac{1}{4}$ to 8 ounce contents.
Maximum diameter of finish	- $1\frac{3}{4}$ inches.

Maximum body diameter of article	- 2 inches.
Height under finish ring holder	- $4\frac{3}{4}$ inches. If the neck ring holder is turned over or reversed then it is possible to increase this to 6 inches.
Feeding the machine	- by automatic feeder, not recommended by hand gatherer.
Speed	- up to 45 articles per minute. See speed graph.

Services required are;

Cooling air requirements	- approximately 125 cu. ft. of free air per minute at 40 psi at the machine.
Cooling wind requirments	- a minimum of 4000 cu. ft. of free air per minute at a gauge pressure of 4 inches of water, or about 4 inches of fan air.

Just to give you some appreciation of the size of a 'Baby Lynch' the following are the dimensions for boxed up machines for export, along with their gross and nett weights.

	Machine Type		
	Baby Lynch	Lynch 'R'	Lynch '10'
Length	6 ft 10 inches	10 ft 4 inches	11 ft
Width	4 ft 2 inches	5 ft 6 inches	6 ft 4 inches
Height	4 ft 5 inches	5 ft	6 ft 8 inches
Gross weight	4100 pounds	9600 pounds	16 000 pounds
Nett Weight	3000 pounds	8300 pounds	14 000 pounds

Table 8,1. Sizes of boxed machines for export along with their weights

The nett weight of a Lynch 16 - B machine is 32 000 pounds, more than ten times the weight of the 'Baby Lynch'.

Figures 8,22 and 8,23 show the neck ring holder in the normal position and the inverted position.

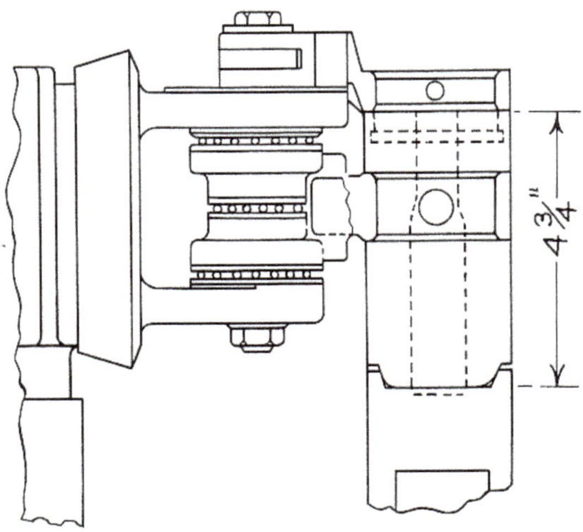

Figure 8,22. Neck ring in the lower position, i.e. the normal position

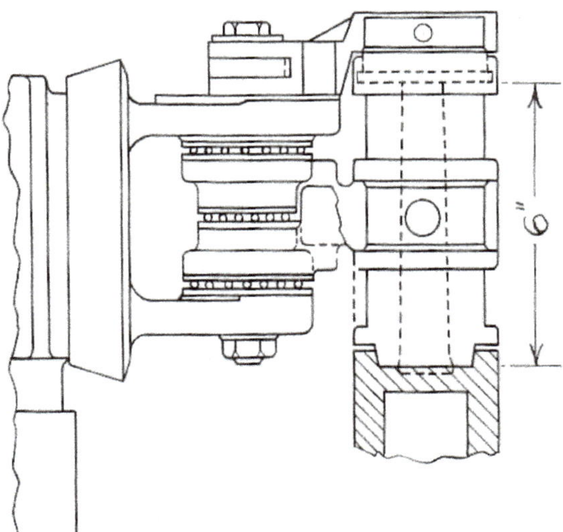

Figure 8,23. Neck ring shown in the upper or reversed position

Figure 8,24. Some of the bottles made by the 'Baby Lynch'

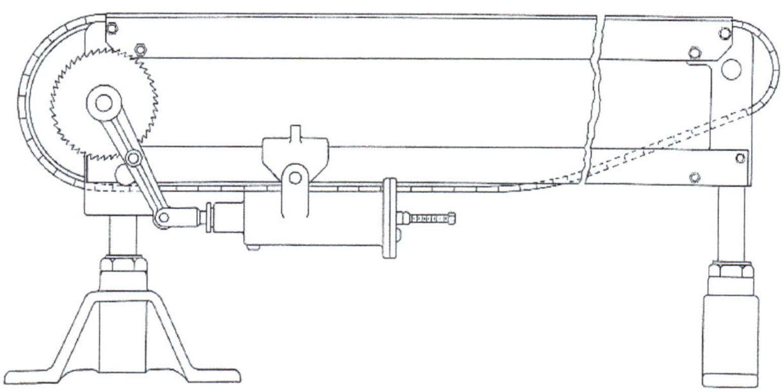

Figure 8,25. A conveyor which will unload most of the production

The conveyor shown above is the same air operated conveyor that is used for the Lynch 'L-A' and the Lynch 'R' machines. It can be used with the Lynch 'B' machine but its maximum speed is around 35 pieces per minute. The speed curve that follows shows that over the machines weight range, the speed ranges from 30 to 45 pieces per minute. If the ware to be made will always be in the $4\frac{1}{2}$ to 7 ounce weight range then this conveyor will suffice, if a wide range is wanted then an alternative conveyor needs to be found.

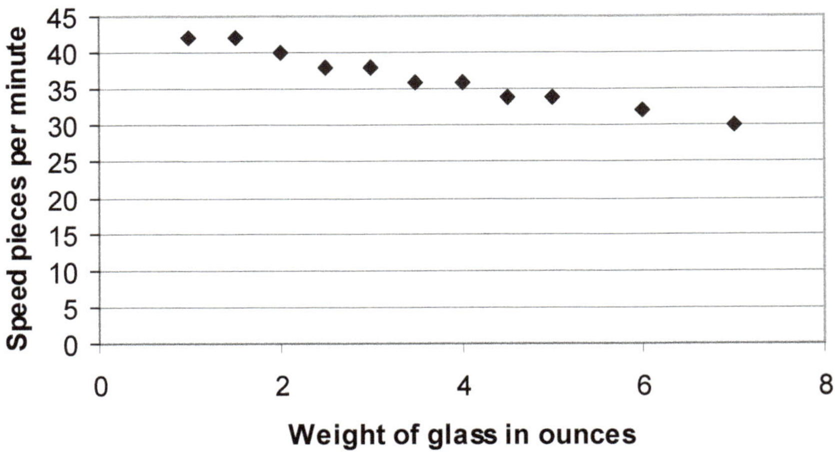

Figure 8,26. Speed curve for the Lynch 'B' model

The machine will manufacture bottles that are less than one ounce in weight, the speed quoted for all of these fractional weights is 45 pieces per minute.

The minimum amount of mould equipment required to make the machine work satisfactory is;

6 Blow moulds and bottoms, 6 Blank moulds,
6 pairs of finish rings, 1 Plunger,
1 Baffle plate, 1 Guide ring, or funnel,
2 Blow-heads.

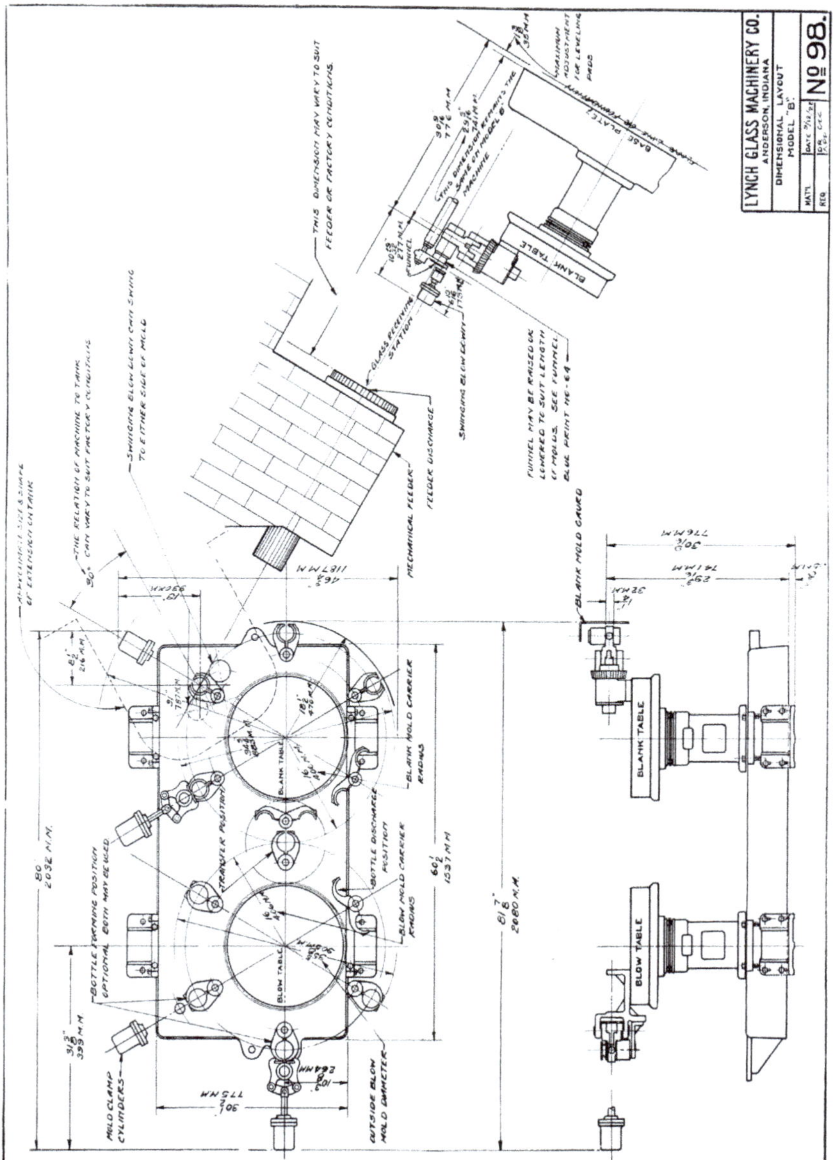

Figure 8,27. Layout plan for the Lynch 'B' machine

The following Figure 8,28 is probably that of a 'B' machine but it could be an 'F' machine which came out that year.

British Glass Industry Directory. 197

AUTOMATIC

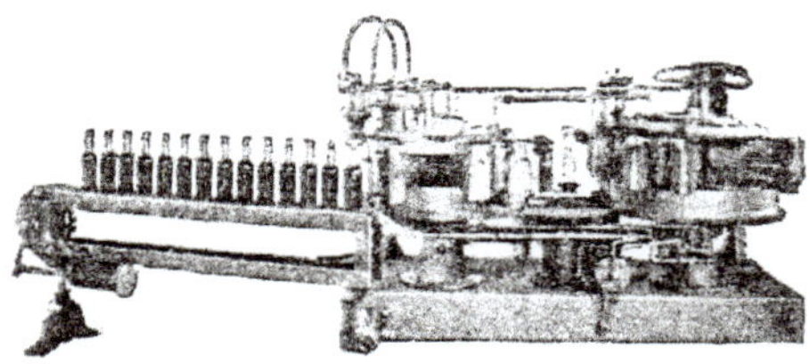

May be Worked on Floor Level—No Pit Needed

NARROW NECK AND WIDE MOUTH

Bottle Blowing Machine

No Transfer Boy   No Take-Out Boy

Speed 10 to 32 Bottles per minute ranging in size from ½ oz. to Half-Gallon

STRICTLY AUTOMATIC

When Operated With Mechanical Feeding Device

May also be worked by Hand Gatherers

Lynch Glass Machinery Co.

Anderson, Indiana, U. S. A.

CABLE ADDRESS "LYNCHNOBOY" ANDERSON

CODES A. B. C. 5th ED. BENTLEY'S WESTERN UNION

Figure 8,28. Taken from the British Glass Industy Directory 1923 edition of an advertisement placed by Lynch Glass Machinery Co.

The next machine that we have a record for is the 'R' machine which was first introduced in 1925. The series of machine mentioned here are similar in design and construction to the Lynch 'L-A' except that they is built heavier, and are capable of producing more bottles and jars of the larger sizes.

The 'R' machine is shown in Figure 8,29. All of the machines are of the twin-table type, and the series consists of;

Model 'R'	- 6 blanks and 8 finishing moulds.
Model 'RS'	- 6 Blanks and 6 finishing moulds.
Model 'R-A'	- 8 blanks and 8 finishing moulds.
Model 'R-00'	- 6 blanks and 8 finishing moulds.

The specifications for the Lynch 'R' are;

Class of glassware	- narrow mouth, wide mouth and semi-wide mouth bottles and jars. Such as Beer, Wine, Whiskey, Flasks, Sodas, Pickle, Olive Oil, Catsup, Vinegar, Fruit Jars, Packers' Jars, and other glass containers.
Capacity	- from 3 ounce to 70 ounce capacity, or two litres.
Finishes	- suitable for all kinds.
Maximum diameter of finish, with special wide mouth neck-ring holders	- $2\frac{5}{8}$ inches.
Maximum diameter of bottle	- 5 inches.
Minimum height under finish, with special blank-mould cages	- 2 inches.
Minimum height under finish, with standard equipment	- $3\frac{7}{8}$ inches.
Maximum height under finish	- 11 inches.
Maximum height under finish, with extra height equipment	- 13 inches.

Speed	- up to 30 pieces per minute. See speed curve.
Services required;	
Compress air requirements	- 200 cu. ft. of free air at 40 psi. gauge pressure at the machine.
Cooling wind requirements	- 13 000 cu. ft. of free air is required per minute at a gauge pressure of 8 inches at the machine. This is split as 11 000 cu. ft. to the machine and 2000 cu. ft. to the conveyor.

Figure 8,29. The Lynch Model Number 'R' machine

The average speed curve for the 'R' type machines is;

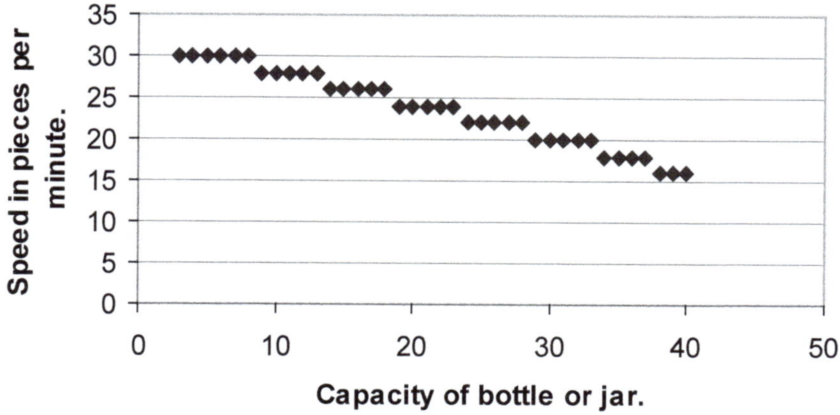

Figure 8,30. The Average speed curve for the Lynch 'R' machine

In Table 8,2 you can see the various combinations of moulds and blanks, the 'R-OO' is a special machine a description follows later.

	Machine Type			
Mld. Equipm't	'R'	'RS'	'R-A'	'R-OO'
Blanks	6	6	8	6
Blow-moulds	8	6	8	8
Baffle plates	1	1	1	2
Blow-heads	2	2	2	2

Table 8,2. Mould and blank combinations for the 'R' series of machines

The following two slides give some indication as to what narrow neck, wide mouth and semi-wide mouth bottles and jars can be made by this family of machines.

Figure 8,31. Some of the range of narrow neck bottles that have been made on these machines

Figure 8,32. Some of the wide mouth and semi-wide mouth bottles and jars that can be made on this family of machines

The 'R' Double Zero machine, a schematic shown below in Figure 8,33, is a machine modification to the Model 'R'.

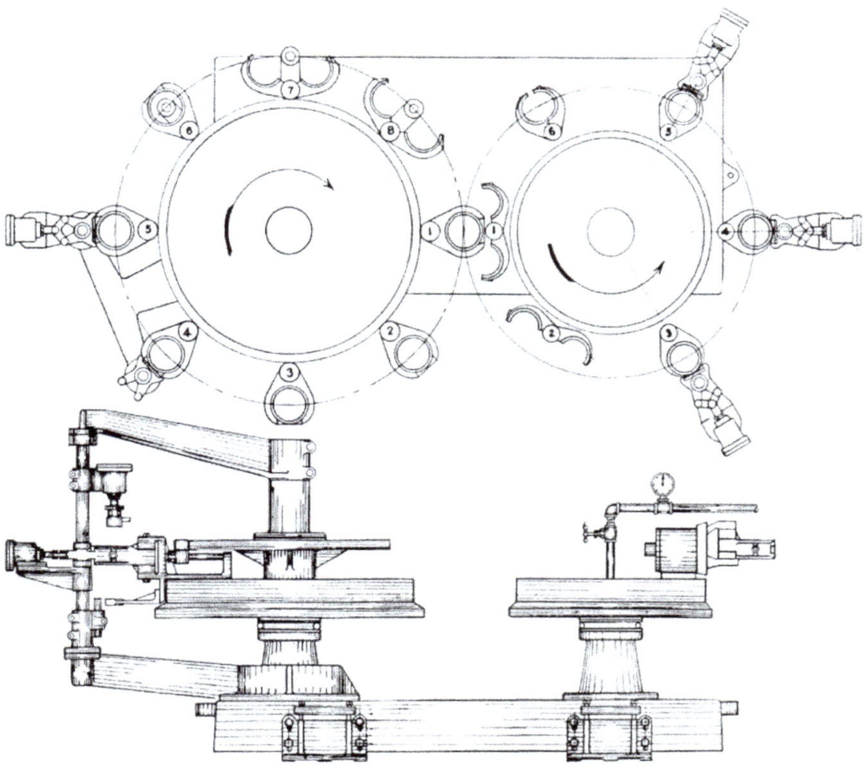

Figure 8,33. Schematic of the 'R-OO' type of machine

The blank-mould table carries 6 blank moulds, and the stations are listed as follows;
Station No. 1 is the transfer station.
Station No. 2 is the idle, or swabbing station.
Station No. 3 is the loading, or charge station.
Station No. 4 is the first baffle, or counter blow station.
Station No. 5 is the second baffle, or counter blow station.
Station No. 6 is where the blank-mould is partially opened as shown.

There are no changes to the blow mould table.

The 'R' Double zero is a two table bottle machine, carrying 6 blank-moulds on one table and 8 blow-moulds on the other, the same as the standard 'R' machine. The cycle of operation begins with the loading station, counter blow, counter blow a second time, opened blank-mould, and then transfer.

The principle change is in the cycle of operations. The blank mould is charged two stations earlier, and two baffle stations are employed in the place of one; also opening the blank-mould before reaching the transfer station.

The blank-mould is inverted between stations 2 and 3 to bring the finished end of the blank-mould to the bottom, or in an upside down position, where the glass is dropped into the blank-mould at the loading, or charge position.

This blank-mould is inverted between stations 3 and 4 to bring the finished end to the top and the counterblow of the parison is down at baffle stations 4 and 5. The blank-mould is partly opened on the run between stations 5 and 6 and is fully opened between stations 6 and 1.

There are eight changes required when changing a standard 'R' machine to a 'R-OO' machine.

1. Requires a base extension plate, which is bolted onto the machine base.
2. Requires a finish forming assembly to be installed at station 3, formerly shown at station 5.
3. Requires a new baffle assembly which will be installed at station 4.
4. Requires a new 00-85 cam.
5. Requires a new finish-ring opening cylinder assembly.
6. Requires six finish-ring holder lock assemblies. These parts required to keep the finished rings locked until they reach the transfer point.
7. Requires a parison patter (or adjusting) cylinder assembly. This unit is installed beneath station 6.
8. Requires six pairs of neck-ring holders, with lugs thereon. These parts work in connection with the finish-ring holder locks mentioned in item 6.

There is no advantage of using the Double Zero cycle of operation, unless it is used with 8 blow-moulds on the blowing table.

To make really small bottles on the Lynch 'R' machine it is necessary to use special blank-mould cages. To make really tall bottles sit is necessary to use the Extra High, or Goose-neck equipment. To make wide mouth bottles and jars, it is necessary to purchase wide mouth neck-ring holders only.

While the machine is designed to make bottles up to 70 ounces in capacity, it is possible to make items up to 128 ounces in capacity by making the blow-moulds with hinges attached. To accomplish this it requires a few changes to the machine as well. To make push-up bottom bottles, it is necessary to purchase a push-up cylinder for the transfer position, and other necessary equipment.

Now we come to the tenth machine that they made; the popular Lynch 10, their catalogue states that it was a two table, indexing machine with 6 blank-moulds on one table and 6 blow-moulds on the other, capable of making bottles and jars from ½ ounce to 64 ounce capacity which is nearly two litres see figure 8,35.

Figures 8,36 to 8,38 show some of the range that can be made on this machine. The first, Figure 8,36, shows a collection of small intricate shapes such as perfumes and toiletries, and tableware such as vinegars and other such items that need to be poured at the table from a bottle. The second, Figure 8,37, shows a collection of larger narrow necked bottles such as whiskies and beers. Finally Figure

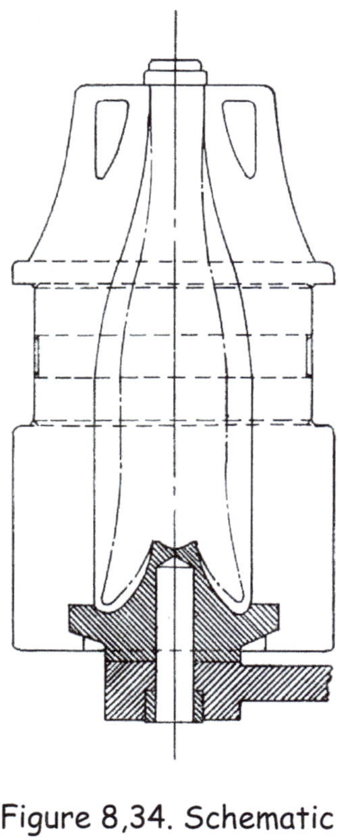

Figure 8,34. Schematic of parison mould showing the push-up arrangement

A 6 MOLD MACHINE . . .

for GLASS BOTTLES *and* JARS
from ½ to 64 Ounces

Figure 8,35. The new improved Lynch 10 machine

These bottles were made on the Lynch Model "10" Machine. While many other intricate sizes and shapes can be made as well as those shown here, space will not permit the showing of all bottles within the range of the machine.

Figure 8,36. Illustration of small bottles

8,38 shows a collection of wide mouth and semi-wide mouth jars and bottles, to make these containers it is necessary to purchase wide mouth equipment.

Beer bottles, Whiskey, Wine, Brandy, Catsup, Vinegar, Soda, Syphon, etc., are other examples of glassware produced on the Lynch Model "10" Machine.

Figure 8,37. Illustration of common narrow necked bottles

The Lynch 10 is ideal for producing champagne bottles with push-up bottoms. The depth of the push-up is made in the baffle plate. And the push-up operation takes place at the counter blow

Figure 8,38. Examples of some of the wide mouth and semi-wide mouth bottles and jars that have been made on the Lynch 10

station where the parison is blown up around the push-up portion in the baffle.

Like the model 'R' the range of jobs that can be made by the Lynch 10 machine this is done by using special equipment. These are namely;

1. The machine will make up to 64 ounces capacity using blow-mould cages, if the hinges are attached to the blow-moulds then this will increase this up to 128 ounce capacity, making it suitable for gallon jugs.
2. The machine will make wine and champagne bottles with push-up bottoms no additional equipment is required. The push-up

is made in the baffle plate so this feature is considered as mould equipment.

3. When making small bottles it is necessary to have the Short Height Equipment.
4. When making tall bottles it is necessary to use the Extreme High Equipment.
5. When making wide mouth and semi-wide mouth bottles and jars the Wide Mouth Equipment is required.
6. When making medium or a general run of bottles the Standard Equipment is required.

The specifications for the machine are;

Class of Glassware	- Narrow mouth, Wide mouth and Semi-wide mouth galss bottles and jars.
Capacity	- From ½ ounce to 64 ounce capacity inclusive.
Maximum diameter of finish	- With Seamless Top: 3½ inches. - Without Seam-on Top: 4 inches.
Maximum body diameter	- 5 inches.
Minimum length under finish	- With Special Blank Holders: 1 inch.
Maximum length under finish	- With Standard Equipment: 12½ inches.
With Extra High Neck Ring Holders	- 14½ inches.
Inside diameter of Blank mould holders	- 5 inches.
Inside diameter of Blow mould holders	- 6 inches.
Speed	- up to 45.2 bottles per minute (1 ounce). Fractional weights (under 1 ounce) are 48 bottles per minute. See speed curve.

Services required;

Compressed air	- 200 cu. ft. of free air per minute at 40 lb. gauge pressure at the machine.
Cooling air	- 12 000 cu. ft. per minute of free air at a gauge pressure of 8 inches of water when measured at the inlet valve to the machine.
Vacuum	- at least 125 cu. ft. per minute piston displacement rated at 26 inches of mercury vacuum at sea level.

This information was taken from a catalogue published around the 1940's and it was for an improved Lynch 10 machine. One of the things that it also gives is how to convert the old model to the new blank-cracking method as fitted to the improved machine.

All that is required is to remove one turn-over assembly at the loading station, remove the taper pin from the cam, then shift the cam round about 13 degrees. The cam must be moved around in the opposite direction to the table rotation. After shifting the cam, it is necessary to re-drill a hole in the cam and column, then taper pin the cam again. By doing this it cracks open the blank-mould about $\frac{3}{4}$ of an inch at the station before transfer, instead of past this station.

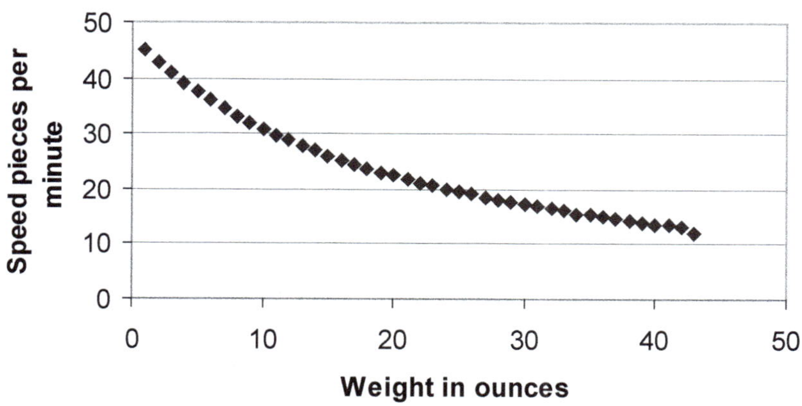

Figure 8,39 shows the speed curve for the machine

In the late 1940s the Lynch turned the Lynch 10 into a dual cavity machine, and all of the existing models could be converted to the dual operation. Because the machine runs slightly slower it improves the quality of the bottles giving in some instances a higher pack efficiency. The machine is shown in Figure 8,40

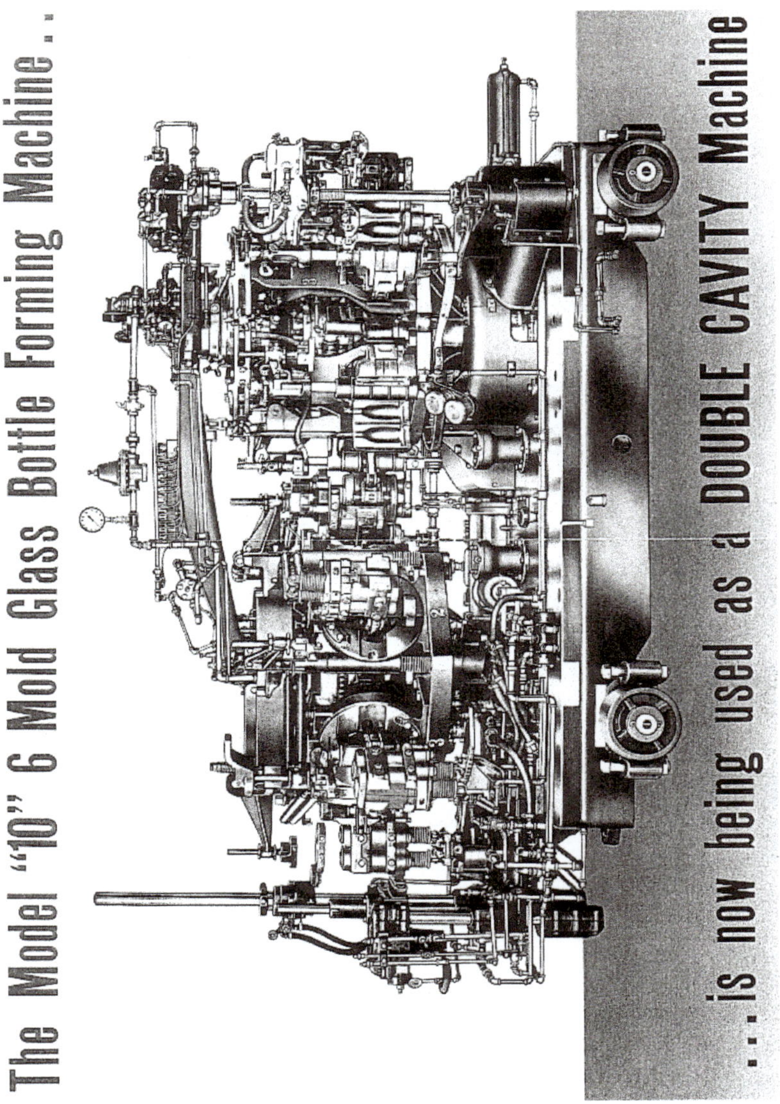

Figure 8,40. The Lynch 10 dual cavity machine

Figure 8,41. Photograph showing the dual blow moulds and take-outs

Figure 8,42. Showing the blanks and their mountings

along with the takeout mechanism and the mounting of the blank moulds, shown in Figures 8,41 and 8,42.

Figure 8,43. Examples of bottles made on the Dual Lynch 10

In late 1939 - early 1940 Lynch marketed their Model 16-b machine, see Figure 8,44.

Figure 8,44. The Lynch model 16-b

The machine carries eight blank-moulds on one table and eight blow-moulds on the other, the blank table rotates to the left (anti-clockwise) and the mould table rotates to the right (clockwise). The machine is capable of manufacturing Narrow Mouth, Wide Mouth and Semi-wide Mouth Glass Containers, from a minimum capacity of 8 ounces to a maximum capacity of 128 ounces. It is suitable for bottles and jars with finishes that do not exceed five inches in diameter, and it is ideal for making wine bottles with a push-up portion in the bottom.

The Machine specifications are;

Class of glassware	- narrow mouth, wide mouth and semi-wide mouth glass bottles.
Capacity	- from 8 ounce minimum to 128 ounce maximum inclusive.
Maximum diameter of finish	- with seamless top - $3\frac{1}{2}$ inches. - with seam-on-top - 4 inches
Maximum diameter of finish	- using Special Equipment - 120 mm., or for finishes that do not exceed 5 inches.
Minimum height under finish	- with Standard Equipment - 6 inches.
Maximum height under finish	- with Standard Equipment - $12\frac{1}{2}$ inches.
Maximum height under finish	- with Extra High Neck Ring Holders - $14\frac{1}{2}$.
Maximum body diameter	- with mould cages - $5\frac{1}{2}$inches. - with hinged moulds - $6\frac{1}{2}$ inches.
Speed	- up to 40 bottles per minute.
Services required;	
Compressed air	- 200 cu. ft. of free air per minute at 40 pounds gauge pressure at the machine.
Cooling wind	- 10 000 to 12 000 cu. ft. per minute with static pressure of 10 to 12 inches at the outlets.

	Or 6 to 8 inches when measured by water in a U-tube.
Vacuum	- requires a pump capable of supplying 125 cu. ft. per minute cylinder displacement at 26 inches.
Motor	- $7\frac{1}{2}$ HP motor at 12 000 r.p.m.

Mould equipment required to operate the machine no spares. See Table 8,3.

Blank - mould Table	**Blow - mould Table**
8 Blank-moulds	8 Blow-moulds
8 Plungers	8 Bottom plates
8 Pairs of Finishing rings	8 Blow-heads
8 Baffle plates	2 Pairs of Take-out jaws
1 Guide ring, or Funnel	

Table 8,3. Minimum amount of mould equipment required to operate the Lynch '16-b' machine

Figure 8,45. 12 valve timer assembly, PIV gear and vertical feeder shaft, and 8 valve auxiliary timer assembly for the pneumatic feeder operation

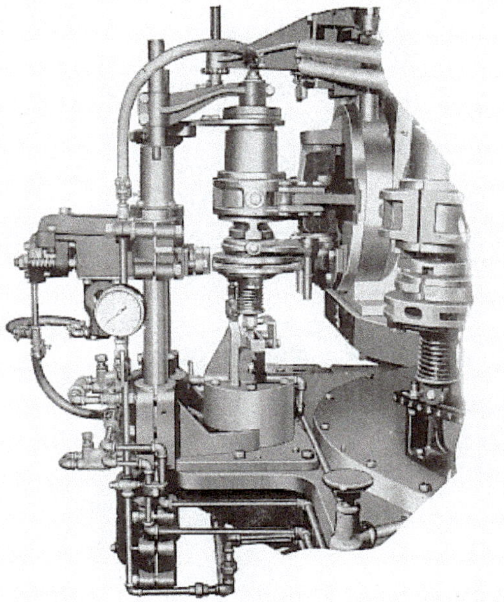

Figure 8,46. Spiral blow-down and Vacuum head

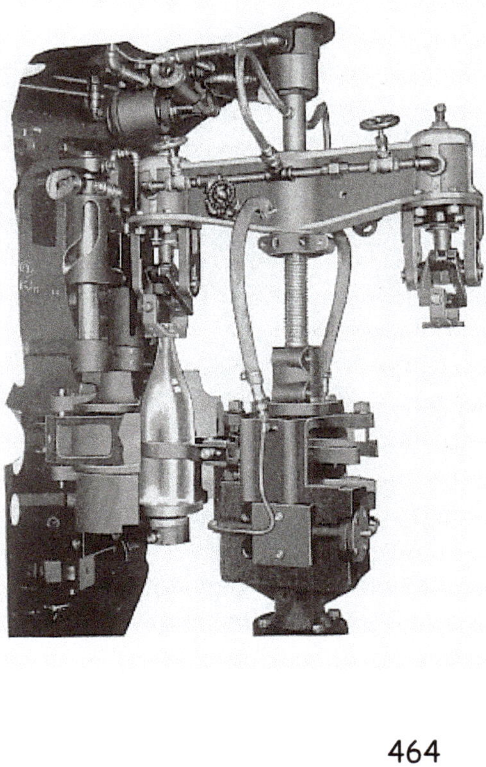

Figure 8,47. Take-out mechanism, same as on Lynch 10 but much stronger

Figure 8,48. Narrow mouth bottles made on the Lynch 16-b machine

356 Glass-blowing Machine. H. Fairbrother, London (Brit. Pat., No. 160366 (Lynch Glass Machinery Co.; Jackson Street, Anderson, Indiana, USA), April 6th, 1920). - Relates to machines for blowing glass hollow-ware of the type in which a number of turnover parison moulds and an equal number of finishing moulds are carried on two separate rotating tables on the same level, the transfer of the blank from the parison mould to the corresponding blowing-mould taking place where the mould paths meet.

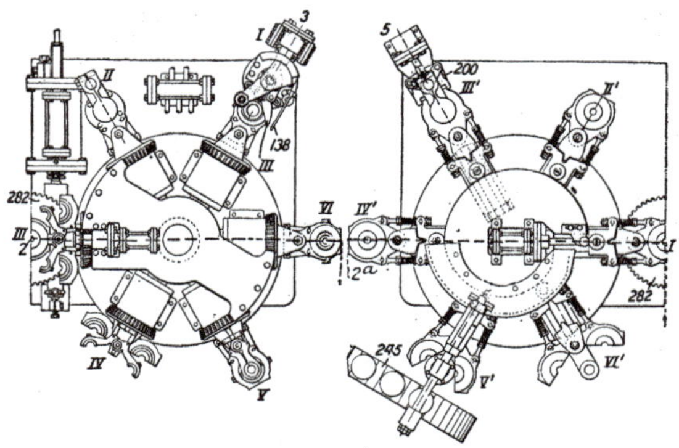

Abst. 356 Vol. V 1921 Brit. Pat. By Fairbrother re: Lynch machine (plan)

Figure 8,49a. Figure 162 of abstract

The machine is pneumatically operated and is arranged so that, in charging the parison mould, the gatherer strikes a valve lever with the gathering-iron, after which all operations are automatic, including the delivery of the finished article. The machine is shown in Figure 162, and in vertical section on the line 2-2^{a}, in Figure 164.

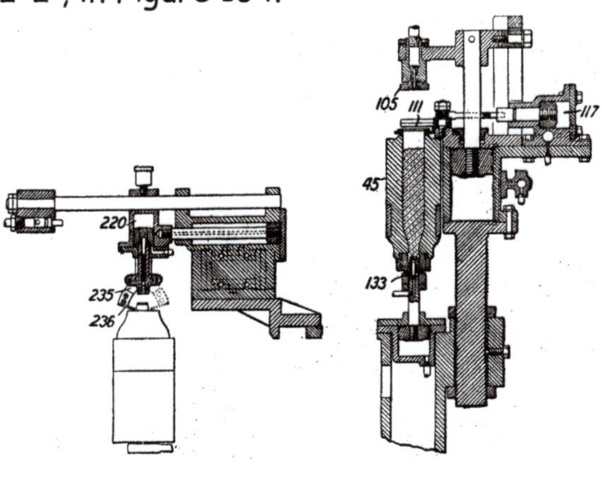

As above but special sections

Figure 8,49b. Figure 163 of abstract

The parison moulds, 45, are charged by hand at station I. The gatherer depresses the lever, 138, which admits air to the cylinder, 117, operating shears, 111, which cut off the charge. An air-cooled neck-rod, 133, rises and a cover, 105, descends, air being admitted through the latter to compress the charge. At station II the blank is partially blown. In passing from station II to station III, the parison mould is inverted and at the latter station the blank mould opens and releases the blank, and the finishing mould closes on it.

At station III', the blank is blown and at V' it is lifted from the open blow mould on to a conveyor, 245, by pneumatically-operated grippers, 235, 236, mounted on a slide, 220. Pneumatically-operated grippers insure the firm closure of the blow moulds at station III'. The mould tables rotate intermittently in the same direction by a toothed wheel, 282, which engages teeth, 281, on the tables. They are locked during periods of rest by bolts, 296, 297.

As above but vertical sections through machine.

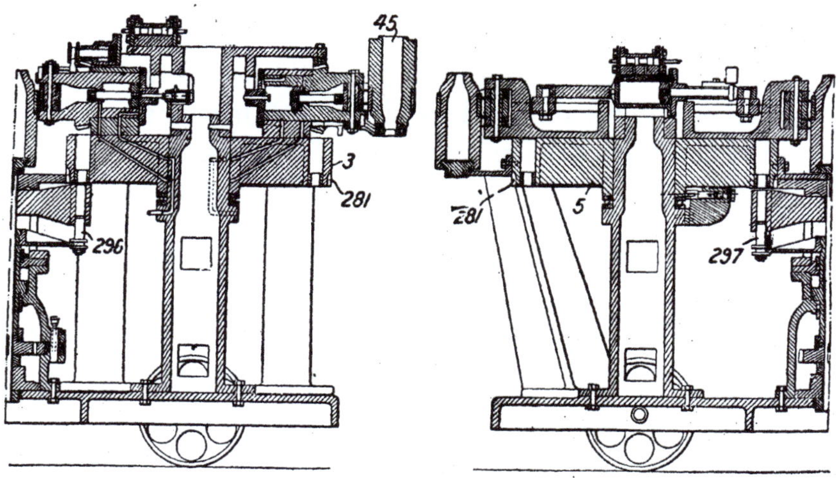

Figure 8,49c. Figure 164 of abstract

Chapter 9

The rise of the 'in-line' machine

This chapter is devoted mainly to the IS machine, which is commonly called the 'individual section machine' but in the machines formative days the IS was said to stand for Ingle (Henry W.) and Smith (Francis Goodwin). Ingle being the machine's designer and Smith the company President - the company was the Hartford-Empire Company of Hartford, Connecticut in America.

Originally all of these machines were made by the Hartford-Empire Company of America, but since the end of the Second World War Emhart's have also being manufacturing them at Sundsvall in Sweden. As with all good things, when the patents expired, others started to manufacturer in-line machines themselves, some were successful while others failed.

Some manufacturers tried a different design such as the Lynch Corporation with their '44' machine, the O'Neill Machine Company with their '55' machine and Mitchell with his single unit machines, but none of these alternative machines made any impression on the use of the original IS machine.

They are probably the nearest to the 'Universal Machine' that has ever been built. The scenario that they conjure up is a machine that can work any combination of sections from one to twelve, and the number of containers that can be made by a section at anyone time is from one to four. If twelve sections are not enough, a twelve section machine is the maximum number of sections before the outside one give delivery problems, then two machines can be put side to side (end to end called being in tandem) with the output going down one single conveyor to a common lehr. At one time many glass manufacturers did not like to use more than a ten section machine because of this perceived delivery problem, but 12 section machines are now becoming more common.

Working in tandem can be a very useful way to use up your old machines, for example two six's make a twelve section machine, two eight's make a sixteen section machine, and two ten's make a twenty section machine. Obviously the size of the orders has some influence on the subject, no manufacturer would want to be changing the job every two or three days.

Here once again just like Owens, O'Neill and Lynch, we have a company that has done so much for the glass industry, it developed over time to be a multi-faceted company but we will in this biopic deal with only the side which affected the glass industry. Some of this has been mentioned before in the chapter on Forehearths.

William Honiss and William Lorenz, Hartford-Fairmont Company Ltd., Hartford-Empire Company, Emhart Manufacturing Company

The Honiss and Lorenz Partnership

Around about 1880 William H. Honiss (1854-1940) and William A. Lorenz (??-1928) formed a partnership who's working brief was to design and perfect various types of machinery. The offices were located at 380 Asylum Street, Hartford, Connecticut, USA, and their first project was to develop a automatic machine for the manufacture of paper bags.

The two men had both had had practical training and experience as machinists with various engineering companies, an asset which was to be highly useful when designing glass works machinery. William Lorenz had come to Hartford from Philadelphia. He had graduated from an intermediate school in Germany and continued his education in America. After an engineering apprenticeship he became a Master Machinist at Pratt and Whitney where he eventually became Chief Engineer.

William Honiss who had trained as a machinist started work firstly in Middletown in Connecticut and then at Pratt and Whitney where he met Bill Lorenz. He also came under the influence of Stimson Walker who was an authority on patents. Honiss was then able to say what was patentable which was to be very useful when they patented their ideas. It was not just the patenting of their ideas but then protecting those ideas from others, i.e. protecting their rights as the patentees.

Three other engineers joined the group the first was William Lorenz's son Edward H. Lorenz who had worked in his father's business. Ed knew Karl Peiler a fellow classmate who he had met at the Massachusetts Institute of Technology in 1904, who in 1910 joined them, they became known as the 'Hartford Engineers'. Finally there was Joseph Merritt who was in charge of installation work.

Karl Peiler was the sole inventor in 147 U.S. patents relating to the glass machinery business and had many foreign patents in 17 countries. These inventions related to feeding molten glass, conveying molten glass containers, heat treating of glass, forming glassware, glass tube production, annealing lehrs and the making of ceramic refractories. He is best known as the inventor of the Hartford Single Feeder.

Their entry into the glass industry can probably be traced back to a chance or off-hand remark from Bartlet Arkell the President of The Beech Nut Packaging Corporation of Canajoharie, N.Y. who enquired if Lorenz and Honiss who were now trading as Mechanical Engineers and Patent Attorneys could do something about the shear mark that was prominent and visible on their glass jars. He wanted to vacuum pack his products but was unable to do so due to leakage around the seals.

They advised the establishment of a new glass works to make the jars on machines that they would build and by methods that they would design. Accordingly the Monongah Glass Company was formed in 1904 with a factory located at Fairmont, West Virginia. Many of the existing forming machines were tried but they were not good enough, an Owens machine license was not obtainable so the company decided to undertake a special research and development programme into the automatic feeding of glass to bottle making machines.

They had input into the factory design. The specially designed moulds, tools and gauges were shipped out to the Monogah Glass

Company. The Monogah staff were enthusiastic regarding the project and the newly designed Beech Nut jars gave a better seal but still lacked the uniformity in both weight and contents.

In 1910 Karl Peiler joined the company and soon after (1911) was assigned to work on the shear mark and other production problems. The Beech Nut Packaging Company and the Monogah Glass company funded Peiler to carry out a survey of the glass industry, the study was to design ways of improving the forming machinery.

When he reported back he said that he had found that the industry was disinterested, what was holding their attention was the new revolutionary automatic machine owned by Owens. The machines used more glass than the hand made ones but they still suffered with variations in weight and contents.

Peiler reported that factories would not improve their semi-automatic forming until they had an automatic feeder to compete with the Owens suction machine. His employers after discussions with the people would funded his survey assigned him to think of ways of feeding glass to semi-automatic machines. This topic has already been dealt with in the relevant Chapter 6, The Development of the Feeder and the Forehearth.

The Hartford Fairmont Company

A new company was formed to design and build the new and improved glass-working machinery, it was incorporated in the spring of 1912 as the Hartford-Fairmont Company. Hartford being the location of Honiss & Lorenz and Fairmont that of the Monongah Glass Company. Seven directors were appointed and a capital stock of $60 000 issued. The seven directors were;

President	- Bartlett Arkell (President of the Beech Nut Packaging Company, Canajoharie, New York)
Vice President	- William Lorenz
Director	- Walter Lipe (Secretary and Treasurer of the Beech Nut Packaging Company)
Director	- H. L. Heintzelman (President of the Monogah Glass Company of Fairmont, West Virginia)

Director - R. T. Cunningham (Secretary and Treasurer of the Monogah Glass Company)
Director - William Honiss
Director - Karl E. Peiler

In 1913 using a paddle-needle feeder to feed a Hartford press the combination of the two enabled the development team in Hartford to produce the Beech Nut Jar, see Figure 9,1, - the glass was cloudy, full of seeds and blisters, crizzles and lap marks. This was a reflection of the conditions at Hartford in the mixing and melting of the glass.

Figure 9,1 shows a photograph of one of the original bottles that was made using a Paddle-Needle Forehearth and a Hartford Press. The bottle was signed 'Made August 6, 1913, Karl E. Peiler signed August 7th 1913'. In the photograph the jar is shown set between two modern machine made jars that have been made by the press and blow process on a modern IS machine.

In the spring of 1912 premises were rented at 41 Arch Street, the site of an unused bayonet factory, to house a Glass

Figure 9,1. One of the original made jars, signed by Peiler

Laboratory, which was immediately named 'the glass factory'. A furnace was built and the original development of the 'paddle feeder' was carried out there, as was future development works. It was in this factory that the pressed jar was made.

Figure 9,2. Group of pioneers at the Lorenz Laboratory at 41 Arch Street, Hartford. This photograph of some of the men that worked in the 'Glass Factory' was taken 13th July 1914. William Lorenz is the second from the left and Edward Lorenz, Williams son, and Karl Peiler are third and fourth from the right.

The next event in the history of the company was in 1915 when Lorenz and Peiler started the development of a rotary automatic press and blow machine for the production of milk bottles, Also at that time the Hartford Engineers felt that they needed a

General Manager to deal with the business end of the company especially the finance and the licensing. A list of requirements for the new position was drawn up and the engineers eventually came to the conclusion that no such man existed, but eventually the choice came down to a man who some had known since childhood.

His name was F. Goodwin Smith and he was appointed as Vice President and General Manager later that year. Goodwin Smith was to have a major influence on the company from 1927 when he became President and Chairman of the company. He was born in 1881 in Hartford and after graduation he gained experience in a number of local industries before being appointed as Hartford Fairmont's new General Manager.

To quote from John Edgington's book recording the history of Emhart's;

> "Mr. Smith was a mild-natured man who dealt with his staff using the well known principle of 'firm but friendly'. His major contribution to the success of Hartford Empire can never be fully recorded. He led the company from loss to profit, through the tricky anti-trust suit and from receivership to a profitable, reliable supplier to the world-wide glass container industry."

The following is a record of some of the founders of the company.

General Manager and then from 1927 the Company's President and Chairman. Retired in 1954 after 38 years service. Board of directors until his death in 1956.

Figure 9,3. Francis Goodwin Smith

Figure 9,4. Mr. William A. Lorenz

One of the two partners in the original company the Honiss and Lorenz Partnership. Later to be the vice President of the Hartford- Fairmont Company. He died in 1928 after 16 years working for the Hartford Fairmont and later the Hartford Empire Company.

Figure 9,5. Edward Lorenz

William's son worked for his father before the partnership was founded. He worked for company for 33 years.

Director of Hartford Fairmont Company. Became Chief Engineer when they merged with the Empire Company, became one of the companies Vice-Presidents in 1936. Worked for company 42 years. In 1949 awarded a Honorary Degree of Engineering.

Figure 9,6. Karl E. Peiler

Figure 9,7. Some of the company's management group, Goodwin Smith in discussion with some colleagues. Left to right: Karl E. Peiler, Arthur Sandford (Company Secretary), Goodwin Smith (President), and S. Parham (Legal)

The next two figures show the payments for the month of January 1923 and a listing of the Directors salaries for 1923, dated 4th February, 1923.

P. Goodwin Smith Special Hartford-Empire Company Salaries

January 1923

1923					
Jan		Deposited to this account check No			7424
					200
Check No 1	31	P Goodwin Smith	for January	1250	
	2	Geo E Howard	"	750	
	3	William H Howiss	"	500	
	4	Karl E Peiler	"	666	
	5	E O Hiller	"	500	
	6	Virgil Mulholland	"	525	
	7	J P Green	"	400	
	8	Herbert Knox Smith	"	450	
	9	Alfred R Hunter	"	300	
	10	Fritz Kemp	"	300	
	11	A. P Northend	"	300	
	12	R. M Eldred	"	275	
	13	Henry W Ingle	"	275	
	14	T. C. Champeau	"	275	
	15	William A Lorenz	"	200	
	16	Robert W Canfield	"	458	
	17	H M Stoner - cancelled		000	
	18	Robert W Canfield Balance	"	83	
	19			7507	7624
		Balance to February		117	
				7624	7624

Figure 9,8. January's 1923 wages bill for the management of Hartford Empire

In 1985 the Emhart Glass Machinery Group gave or took part in a symposium on Glass Manufacture at Cairo in Egypt. In one of the papers given by E. P. Munz, Sale Engineer, the was an

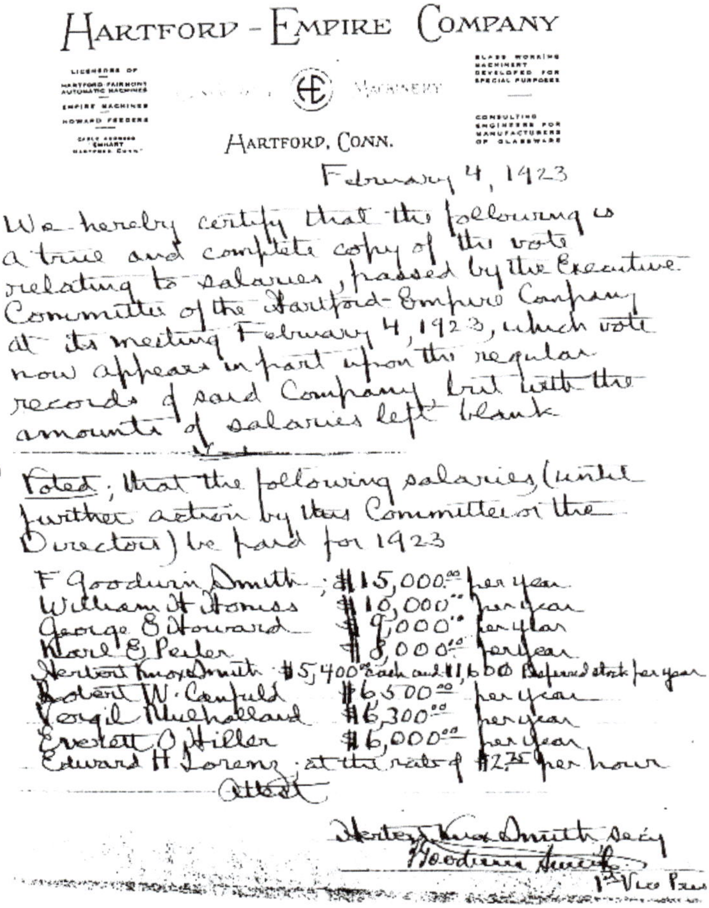

HARTFORD-EMPIRE COMPANY

LICENSORS OF
HARTFORD FAIRMONT AUTOMATIC MACHINES
EMPIRE MACHINES
HOWARD FEEDERS
CABLE ADDRESS "EMHART" HARTFORD, CONN.

GLASS WORKING MACHINERY DEVELOPED FOR SPECIAL PURPOSES
CONSULTING ENGINEERS FOR MANUFACTURERS OF GLASSWARE

HARTFORD, CONN.

February 4, 1923

We hereby certify that the following is a true and complete copy of the vote relating to salaries, passed by the Executive Committee of the Hartford-Empire Company at its meeting February 4, 1923, which vote now appears in part upon the regular records of said Company but with the amounts of salaries left blank.

"Voted; that the following salaries (until further action by this Committee or the Directors) be paid for 1923

F Goodwin Smith ; $15,000.00 per year
William H Honiss $10,000.00 per year
George E Howard $9,000.00 per year
Karl E Peiler $8,000.00 per year
Herbert Knox Smith $5,400.00 each and $1,600 Preferred stock per year
Robert W. Canfield $6500.00 per year
Virgil Mulholland $6,300.00 per year
Everett O Hiller $6,000.00 per year
Edward H Lorenz at the rate of $2.75 per hour

Attest

Herbert Knox Smith Secy
F Goodwin Smith
1st Vice Pres

Figure 9,9. Wage structure for the Committee of Directors. Signed by Herbert Knox Smith the Company Secretary

addendum to the paper which listed the development of the IS machine right from the beginning up to the date at that time (i.e. 1985 although the date on the chart says 1981). Later with the publication of the book "Emhart Glass And The Story Of Glass Packaging" the time line was updated and it is these timelines that we will use.

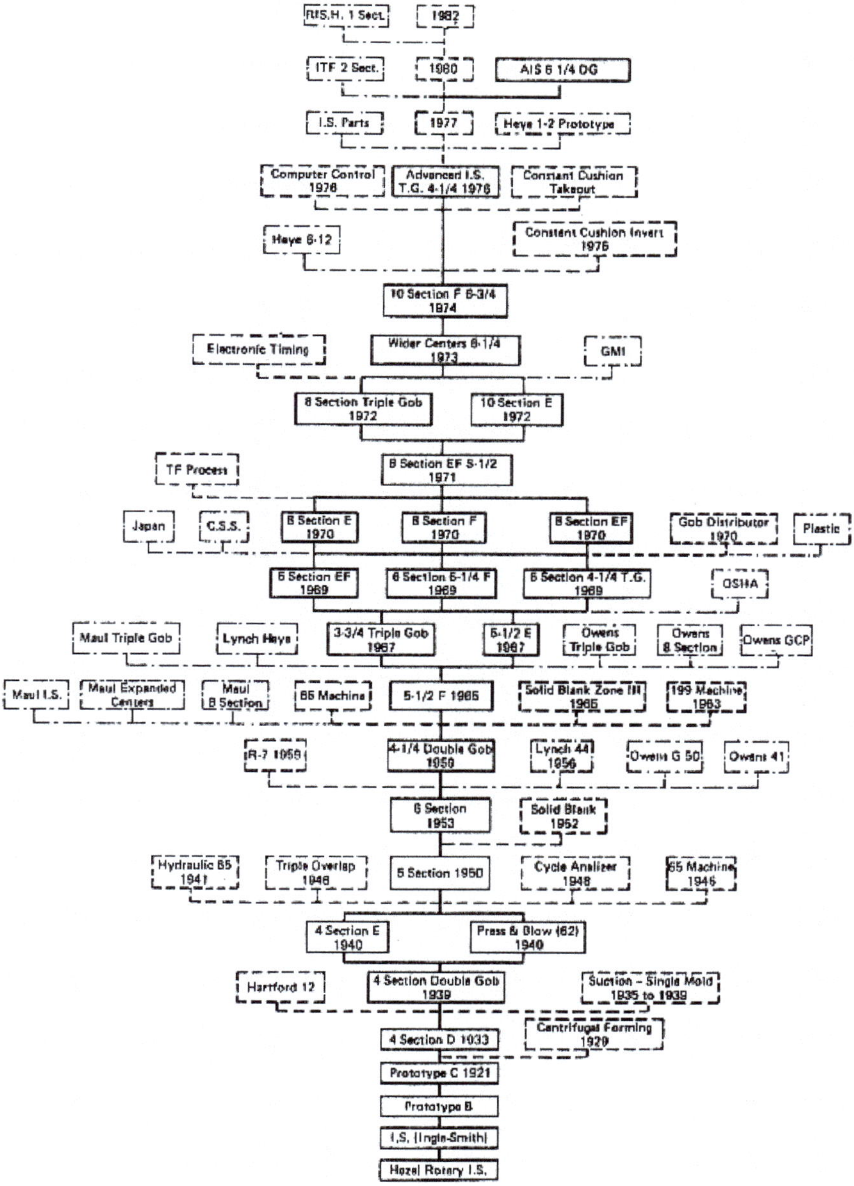

Figure 9,10. 1984 timeline for the IS machines and some of their competitors

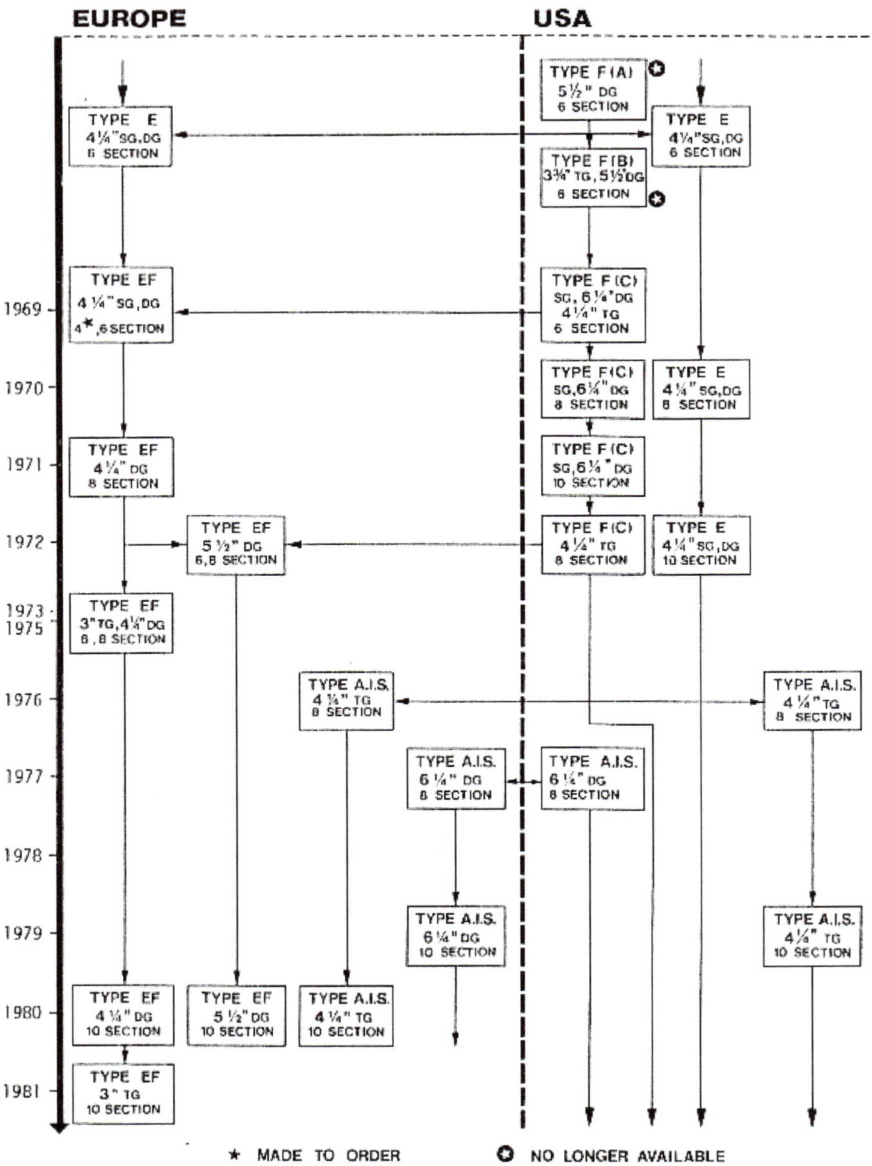

Figure 9,11. A more detailed look at the timeline between the later 1960s and the early 1980s. Comparing the different types of IS machines in the United States and those for the rest of the world

The second set of timelines, circa 1991, which we will use, are;

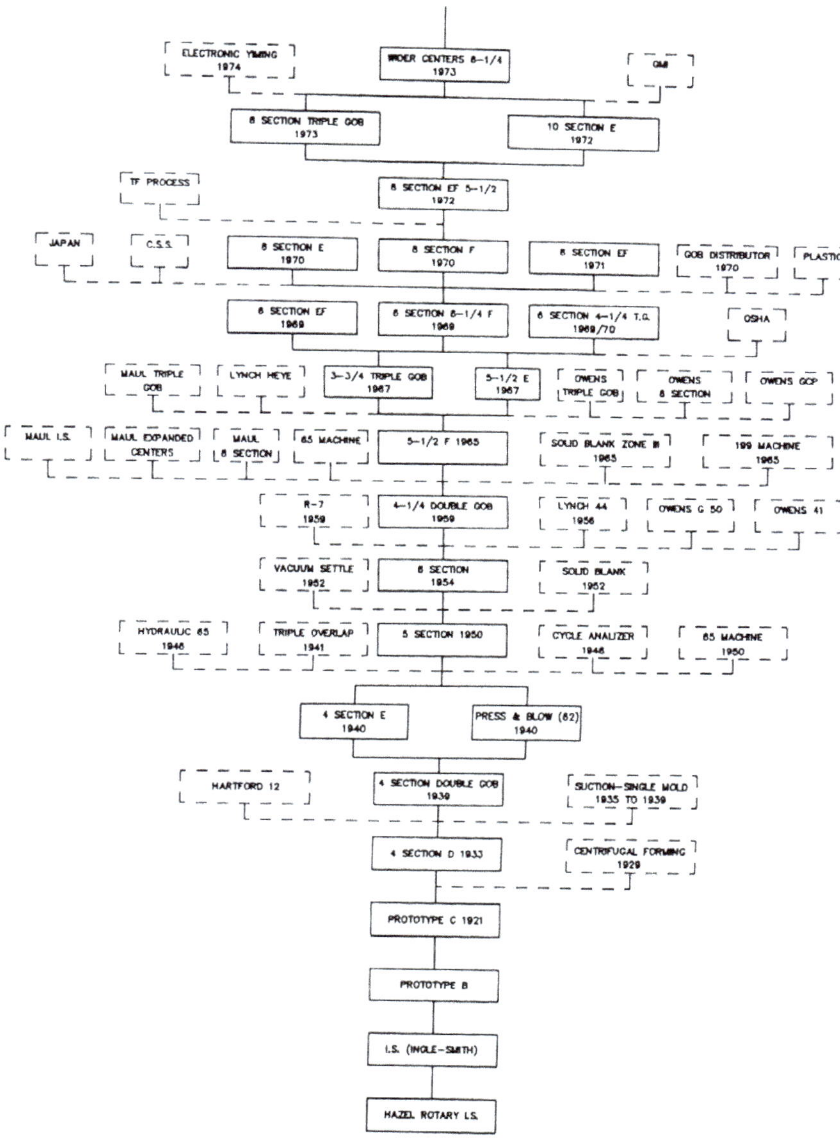

Figure 9,12. The development of the IS machine up to 1974

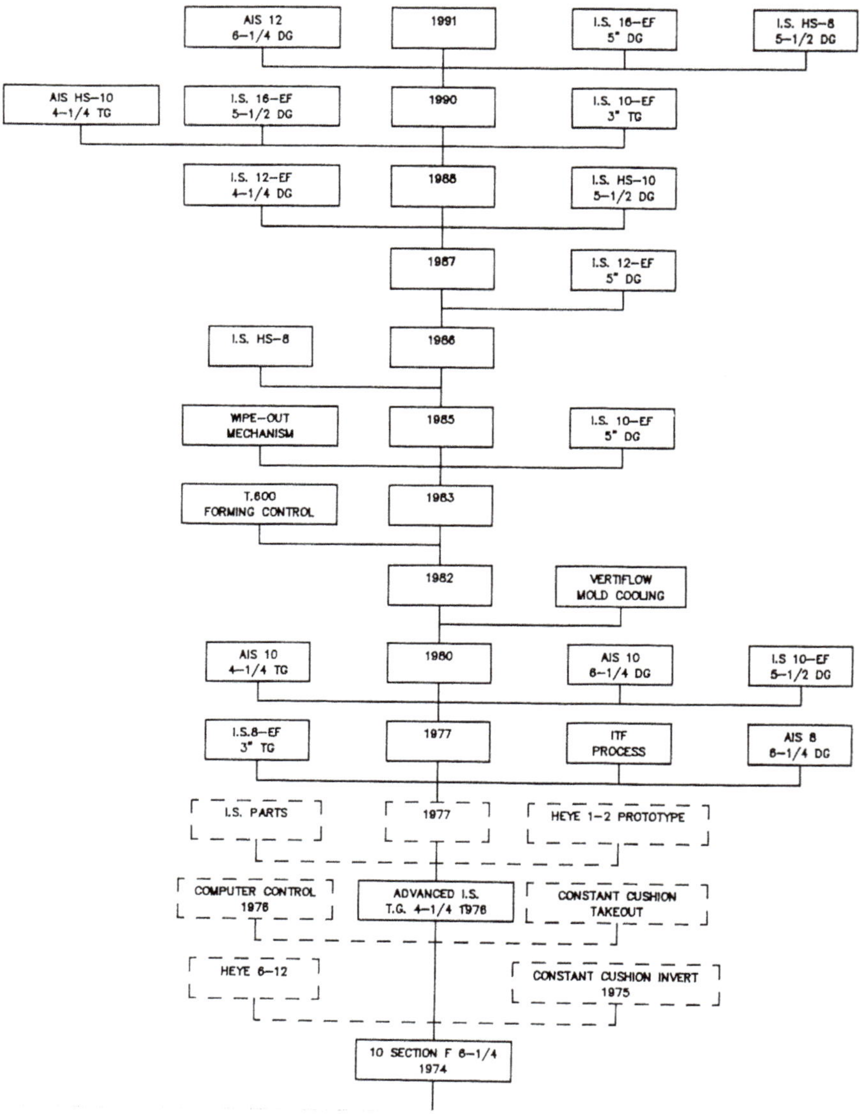

Figure 9,13. The development of the IS machine from 1975-1991

The development of the IS machine was not as straight forward and direct as the aforementioned charts indicates. There were many blind alleys for instance in 1950 the so-called Type 65 IS machine with $6\frac{1}{2}$ C.D., D.G., with either three or four sections did not take off due to it being too big. The same applied to the Type 199 IS machine introduced in 1960, this machine was a speciality machine which had five sections each of $5\frac{1}{2}$ C.D., D.G. P. & B., this machine occupied more space than a six section machine.

According to the listing in Figure 9,12 the saga started with the Hazel Rotary IS machine and the machine proper began in 1921. The design of the machine was revolutionary the machine stood still and it moved the glass around during the bottle making process, unlike all of the other machines which moved the machine around while making the bottle, which reduced the amount of power that the machine required.

These individual machines could be worked "side by side" but a fault on one of the machines did not affect the others. Unlike the rotary types which often had to be stopped while a fault on one head was rectified. When the sections were operating side by side it meant that ultimately the machine would be able to manufacture different containers provided that they were all the same weight. Due to it's "none moving or stationary" mode of operation it meant that machine speed could be increased for the same sized bottle and the quality of the container improved.

The next important step was in 1920 when Henry W. Ingle joined the company as Designer and took over the IS programme. In 1923 along with Peiler's single feeder and a single section machine meant that an automatic production line could be offered. A record in Henry Ingles record book shows that a single unit was installed in at the Salem Glass Works, in Salem, New Jersey, See Figure 9,14. In the same year the two section was made and four of these units were sold to the Hocking Glass Company of Lancaster, Ohio, see Figure 9,25 but the date may have been in 1924 as the single section trial at Salem did not take place until December 1923. Four years later this has graduated to a four section machine the first experimental one being installed in the Carr-Lowrey Glass Company's Baltimore factory in Baltimore, Maryland. See Figure 9,16.

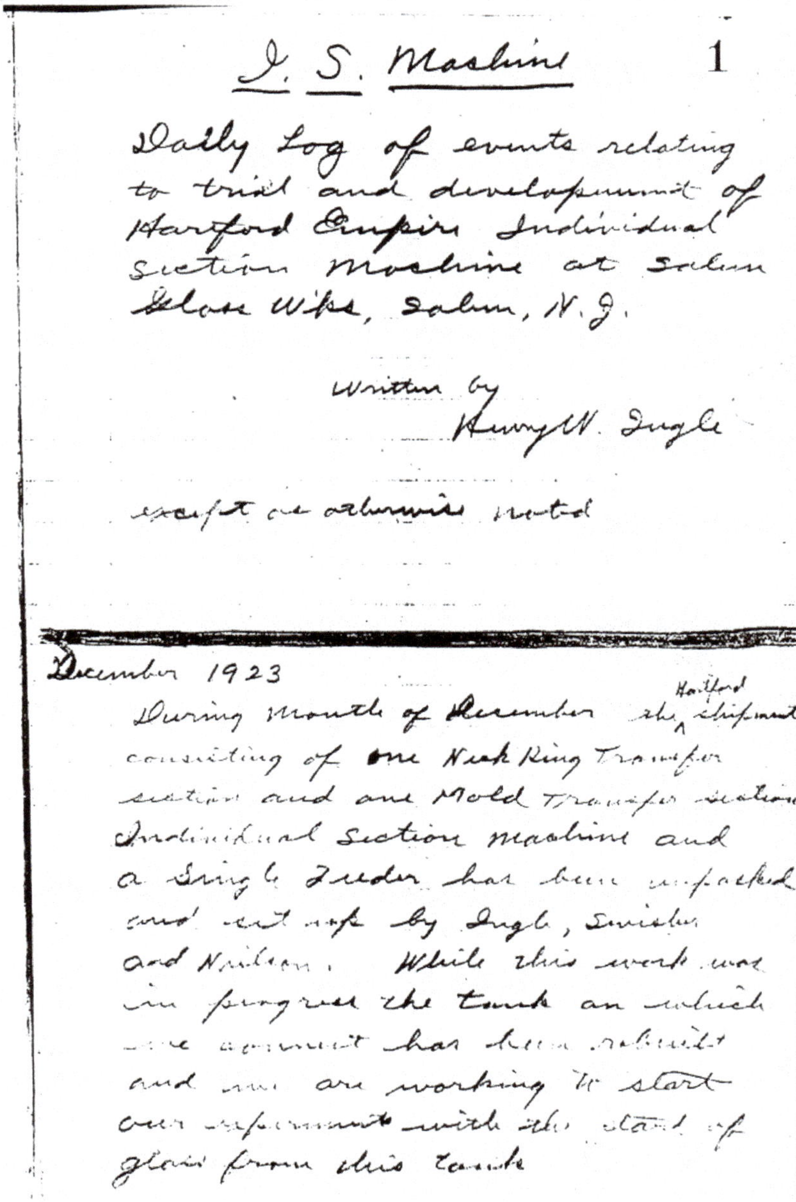

I. S. Machine 1

Daily Log of events relating to trial and development of Hartford Empire Individual Section Machine at Salem Glass Wks, Salem, N.J.

Written by
Henry W. Ingle

except as otherwise noted

December 1923

During month of December the Hartford shipment consisting of one Neck Ring Transfer section and one Mold Transfer section Individual Section machine and a Single Feeder has been unpacked and set up by Ingle, Swisher and Neilson. While this work was in progress the tank on which we experiment has been rebuilt and we are working to start our experiments with the start of glass from this tank

Figure 9,14. Page Number 1 of Henry Ingles report book of the trial at the Salem Glass Works, Salem, N.J.

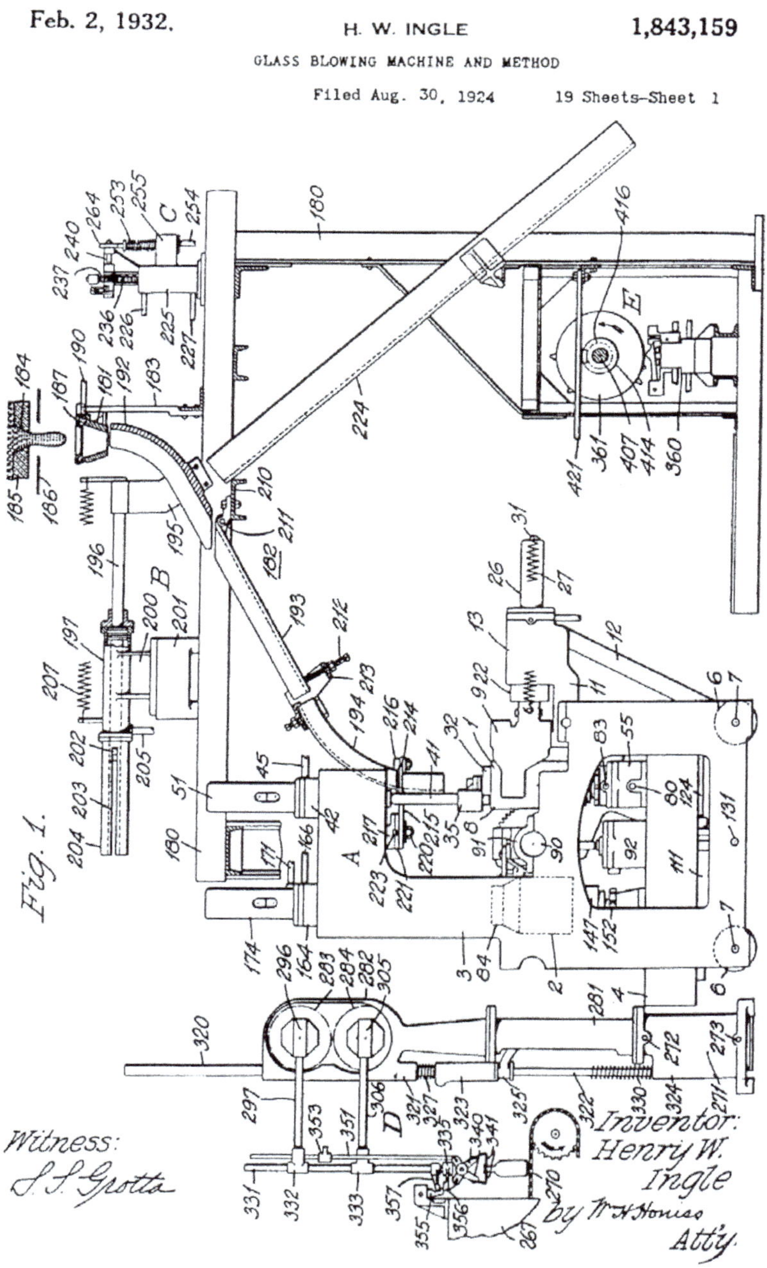

Figure 9,15. Drawing of Henry Ingles' IS machine, Sheet 1 out of a total of 19

Another important development in 1922 was the formation of the Hartford-Empire Company, this came about due to patent interference in 1916. The dispute was between B. D. Chamberlain an employee of the Empire Machine Company and the Hartford-Fairmont Company regarding the paddle feeder that Chamberlain said that he invented. This was finally resolved by Hartford-Fairmont merging the Empire Machine Company from its owners Corning into the company, Corning being able to use the benefits of the new companies developments to further their own business. See Chapter 6 for more information on this subject.

The first experimental 4 section machine was installed at Carr Lowrey in 1927, Figure 9,16 shows the take out side of the machine, including the conveyor and the tong stacker. Photograph taken 19th September 1928, and Figure 9,17 shows examples of the small ware made.

Figure 9,16, IS machine and conveyor.
Taken September 19th 1928

Figure 9,17. Examples of bottles made on the first experimental machine

The Anti-Trust Suit. The United States-versus-Hartford Empire Company. 1938-1945

According to John Edgington a perusal through the court papers showed where the Government was coming from. In 1916 the Owens Bottle Company acquired the Graham Glass Manufacturing Company, because it appeared to be on the verge of manufacturing an automatic gob feeder. They also bought up the rights to the Homer Brooke feeder. These were to safeguard their interests in the Owens suction machine of which they had a monopoly.

After the merger in 1922 between Hartford Fairmont and the Empire Machine Company, Owens sued Hartford Empire, maintaining that Peiler's invention infringed upon the Graham patents. The two companies became locked in bitter legal battles until 1924 when they settled out of court. Owens turned over its machine patents to Hartford Empire and paid them royalties on its own machines, in return it was given the use of forty machines without royalty charge, and approximately half of Hartford's net profit from licensing. Owens was also given the right to veto any licensing application for the next eight years.

Because Hartford Empire did not manufacture any glass themselves this agreement worked quite well. They controlled the areas of manufacture by restricting entry of manufacture to confined geographical areas. Hartford Empire acted as an unofficial regulator to the industry using its powers to discourage over production and price wars. They had the market all sewn up, they held patents for nearly 700, so virtually looking up the production of automatic machines.

The great depression and the Prohibition Act did not affect the glass industry and they undertook a period of growth. Then in December 1938 the Temporary National Economic Committee (TNEC) decided to investigate the glass container industry and all people and companies who had been most instrumental in its bottle manufacturing over the last 20 years.

The purpose of the hearings was to investigate the relationships existing in actual practice, between the patent monopolies and the competitive conditions within the industry, to ascertain to what extent the patent laws circumvented the anti-trust laws and to present this data to Congress with the ultimate view of drafting new patent legislation, that would maintain a free and open market.

The Hartford Empire was essentially a monopoly on the gob feeding method as will be seen in the breakdown for 1938;

Made under Hartford licences	67.4%
Made by Owens machines	29.2%
Made by other methods	3.4%

A total of 43 organisations where involved in the hearings including the Owens Illinois Glass Company, the Hazel Atlas Glass Company, Ball Brothers Corporation, Knox Glass Associates, Amsler-Morton, Corning Glass and The Glass Bottle Blowers

Association - a trade union. The Department of Justice drew up a chart showing the relationships between the various glass companies for the TNEC to use during their enquiries.

The TNEC reported back and after consideration, the government brought a case against members of the glass industry, the charges being of violating the Sherman Anti-Trust Act and the Clayton Act, by forming monopolies and conspiring to monopolise with respect to patents for the manufacture of glassware.

Just before the trial commenced two incidents were reported;

- a company lawyer named Hatch was ordered to investigate patent infringements at certain plants, which he did using many devious methods. One of which was to seek employment on a daily basis to look at equipment being used, and another where he climbed up onto the roof to take photographs and was shot at by the companies security staff, fortunately he was not hit.
- some of the independent glass plants decided to sue Hartford Empire claiming damages caused by the companies licensing system. During these investigations it was found that an employee of Hartford Empire had bribed a Union official. This was now considered to be a case of fraud which made a trial under the anti-trust laws inevitable.

The trial was held in the Federal Court of Toledo, Ohio, before District Judge Frank L. Kloeb, it started on the March 3 1941 and ended on the July 3 1942 a total of 112 days of testimony. The records of the trial amounted to over 12 000 pages including 3300 exhibits.

The Judge found a "violation of the anti-trust laws ... as deliberate as any that I can find in a review of anti-trust cases". He found violations;

- in the agreement in 1916 between Hartford and Empire which gave Corning its exclusive position in the non-container field.
- in the agreement between Owens and Corning in 1924.
- and in the dealings with Hazel Atlas, Ball, and Lynch.

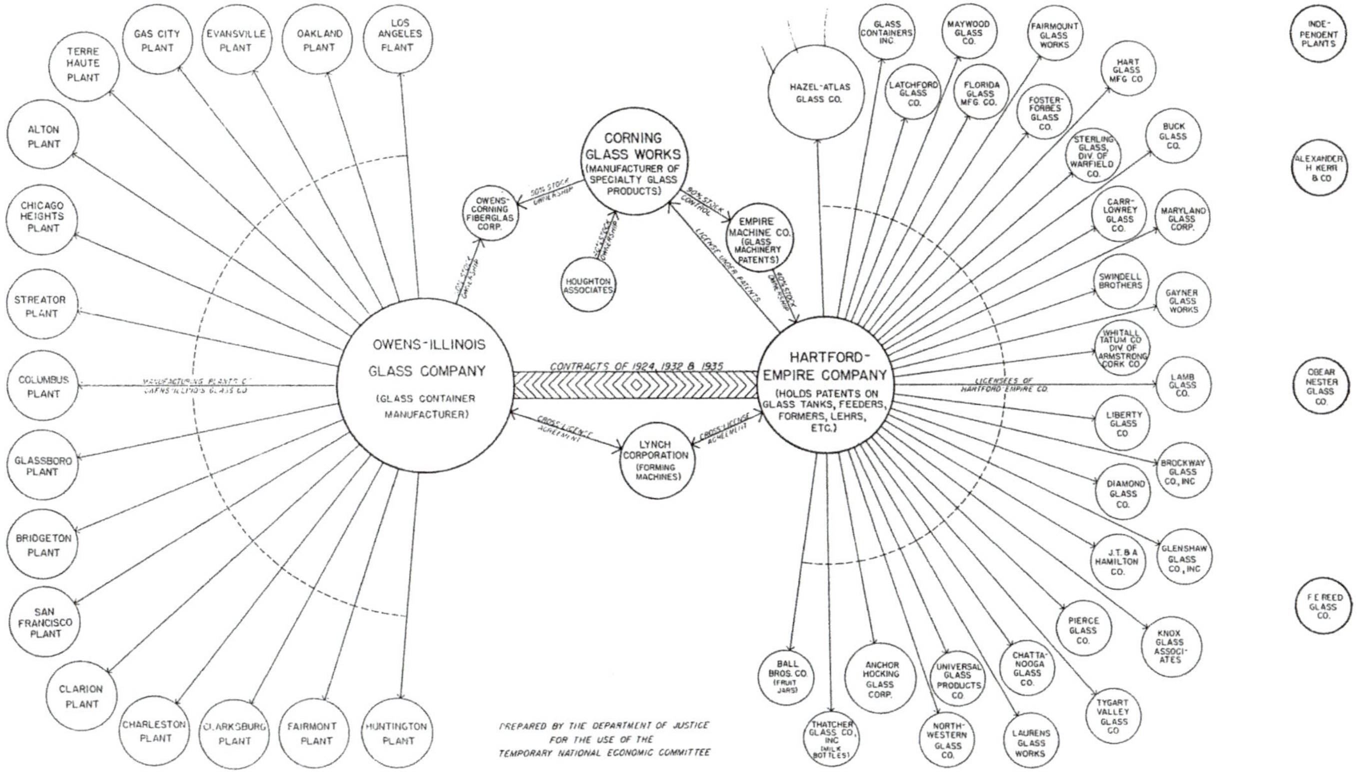

Figure 9,18. Relations within the Glass Industry

The decree was both severe and complex, it included the dismembering of the Hartford Empire Company, it gave an order to all defendants to grant royalty-free licenses under all current patents and to license future patents at reasonable royalties.

The Supreme Court affirmed the findings but softened the decree, it said that the findings did not warrant compulsory royalty-free licensing, which amounted to confiscation of the patent. The defendant was ordered to license at reasonable royalties.

At the end of the trial Hardford Empire was put into receivership, James Shanley being appointed as receiver. There followed a number of appeals against the final judgement and the consent degree. Then on the 8th January 1946 the court handed down its decision. Receivership was discontinued, and Hartford Empire was ordered to agree to royalties that the company could charge for use of its patents from companies manufacturing glass containers.

Finally in 1947 a decree was signed, in the Toledo Court by Hartford Empire, who were now under orders to give customers the option of buying or leasing its forming machines, and they were to license all applications from independent machinery manufacturers at agreed fees.

The TNEC hearings, the anti-trust suit, and the final consent decree paint the company in a very bad light, Dr. F. W. Preston in his summary of the Hartford affair highlights some valid points;

- The Hartford licensing policy kept many small business's alive and well. These would most probably have collapsed if the feeder patents had been used to put Hartford in the glass manufacturing business.
- The composition of the industry had not been static with new companies coming in to the business.
- Over the years in question the glass industry had made great strides in efficiency.
- During the depression the glass industry had maintained steady employment throughout.

In his final summary his quote is worth repeating:

"We believe that the glass container industry has contributed greatly to social stability in the last decade (The

Great Depression Years), and this has been due to the steadying influence of two organisations, Owens and Hartford, which the Department of Justice was doing its best to impeach".

Before we go on to the various machine specifications let us look at the engineering history of the Hartford people. The following listing is of the glass making machines and their accessories which came from there, these exclude the 'in line' machines. The 'in line' machines come after these.

Developments by the various companies that eventually became known as the Glass Division of the Emhart Corporation from 1910 the 1990s are:

The 1910s

1911 Peiler started his research into feeders.

1912-16 Designed and built an automatic press machine for vacuum-sealed jars at a cost of $65 000. It was far ahead of its time and it was also too big and expensive for general use.

1913 In their 'Glass Factory' the Hartford engineers successfully made jars with the 'Paddle' feeder, but the glass was of poor quality. On the second attempt it failed. This is when the company learnt that deviation from the normal glass making parameters would not be tolerated. This would be put into operation in later years.

1913 Peiler started his research into the 'Paddle and Needle' feeder.

1914-15 Trial with new feeder and mould equipment started at the Monongah's Glass factory.

1915 Lorenz and Peiler started to design an automatic press and blow rotary machine for the production of milk bottles. Its development during 1916-17 cost about $87 000. The first machine was put into the J.T. & A. Hamilton plant. This intermittently rotating machine produced 'disc-seated' top milks.

1916 Commercial use of the 'Paddle' feeder. The results of the 'paddle' feeder were found to be generally unacceptable due to its limitations in gob shaping for narrow neck ware. It was generally acceptable for press and blow but a failure for blow and blow.

1916 Started work on an automatic machine to manufacture 'narrow neck' ware.

1918 Commercial use of the Paddle and Needle' feeder.

1919 Work on a Hartford press and blow machine for other types of wide and semi-wide mouth jars. First unit completed in 1920 went to work in Baltimore. The direct development costs were $180 000.

1919 Hartford 'Narrow Neck' had already passed fits cold trials. Started the hot trials. Great difficulties ensued with expansion causing a redesign of many of the parts. The redesign was successful and the machine was put onto the market.

I have no other information regarding this machine but could this have been the Hazel Rotary IS machine.

The 1920s

1922 First successful trial took place of the single feeder in November. Later to be called the Hartford 144 feeder.

1922 Hartford branched out into lehrs with the new Hartford Type A lehr.

1923 Improved paste mould process.

1924 New woven type lehr belt introduced on the instigation of Henry Ingle. First one of the new Type C lehrs with woven belt installed at Glenshaw Glass.

1924 Introduction of the 'Tong' lehr stacker. Bottles were still transferd to the lehrs until speeds exceeded 40 containers per minute.

1926 Introduction of the Model 133 lehr.

The 1930s

1930 Introduction of the H 21 paste mould machine.

1931 Introduction of the H 12 machine for 'narrow neck' production. It was an 8 mould gob fed continuously rotating machine.

1932 Introduction of the H 28 paste mould machine. Final version made available in 1939.

1937 Introduction of the Type M annealing lehr.

1938 The Type MD decorating lehr put onto the market.

1939 The Type 61 Batch charger introduced.

The 1940s

1941 The Tri-angular doghouse design for feeding batch into the furnace introduced.

1943 The H-61 Tempering process introduced.

1944 Introduction of the KU (16" wide) and KW (26" wide) gas fired forehearths. These were an advanced version of the oil fired Z type which had been introduced earlier. In later years they where followed by the K 36 and the K 48.

1944 Introduction of the HE-73 and the HE-73B. Due to resistance from the glass producers all tempering trials stopped in 1947.

1945 The Type 74 Check Inspector entered the market.

1946 Automatic glass level control put on the market.

1946 The Enfolding Type 90 Batch Charger for use on Unit Melters introduced. By 1959 some 40 were in operation and by 1962 the total had climbed to over 100.

1947 The Seedscope introduced.

1948 Development of the Type 83 Hartford-Empire Gang stacker. It lifted a complete row of bottle from the cross conveyor onto the lehr belt.

1949 Introduction of the HE-84 Glass Spike Detector.

Other developments during this period were;

The Type M and MD lehr improvements

Development of a treatment to protect glass surfaces

IS Mould Gauges

Container ware comparators and gauges

The 1950s

1950 Introduction of the Hartford 81 Turbex feeder.

1951 Introduction of the 113 feeder. Used with the HE 111 High Torque air shears used for television production. Introduced in 1951.

1951 Introduction of the 115 feeder for extended range of double gob operation.

1953 Sundvall plant shipped its first IS machine. It was a 5 section machine which was delivered to Redfearn Bros., Barnsley, England.

1954 Development of the HE-127 Automatic Finish Check detector was completed. Original for the 'baby food' market.

1954 The Thermal Shock Machine.

1955 Introduction of the single-liner for the lehr end.

1957 Successful trials of the HE-X131 servo-feeder mechanism but not marketable.

1958 The Type 192 Choke Detector put onto the market.

During this period the 503, the 513 and the 515 were developed from the 81 feeder for use with the wider forehearth required by tonnages of up to 150 tonnes per day.

The 1960s

Early 60s Introduction of the Type 162 Batch charger. Improved Type 61.

1962 The Check and Mark detector put onto the market.

1962 Introduction of the 194 feeder, basically a 144 but with enlarged spout for higher tonnages.

The 1970s

1970 Last lehr to come onto the market in the form of the 566.

1978 Introduction of the straight-line shears to the glass industry.

During this period the Type 540 advance forehearth system was introduced. It provided an improved cooling and heating system along with sophisticated temperature control giving rise to greater temperature homogeneity.

The 1980s

1985 Introduction of the 555 feeder. This was micro-processor controlled with a servo-plunger mechanism and a servo-electric parallel shear mechanism. This made for many applications for computer control leading to the Flex-Line system.

The 1990s

In this period the following were introduced;

Major developments in automatic inspection
An improved 555 Feeder System
Hot-end inspection
Improved 740 Forehearth Control System

The highlights of the growth in the IS glass forming machines by the various companies which eventually became the Glass Division of the Emhart Corporation from 1920 to 1995 are:

The 1920s

1921 Started development of the IS machine.

1921-22 Henry Ingle joined the company.

1923 Trials with a single section Blow and Blow IS machine took place at Salem Glass Works, Salem, New Jersey in the December.

1924 Ingle filed his patent application. Accepted in 1932.

1927 First commercial single section put on the market.

1927 First experimental 4 section machine installed in the Carr-Lowry Glass at the Company's Baltimore factory.

1928 First 2 section machine put on the market.

The 1930s

1932 On February 2nd Ingle granted his patent for the IS machine.

1939 Introduction of D.G. feeder.

The 1940s

1940 Development of the 'E' Type machine started.

1940 Introduction of the P. & B. Process.

1940 Introduction of D.G. machine (first with $3\frac{5}{8}$ inches C.D. later by $4\frac{1}{4}$ inches C.D.)

1941 An IS 4 section Double Gob machine installed at Anchor Glass.

1941 After trials the Hartford 65 machine entered the market. This was a variation of the IS principle of manufacture. Had limited sales success.

1947 First Emhart Type 'E' IS Machine (later Sundsvall built over 4000 sections).

1948 Up to this time all of the Hartford Empire machines sold had been sub-contracted out for manufacture. One of the main ones being the Hartford Special Machinery Company who were situated next door to Hartford Empire. The following year, 1949, Hartford bought the Henry and Wright Company who specialised in high speed metal forming machines and started to manufacture their own.

The 1950s

1950 Introduction of the five section machine

1952 Development of vacuum settle.

1954 Introduction of the 6 section machine.

The 1960s

Early 60s Introduction of the Hartford 199 press and blow machine. Just like the Type 65 the machine it followed the IS principle. Had limited sales success. Only four were built.

1968 Introduction of the triple gob machine.
1969 First Type 'EF' machine.
Also marketed during this period were;
A check and mark detector
IS machine Type F $5\frac{1}{2}$" mould centres
IS machine Type F $6\frac{1}{4}$" double gob and $4\frac{1}{4}$" triple gob.
First 8 section machine installed

The 1970s
1970 The introduction of the gob distributor.
1970 Introduction of the 8 section machine.
1974 The application of electronic control put on the market.
1974 The development of 'narrow neck' press and blow.
1977 Introduction of the AIS machine with parallel mould motion.

The 1980s
1984 Application of advanced forming control system.

The 1990s
1990 IS 10 sections Type EF 3" triple gob came to market.
1990 IS 16 sections Type EF $4\frac{1}{4}$" double gob became available on the market.
1990 IS 16 sections Type EF $5\frac{1}{2}$" double gob came onto the market.
1990 AIS-HS 10 sections entered the marketplace.
1991 AIS 12 section $6\frac{1}{4}$" was introduced.
1991 IS 16 section 5".
1991 IS HS $5\frac{1}{2}$" double gob came onto the market.
1991 Improved version of the electronic control system put onto the market.

1992-1995
During this period the following were marketed;
Servo-electric parallel shears
Mould Open/Close mechanism in oil

The T.600 LAN Forming Control System
A Pantograph Baffle mechanism
The 560 Sevo-electric Pusher
New Funnel and Blowhead mechanisms
New Pocket Air and Fingers for pusher mechanism
Powernet Process Management System

Professor W. E. S. Turner in his 1937 Presidential Speech listed 47 tables, of which I have used eight, either in their entirety or as a shortened version, this table is one of them. It lists the first sales of the then Hartford-Empire Machinery Co. that took place outside of America during the period 1916 to 1936.

First non-USA sales of Hartford-Empire machinery 1916 to 1936

Paddle Feeder Date & Country	Paddle Needle Feeder Date & Country	Single Feeder Date & Country	IS Machine Date & Country
27/8/1917 Canada	30/10/1918 Australia	22/3/1923 Japan	23/3/1928 England
29/4/1918 England	14/3/1919 Holland	19/9/1923 Argentine	12/10/1929 Brazil
30/4/1918 Scotland	22/1/1919 Cuba	8/10/1923 England	26/6/1936 Australia
	2/12/1920 Argentine	21/12/1923 Australia	11/9/1936 U.S.S.R.
	6/12/1920 Brazil	26/8/1926 Canada	
	17/2/1925 Canada	21/11/1927 Chile	
		4/10/1929 Brazil	
		9/12/35 China	
		6/10/1936 U.S.S.R.	

Table 9,1. The First Non-USA Sales of Hartford-Empire or Hartford-Fairmont Machinery by Country

Since this time Emhart have marketed their version of the 'Quad' machine it has 95 mm centres and was introduced in the mid-2000s. A report by Emhart in Glass International in 2007 says that "currently three NIS 12 section QG machines are setting new benchmarks".

Looking at the published brochures for the family of $4\frac{1}{4}$", 5", $5\frac{1}{2}$" and $6\frac{1}{4}$" Emhart machines we can see that the only differences are due to size. Most of the dimensions for each size machine remains the same, i.e. for the 6, 8 and 10 section machines, the 12 section is different due to the delivery system.

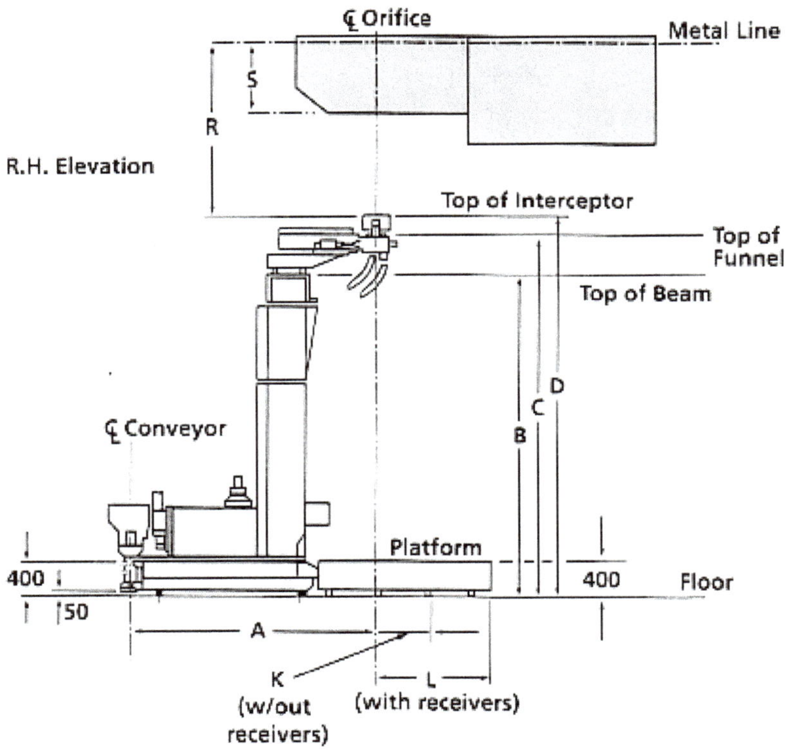

Figure 9,19. Showing an end view of a $4\frac{1}{4}$" C.D. IS machine

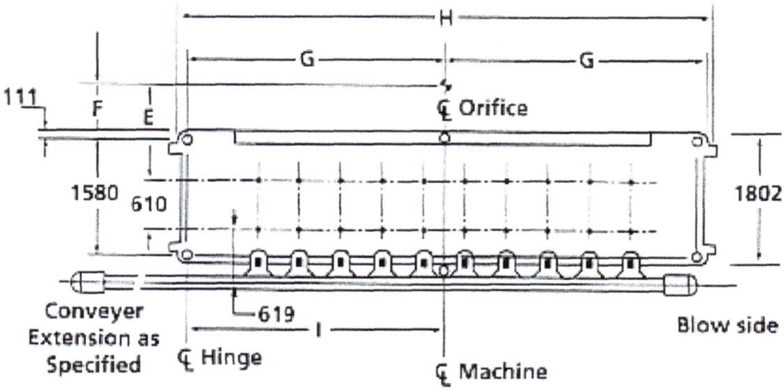

Figure 9,20. Showing a plan view of the same machine

The dimensions for the forehearth/feeder are the same for each model of machine. They are listed in Table 9,2.

	S (mm)	**Min. R (mm)**	
Feeder & Spout		**Arcurate Shear**	**Parallel Shear**
81 Std	324	708	737
81 Deep	375	759	788
503	414	822	827
515	477	889	890
555 Std	414	N/A	827
555 Deep	477	N/A	890

Table 9,2. Forehearth /Feeder dimensions

Of the machine dimensions some of them are common irrespective of the machine centres while others vary, for example, the distance between the blank mould and the blow mould positions on the $4\frac{1}{4}$" machine is 610 mm this increases to 629 mm for the 5" machine, to 673 mm for the $5\frac{1}{2}$" machine and to 711 mm. for the $6\frac{1}{4}$" machine. Likewise the distances from the blow mould to the centre of the conveyor are 619 mm. for the $4\frac{1}{4}$", 629 mm for the 5", 724 mm for the $5\frac{1}{2}$" and 724 mm for the $6\frac{1}{4}$" machines. All other marked dimensions in both Tables 9,3 and 9,4 are the same for all machine types.

Dimension Specifications

Ctr, Dist.	4¼"		5"		5½"		6¼"	
No. Sections	8	12	8	12	8	12	8	12
	mm	mm	mm	mm	mm	mm	mm	mm
A	2398	2658	2417	2677	2566	2826	2604	2864
B - D.G. ¬	3175	3910	3302	3910	3232	3980	3232 ¬	3980 ¬
B - T.G. [2]							3258 [2]	4045 [2]
C - S.G.	3525	4260	3652	4260	3582	4330	3662	4409
C - D.G.	3604	4339	3731	4339	3661	4409	N.A.	4427
C - T.G.	3552	N. A.	3679	4263	N. A.	4427	3635	4398
D - S.G.	3677	4412	3804	4412	3734	4482	3813	4561
D - D.G.	3756	4491	3883	4491	3813	4561	N. A.	4595
D - T.G.	3713	N. A.	3840	4425	N. A.	4595	3796	4559
E	1168	1429	1159	1419	1168	1429	1168	1429
F	651	911	651	911	618	879	637	898
G	2772	3839	2772	3839	2772	3839	2772	3839
H	5797	7931	5797	7931	5797	7931	5797	7931
I	2804	3871	2804	3871	2804	3871	2804	3871
K	815	555	815	555	842	575	815	568
L	1264	1022	1264	1022	1295	1029	1275	1022
Wt. Inc m'lds (tonnes)	35	51	35	51	37	55	38	56

Table 9,3. Specifications for selected common designed Emhart machines

Emhart IS Machines

Centre distances of various machines and the number of bottles that they can manufacture at any one time

	S.G,	**3"**	**85 mm**	**95 mm**	**4¼"**	**5"**	**5½"**	**6¼"**
4¼"	S.G.	T.G.	---	---	D.G.	---	---	---
5"	S.G.	---	T.G.	---	---	D.G.	---	---
5½"	S.G.	---	---	---	---	---	D.G.	---
6¼"	S.G.	---	---	---	T.G.	---	---	D.G.
AIS	---	---	---	---	T.G.	---	---	D.G.
NIS	---	---	---	Q.G.	---	T.G.	---	D.G.

Table 9,4. Bottle centre distances for various Emhart IS machines according to type

N.B. Some confusion arises here because Figure 9,11 states that the Type F(B) could manufacture 3¾" (95.3 mm) triple gob.

Emhart ISMachines - Maximum Body Diameter

Centre distance	**Single Gob**	**Double Gob**	**Triple Gob**	**Quad. Gob**
4¼"	156 mm	76 mm	51 mm	----
5"	156 mm	95 mm	60 mm	----
5½"	156 mm	102 mm	----	----
6¼"	156 mm	121 mm	76 mm	----
AIS	----	121 mm	80 mm	----
NIS	----	121 mm	90 mm	65 mm

Table 9,5. Maximum body diameters for various Emhart IS machines according to the machine centres. All machines equipped with Vertiflow

Comparative Outputs of Different Sizes of IS machines

No. CTRS Sections	$3\frac{5}{8}$" S.G.	$4\frac{1}{4}$" S.G.	$5\frac{1}{2}$" S.G.	$5\frac{1}{2}$" to $6\frac{1}{2}$" D.G.	7", $7\frac{1}{2}$" to $8\frac{1}{2}$ T.G.
5 Section	100	---	---	---	---
6 Section	115	120	135	220	(300)
8 Section	---	165	180	300	415
10 Section	---	200	220	360	(505)
(12 Section)	---	(240)	(265)	(440)	(615)
(16 Section)	---	(320)	(360)	(600)	(830)
(20 Section)	---	(400)	(440)	(720)	(1005)

Table 9,6. The comparative outputs from various IS machines rounded to nearest 5. Information as circa 1975 but amended to cover the newer machines.
N.B. Figures brackets (---) are estimated. 16 section machine taken as two 8 section machines and the 20 section as two 10 section machines

Some of Emhart's current machine sales catalogue data - circa 2006, is;

IS machines - traditional pneumatic forming machines. Sizes;

single gob;
5½" double gob;
85mm. Triple gob;
4¼" double gob;
6¼" double gob;
4¼" triple gob;
5" double gob;
3" triple gob.

AIS machines - traditional pneumatic forming machines with parallel mould and blank opening. Sizes;

single gob;
4¼" triple gob;
6¼" double gob;

NIS machines - the first fully servo electric forming machine. Sizes;

single gob;
$6\frac{1}{4}$" double gob;
5" triple gob;
95 mm. Quadruple gob.

These machines for container forming have sizes varying from 4 to 20 sections. The range is comprised of 4, 5, 6, 8, 10, 12, 16, and 20 section machines. The 16 and 20 section machines are fed from twin orifices and are called tandem machines.

Other specialist forming products consisted of InVertiFlow neck ring cooling; pantographic baffle mechanism; servo electric invert; flex pressure system; Vertiflow blank and blow mould cooling system.

For delivery equipment there is a choice of the Deflector Series 1800, which consists of scoops, troughs, and deflectors or the 535 servo gob distributor.

The hot end handling equipment consists of either constant cushion takeouts, or servo electric takeouts; 8000 series main conveyors; flexpushers; cross conveyors; push bar stackers; and choice of 178, 478 or 878 ware transfers. The company also makes a vast range of spare parts.

A modern IS machine consists of, as a minimum; a servo electric gob distributor, servo electric pushers, a T 600 LAN Forming control, a ratio programmable drive and quick change plunger mechanisms and accessories.

Now we will go through the archives to see what photos we have, but before we do that we will include an abstract #147 from the 1927 edition of the Journal of the Society of Glass Technology. This is the first entry for this machine and is quite informative.

#147. An Automatic Bottle-making Machine. P. C. Ruchen, London (Hartford Empire Co. Connecticut, USA) Brit. Pat., No. 257 637, March 7th, 1925; from Ill. Official J. Pat. No. 1 971, October 27th, 1926.

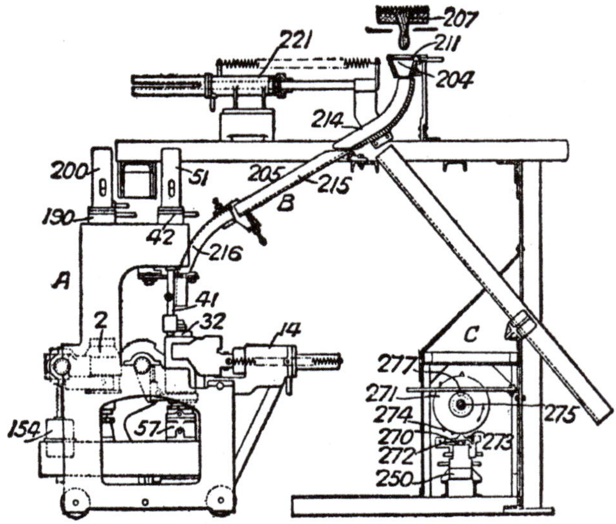

Figure 9,21a. 1927 Patent application for an IS machine Figure 20

FIG. 20.

- It is an apparatus for making bottles or similar glassware comprises a number of independent forming machines, A, Figs. 20 and 21, a delivery mechanism, B, for conveying charges from a feeder to each of the machines, and a timing-device, C, for controlling the operations of each machine in proper timed relation to one another and to the feeding-device.

The machines, A, are preferably arranged side by side and each comprises an inverted parison mould, 1, Fig. 22, a finishing mould, 2, and a ring mould, 3, which is adapted to swing about a horizontal axle, 97, to transfer a parison from the parison mould to the finishing mould, 2. The operation is so arranged that a parison is formed in the parison mould at the same time as the previously formed parison is being blown to its final form in the finishing mould.

The parison mould sections are hinged on a pin, 11, Fig. 22, and are opened and closed by an air cylinder, 14. The open end of the mould is closed by a cover or baffle plate, 32, carried by an arm,

35, adjustably secured to the piston rod, 41, of an air cylinder, 42. The other end of the rod had a pin which engages in a cam slot, 52, formed in a tubular extension, 51, of the cylinder, 42. Thus, as the cover is lifted from the mould, it is swung to one side out of the path of the ring mould, 3.

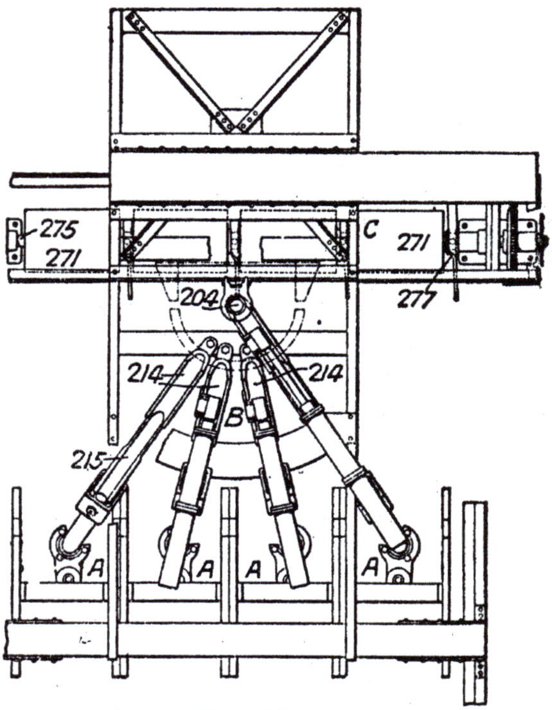

FIG. 21.

Figure 9,21b. 1927 patent application Fig. 21

During the parison-forming operation, the ring mould, 3, is positioned below the parison mould, 1, and a mouth-forming plunger, 71, Fig. 22, is projected into the ring by an air cylinder, 57. This cylinder is provided with two pistons, one of which, 60, has a tubular piston rod, 61, carrying a head, 63, that seats against the underside of the ring mould. The other piston, 67, has a piston rod, 70, which slides through the rod, 61, and carries the mouth-forming plunger, 71.

Suction is applied to the ring mould, through a pipe, 80, which communicates with the space between the two pistons and thence through passages with the interior of the ring mould. Pressure is then applied through the pipe, 80, and this forces the piston, 67, downwards and withdraws the plunger, 71, from the ring mould. At the same time the pressure is transmitted to the ring mould

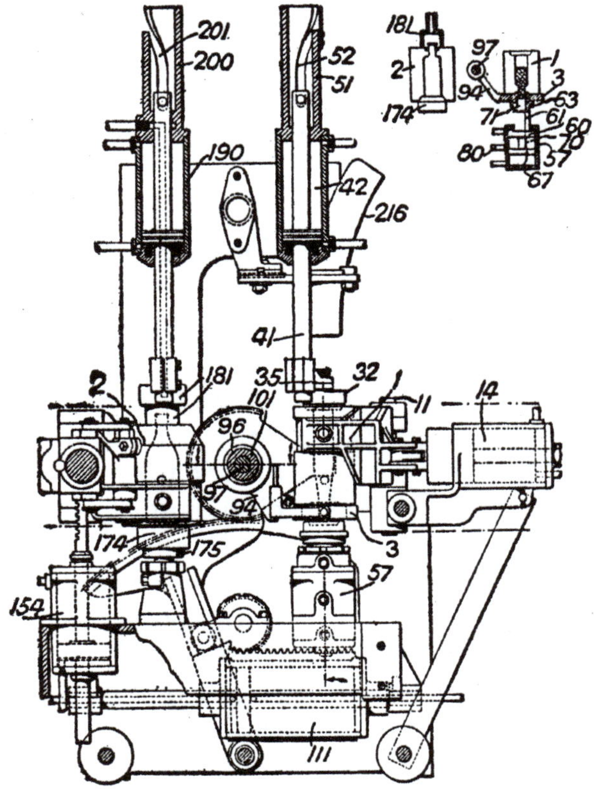

FIG. 22.

Figure 9,21c.
1927 patent appl.
Figure 22

and counterblows the charge to for the parison.

When the parison is formed, the mould, 1, is opened and the axle, 97, Fig. 22, is rocked by a cylinder, 111, in order to turn over the arms, 94, on which the ring mould is mounted and thus transfer the parison to the finishing mould, 2. The arms, 94, each carry a section of the ring mould and each is carried by opposed cylinders, 96, slidably mounted on the axle, 97, and pressed towards one another along a common piston, 101, by springs. When air is admitted to the cylinders, they are forced apart and open the ring mould.

The finishing mould halves are opened and closed by a cylinder, 154, and the bottom plate, 174, is carried by a holder adjustably mounted on the baseplate of the machine. A blowhead. 181, is raised or lowered by the piston of an air cylinder, 190, which is provided with an extension, 200, having a slot, 201, to swing the blowhead to one side as with the baffle plate, 32.

The means for delivering the mould charges to the different forming machines comprises a funnel, 204, Figs. 20 and 21, which receives the charges from the feeder, 207, and a number of guideways, 205, which extend from the funnel to the machines. The funnel has its inner surface lubricated by air or steam projected from an annular chamber, 211, mounted over its upper edge. Each guideway comprises a trough section, 215, a deflector, 216, and a reciprocating distributor, 214, which is carried by the piston rod of a cylinder, 221. The cylinders are operated in succession, so that the distributors, 214, are brought in turn beneath the funnel and deflect the charges in succession to the troughs, 215.

The pressure-controlling mechanism, for timing the operations of the different parts of the machines and of the delivery mechanism, comprises a valve chest, 250, which is connected to the various cylinders by air pipes. Air is admitted to these pipes by valves actuated by levers, 272, Fig. 20, which are operated by studs, 270, adjustably secured to a constantly rotating drum, 271. The valves are held open by latches, 273, until they are tripped by studs, 274, on the drum. A number of drums are mounted on a shaft, 275, which is driven in timed relation with the feeder, and each drum may be disconnected from the shaft by a clutch, 277.

Now, as promised, we turn to the first of our photographs which is of a single section machine with its take-off plate from which the bottles could be loaded by hand into the lehr. This machine would have been similar to the single section trailed at the Salem Glass Works, Salem, New Jersey.

This is followed by a two section machine, once again it has a take-off plate for manual lehr loading, signifying an earlier model. The first four of this model were sold to the Hocking Glass Company of Lancaster, Ohio.

Figure 9,22. A single section IS machine

Figure 9,23. A two section IS machine, note the take-off plate the one services both sections

Next we have a four section experimental machine in the machine shop, there are two views of the completed machine prior to its installation. Note the date as it was probably the one installed at the Carr-Lowrey Glass Companies factory in Baltimore, Maryland, USA in 1927.

Both of the photographs were taken in 1925, Figure 9,24 shows a partly completed machine from the timing drum side and Figure 9,25 shows the front of the machine including a take-off conveyor end you can see the back of the scoop mechanisms.

Figure 9,24. Photograph of a partly assembled machine. Dated September 12, 1925

Figure 9,25. End view an experimental machine, dated September 1925

Finally on this four section machine is a front view of a four section machine and it is dated July 15th, 1927 which is the year when the first experimental four section machine was installed in the Carr-Lowrey plant in Baltimore. Maryland, USA. The machine ready for dispatch in 1927 is seen in Figure 9,26.

Figure 9,26 (above) and Figure 9,27 (below) a 5 section working at Foster's Glass

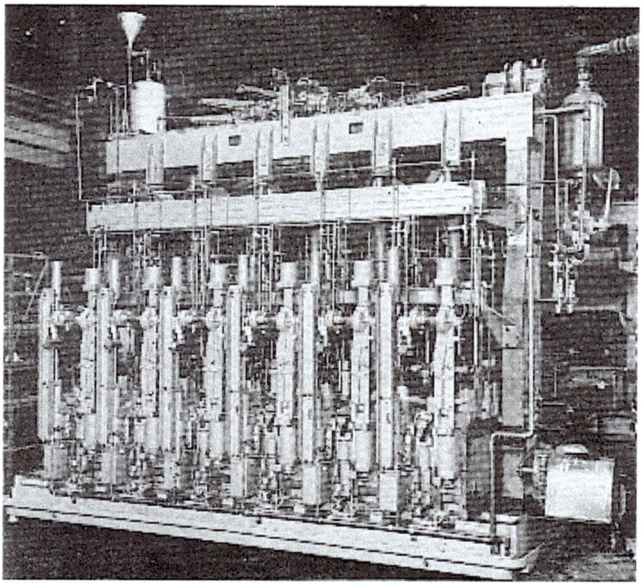

Figure 9,28. View of a 6 section IS machine and below Figure 9,29 is the 1000th machine made at the Sundsvall site. It was an 8 section machine made for Verriere Ouvrerie d'Albi of France. Photograph date 10th October 1974

Figure 9,30. A 12 Section double gob machine. In this case it was a 12 oz. Container Cycle Analyser B machine

Figure 9,31. A 16 section IS machine. Notice the twin delivery systems. The machine uses two feeder mechanisms from the same forehearth

That finishes the Emhart biopic but although they put much into the development of the IS machine after the early 1950s other companies started to make their version of the IS machine. Some companies like Maul Technology manufactured and sold their machines on the open market while others such as Owens-Illinois Inc. made their version of the machines and sold them to their own company, their affiliate companies, or to companies that had technical agreements with them. Often their machines would be marketed or used by themselves before the Emhart version came into being, a typical example of this is the quad machine, Owens-Illinois and their affiliates where using the quad gob machine as early as 1986, (see the BHF bioptic in Chapter 6), whereas Emhart sold their first three NIS quad machines in 2005.

Using the Emhart timeline some of the earlier contributions made by others manufacturers are;

In 1939	Emhart marketed their 4 section double gob machine, this looks like it was preceded in 1937 when Owens announced their double cavity mould. This is the only information that I have but I think that it must be for a gravity fed machine as their suction machine had been working as multi-cavity machine for years.
From 1953 to 1959	there was the introduction of the Roirant R7, the Lynch 44, the O'Neill 55, the Owens G50 and the Owens version of the 'solid blank'. Emharts had introduced this concept in 1952.
From 1959 to 1965	Maul introduced its Maul IS machine, the Maul expanded machine centres, and the Maul 8 section machine.
From 1965 to 1969	there was the Maul triple machine, the Lynch-Heye machine, the Owens triple machine, the Owens 8 section machine, and the Owens GCP.
From 1970 to 1974	saw the introduction of the CSS pushers, and the gob distributor.
1974	saw the introduction of electronic control or timing. The Owens-Illinois name for their system was the ComSoc.

From 1974 to 1976	saw the introduction of the Heye 6-12 machine and Constant Cushion Inverts.
1976	saw the introduction of Computer Control and Constant Cushion Take-outs.
1977	saw the introduction of IS Parts and the Heye 1-2 prototype.
1982	Vertiflow was introduced this was matched by Owens-Illinois Axial mould cooling system.

Obviously there is a 'time lag' between when a machine was first introduced to the market and the date of its installation. According to the records the first $4\frac{1}{4}$" dual machine was introduced to the American market in 1959 and by 1962 30.5% of all attempted hours was by this method of manufacture. By 1968 dual operation of IS machines were 53.0% America and 50.7% the international market (Zones II and III).

The same year the first IS triple gob operation was recorded which amounted to 0.1% of machine hours reported. By 1970 the machine hours attempted were 2.7% for America and 0.04% for the international market.

Percent Double and Triple Gob Operation

	Domestic		International	
	D.G.	T.G.	D.G.	T.G.
	% Att. Hours	% Att. Hours	% Att. Hours	% Att. Hours
1968	53.0	0.1	50.7	
1969	57.6	1.0	51.3	
1970	60.8	2.7	53.8	0.04
1971	66.3	3.3	62.0	0.16

Table 9,7. Percent Double and Triple Operation for selected years

In 1983 Emhart in their reporting system quoted for Zones II and III an IS 6 quad machine which worked that year for 146 days, and in 1985 there was a IS 5 machine reported.

It must be noted here that every manufacturer did not contribute their production statistics to the Emhart Machine Performance Analysis - formerly called Machine Production Rates. So although they are a good guide they are not 100% perfect.

Let us now look at the development of the 'quad' machine, much of it is anecdotal and is from peoples recollections. I believe that the first attempt at a quad operation took place in France in the late 1960s/early 1970s by St. Gobain Desjonquères where they converted a 5 section single IS machine to a $2\frac{1}{8}$ inches centre distance quad machine. A single cylinder using a candelabra arrangement operated on each machine section.

The next 'recollection' that I have obtained is that in the late 1970s Bormioli converted a $4\frac{1}{4}$ inch double gob machine to a quad operation, each of the cylinders were converted to a twin candelabra arrangement. Owens-Illinois started their investigations into quad gob operations at the start of the 1980s and by the early 80s they had their first machine in operation. Here the centre distances were not equal, the centre one being larger than the other two, the moulds that the machine used was two 'block' moulds.

Figure 9,32. A B-HF quadruple shearing system

In 1986 BSN in France announced that the company had achieved a record daily production rate of 720 000 25cl. beer bottles from one line, it was a quadruple machine that used an Owens-Illinois parallel shearing system that had been modified to incorporate a B-HF non-arcular shearing system. I believe that the machine was an Owens-Illinois machine although BSN had been trying to design a machine with larger section boxes to give larger centre distances. Although this photograph has been shown before Figure 9,32 shows the quad shearing system while Figure 9,33 shows a quad sweep out system.

Figure 9,33, An Emhart quadruple sweep-out arrangement

There are other references to quad machine before the 2005 launch by Emhart, in 1997 B.D.F. of Italy were offering two to twelve section machines which ranged from single to quadruple gobs.

There are a number of machine sellers and repair specialists, there are too many to record here but the following table lists some of them, showing the machine types and sizes, and the mechanical/electrical options that are available.

Guide to some Machine Suppliers other than Emhart

Machine size	Single gob								Double gob								Triple gob							
	BDF	Bottero	Glass Eqip.	HEYE	Maul	MAVSA	Shamvik	Sklostroj	BDF	Bottero	Glass Eqip.	HEYE	Maul	MAVSA	Shamvik	Sklostroj	BDF	Bottero	Glass Eqip.	HEYE	Maul	MAVSA	Shamvik	Sklostroj
Single	☐	☐		☐	☐	☐	☐	☐		☐		☐	☐	☐	☐	☐		☐		☐	☐			
2 section	☐	☐			☐	☐	☐		☐	☐			☐	☐	☐		☐	☐			☐			
4 section	☐	☐		☐	☐	☐	☐	☐	☐	☐		☐	☐	☐	☐	☐	☐	☐		☐	☐			
5 section		☐			☐	☐	☐			☐			☐	☐	☐			☐			☐			
6 section	☐	☐	☐	☐	☐	☐	☐	☐	☐	☐	☐	☐	☐	☐	☐	☐	☐	☐	☐	☐	☐		☐	☐
8 section	☐	☐	☐	☐	☐	☐	☐	☐	☐	☐	☐	☐	☐	☐	☐	☐	☐	☐	☐	☐	☐			☐
10 section	☐	☐	☐	☐	☐		☐	☐	☐	☐	☐	☐	☐		☐	☐	☐	☐	☐	☐	☐			☐
12 section	☐	☐	☐	☐	☐		☐	☐	☐	☐	☐	☐	☐		☐	☐	☐	☐	☐	☐	☐			☐
6+6 section	☐	☐	☐	☐	☐			☐	☐	☐	☐	☐	☐			☐	☐	☐	☐	☐	☐			
8+8 section	☐	☐	☐	☐	☐			☐	☐	☐	☐	☐	☐			☐	☐	☐	☐	☐	☐			
10+10 section	☐	☐		☐	☐			☐	☐	☐		☐	☐			☐	☐	☐			☐			
Centre distance																								
3 in.																	$3^{5}/_{16}$	☐	☐		☐		$3\frac{1}{2}$ in.	
$3\frac{1}{8}$ in.																								
$3\frac{7}{8}$ in.																								
$4\frac{1}{4}$in.	☐	☐	☐	☐	☐	☐	☐	☐	☐	☐	☐	☐	☐	☐	☐	☐		☐	☐	☐	☐		☐	
5 in.	☐	☐	☐	☐	☐		☐	☐	☐	☐	☐	☐	☐		☐	☐				☐				☐
$5\frac{1}{2}$ in.	☐	☐	☐	☐	☐			☐	☐	☐	☐	☐	☐			☐								
$6\frac{1}{4}$ in.		☐		☐					☐	☐		☐	☐		☐	☐	☐			☐				

Table 9,8. Various machine suppliers. Legend:

BDF Industries SpA. Vincenza, Italy. Email: bdf@bdf.it Web site: www. bdf.it
Bottero SpA. Cuneo, Italy. Email: hollowglass@bottero.com Web site: www. bottero.com
Glass Equipment (India) Limited, Bahadurgarh, (Haryana), India. Email: bsomany@hngil.com Web site: www.hngil.com
HEYE International Gmbh, Obernkirchen, Germany. Email: info@heye-international.de Web site: www.heye-international.de
Maul Technology (A Division of VHC Ltd.), Sunderland, UK. Email: sales@ maultechnology.co.uk Web site: www.maultechnology.com

Maquinas Argentinas Del Vidria SA, Buenos Aires, Argentina. Email info@mavsa.com.ar Web site www.mavsa.com.ar
Shamvik Glasstech PVT Limited, Mumbai, India. Email vsm@shamvikglass.com/office@shamvikglass.com Web site www.shamvikglass.com
Sklostroj Turnov CZ, s.r.o., Turnov, The Czech Republic. Email sklostroj@sklostroj.cz Web site www.sklostroj.cz

Various choices of machine equipment

Company		BDF	Bottero	Glass Equip	Heye	Maul	MAVSA	Shamvik	Sklostroj
Forming process	Blow-blow	x	x	x	x	x	x	x	x
	Press & blow	x	x	x	x	x	x	x	x
	NNPB	x	x		x	x		x	x
Machine driving mechanisms	Pneumatic	x	x	x	x	x	x	x	x
	Electric	x	x	x	x	x	x	x	x
	Hydraulic					x	x		
Type of gob distributor	Mechanical	x	x			x	x	x	
	Servo	x	x	x	x	x	x	x	x
Machine timing	Electronical	x	x	x	x	x	x	x	x
	Mechanical	x	x	x		x	x	x	
Refurbishment services		x	x	x	x	x	x	x	x

Table 9,9. The options of machine equipment available from different suppliers

Other manufacturers and/or refurbishers are;
Emhart Glass, Cham, Switzerland. Email info@emhartglass.com Web site www.emhartglass.com
Owens-Illinios, Inc. Toledo, USA. Web site www.o-i.com
IS Maintenance Ltd., Barnsley, UK. Email sales@ismaint.co.uk Web site www.ismaint.co.uk
GPS Glasproduktions-Service Gmbh. Essen, Germany. Email s.howells@gps-essen.de Web site www.gps-essen.de
Inter-rep Associates, Inc. New Jersey, USA. Email bleafy@inter-rep.com Web site www.inter-rep.com
Toyo Glass Machinery Co., Japan. Email fjp55036@nifty.com Web site www.toyo-glass.co.jp

In 1997 BDF were advertising that you could purchase from them, IS machines that ranged from two to twelve sections in length with options to manufacture either single, dual, triple or quadruple gobs. At the same time Bottero had just sold a twelve section to Nippon Glass in Japan and GPS had already sold five twelve section machines to different Thai glass manufactures. In the early 1990s Emhart themselves had sold 12 section machines to Saudi Arabian Glass in Jeddah and in Dubai. The following is a $5\frac{1}{2}$", 16 section machine that was sold to St. Gobain and was installed in their St. Romain de Puys factory and commissioned in December 1996.

Figure 9,34. A Bottero $5\frac{1}{2}$", 16 section machine built to St. Gobain's specifications

Next we come to the **Lynch 44**. This machine can produce containers by the blow and blow and the press and blow processes. It was designed for the making of high quality production of short run length. The machine took very little time to change from single to dual operation as well as changing from blow and blow to press and blow operation. It can be used as a twin unit machine, a three unit, a four unit, a six unit working back-to-back and possibly an eight unit working back-to-back.

The Lynch 44 is a multipurpose machine, receiving gobs by the well known and proven Hartford feeder principle, i.e. off set scoop, trough and deflector. It is a much larger machine than the IS machine, the footprint of a four section machine has a table depth of about 1.5 metres and a length of approximately 5.6 metres. The equivalent IS machine dimensions for a four section machine would be approximately; table depth 1.45 metres, machine length 2.9 metres. Which equates to an increase in area of nearly 70%, even allowing for the claimed increase in speed it is a vast footprint when access and working conditions are at a premium.

According to the manufacturers brochure the unique turret action design of the Lynch 44, which employs three finish ring arms, makes possible;

- simultaneous invert and revert action
- maximum overlap in the glass forming cycle, creating higher production speeds
- fewer mechanical actions per cycle
- elimination of take-out operation, reducing finish defects
- holding of the finish by a finish ring throughout the cycle, the ring opening only top lace the ware on the deadplate
- positive register of the parison in the blowmould
- final blowing through the finish ring to eliminate bulged or distorted finishes

The Lynch 44 turret rotates the glass through three simultaneously operating positions - the blank station, the blow station and the dead plate station. Through the use of three finish ring arms, all three work concurrently, accomplishing maximum overlap in the glass making cycle. The overall machine time is reduced by the elimination and combination of mechanical actions. The mechanical speed of this machine is up to 25-30 cycles per minute per section.

The Lynch 44's design shortens the travel time required to carryout the cycle, creating for the operator more glass forming time and the cooling system more heat extraction time. This allows him more blank time, reheat time and blow mould time to give better control of the glass forming process.

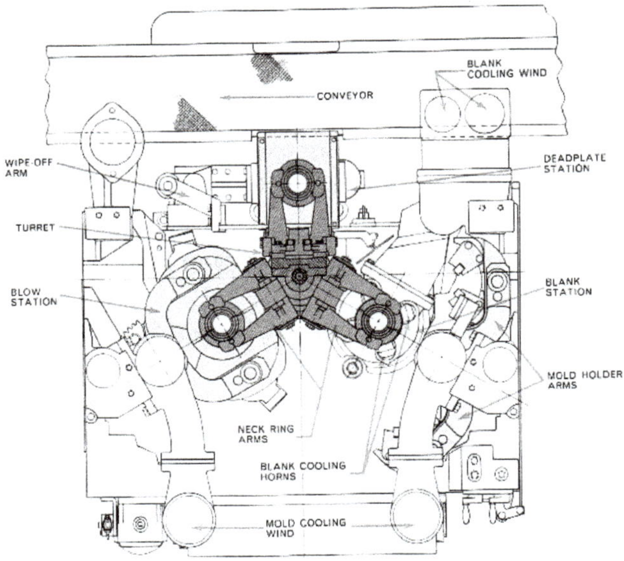

Figure 9,35. Plan view of the Lynch 44 machine

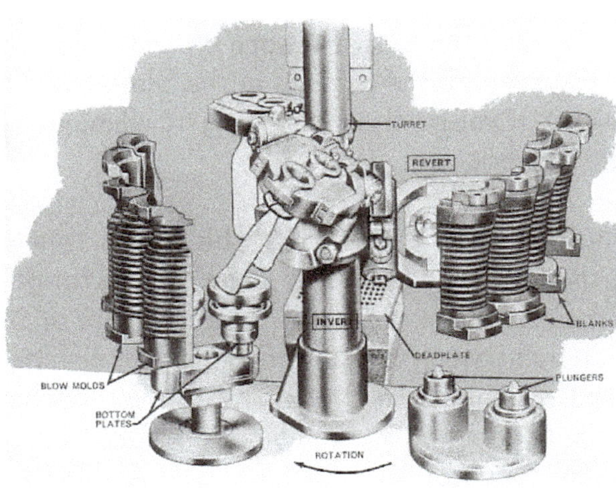

Figure 9,36. Schematic of the turret action

By the use of vacuum immediately the glass is pulled into the finish, the counterblow then forms the parison. Since the machine uses vacuum instead of settle blow there is little evidence of a settle wave which means that the container can be made at a lower glass weight. Hence increasing the manufacturing speed and giving a cost saving on the glass.

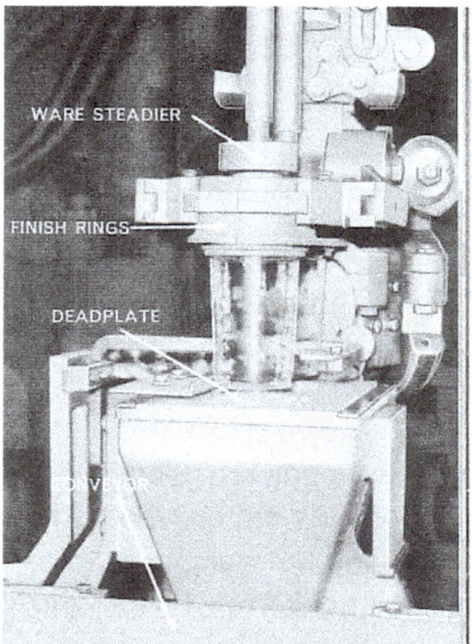

Figure 9,37. View of the dead plate area of the machine showing that there are no take out tongs

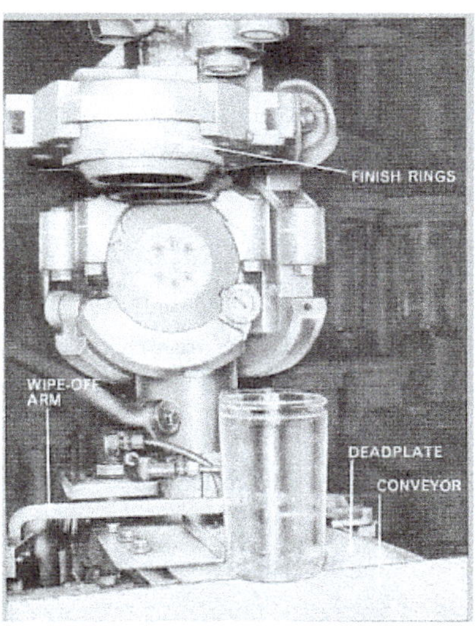

Figure 9,38. View of dead plate area showing wipe out arm and finished rings after releasing the container

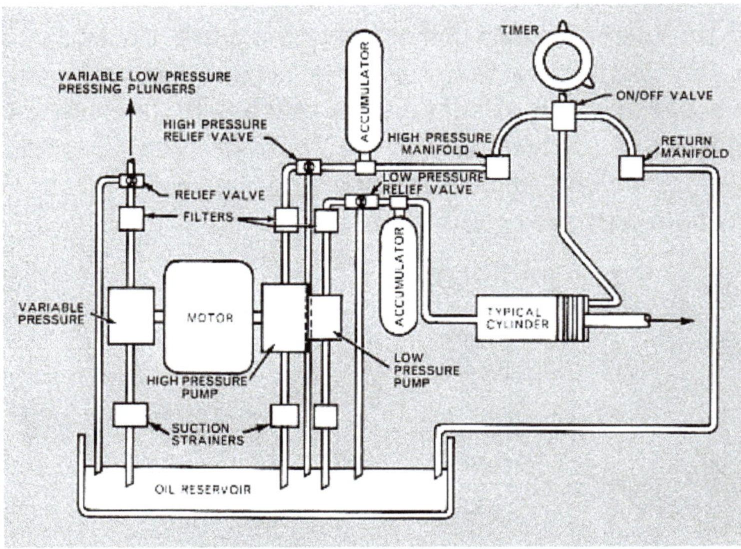

Figure 9,39. The Lynch hydraulic system

The Lynch 44 uses a hydraulic system to power its operation, the hydraulic system uses pistons and valves except that they are operated with a non-flammable fluid and not compressed air. According to the brochure some of the benefits of the Lynch 44's hydraulic system are;

- higher speeds with precision timing are obtained by controlled pressure of non-compressible fluid
- better pressing action
- reduction of friction and wear, resulting in lower maintenance costs
- turret action that can be stopped by hand for safety
- infinitely quieter operation increases personnel efficiently
- appreciably lower operating costs than that of a pneumatic system.

The brochure then claims that these features and many others give 20-50% HIGHER PRODUCTION.

The precision mould cooling system is designed to remove heat quickly and precisely form the moulds. The fast forming requires close and controlled wind application, by directing cooling through the mould holders and wind horns, maximum wind utilisation is achieved. The cooling wind is always available to the mould holders whether open or closed, and to the wind horns which travel up and down with the solid blank. Additional control is supplied to give timed wind control, the blank, blow mould and bottom plate

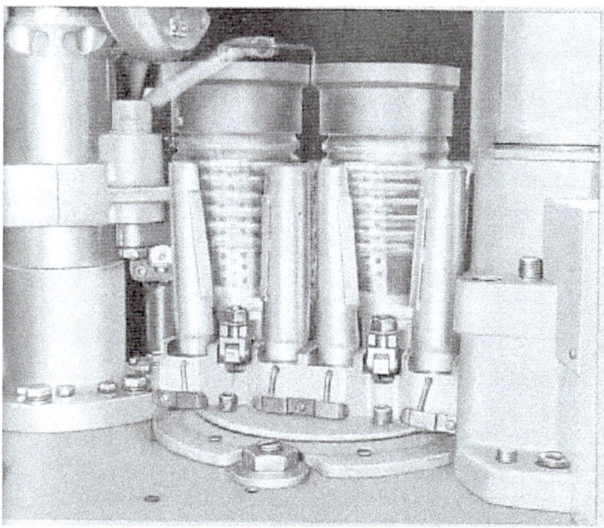

Figure 9,40. Solid blanks

Figure 9,41. Split blanks and blow moulds

wind are turned on and off by timer adjustments. The volume of each wind application can also be manually set.

The cooling wind is placed $\frac{1}{2}$ inch from moulds - gives maximum wind utilisation for fast glass forming cycle and high quality ware.

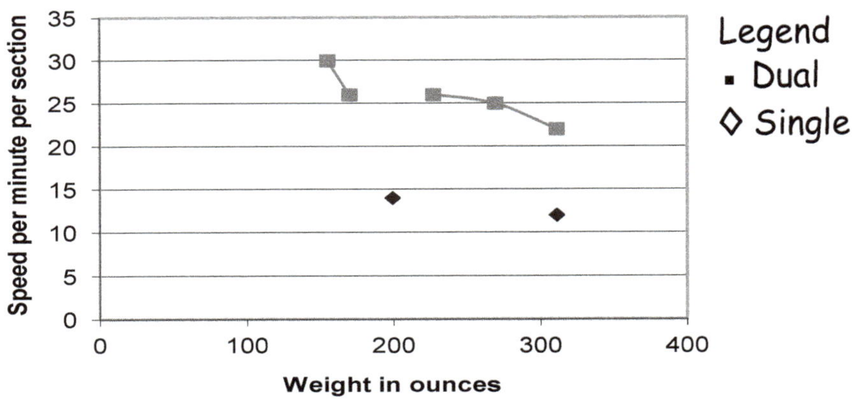

Figure 9,42a. Speed curve for Lynch 44 - Press and Blow. Manufacturers information. Circa 1970

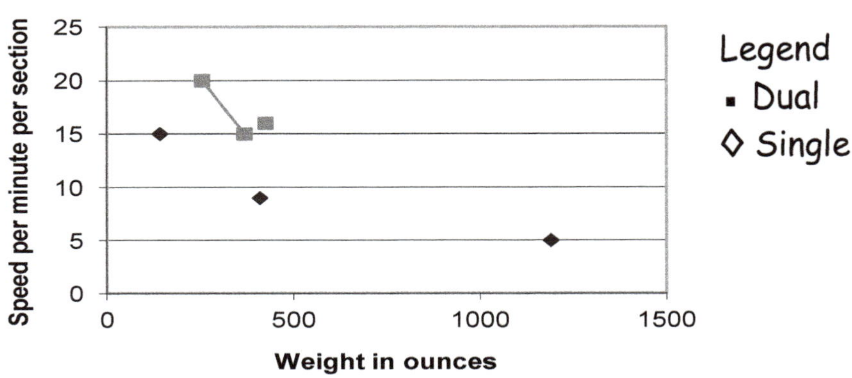

Figure 9.42b. Speed curve for Lynch 44 - Blow and Blow. Manufacturers information. Circa 1970

Lynch 44 Maximum Ware Sizes

	Single Operation	
	Blow & Blow	Press & Blow
Height to base of finish	$14\frac{1}{2}$ inches or 368.3 mm	$9\frac{1}{4}$ inches or 247.65 mm
Body diameter	$6\frac{3}{4}$ inches or 165.1 mm	$6\frac{3}{4}$ inches or 165.1 mm
Finish diameter	4.5625 inches or 120 mm	4.5625 inches or 120 mm

	Dual Operation (Mould centre distance 5.0 inches or 127 mm)	
	Blow & Blow	Press & Blow
Height to base of finish	12 inches or 304.8 mm	$8\frac{3}{4}$ inches or 215.9 mm
Body diameter	4.0 inches or 101.6 mm	4.0 inches or 101.6 mm
Finish diameter	$3\frac{1}{8}$ inches or 89 mm	$3\frac{1}{8}$ inches or 89 mm

Note: Push-up bottoms of 2 inches, or 50.8 mm maximum depth can be made 'blow and blow' single or dual cavity.

Table 9,10. Maximum ware sizes that can be made on the Lynch 44 machine

Ware sizes can be made from fractional ounce to 160 ounce capacity. Five inch mould centres makes for dual cavity operation up to 4 inches in diameter.

Next we come to the O'Neill 55 machine of which there are only two references to this machine the first is in Glass Machines by Giegerich and Trier who say that it was marketed by the O'Neill Machine Company Ltd. of Burton Road, England. The actual address was Dormiston Trading Estate, Burton Road, Upper Gornal, Nr. Dudley, Worcestershire, England, and the second in the Society's Journals and publications. The only picture of this machine is an advertisement taken from a 1961 Glass Technology. I do not know of any sales of this machine.

Figure 9,43. The O'Neill 55 Machine

According to the advertisement the machine could be up to six unit long and if like the Lynch 44 it could be worked 'back to back' then that would make a total of 12 units.

The concept of 'back to back' working is not new, it probably started in the early 20th century, most likely after the introduction of the forehearth. For example there are records of some manufacturers utilising two, four, or five and even six

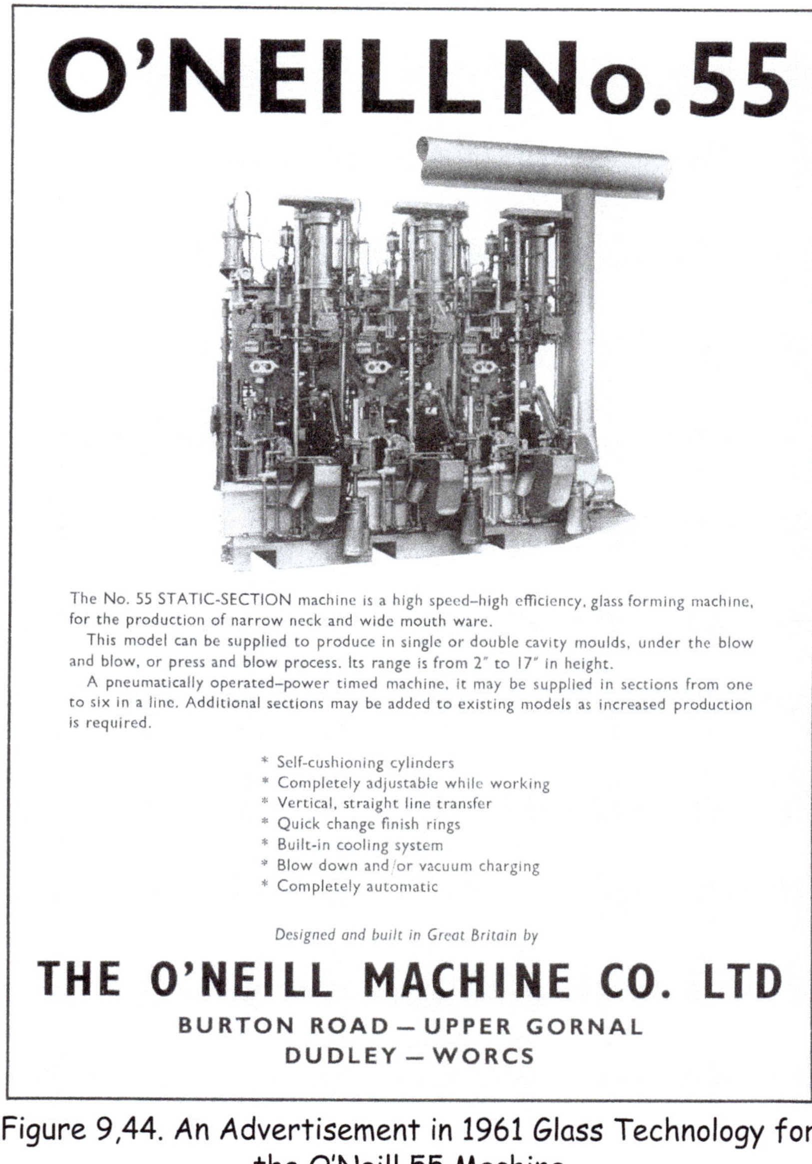

Figure 9,44. An Advertisement in 1961 Glass Technology for the O'Neill 55 Machine

section machines as a cheaper way of increasing their machine output. It had the added advantage of using up machines what would otherwise have been made obsolete.

Chapter 10

Why Machines Became Bigger

This chapter gives some idea of:

- The development of the glass making machines and why they all now cease to exist except for the IS machine.
- The consolidation of the worldwide glass container industry that has given rise to three international glass manufacturing giants.

Much of this information comes from America because their historical records are much better than most other countries.

Changes to the machine types due to changes in the market

The changes or the speed of the changes that occurred in the United States of America vis that in Europe can be put mainly down to market size. America is a large single country and so has a large market, whereas Europe, before the European Union, was comprised of individual countries but, in total, the market is nearly as large as the market of the USA.

Both sides of the Atlantic went through the same teething problems in the size of the markets of individual bottles. The bottle maker has always wanted all of similar ware to be alike i.e. all bottles would be classed as 'general trade', whereas the purchaser of the bottles has always wanted his bottles to be of an individual shape, i.e. one which can be uniquely identified with his product. So this leaves each party at opposite end of the spectrum. Even as early as the early 1920s just before his death Michael Owens was trying to get buyers to standardise their designs. This was a company objective as the Owens machine was good at manufacturing large quantities but could not compete with the short orders that the hand and semi-automatic makers made.

As time went by the glass market became more 'rationalised' due to mergers between the glass company's customers and the increases due to the natural growth in the market place, which meant that orders became larger and the ' making and design' of containers became closer together.

More and more people became used to buying products in glass but more were bought in glass per capita in the USA than in Europe. Table 10,1 is a selected extract from Moody which was published after 1961 showing the consumption of container glass for 1961 by country, the table shows only the top 10 countries.

World Consumption of Glass Containers

Country	Metric tonnes consumed (in thousands)	Number of containers used per person
USA	6800	129
Great Britain	1809	86
West Germany	1229	65
France	846	less than 50
Italy	352	20-25
Belgium	115	less than 20
Portugal	42	less than 20
Norway	(22)	less than 20
Japan	531	less than 20
Yugoslavia	54	less than 20

Table 10,1. The highest container consumptions per country

N.B. There is incomplete information for some countries container consumption for example; Austria used 64 000 tonnes, Denmark 56 000 tonnes, Finland 36 000 tonnes, and Switzerland 48 000 tonnes, all of these countries have small populations so probably they would have been included in the above table.

It shows that consumption of glass in the USA is 50% greater than Great Britain, which is the second largest, and is approximately double the consumption in Western Germany.

The United States of America was the biggest market and it did not import glass bottle making machines in any quantity, so the changing of the forming line (i.e. upgrading it) was different to Europe.

In America there were six main types of automatic machine, the Owens from 1904 to 1983, these were supplemented by the O'Neill and the Lynch machines, and the W. J. Miller blow and blow machine, the W. J. Miller and E. Miller their presses and press and blow machines, and finally the IS machine used from 1927 until now. There were other machines such as the Knox but these were only used by Knox Glass until the patents ran out, when they were then put on sale for others to purchase - but there were very few takers.

Prior to 1901 the records show that the following glassworks were using some form of machine to manufacture their wide mouth bottles and jars; the 'Butler' was being used by the Redkey Glass Co. of Indiana, the 'Pyle' at Hollweg and Reese at Greenfield, Indiana, the 'Hayley-Bridgewater' at the Cumberland Glass Co. of Bridgeton, New Jersey and the Illinois Glass Co. of Alton, Illinois, and the 'Powers' at Swayzee Glass Co. of Swayzee, Indiana and prior to that at the Gilchrist Jar Co. of New Jersey. Some glass-makers used machines made by O'Neill while other manufacturers such as Atlas, Flaccus, Ball Bros., the Marion Fruit Jar and Bottle Co., and the Rochester (NY) Glass Co. used, in some cases, in-house designed machines to manufacture their wide mouth jars.

After 1901 other machines were installed, Glenshaw Glass Co. of Glenshaw, installed Ashley type machines in 1908, the Olean machines were used at the Olean Glass Company (Works) and the Brockway Glass Company in 1910, the Root or Red Devil Machine was worked at the Root Glass Company in Terre Haute, Indiana, circa 1912, and by 1918 there were three Twenty Century machines at Clarion, PA.

In addition to the above there were a number of semi-automatic machines, the use of which gradually became less and less. Nowadays they are only used for very special 'short order' items which would be uneconomic to make on a automatic machine, although there are still some Lynchs around, which could be used viably in between the semi-automatic and an IS machine.

Before we go much further because of the difference in terminology between the United States and Europe it maybe a time to define what we mean when we talk about the different types of machine, they are shown in Table 10,2. Holscher in his booklet "Hollow and speciality glass" for Owens-Illinois described the progress in machine development slightly different, he describes the changes as firstly using the fundamentals learned in the late 1880s early 1890 came the first machines which were single head and looked something akin to a modern pillar drill, all functions were hand operated and gatherer loaded these he called or recorded as semi-automatics. See Figure 10,1, views a to d below, which shows a similar type machine going through its cycle. Here I must apologise for the quality of the photographs, I took them at the John Goodchild Archives and later realised that the plastic protection sleeves that they were in had caused reflection from the overhead lights. I do not know what machine this is other than it came from the George Bradley archives and is possibly one from either Castleford or from the West Riding of Yorkshire. George Bradley was a solicitor who practiced in Castleford.

The next step was to mount these individual machines on to a table and drive them by pneumatic or electrical power still termed as semi-automatics but once the use of a more reliable feeding mechanism they became automatic machines. This is the debatable part as other people referred to 'hand fed' table mounted machines as automatic machines. We have described the two systems as follows;

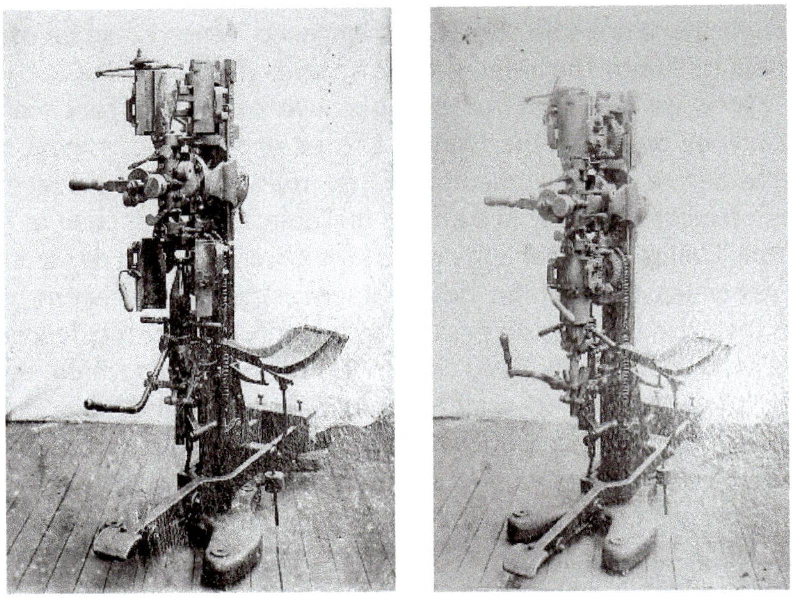

Figures 10,1 a & b. Showing the first two simulated stages of a semi-automatic machine

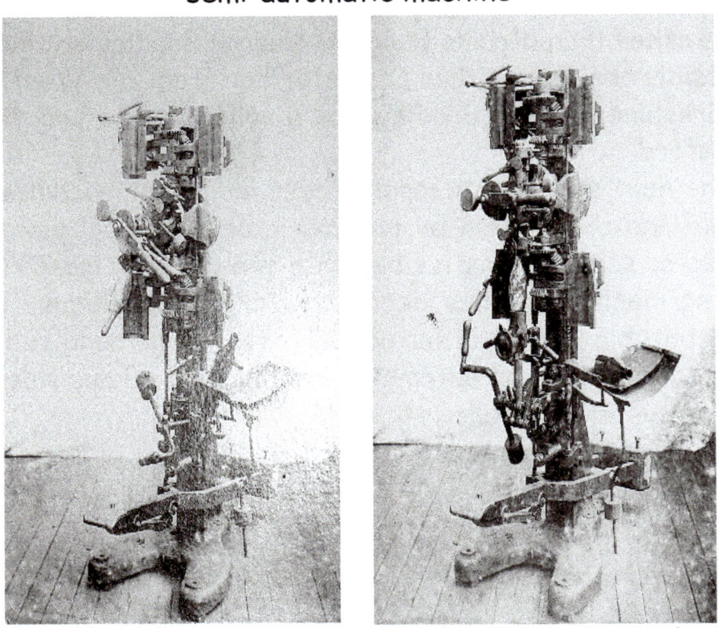

Figures 10,1 c & d Showing the third and fourth simulation stages of a semi-automated

	USA	Europe
Hand operated		All functions are operated by foot or hand.
Semiautomatic	All functions are hand operated	The machine is fed by a gatherer and then started manually before it makes a bottle, often automatically.
Automatic	The machine can work automatically irrespective of how the glass is fed to it.	The glass is loaded to the machine automatically while the machine runs continuously.

Table 10,2. Definitions of machine types USA vis Europe

The following statistics for the years 1898 to 1924 inclusive show the rise and fall of the semi-automatic machine in America. It reached its peak in 1916, after which came a more reliable automatic feeder. The figures marked with an asterisk are from union records so they only record places which were unionised.

Year	1898	1899	1900	1901	1902	1903	1904	1905	1906
No. of machines	50	60	80	90	100	150	200	250	168*
Year	1907	1908	1909	1910	1911	1912	1913	1914	1915
No. of machines	191*	223*	224*	216*	225*	266*	297*	312*	458*
Year	1916	1917	1918	1919	1920	1921	1922	1923	1924
No. of machines	459	428	417	417	315	288	164	130	72

Table 10,3. The rise and fall in the usage of semi-automatic machines in the USA Here the term semi-automatic refers to a machine where all of the functions are manually operated; they do not refer to an automatic machine that is fed by a gatherer

Shortly after the installation of the first Owens machine in 1904 the first narrow-neck semi-automatic machines to be sold (as against in-house developments) appeared on the market. The first was actually an import from Britain known locally as the John Bull, this machine was built on the pattern of the Ashley machine, it possibly could have been the Horne version of it, as he made many of the original ones and his patent application was made in 1902. According to Lockhart in 1908 there was at least one Ashley machine installed at Glenshaw Glass Company of Glenshaw, PA (or it possibly could have been either the Horne or the Breffit-Horne, both improvements on the original Ashley Machine), or it could have been a Horley machine who's patent application was made around the same time. See Chapter 6 - Horleys biopic taken from the Society's journals.

This machine had other names which were, the English or the United, after the company who was making them in the United States. As stated earlier F.H. Pierpoint and W.J. Demming from the P.B.M. syndicate had the licences - possibly for the United States only. I believe that this machine was the 'Horley', which were installed in the United States between 1908 and 1916 by Ronald Hirst. See Chapter 6. The United machine was licensed for seven years at $500 per year, and their popularity can be seen in Table 10,4 below.

Year	1909	1910	1911	1912	1913	1914	1915	1916	1917	1918	1919
No. of machines	14	38	71	93	94	113	114	94	83	19	10

Table 10,4. The number of 'United' machines working in the USA by year

Other machines were soon developed in America such as the 'Jersey Devil' a two man operated twin table machine, in its crudest form it was a twin table machine with manual transfer from the blank table to the blowing table and again manually from machine to lehr. Each table usually had either two or four moulds with the blank moulds which incorporated a compressed

Figure 10,2 . The Jersey Devil

air actuated plunger being positioned nearest to the furnace, and the blowing table having a compressed air blowing valve.

The Teeple-Johnson machine, another twin table machine was quite popular (see Chapter 8) while other American developed machines were the South Millville and the Turner machines.

Year	1903	1904	1905	1906	1907	1908	1909	1910	1911	1912	1913	1914	1918
No. of machines	1	1	6	8	22	46	51	71	103	133	161	187	194

Table 10,5. Growth of Owens machines in the USA

Up to the advent of the forehearth in 1915 - 1920 era the Owens machine 'ruled the roost', there were no automatic versions of gravity fed machines available and by this time Owens had sold 200 machines, see Table 10,5. In 1917 90 to 95% of all bottles were made by machines of which 45% were made by Owens machines.

	Semi-automatic Machines			
	Wide Mouth	Narrow Mouth	Total	Owens Machines
1906	168		168*	8
1907	191		191*	18
1908	223		223*	34
1909	205	19	224*	46
1910	216		216*	61
1911	173	52	225*	103
1912	170	96	266*	124
1913	201	96	297*	152
1914	210	102	312*	164
1915	193	265	458*	176
1916	167	292	459	187

Table 10,6. The growth of semiautomatic and Owens machines over the period of 1906 to 1916 inclusive. The totals marked with an asterisk for semiautomatic machines are for unionised manufacturing units only, there would have been a number of extra machines

Table 10,6 shows for the period 1906 to 1916 the rise in 'machine made' glass containers via the number of machines that

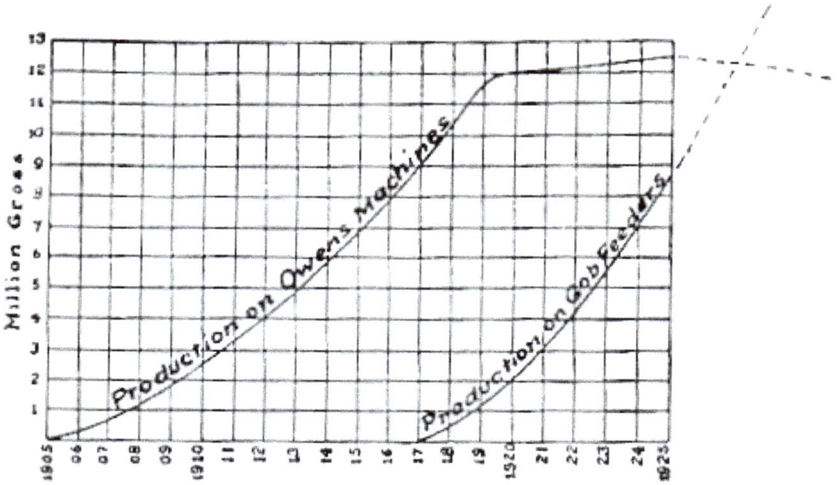

Figure 10,3. Comparison between the number of bottles made by Owen's machines vis gob fed machines

were in operation. It shows quite markedly for the semi-automatic machines that the initial growth came from wide mouth making side of the business.

The earlier gravity fed machines were automatic machines that either used a feeder such as Homer Brooke's flow device or the later Peiler versions of the 'Paddle' or the 'Paddle and Needle' feeder, or they were semi-automatic machines - meaning that they were fed from a gathering iron and then after charging the mould the gatherer would set the machine in motion by knocking a switch.

Once Peilers automatic needle feeder was marketed in 1922 and by the time that the IS machine was introduced in 1927 the dominance of the Owens machines was already in decline, Figure 10,3 shows this in detail. This was an original graph updated manually and shows that after 1927 more bottles were made from gob feeders than those made by Owens machines.

In Figure 10,3 the gob feeder started in 1917 with the advent of the Hartford Paddle feeder, obviously there were machines that were fed by gatherers or by the 'flow method' of feeding.

Since 1917 many feeder/forehearths came onto the market, many of which differed only slightly in design giving rise to many battles in court.

At this time it is probably best to define the forehearth types, first was the single forehearth feeding a single machine, and second was a multiple feeder usually incorporating a paddle-needle feeder. This device can feed two to four machines, when this is the case there is a trough for each machine which locates underneath the orifice, in its turn, and that gob is fed via the trough to the machine.

This increases the flexibility of the ware being made, each machine can make a different job but the downside is that they all have to be the same weight. Many people tried to make a gob feeder charge gobs of different weights, two such people to patent their ideas are - J. H. O. Bunge and the Hartford-Fairmont Syndicate Ltd granted 23rd, January 1920 for a variable weight paddle feeder, and W. A. Lorenz as assignor to the Hartford-Empire Co. filed on the 4th April 1922, in the USA.

The Bunge application was in the 1921, Volume V, abstracts in the Journal of the Society of Glass Technology and the Lorenz application was in the 1925 edition of the journal. The abstracts are as follows;

#363. Device for Varying the Weight of Charges by Glass-feeding Machines. British Hartford-Fairmont Syndicate, Ltd., and J. H. O. Bunge, 148, Audley House, Ely Place, Holborn, London (Brit. Pat., No. 161 001, January 23rd, 1920). - Relates to a modification of the apparatus described in Specification 105 672 which adapts it to automatically charges of molten glass to two or more machines requiring charges

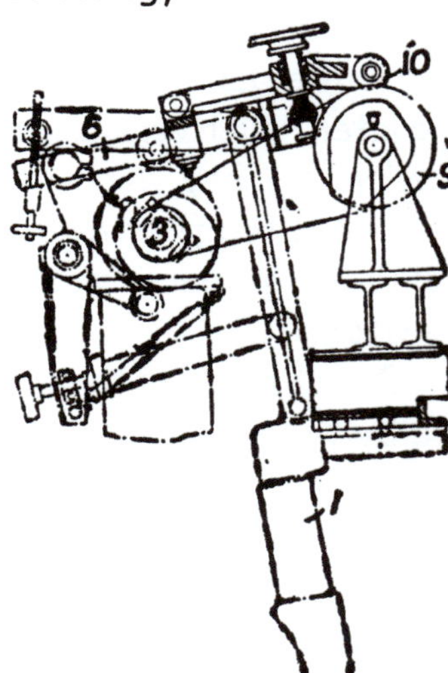

Figure 10,4. Bunge's application

of different weights. According to the invention, an auxiliary cam, 9, positively driven from the main cam shaft, 3, is provided. A roller, 10, adjustably mounted on the lever, 6, is engaged by cam, 9, thus lifting the lever, 6, and reducing the immersion of the paddle, 1, in the molten metal and hence the weight of charge delivered by that particular stroke. The frequency of these reduced impulses can be varied by altering the ratio of gearing between the shaft, 3, and cam,9.

The W. A. Lorenz abstract is from the 1925 edition of the Journal of the Society of Glass Technology regarding a U.S. patent filed on April 4th, 1922. The abstract is as follows;

#616. An Automatic Feeding Device. W. A. Lorenz, Assignor to The Hartford-Empire Co., Hartford, Conn. (USA Pat. No. 1 525 189, Feb. 3rd, 1925. Filed April 4th, 1922, No. 549 448). - The invention relates to methods of feeding glass automatically to forming-machines and its object is to feed charges of sizes and weights in a desired order of succession. The method of producing the different-sized charges, according to the invention, consists in severing a stream of glass at different levels in a suitable order.

Figure 10,5 shows a stream of glass discharging from the outlet, 10, of a forehearth, 11, and charges of glass of different sizes, are obtained by severing the stream alternately in the planes a-a and b-b. The successive figures show the position of the flow, at successive intervals of time. USA Patent 1 525 190, Feb. 3rd, 1925, describes mechanism by which the successive cuts in different planes are made by two pairs of shears operating alternately. USA Patent 1 525 191 describes mechanism in which the cuts are made by a single shearing-device operating alternately in two plane a-a and b-b.

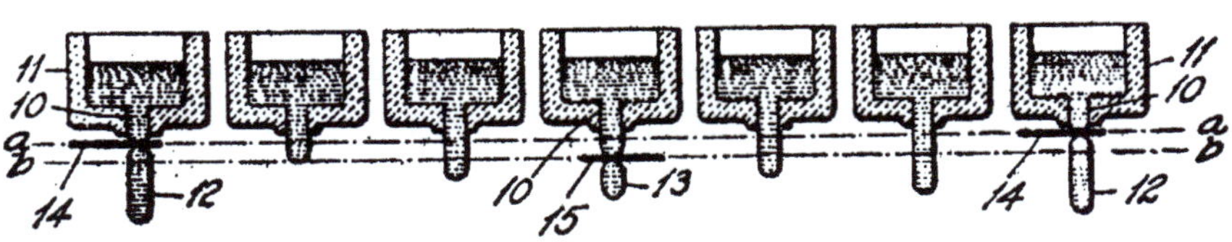

Figure 10,5 Schematic of Lorenz's feeder

I do not know if these inventions were ever successful, but at least the facility of working two different weights from a single orifice was available for the bottle-maker of that time.

	Number used during the 1916-17 season	**Number used during the 1917-18 season**
3-man machines	43	9
2-man machines	113	97
1-man machines	272	301

Table 10,7. Reduction in shop floor workers between 1917 and 1918 seasons

The order in manning changes due to mechanisation saw the "three-man, two-boy" machine being giving way to the "two-man, two-boy" machine, which in turn was superseded by the "one-man, one-boy" and finally the "one-man, no-boy" machine, i.e. it had become very nearly a fully automatic machine. Some of these machines still required the services of a gatherer but once a feeder was installed then the fully automatic machine became a reality.

The no-boy machine was introduced by both the O'Neill and the Lynch machine companies in 1917, on August 8th and November 17th, respectively. Prior to this time in 1913 Hartford-Fairmont had installed their first 'paddle feeder' at Fairmont. See later for the manning levels for different machines.

By this time the industry was dividing its self into three main types of manufacturing processes, mostly they were one of these types or they fell into two of these categories which were;

1. The hand process limited to very small orders and oddly shaped containers.
2. Used the Owens automatic machine, which required large orders; - very large orders for the larger sized machines.
3. The 'feeder and flow' automatics used both for middle sized and larger orders but having a main use in the mass production of slightly smaller orders - i.e. 'stock' or 'general trade' bottles.

Hypothetical question - What type of machine would I have

chosen?. Obviously it depends upon the size of the orders, the larger the machines are, the larger the order needs to be to make them economical. Around about the mid 1920s the minimum order for a large automatic machine was a minimum run of 1000 gross. The cost of making and repairing the number of moulds required by these larger machines, the time taken in the setting and the adjustments of them and the changing of moulds had a influence on the size of the minimum order.

The smaller six headed 'feed and flow' type machine required a minimum order of 250 gross to make them competitive while containers made by the semi-automatic and hand processes manufactured orders less than these. They would be used for the manufacture of perfumeries, toilet ware and to the manufacture of 'oddly shaped' bottles that may or may not have been used for advertising purposes.

There were some mitigating circumstances regarding the size of the orders; Owens when they sold and licensed a machine they dictated at the outset what type of product the machine could manufacture, i.e. the type of ware that the machine had been purchased for. This put a massive constraint upon the manufacturer - the downside - but the upside was that they had that total rights to the manufacture of that type of ware amongst Owens licensees which increased the size of the orders dramatically considering the size of the country. If the market was large enough then two licenses would be issued, as were the ones for whiskeys given to Chas. Boldt and the Illinois Bottle Company.

An example of this was that the Illinois Bottle Company took out licenses for three machines, the first on June 11th, 1910 for perhaps branded whiskies, on January 18th, 1911 for pharmaceuticals although this is not 100% clear and one in 1914 for the manufacture of 5 to 13 gallon carboys. This licensing agreement meant that no other company could make any of these products on an Owens machine. The Illinois Glass Company was in the process of rebuilding its plant to use the Owens machines and it is reported that the following year that 22 machines had been installed. Whether these were all Owens I do not know.

Between 1904 and 1906 Owens licensed his No. 5 machine to important customers such as the Thatcher Manufacturing Company (who were issued the first licensee for an Owens machine to manufacture milk bottles), the Ohio Bottle Company and the American Bottle Company. Then from 1906 to 1911 he licensed his No. 6 machine, which was now called the "A Type", to Hazel-Atlas, Charles Bolt, Illinois Glass Company and the Whitney Glass Works.

In the following tables the manning is given for a number of different machines and the average gross per hour production for a number of different jobs in the mid 1920s. This makes the order size easy to calculate, and it gives some indication of the size of orders needed, for example, in Table 10,9 for the manufacture of an 8 ounce oval prescription, a 6 arm A.E. machine made 6.05 gross per hour as against the 10 arm C.A. machine working in the double-triplex mode, which made 38.33 gross per hour.

So the 6 arm A.E. machine would need 166.7 hours to manufacture 1000 gross but the C.A. would only require 26 hours, a situation which is totally untenable, the order for the C.A. machine to make it comparable to the 6 arm A.E. machine would have to be 7000 gross. If we consider the smaller prescription then to keep the order size the same as the 8 ounce - i.e. at a one week run, then for the half ounce medical the order size need to be in the region of 13 000 gross, which was a lot of bottles in those times.

The table is the yearly averages, in gross, for all machines of that type and if they are broken down into monthly figures there are variations. The figures allow a speed to be calculated, efficiency probably would be no higher than 75%, and a curve for each type of machine. Remember the sizes given are for contents they are not the weight. Total grossage made in 1925 was, 115 849 half ounce, 190 405 two ounce, 168 389 four ounce and 104 732 of eight ounce.

An odd thing in these statistics in Table 10,10 is the Whiskey Dandies, these figures come from the middle of the prohibition era so it is odd that these are quoted in government statistics.

Machine Type	Blowers	Finishers	Mould boy	Cleaning-off boy	Snapping-up boy	Carry-in boy	Chief foreman	Machinist	Chief operator	Machine foreman	Feeder operator	Machine operator	Peanut roaster boys	Hot ware inspector	Gatherer	Presser	Transfer boy	Take-out boy	Swing man
Hand	2	1	1	1	1	1													
Semi auto Jersey Devil						0.5				0.16			1		1	1	1	1	
Semi auto Teeple-Johnson						1		0.25		0.08					1		1	1	
Semi auto one man						1							1		1		1	1	
Semi auto Lynch						1		0.25		0.08					2				
Semi auto O'Neill + feeder						1		0.16				1	1				1		
Auto O'Neill + 3 x feeder						1				0.40	1	1	1						
Auto Lynch + feeder						1			0.16			1							
Auto Owens A.N. Double						2.33				0.3		1		0.11					
Auto Owens A.N. & A.V. S., D. + con.							0.16			1		1							
Auto Owens C.A. Doub.-triplex + con.							0.16			1		2							
Auto H-E triple unit + P N. feeder						6				0.5		3		0.66					1

Table 10,8. Typical manning arrangements for various machines. As the weight of the jobs increased then the manning would also

During this era it was possible to obtain whiskey through your doctor if prescribed for medicinal purposes, he had to prescribe it and then fill out a specially designed government prescription form which would then be filled at your pharmacy.

When stocks of whiskey were getting low then the government would issue more. The system was open to abuse much being 'siphoned off' by 'bootleggers' and by others for their own use. Another alternative is that some were made for export such as to Canada or to Mexico.

Just what the Whiskey Dandies were made for I do not know, but only 20 711 gross were made. The totals made for the other beverages were 15 108 gross of pint beers, again affected by the prohibition law and 151 223 gross of sodas.

The following table, Table 10,11 lists the output for the different Owens machines for pint beers with the following notes. N.B.1. The "AR" and the "AQ" were designed to manufacture large ware, while the "AN" and the "AV" were designed to manufacture small ware.

Prescription Ovals	Output per hour in gross			
Machine configuration	One Half oz.	Two oz.	Four oz.	Eight oz.
Hand (actual)	3.48	2.86	2.00	1.59
Semi auto two man Jersey Devil		4.82*		3.62*
Semi auto one man		4.69*	3.59*	3.22*
Semi auto O'Neill + feeder		6.70		
Auto O'Neill + single feeder		6.93	6.90	5.03
Auto O'Neill with triple feeder	19.12			
Auto Lynch and single feeder	11.81	10.40	8.19	7.13
Auto Owens A.E. 6 arms single				6.05
Auto Owens A.N. 10 arm single				11.42
Auto Owens A.N. 10 arms single + conveyor			15.46*	
Auto Owens A.N. 10 arms double	33.51	30.11	23.63	
Auto Owens A.N. 10 arm double + conveyor	33.01			
Auto Owens A.R. 10 arms double				18.33
Auto Owens A.R. 10 arms double + conveyor				22.16
Auto Owens A.V. 15 arm single + conveyor	21.58*	20.90*	21.48*	19.54
Auto Owens A.V. 15 arm double + conveyor	43.37	39.74	35.13	25.52
Auto Owens C.A. 10 arm double triplex + con.	79.54	79.54	69.75	38.33

Table 10.9 Average hourly production, in gross, of various sizes of oval prescriptions. All semi automatic machines without a feeder includes a gatherer. Entries marked with a * are for earlier years

Various beverages	Output per hour in gross		
Machine configuration	Half Pint Sodas	Pint Beers	Pint Whiskeys
Semi auto two man Jersey Devil machine	2.86*	2.63*	2.88*
Semi auto Teeple-Johnson machine	3.37	3.36	3.35
Semi auto O'Neill	4.37		3.04*
Semi auto Lynch	4.50		3.66*
Semi auto O'Neill + feeder	6.57		5.13
Auto O'Neill + single feeder			5.74
Auto Hartford-Empire + triple feeder (P&N)	20.15	18.60	
Auto Lynch and single feeder			5.74
Auto Owens A.E. 6 arms single			7.04
Auto Owens A.E. 6 arm single + conveyor	8.88*	9.44*	
Auto Owens A.R. 10 arms single			9.44
Auto Owens A.R. 10 arms single + conveyor	11.83*	13.46	
Auto Owens A.Q. 15 arm single + conveyor	15.47*	18.65*	

Table 10,10. Mid 1920s outputs for various beverages, the ones marked * are for earlier years. Semi automatics include a gatherer except for the O'Neill

N.B.2. All of these machines could work single gob any weight provided that it was within its capability - see below for examples of the differences between the "AQ" and the "AV" machines. All of these machines can use either dual or triple moulds, dual from 4 to 13 ounces in weight and triple less than 4 ounces.

For the "CA" and "CB" machine you would double the number of cavities on the machine, e.g. for a "CA" machine making ware less than four ounces then the number of cavities would be 60 and if the speed of the machine was four revolutions per minute then 240 bottles would be made per minute.

Kind of Machine		Year it was Introduced (Approx)	Approximate Daily Output (in gross)	
Type	No. of Arms		Pint Beers	Other Bottles Depending on Their size
"A"	6	1903-1904	125	100-200
"AC"	6	1908	150	?
"AE"	6	1909	200	120-240
"AR"	10	1912	400	120-450
"AN"	10	1912	400	200-600
"AQ"	15	1914	500	180-600
"AV"	15	1917	625	300-900

Above Table 10,11. Approximately the Average Outputs of various Owens Machines

Packer Jugs	Half Gallon	Gallon
Hand	1.25	1.00
Semi auto two man Jersey Devil	1.49*	1.15*
Auto O'Neill and feeder	3.77	
Auto Owens AL 6 arms single	3.74*	3.64
Auto Owens AR 10 arms single	3.79*	
Auto Owens AQ 15 arms single	10.59	8.35

Table 10,12 For packer jugs, * are for earlier years. Quantities are in gross/hour

Finally for our voyage through the land of statistics we have the milk bottle. The milk bottle market was an important market and the total number manufactured for the one pint and the quart milk bottles was 163 565 gross of the pints and 279 283 gross of the quarts. Remember that these bottles were returnable, i.e. multi-trip, a survey in the early 1900s by the United States Department of Agriculture stated that the average life span was 22.5 trips, with a range from 6 to 60 trips. In 1935 Owens-Illinois and the Thatcher Manufacturing Company accounted for just over half of a 300 000 000 total milk bottle market i.e. they made more than 1 042 000 gross between them. The 300 million bottles covered growth and breakages.

Milks	Pint Milks	Quart Milks
Hand	2.50	2.00
Semi auto Teeple-Johnson + gatherer	3.08	2.58
Semi auto Miller + gatherer	5.20	4.56
Auto Miller + feeder + conveyor	4.50	3.49
Auto H-E + double feeder	13.38	11.17
Auto H-E + double feeder + conveyor	11.41	8.98
Auto Owens A.E. 6 arms single	5.49	5.05
Auto Owens A.R. 10 arms single	6.83*	5.76*

Table10,13. Mid 1920s statistics for milk bottle manufacture. The ones marked * are for earlier years. Quantities are gross/hour

By the mid 1920s milk bottle manufacture by Owens machines had been phased out, now most of the milk bottle being made were by specialist milk bottle machines such as those marketed by Miller and the Hartford-Empire companies. Figure 10,6 shows a twin table Hartford-empire bottle making machine. Even Owens had stated that these machines were more efficient and they had already purchased replacement machines for one of their important markets.

Figure 10,6. A Hartford-Empire twin table milk bottle making machine

The Hartford-Fairmont twin table milk bottle machine, first used a paddle feeder and then, in 1919, a paddle and needle feeder and finally after 1922 a single needle feeder. Figure 10,7

Figure 10,7. A triple distribution system

shows from the front a paddle and needle feeder with a triple distribution system.

The following curves in Figure 10,8, show the speed curves for various methods of manufacturing glass containers, they range from hand working, to semi-automatics, through to feeder fed intermittent rotary machines and suction fed automatic machines. The suction machines range from single, to dual, and on to dual arms with triplex moulds. The machines marked AN are for AN and AR machines and those marked AS are for AQ and AS machines.

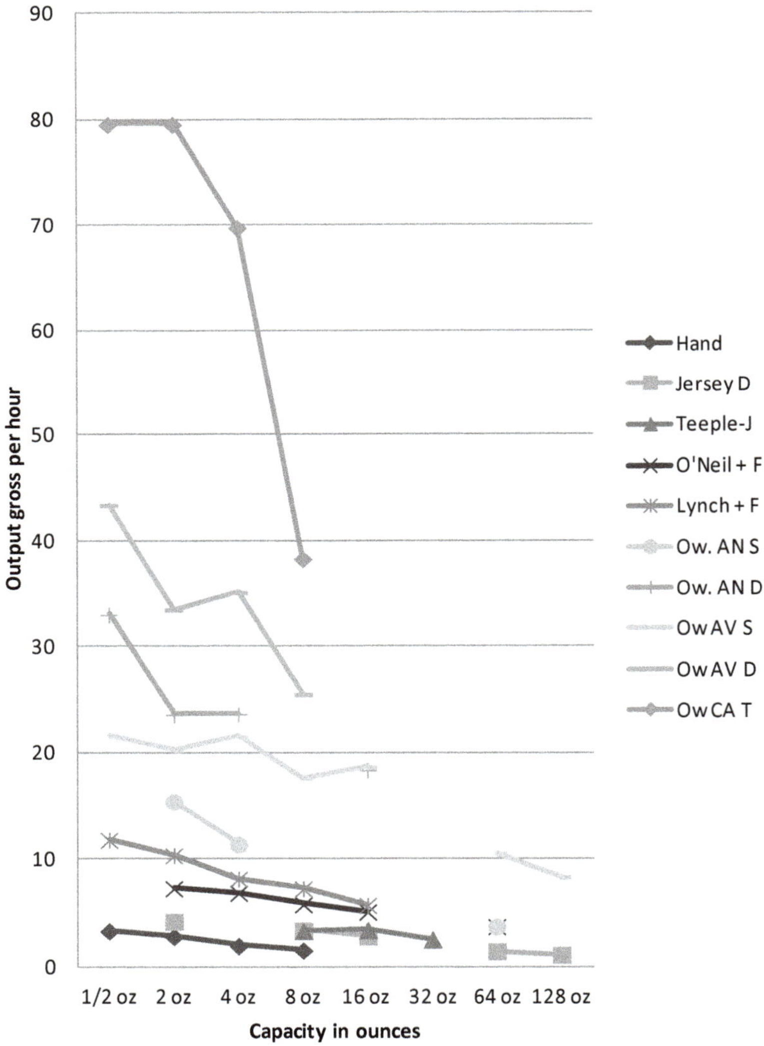

Figure 10,8 Average speed curves for different methods of manufacture circa the mid 1920s

Changes to the bottle making machine line-up as used in the USA, with time

The first thing that is apparent is that the curve, although shown continuous, is not, and that there is a big gap between 1963 and 1999. Even so it gives a good indication of the trends and it

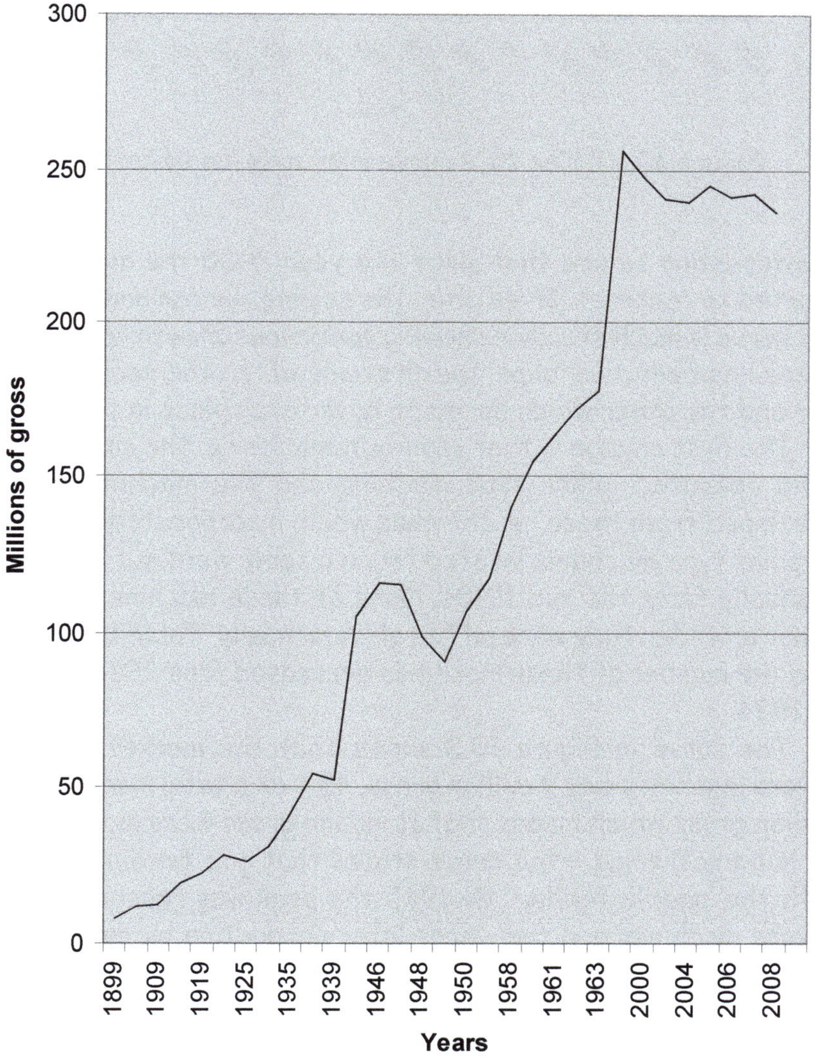

Figure 10,9. Curve showing the trend in the USA container market. The figures are for gross manufactured - not for gross sold

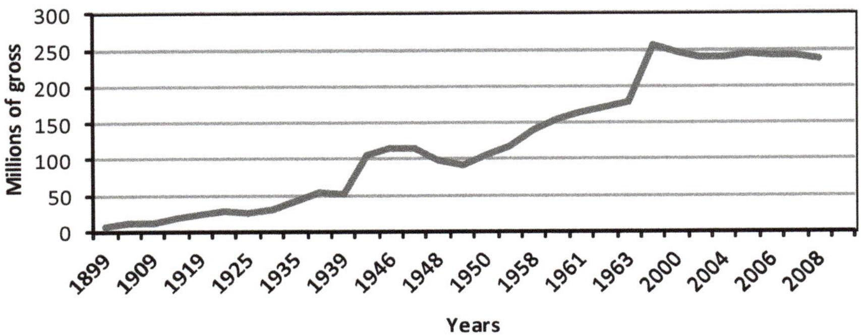

Figure 10,10. The 10,9 curve with axis modifications

is interesting to see that since the year 2000 the market has started to contract. If we alter the scaling, we can now see that the curve is much steadier showing continuous growth up to 2000, there has been two 'blips' the first one after the second world war and the other which seems to have taken place in the 1980s.

The first change is that semi-automatics i.e. the single head, hand operated, pillar type machine and the machines which developed from these i.e. the ones which incorporated a number of pillar type machines located round a table went out of favour gradually from the mid 1910s. None of these machine were put under a feeder they were all fed glass manually. Table 10,3 shows how the number of these machines decreased from 1916 through to 1924

The curve in Figure 10.3 shows that the market made by Owens machines was 9 million gross, 45% of a total market of 20 million gross which means that 11 million gross were made mainly by hand gathering - the curve states that gob feeding started with the 'paddle feeder'. By 1925 the peak was reached by the Owens machines and two years later production by suction and by flow where equal.

We are going to deal with mainly the flow machines because as we have already said suction machines started to lose their dominance of the market after 1927. So which type of machines were used first; obviously the machines that were already in use continued to be used. When these machines were replaced then, for wide-mouth ware initially machines such as the press and blow machines that had been manufactured by W. J. Miller or Ed. Miller would have been preferred although F. O'Neil did make some in his earlier career, while for narrow-neck ware the machines supplied by F. O'Neil, W. J. Miller and J. Lynch would have been used.

Figure 10,11 shows the speed curves for Ed. Miller's press and blow machines while Figure 10,12 shows the speed curves of the earlier Lynch machines.

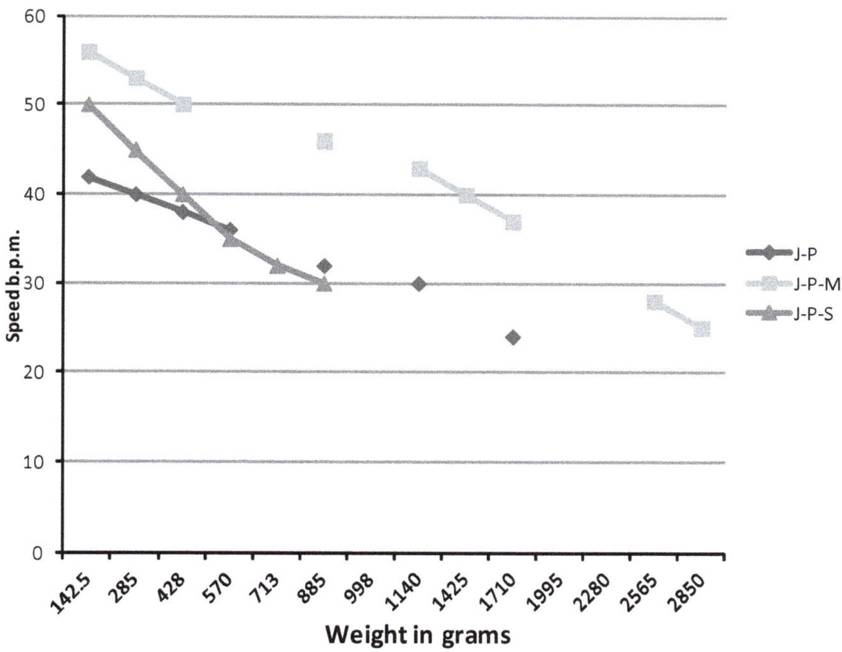

Figure 10,11. Speed curves for the Ed. Miller J-P series

In the beginning these single and two table machines cornered the market as the IS machine did not enter the market until roundabout 1928, when both the two and the four section machines were made available but these could only manufacture narrow-neck bottles, the wide-mouth version did not reach the market until 1940. This meant that the single and twin table press

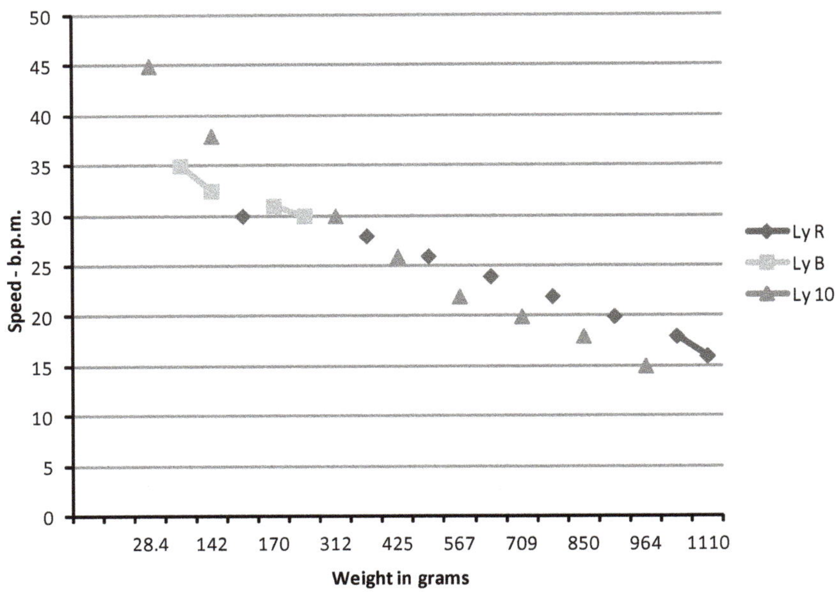

Figure 10,12 Speed curves for early Lynch machines. The Lynch B (six blanks & six moulds) was introduced in 1921, the Lynch R (six blanks & eight moulds) in 1925 and the Lynch 10 (six blanks & six moulds) in 1930

and blow machines did not face any competition until at least 1940.

The blow and blow machines faced competition much earlier although there was no significant swing in favour of the IS machine until 1939 when the first dual operation was unveiled and then in 1950 when the five section machine was introduced to the market, here we are talking about single operation. In the following curve Figure 10,13 shows the first four machine types of the IS machine, namely the two, four, five and the six section machines, the latter was introduced in 1954. Included in this graph is a speed curve for a Lynch 10 machine which shows that it was completive with the four section but it was outclassed by the five section machine in terms of output.

In 1940 the IS press and blow system came on the market the next figure, Figure 10,14 shows the outputs for the 'as then' current four section IS machine. If we then superimpose the

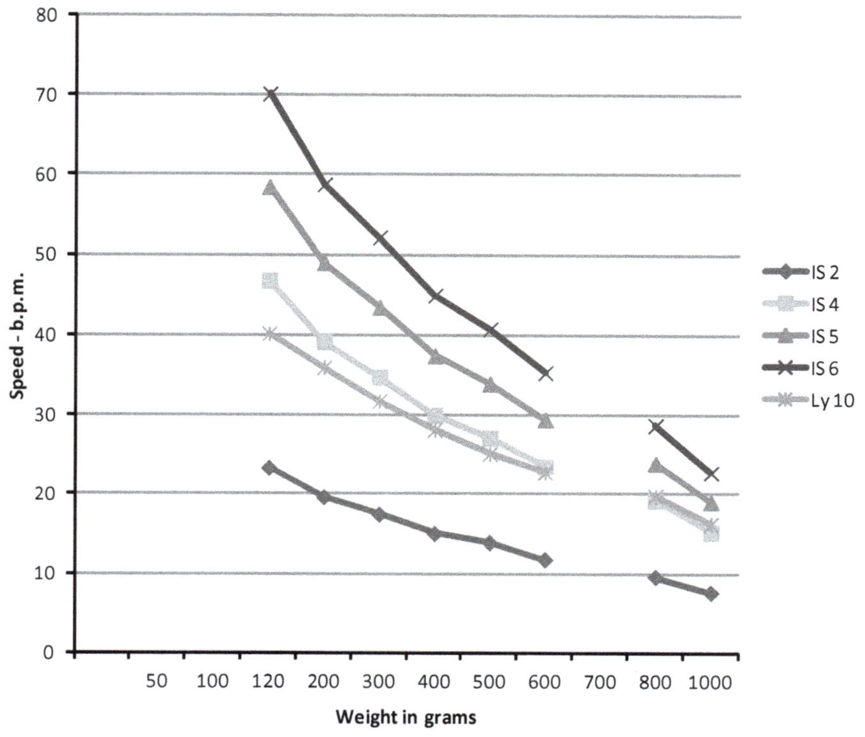

Figure 10,13. Speed curves for various IS machines and the Lynch 10

fastest Miller machine and the Lynch 10 on to the same graph, see Figure 10,15 we can see that the Miller is the fastest press and blow machine and except for weights less than four ounces the Lynch compares favourably with the IS machine. So there is no incentive to change to the IS machines yet.

As we said previously in 1939 Hartford Empire introduced their 'double gob' process onto the market which changed the balance in favour of the double gob four section machine. Figure 10,16 shows the increase in output between single and double gob operation. Lynch replied by marketing their double gob machine in the late 1940s, the percentage increase would be about the same as the IS machine, the Miller press and blow machine would stay the same until 1940 when Hartford Empire introduced their

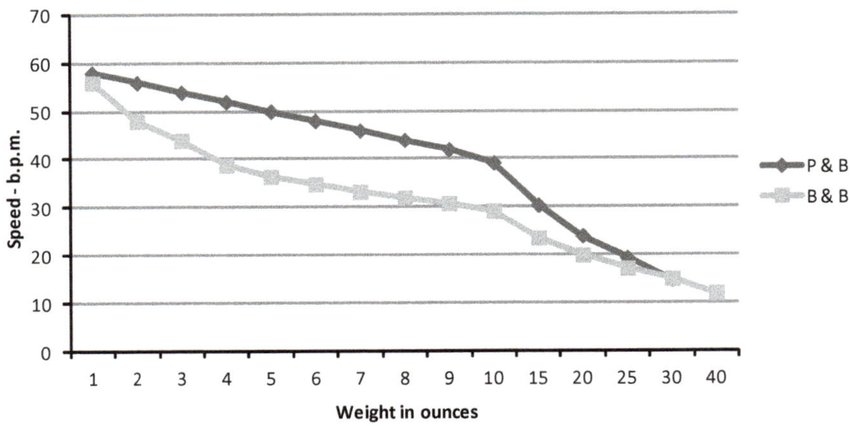

Figure 10,14. Speed curves for a four section IS machine

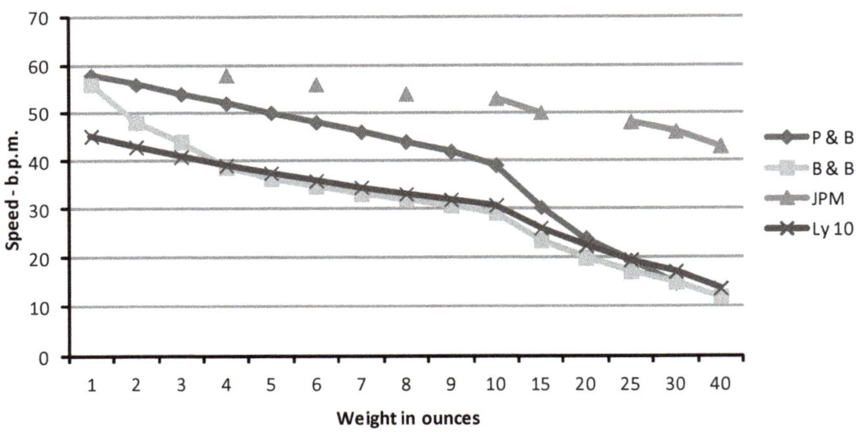

Figure 10,15. Speed curves for a four section IS machine plus those for a JPM Miller press and blow machine and a blow and blow Lynch 10 machine

'press and blow' process.

The centre distances of the first dual IS machine at $3\frac{5}{8}$" were much smaller than the current larger machines making it not as versatile as say the $4\frac{1}{4}$" machines so the maximum bottle diameters would be limited to around 2 inches so confining it to the lesser weight bottles. If you look at Figure 10,16 you will see

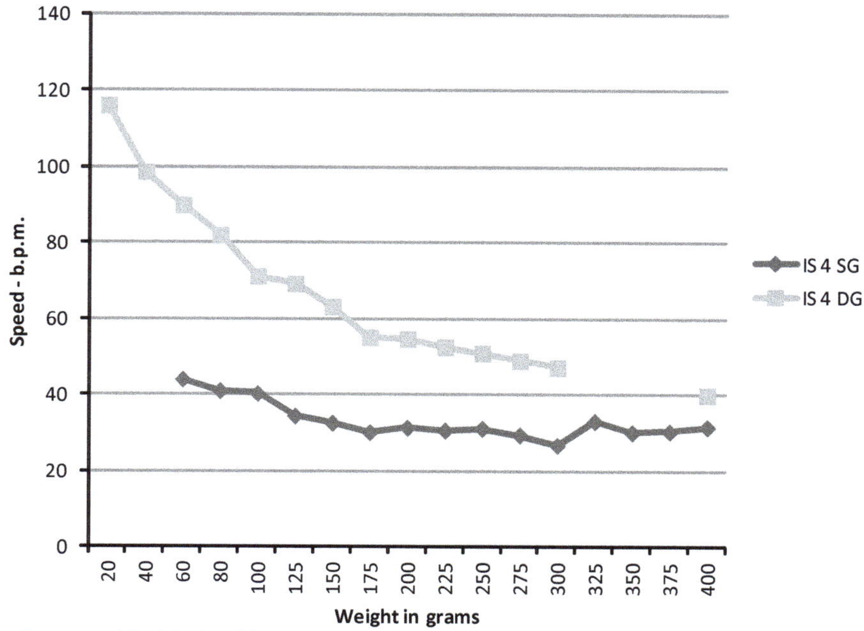

Figure 10,16. Differences between the output of a single four section machine and a dual one

that there is a distinct change in the slope at around 175 grams, bottles made at more than this weight must have been taller with all of the resulting bottle making and handling problems.

In 1954 the six section was introduced, again the dual version was $3\frac{5}{8}$" centre distances but this was upgraded to $4\frac{1}{4}$" centres in 1959. This increased the scope of the dual manufacturing range, and by 1965 the $5\frac{1}{2}$" centre distance 'F' machine was introduced to the market.

In the 1966 Glass Industry International Directory for the United States of America out of the 59 plants that gave detailed information the split was about 80% for IS machines to 20% for Lynch 10 machines this is for 'flow operation' only as Owens still had a considerable percentage of their production made by the suction process.

By 1970 the Emhart returns indicated that approximately 95% of container manufacture was made by IS machines - again this maybe an over estimate as there is no information regarding the number Owens machines of which were still operating in the United States at that time but I estimate that a total of at least 30 to 35 were still in operation. You will remember that the last Owens machines ceased production on the 17th December 1982 so some of them were working long after 1970, the crunch came with the fuel shortages in the late 1970s when most of the furnaces with Owens machines were shutdown due to the amount of fuel used to heat the revolving pots. In the same returns (1970) Lynch 10 machines only accounted for 2.7% of the total attempted hours.

By the end of 1983 all of the glass containers in the USA were being manufactured in one form or another by an IS machine. The number of machines had increased in both the number of sections that the machine had and the number of cavities per section. In 1967 Emhart had introduced their $3\frac{5}{8}$" triple machine followed in 1969-1970 by the $4\frac{1}{4}$" version. Mauls introduced their triple machine and their eight section machine in 1965 and Owens introduced to their company and affiliates, their version of the triple machine and the eight section machine in 1967.

Emhart introduced their eight section machine in 1973. The ten section machine was introduced in 1972 - $4\frac{1}{4}$" C.D. E type machine, followed in 1984 by the 5" C.D. machine and the 12 section 5" machine in 1987 and in Europe the 16 section EF machine. Round about 1984 Owens introduced their 'quad' machine.

The growth of the IS machine has seen a steady improvement in its output/section, from the sale of the four, two section single gob machines to Hocking Glass through to the introduction of the 12 section quad gob machine, and the Heye 18 section triple machine.

We will try and illustrate the growth in the following way:

The progress in Europe followed much the same pattern but with a different mix of machines. The machines developed 'inhouse' were in Great Britain the Ashley, Windmill, Rylands, Dobson, Hayley, Clegg, Horley, Foster, Simpson & Bradshaw, Horne and Breffit-Horne, Sweeting-Hardman, Harrison and

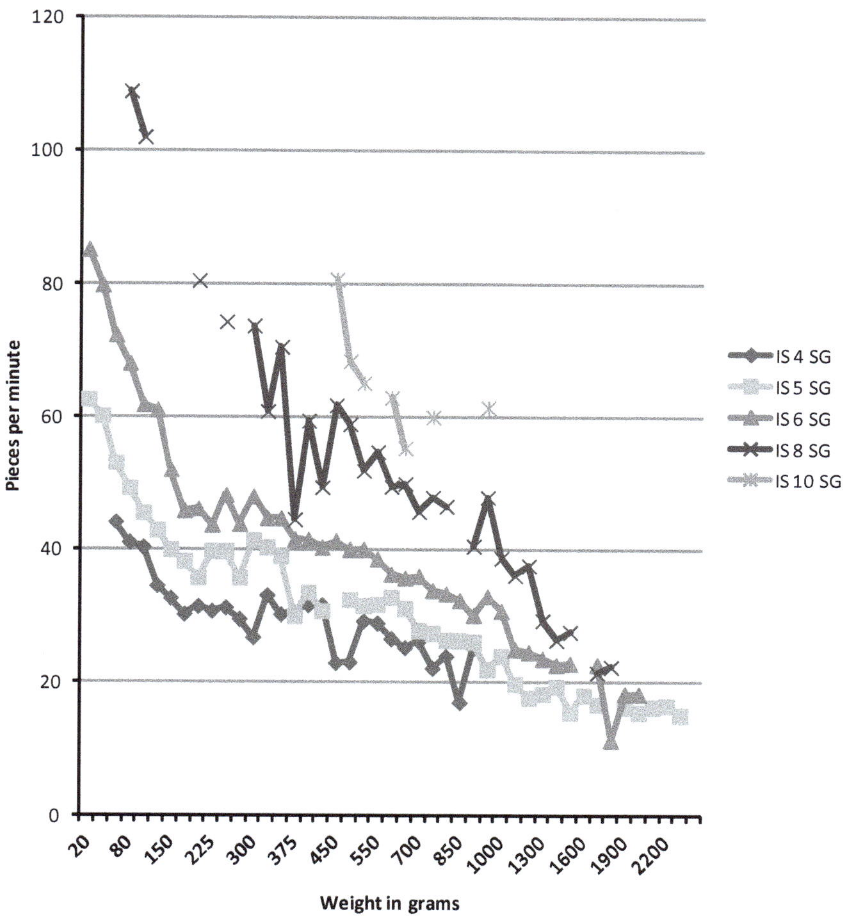

Figure 10,17. Comparison of speed curves for different sized IS machines

the Knowle-Inman machines. In France there was the Verney, Bouchier, Tourres and Barrez machines, Germany had the Hilde, Grote, Severin, Schiller, Wolf, Kutscher and Mainzer machines while in Austria there was the machine by Heerdt.

Moving to the automatic machines, this listing is far from being comprehensive, it has a bias towards machines of USA origin and it does not included machines designed and built by the country named for example the Redfern machine made and designed in

Great Britain and worked at Castleford in Yorkshire. Some of the machines recorded are; in Great Britain there was the Owens machine from 1904, there were 4 by 1908 and by 1919 there were 19 in operation and with another 7 on order. In total there were 7 - 6 head machines and 19 - 10 head machines, other machines in operation were those made by W. J. Miller and F. O'Neil from 1917, E. Miller and Lynch from 1919, the IS machine by 1928, the Monish from 1929 and the Roirant by 1932. In Germany the Owens machine was first commissioned in 1907 with 24 machines operating by 1913, other machines were the Roirant from 1923, E. Miller and Lynch from 1924 and the Monish from 1932.

The first Owens in the Netherlands was in 1908, there were 2 by 1913, and there are records of the Monish being in operation in 1932. The first Owens to be installed in France and Sweden, of which there were two, was in 1910, other machines were, in France 1919/1920 the Lynch machine, E. Miller in 1926, the two section IS machine made by the British Hartford Syndicate in 1930 and the Monish in 1931, while in Sweden the first E. Miller machine had been installed in 1929. In 1912 Owens machines were in operation in Denmark, two, and in Italy were there were E. Miller and F. O'Neil machines installed in 1920 and Roirants in 1926.

Owens machines were operating in Norway 1914 plus machines by Lynch in 1927 and E. Miller in 1932, while Hungary followed them in 1915. Other countries with none indigenous machines were; Russia were they had O'Neil machines by 1920, Roirants by 1923 and IS machines by 1936. Belgium had O'Neil machines by 1919 and Monish in 1932, Switzerland had a British built two section IS Machine in 1930 while Rumania had the Monish machine by 1930.

To make a bit of sense about these different machines it would be better to look at some speed curves. These can be seen in Figure 10,18.

The legend for the machines below with their dates of introduction to Europe is, at 550 grams weight, starting from the bottom, single section Roirant B (1922) & single section F machines (1929), I.S 2 section (1929), Owens AE6 single (six arms and

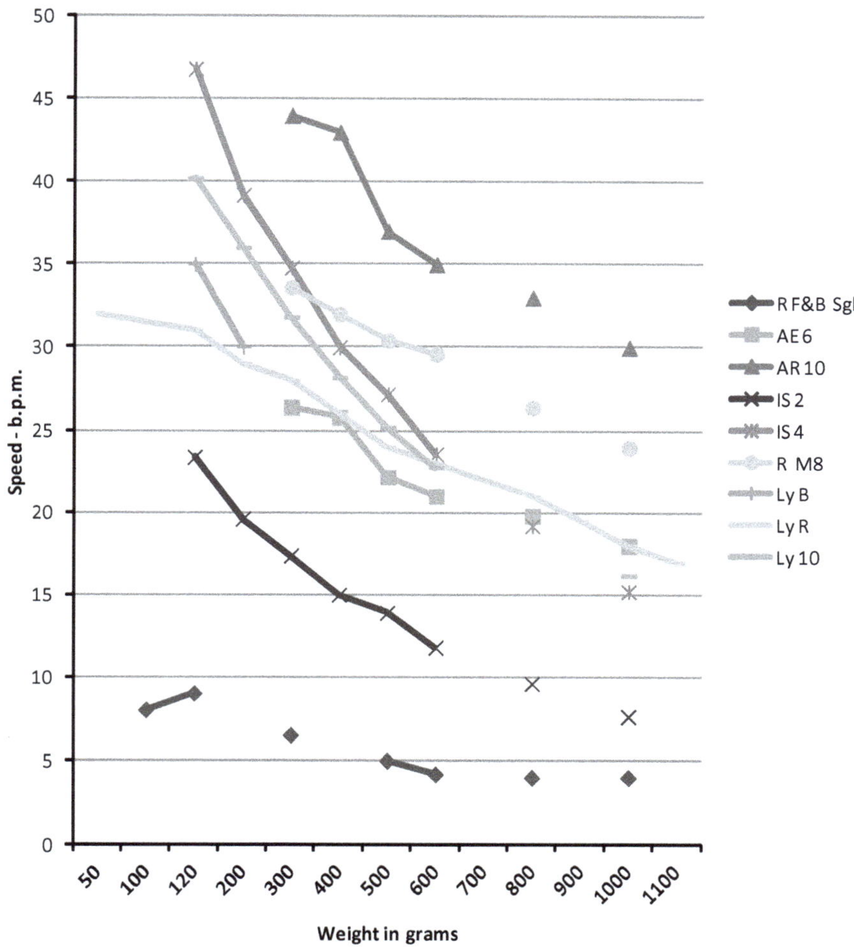

Figure 10,18 Speed curves for glass making machines

after 1910), Lynch R (1925, six blanks and eight moulds), Lynch 10 (1930 six heads), IS 4 sections (1929), Roirant M8 (about 1930 with eight heads), and Owens AR 10 single (ten arms and after 1912). The six head Lynch B machine which makes small ware only is the line between 35 and 30 bottles per minute.

All of the comments regarding the American market hold good for Europe except that there were more types of machines and initially the market was much smaller. The first complication was the relationship between the outputs of the four section IS machines and that of the Roirant M8, the slopes of the speed curves are significantly different below approximately 350 grams the four section IS machine was the quickest but at heavier bottle weights than those then the Roirant M8 machine was the fastest, so the weight of the bottles to be made came into the equation.

Another important factor in the, 'What type of machine should I choose?' equation is the cost of installing the machine, as against the buying of it - these comments apply equally to where-ever in the world the IS machine is being installed. The height of the IS 'metal line' was much higher than that of the existing twin and single table machines that it would be replacing, so either the roof of the building must be raised or the machine shop floor needed to be lowered - the increase in the required drop was something in the order of four to five feet and much higher for the eight, ten and twelve section machines which came later.

So looking at the European market prior to say 1960 there was not much money around to change to the IS machine, especially as when Emhart in 1954 introduced their six section machine but Roirant brought out their R7 machine at nearly the same time which had nearly an identical speed curve but with an overall lower installation cost. The S10 was brought out in the early 1960s.

The next two figures 10,19 and 10,20 show the output of the IS and Roirant machines. Figure 10,19 show the relative outputs of the IS 5 section and the IS 6 section machines along side those of the Roirant R7 and S10 machines. The graph shows that the S10 is the most competitive machine, the IS 6 and the R7 are about the same and the IS 5 section is the least. Figure 10,20 shows once again that for weight above 350 gram the IS 6 and the R7 are much the same and below 350 grams the R7 output is higher, while the IS6 D.G. and the S10 D.G. are over the area in which they are comparable show that they are much of a kind.

So in Europe until the eight section IS machine was introduced around about 1970 and the ten section in 1972 the Europeans were going their own way. Just like the IS machine the Roirant

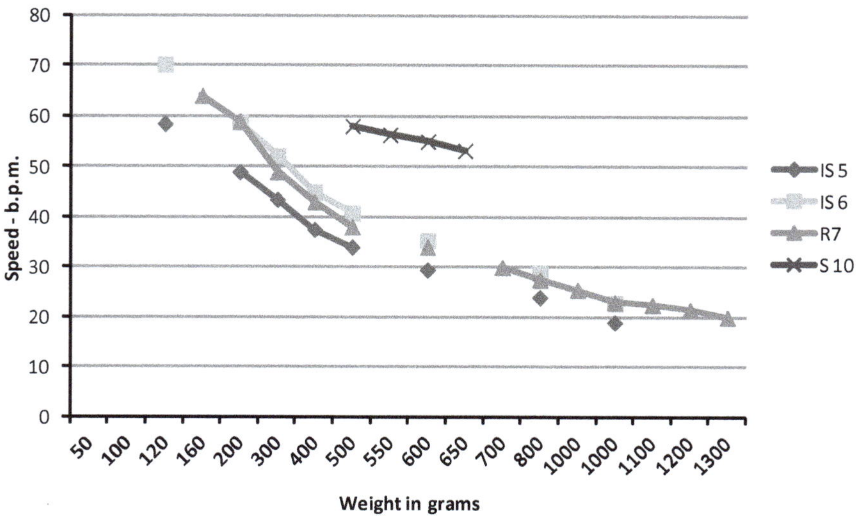

Figure 10,19 Comparison of speed curves for IS and Roirant machines

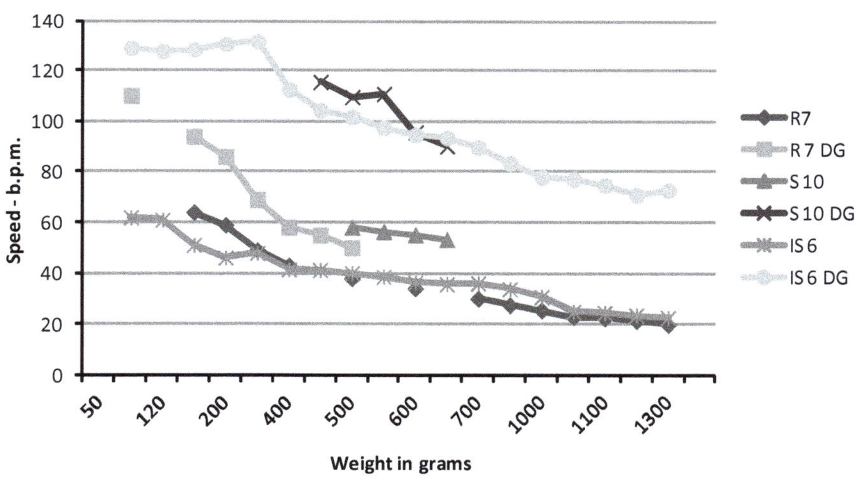

Figure 10,20 Speed curves for IS and Roirant machines both single and double gob

machines gradually increased their centre distance so increasing the range. The centre distances were for R7 - 82 and 92 mm. while the S10 had centre distances of 82, 92, and 111 mm.

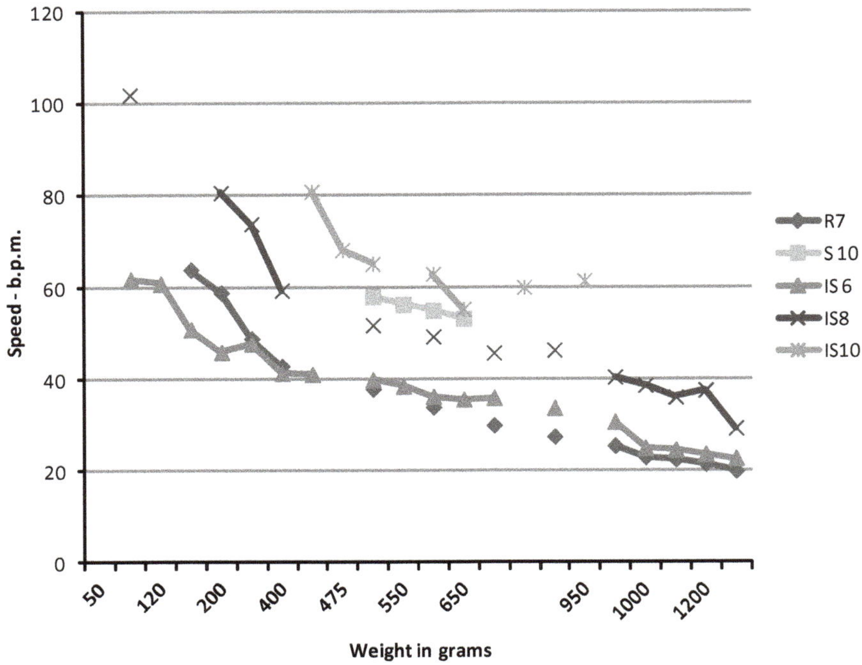

Figure 10,21 Graph showing the speed curve comparison between the IS 10 section and the S 10 single gob machines

Figure 10,21 shows that the IS 10 section has a higher output than the S 10, the problem with machines is often they are only comparable over a short range, the S10 is an example of this because of the speed of rotation there is a lower weight limit, due not to the machine itself but due to the operators ability to swab the moulds - the speed of rotation is too quick, something like 120 b.p.m. would be the maximum speed that the operator could handle. The R7 can go to a much lower weight limit before the same comment applies.

Concurrently there were two significant changes that were taking place in the market place which affected the machine choice, firstly the European Common Market increased its size from the original six countries to nine in 1973, increasing the population of 167 million people in 1957 to 256 million, a then increase of 64 million people. By 1986 three more countries had joined bringing with them another 59 million people, 1990 saw the reunification of Eastern Germany and Western Germany to form the new Germany adding another 16 million and in 2004 a further 10 countries joined increasing the population by 74 million, making a total of 25 countries and finally in 2007 two more joined, bringing with them another 29 million people. So the original population of 167 million in 1957 had increased to 493 millions in 2007, more than double the size of the USA.

Secondly in 1993 all the tariff barriers between the twelve member states of the community were removed allowing access to any part of the union. With the lowering of the trade barriers the individual glass manufacturers found that the competition had suddenly increased as more efficient manufacturers in an adjacent country could supply containers at a price lower than what they were selling them at. This led to an increase in the updating of the bottle making machines and it also led to using the IS machine as standard.

The big advantage that the IS machine had over the other machines was that it could manufacture over the complete range of glass container weights, the only differences in the individual manufacturing lines was in the choice of the feeder/forehearth which were not as flexible as the machine.

So from 1980 there had been much influence on increasing both the size of the machines and the ware that they could make. Again there were differences between the routes taken by the American manufacturers and the European ones, some of the European manufacturers put two machines in 'tandem' and worked them dual whereas the Americans stuck with one machine and worked then triple and eventually as quad versions. Here we need to define what we mean by 'European manufacturers', these are manufacturers who do not fall under the Owens-Illinois umbrella. The putting together of two machines was a convenient way of utilising what would otherwise have been two obsolete machines, the machines used would be either two six section or two eight section.

Turning to the 'glass giants', how did this come about? There are many glass making companies throughout the world but there are only three glass container manufacturers that operate in more than five countries, these are;
Owens-Illinois Inc. of Toledo, Ohio, USA.
St. Gobain of Paris, France.
Ardagh Group of Dublin, Ireland.

The number of factories that they currently owned or part owned as of the end of 2007 were;
O-I Manufacturing which had a total of 87 factories split as North America 24, South America 11, Europe 41, Asia-Pacific 11.
Saint Gobain Packaging who had a total of 54 factories split as; North America 16, South America 5, Europe 32, Asia 1. Although I have seen a news release (headlines only) stating that they have 62 factories in 12 countries - these additional factories could of course be flat glass ones as they are a major player in this field.
Ardagh Group which had a total of 22 factories, all of them located in Europe.

If we look at these three manufacturers in more detail as of 2007, we are not claiming that they are absolute as most of this information comes from trade directories which are often out of date when printed so some errors can be made here, also companies or factories can disappear and then reappear again later.

We will look at them by size so, the first that we will look at is Owens-Illinois Inc., a company that is headquartered in Toledo, Ohio, in the United States of America. The Owens 'modus operandi' was to invest in a company and at a later date buy them out, there are many examples of this especially in the earlier years.

O-I Manufacturing

The Toledo Glass Company was formed in 1896 and renamed The Owens Bottle Machine Company in 1903, but through its Libbey connection its roots can be traced further back, to the New England Glass Company of Cambridge, Maine which was founded in 1818. The Illinois Glass Company was incorporated in 1873. All of the container manufacturing business worldwide was re-named

O-I Manufacturing on April 12th, 2005.

Owens is by far the largest of these 'giants' with factories and other companies who have made Technical Agreements with them worldwide. In 2007 the wholly owned or part owned factories were located in the following areas.

O-I Manufacturing North America (Total 24)
Main office based in Toledo, USA responsible for;
O-I Manufacturing America. Number of factories 18 plus a 50% share with Coors in the Rocky Mountain Bottle Co. of Wheat Ridge, Colorado.
O-I Manufacturing Canada. Number of factories 5.

O-I Manufacturing South America (Total 11)
Co-ordinated through the main office in Brazil, responsible for;
O-I Manufacturing Brazil. Number of factories 2
O-I Manufacturing Columbia. Number of factories 2
O-I Manufacturing Ecuador. Number of factories 1
O-I Manufacturing Peru. Number of factories 2.
O-I Manufacturing Venezuela. Number of factories 3.
O-I Manufacturing Puerto Rica. Number of factories 1

O-I Manufacturing Europe (Total 41)
O-I Manufacturing Europe is based in Lausanne, Switzerland and is responsible for subsidiary companies in the following countries;
O-I Manufacturing Italy. Number of factories 13.
O-I Manufacturing Netherlands. Number of Factories 3
O-I Manufacturing Poland. Number of factories 2
O-I Manufacturing Spain. Number of Factories 2
O-I Manufacturing United Kingdom. Number of factories 2
O-I Manufacturing Hungary. Number of factories 1
O-I Manufacturing Germany. Number of factories 4
O-I Manufacturing France. Number of factories 9
O-I Manufacturing Czech Republic. Number of factories 2
O-I Manufacturing Finland. Number of factories 1
O-I Manufacturing Estonia. Number of factories 1

O-I Russia. Number of factories 1, Joint venture with the OST group.

O-I Manufacturing Asia-Pacific (Total 11)
The headquarters are based in Melbourne, Australia, they are responsible for;
O-I Manufacturing Australia. Number of factories 5
O-I Manufacturing New Zealand. Number of factories 1
O-I Manufacturing China. Number of factories 4
O-I Manufacturing Indonesia. Number of factories 1
O-I Manufacturing South Africa - Number of factories 5. O-I's interests in Consul Glass were sold in 2007 so are not included in total.

Saint Gobain Packaging
Saint Gobain can trace its history even further back than Owens-Illinois. In fact it goes as far back as 1665 when it was founded by charter as the Compagnie des Glaces who were the royal glass makers to Louis XIV. The companys original name was Dunoyer. Now it owns or controls the following companies that have been split into the same areas as before.

St. Gobain North America. (Total 16)
Saint-Gobain, America. Number of factories 14
Saint-Gobain Desjonqueres, America. Number of factories 2

St. Gobain South America. (Total 5)
Saint Gobain Embalagens, Brazil. Number of factories 3
Saint Gobain Embalagen, Chile. Number of factories 1
Saint Gobain Argentina. Number of factories 1

St. Gobain Europe. (Total 32)
Saint-Gobain. Italy. Number of factories 6
Saint-Gobain. Portugal. Number of Factories 1
Saint-Gobain Spain. Number of Factories 6
Saint-Gobain Germany. Number of factories 4

Saint-Gobain, France. Number of factories 7
Saint-Gobain Russia. Number of factories 2
Saint-Gobain, Ukraine. Number of factories 1
Saint-Gobain Desjonqueres,
Saint-Gobain Desjonqueres, France Number of factories 2
Saint-Gobain Desjonqueres, Spain Number of factories 1
Saint-Gobain Desjonqueres, Germany Number of factories 1
Saint-Gobain Desjonqueres, Russia Number of factories 1

Saint-Gobain Desjonquers were sold to the investment company Cougard with SG now holding a substantial interest in Cougard Investments. So they are included in the totals of the relevant areas.

Saint-Gobain Desjonqueres. Asia (Total 1)
Saint-Gobain Desjonqueres, China. Number of factories 1

Ardagh Glass

This is a recently formed glass company circa late 1990s, it was formed out of the purchase of the Irish Glass Bottle Co. Ltd, of Dublin by Transnational Corporation Ardagh Plc.

Ardagh Europe. (Total 22)
Ardagh Glass, Germany. Number of factories 9
Ardagh Glass, United Kingdom, Number of factories 5.
Ardagh Glass, Italy. Number of factories 1
Ardagh Glass, Poland. Number of factories 3
Ardagh Glass, Netherlands. Number of factories 2
Ardagh Glass, Sweden. Number of factories 1
Ardagh Glass, Denmark. Number of factories 1

That was the situation of the three 'glass giants' at the end of 2007, the changes over the next five years were;

O-I Manufacturing.
O-I Manufacturing North America (Total 20)
O-I Manufacturing, America, Number of factories 18, including

the Rocky Mountain Bottle Company. Charlotte was closed in May and Clarion in July, 2010

O-I Manufacturing, Canada. Number of factories 2

Lavington, Scoudouc and Toronto were all closed in 2008

O-I Manufacturing South America. (Total 13)

O-I Manufacturing Brazil. Number of factories 5

On the 1st of September 2010 O-I bought out the Companhia Industrial de Vidros in north eastern Brazil with three factories

O-I Manufacturing Columbia. Number of factories 4

O-I Manufacturing Ecuador. Number of factories 1

O-I Manufacturing Peru. Number of factories 2

O-I Manufacturing Puerto Rica. This company ceased production on the 18th February, 2008

O-I Manufacturing Venezuela's two container plants were nationalised on the 28th November, 2010

O-I Manufacturing Argentina Bought the Cristalerias Rosario company on the 18th March, 2010. Number of factories 1

O-I Manufacturing Europe. (Total 38)

O-I Manufacturing Italy. No. of factories 11. Closed Naples, and Castel Maggiore around the end of 2008. New factories listed are Castellana Grotte and Ottaviano.

O-I Manufacturing Netherlands. Number of Factories 3

O-I Manufacturing Poland. Number of factories 2

O-I Manufacturing Spain. Number of Factories 2

O-I Manufacturing United Kingdom. Number of factories 2

O-I Manufacturing Hungary. Number of factories 1

O-I Manufacturing Germany. Number of factories 4. The Aachen facility was closed sometime between December 2012 and June 2013. It has been included in the total.

O-I Manufacturing France. Number of factories 9. One of the Reims factories was closed and O-I bought the Vergeze (Perrier) facility from Nestle in 2011.

O-I Manufacturing Czech Republic. Number of factories 2

O-I Manufacturing Finland. Number of factories 0 - Karhula

plant was closed in 2009.
O-I Manufacturing Estonia. Number of factories 1
O-I Russia. Number of factories 1, Joint venture with the OST Tara group.

O-I Manufacturing Asia-Pacific. (Total 13)
O-I Manufacturing, Australia. Number of factories 4. Perth closed between 2007 and 2009
O-I Manufacturing New Zealand. Number of factories 1
O-I Manufacturing China. Number of factories 7. In 2010 purchased factories in Cangzhou City (2), Hebei Province, and Suining City, Sichuan Province.
O-I Manufacturing Indonesia. Number of factories 1

The purchase of Malaya Glass in 2010, was a joint venture with the Berli Jucker company of Thailand to purchase the glass holdings of Fraser and Neave in Malaya Glass. Malaya Glass had four factories; in Malaya at Johor Bahru, in Thailand in the Saraburi Province, in China located in the Sichuan Province and in Vietnam in Ho Chi Min City. The Thailand plant to be run by BJC, the China plant by O-I and the other two under joint management. Number of factories which can be attributed to O-I is two. Actually it is three but the China plant is already in their Chinese factilities. The Berli Jucker company are the owners of Thai Glass.

Verallia
In 2010 all of Saint Gobain Packaging operations were renamed Verallia.

Verallia North America. (Total 13, minus 1)
Verallia, America. Number of factories 13.
Saint-Gobain Desjonqueres. Number of factories 2

Verallia South America. (Total 5)
Verallia, Brazil. Number of factories 3
Verallia, Chile. Number of factories 1

Verallia Argentina. Number of factories 1

St. Gobain Europe. (Total 32)
Verallia. Italy. Number of factories 6
Verallia. Portugal. Number of Factories 1
Verallia Spain. Number of Factories 6
The Jerez factory has been closed.
Verallia Germany. Number of factories 4
Verallia, France. Number of factories 7
Verallia Russia. Number of factories 2
Verallia Ukraine Number of factories 1
Saint-Gobain Desjonqueres,
France Number of factories 2
Spain Number of factories 1
Germany Number of factories 1
Russia Number of factories 1

St. Gobain Africa (Total 1)
Verallia Algeria, Number of factories 1
Saint-Gobains in June 2011 acquired the entire capital of Alver Spa.

Saint-Gobain. Asia (Total 1)
Saint-Gobain Desjonqueres, China.Number of factories 1

Ardagh Glass.
Ardagh Glass Europe. (Total 21)
Ardagh Glass, Germany. Number of factories 8
The Shlende factory has been closed.
Ardagh Glass, United Kingdom, Number of factories 5.
Ardagh Glass Italy. Number of factories 1
Ardagh Glass Poland. Number of factories 3
Ardagh Glass Netherlands. Number of factories 2
Ardagh Glass Sweden. Number of factories 1
Ardagh Glass Denmark. Number of factories 1

Ardagh North America. (Total 9)
Ardagh Glass USA. Number of factories 9

In 2012 the Ardagh Group made their first purchases in North America, the first was Leone Industries of Bridgetown, New Jersey closely followed by the Anchor Glass Container Corporation of Tampa, Florida. Leone Industries was a single plant company while Anchor, the third largest operator in America, had eight factories.

In 2013, which is outside of our current timeline, the Ardagh Group made an agreed bid for Saint-Gobain's American holdings, at the present time, November 2013, they are still arguing with the US government re the final outcome as the joint holdings without any loss of manufacturing facilities would make them by far the largest operator in that market.

So to recap the 'so called' three giants are still growing in total but for two of them purchases are less than the closures, only Ardagh in the total factories that they individually control has grown. The differences in factories in which the giants have an interest between 2007 and 2012 are as follows;

Total number of factories controlled or having interest in.

Year	O-I Manufacturing	St. Gobain Packaging	Ardagh Glass	Total number of factories
2007	87	54	22	163
2012	84	52	30	166
Differences	-3	-2	+8	+3

The reason for the closures are different according to the location but the general trend is for the market to be contracting. There is the competition from cans and plastics, the changes in the local market conditions for example, wine is now being transported in bulk from Australia and South Africa to Avonmouth in the United Kingdom and then bottled there, meaning that the market for glass bottles is contracting in those wine producing areas but it is increasing in the United Kingdom. Then there is the increase in the bottle size; one litre, one and a half litres and two litres are much more common than they were say twenty years ago, this means less bottles in the long run. The machine sizes have also gone up, now 18 to 20 section duals or triples or twelve section quads are now quite common which means that fewer production lines are needed with the reduction in the number of furnaces.

To stay in business there has to be consolidation in the glass industry an example of this is what has happened in North America since 1983.

Year	1983	c 1990-92	2000	2010
Number of plants	121	67	55	48
Number of Companies	21	13	8	7

If we remove the single operating companies, some of which are only manufacturing for themselves, for example the Gallo Glass Company manufacturing for E. & J. Gallo Winery or for Anheuser-Busch InBev through the Longhorn Glass Corporation we get;

Number of plants	115	61	51	44
Number of companies	15	7	4	3
Average number of plants/company	7.67	8.71	12.75	14.67

This highlights the consolidation of the glass container industry in Northern America, in the 37 years that the statistics covered the number of plant closures was 73 and the number of companies owning them shrank from 21 to 7, all of which were brought about by either takeovers or mergers. The three companies in the above set of statistics are, yes you've guessed it, the three glass giants.

So to anyone who is still awake and reading this book we have finally come to the end of our journey though time, so, what will the future bring?.

In my opinion, and its my opinion only, is that the only way forward will continue as before, i.e. that is that the glass giants will keep on getting bigger with the possibility of a fourth or a fifth giant appearing. There are some contenders but they would have to expand to at least twice or three times their size to become near to Ardagh's position.

As far as machines go, unless a new revolutionary machine comes on to the market then I think that the IS machine growth will stop at two 12 section quad gob machines working in tandem giving 96 cavities, a viable order to work a machine of this size must be enormous. To put another two sections on a machine will give problems with the delivery system and the number of bottles required to give a this size machine an economical run length must be totally 'mind blowing'.

J. T. Gledhill.

There now follows four Appendices which are additional material to the text, these are;

Appendix 1 Mould iron and other mould making materials

Mainly from articles published in Glass Technology by T. Ensor along with information from other sources.

Appendix 2 Forehearth design and the control of them

Studies flow through an orifice and how important is forehearth temperature control. This is followed by pointers on how to design a forehearth.

Appendix 3 Identifying old bottles and a glass glossary

Notes on how to identify old glass bottles and a historical glossary of bottle making terms, often there will be more than one explanation.

Appendix 4 List of abstracts

These selected lists are from those printed in the first seven years of the Journal of the Society of Glass Technology publications. There were no abstracts printed in the first two years of the Journal so these are for the abstracts printed in the years 1919 to 1923.

Appendix 1

Mould irons and other materials

The use of moulds in glass making is not new, they have been around very nearly from the beginning. A brief history up to the start of the 19th century is as follows, it was around 3000 B.C. that glass was first softened and pressed into a mould to form glass beads then, from around 2000 B.C. amulets were made by pouring melted glass into stone moulds.

For the next 2000 years or so glass was only made for the 'luxury market' i.e. for kings and the very rich people, but by the 15th century (1500 B.C.) the Egyptians had perfected both their sand core method of making hollow ware, the items were usually less than 6" tall and the casting into a mould. Sand core items were designed to hold either perfume or ointment.

Syria came under Egyptian rule in 1468 B.C. and glass was first blown around 50 B.C. possibly in Syria. The original moulds were made of clay. The methods of both blown glass (free blown and moulded blown) and moulded glass manufacture carried on beside each other and sometimes used together. The Syrian glass moulds were made of wood, stone or modelled in clay into which the patterns were either incised or ground

The first century B.C. also saw the development of wood to shape blown ware, and hinged moulds, which were initially in wood, for blowing a certain shape again and again and again. Other than the move to metal moulds both brass and iron this method of manufacturing bottles lasted until the 19th century. Glass was also pressed in metal moulds.

This is based on a review of mould materials taken from Glass Technology; Volume 11, April 1970, and updates published in Volume 19, October 1978 and Volume 31, June 1990, all three by T. F. Ensor of United Glass Ltd. St. Albans, Great Britain. Additional information is also taken from other sources, but mainly from Glass Technology.

In 1938 Glass Industry published a points rating of mould materials by J.S. Vanick which totals 1000. The following is the listing you will notice that costs of the metals do not come into the equation this is because the machining costs far out ways the cost of material;

Good machineability		200
Ability to produce a good surface finish		100
Resistance to heat checking		100
Low thermal expansion		100
High thermal conductivity		100
Resistance to wear		100
Density or homogeneity		100
High graphite particle distribution		100
Resistance to growth		50
Resistance to oxidation		50

	Total	1000

Machineability

The ease of machining is very important, as these are the highest costs when producing a set of glass moulds. It is fair to state that machining difficulties and hence costs will play an important part when assessing a mould material. Grey iron is easy to machine especially in the annealed ferritic condition, even though in the cast and pearlitic condition where the iron is harder but still quite readily machineable.

The addition of alloying elements increases the hardness by solution hardening and the formation of hard eutectic carbides, or by changing the matrix structure or by both means thereby increasing the machining difficulties.

The absence on hard eutectic carbides is of prime importance, which means that the correct annealing process needs to be applied i.e. where the iron is heated up to 900°C and slowly cooling below 700°C.

Ability to produce good surface finish

It is essential that the surface finish of the mould is good, any rough marks, scores, etc. will be imparted to the surface of the finished product. Certain materials lend themselves to taking a high polish and it is generally agreed that a close grain structure at the surface is necessary to obtain this high polish.

With cast iron this means that coarse graphite flakes at the surface are detrimental, the best form being undercooled graphite. Matrix structure is also important, and in cast iron a pearlite matrix gives a higher polish than a ferritic matrix.

Although Vanek listed this as 100 points it is now not thought to be quite as important. Due to better mould repairing techniques there has been an increase in unpolished mould parts being used.

Thermal shock resistance and physical properties

Heat crazing, checking, or cracking on the working surface of the glass moulds is caused by thermal stresses created by alternate heating by the glass gob and cooling. The thermal stresses eventually exceed the tensile strength of the material and rupture or splitting of the mould occurs.

The two main physical properties governing the extent of thermal stressing are thermal conductivity and thermal expansion. Low thermal expansion and high thermal conductivity prevent large thermal gradients being set up by dissipating the heat quickly from the working surface. Also the working face temperature will be lower thus reducing glass adhesion to the moulds as well as reducing oxidation and growth.

Mathematical models have shown that adjustments to the cooling air can have a greater effect than small changes to the composition to increase the thermal conductivity.

Cast iron with a flake graphite structure has a relatively good resistance to fire cracking, but variations in the properties can be experienced by changes in the structure. A coarse flake has a better resistance to thermal cracking than a finer graphite structure.

Resistance to growth

Growth is a permanent increase in volume due to structural changes brought about by the temperatures to which the materials are heated during service. Furthermore, low temperature growth to increase with alternate heating and cooling.

Numerous papers have been published on the measurement of mould temperatures, in 1961 W. Frank gave a paper issued in Glass Technology that reviewed these. This article was about the mould temperatures during the blowing and pressing of glass articles. The main findings were that the temperatures of the working face of a mould varied widely, particularly on different parts of the body. Also that blow moulds ran hotter than parison moulds, but the cyclic variations were not so great. The results show that moulds on an IS machine operated at 425-500°C, those on a Miller press and blow machine at 520-575°C while those on an Owens machine ran at 550-595°C.

At the higher temperatures quoted above, growth would be expected to proceed fairly quickly due to the breakdown of pearlite, but at lower temperatures the breakdown would be slow and growth small. Papers issued by Barton and Gilbert of the British Cast Iron Research Association state that temperature which growth is negligible is about 400°C.

Growth in cast iron is due to a permanent increase in volume which occurs when there is a structure change to a more stable form. For example, pearlite in the matrix will break down to a ferritic matrix and free graphite. Pearlite consists of alternate layers of iron carbide and ferrite, and forms when the cast iron cools at a particular rate after been heated above 720°C - referred to as the critical temperature.

For example part of the annealing cycle involves cooling in the furnace from the annealing temperature - about 900°C - to at least 500°C before the parts can be air cooled. Cooling at a rate of 100°C/h through the 700°C range will result in a matrix structure which is basically all ferrite.

The stability of pearlite depends on the presence of alloying elements such as chromium, molybdenum and vanadium, which stabilise the carbide. Unalloyed cast iron will have a relatively instable type of pearlite, which will progressively break down on the glass machine resulting in a permanent increase in volume.

The pearlite breaks down by the conversion of the iron carbides lamellae to spheroids which in turn change to ferrite and free graphite, see Figures A1.1 and A1,2.

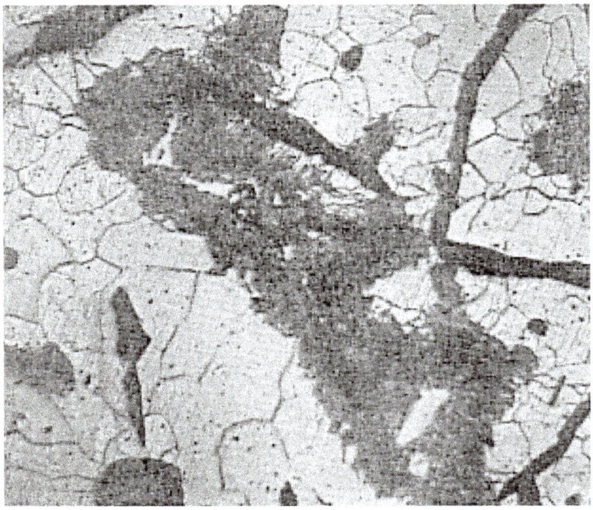

Figure A1,1 above. Lamellar pearlite in cast iron x 400. While below Figure A1,2. Spheroidised pearlite in cast iron x 400. Both etched in 2% nitrone

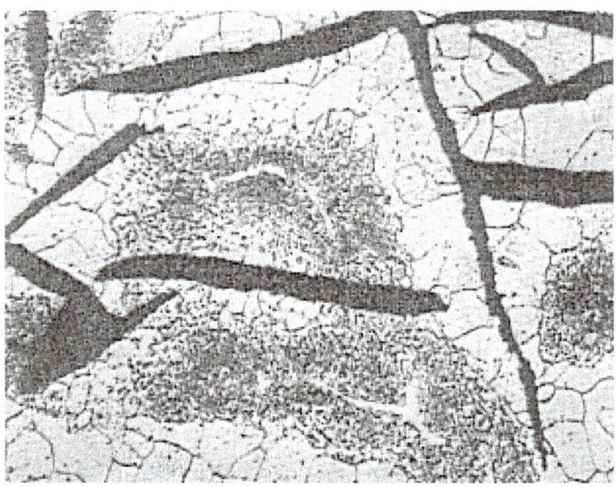

The presence of alloying elements can stabilise the pearlite against complete breakdown, so that pockets of partially spheroidised pearlite are frequently seen in the structure of cast iron moulds, particularly towards the back of the mould where the metal is the last to solidify; these are the areas where the alloying elements tend to segregate during solidification. It is highly unlikely that a 100% ferritic matrix is ever achieved under normal commercial conditions.

Resistance to oxidation

Oxidation of the mould surfaces is responsible for the breakdown of the surface, and it has been found that a combination of wear and scaling is the cause of the largest number of mould failures. Average temperatures of moulds have been reported to be between 425 and 500°C although certain jobs may exceed this.

At temperatures below 500°C. scaling proceeds at a very slow rate but from 500-600°C. large variations in scaling rate can be expected. In cast iron it has been shown that the size and form of the graphite flakes and the presence of alloying elements are important in reducing oxidation.

Additions of chromium, molybdenum and nickel increase the resistance of cast iron to oxidation.

Good wear resistance

It is difficult to describe the type of wear which occurs in the glass forming process, other than to say that it is very abrasive. However, it appears that a pearlite matrix is much more resistant to wear than a ferritic matrix. The corners of mould equipment will withstand the abrasive wear and the burring of sharp edges much better.

Metallurgical requirements to meet properties

Cast iron is an alloy of iron containing principally carbon, silicon and manganese. The maximum solubility of carbon in then molten liquid is about 4.3% and above this amount free graphite (kish) will form when the iron solidifies. Elements such as silicon form graphite and effect the amount of carbon the liquid can hold, hence the carbon equivalent value (CEV) is used to describe the cast iron condition and is calculated as:

$CEV = C + [(Si + P)/3]$

Where C = % carbon, Si = % silicon and P = % phosphorus.

Today with the rise in electric melting the phosphorus content is minimal and contributes nothing to the CEV. However to prevent the formation of harmful kish, glass moulds are usually cast at a CEV value of below 4.3% and are referred to as hypo-eutectic irons. A typical composition would be:

C = 3.3-3.5%
Si = 2.2-2.4%
Mn = 0.5-0.7%
Ti = 0.08-0.1%
S and P are both less than 0.1%

As stated earlier, nickel, chromium and molybdenum, either singly or in combination and up to 0.25% of titanium may be added to promote the formation of the fine flake graphite structures necessary at the glass contact face to resist premature breakdown.

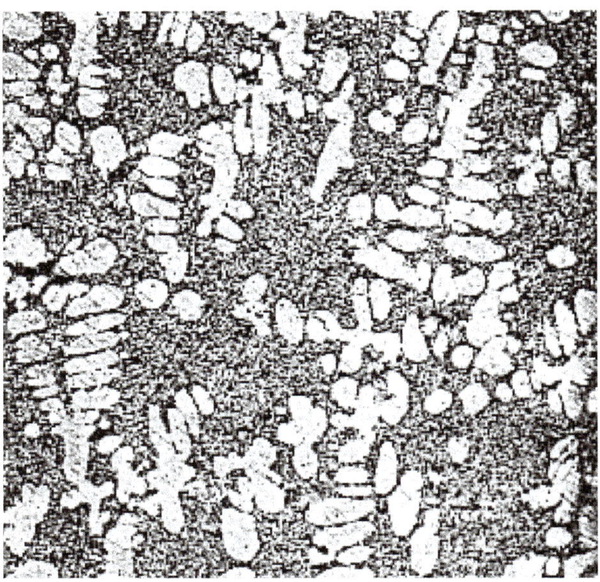

Figure A1,3. Typical structure at the glass contact face of a mould

When the molten iron is poured to form the glass mould the glass contact face is cast against a metal chill to increase the density and encourage the formation of the fine graphite structure. A typical structure at the glass surface is shown in Figure A1,3, and consists of ASTM type D graphite flakes of size 7/8 in a ferritic matrix. However the graphite will coarsen progressively through a series of structures to courser type A flakes size 3/4 at the back of the mould, see Figure A1,4.

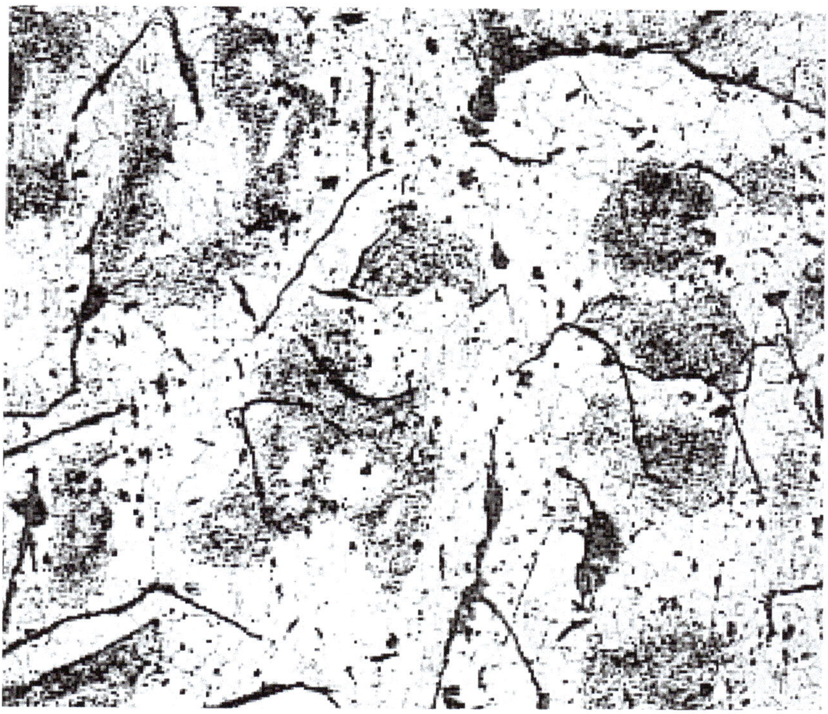

Figure A1,4. Typical structure at the back of a mould

Physical property requirements

The glass gob enters the blank mould at around 1100°C and has to be cooled to below 780°C by take-out from the blow mould. The ability to remove this amount of heat in a controlled manner governs the speed of the machine. The property that governs this conduction of heat is thermal diffusivity which is a measure of how much heat can be removed from a metal in unit time. It takes into consideration thermal conductivity, specific heat and density, which are related to each other by the equation;

thermal diffusivity = thermal conductivity/(specific heat x density)

The next three figures show the relationship of selected variables between room temperature and approximately 500°C, showing that there are significant differences between the variables which gets less has the temperature increases.

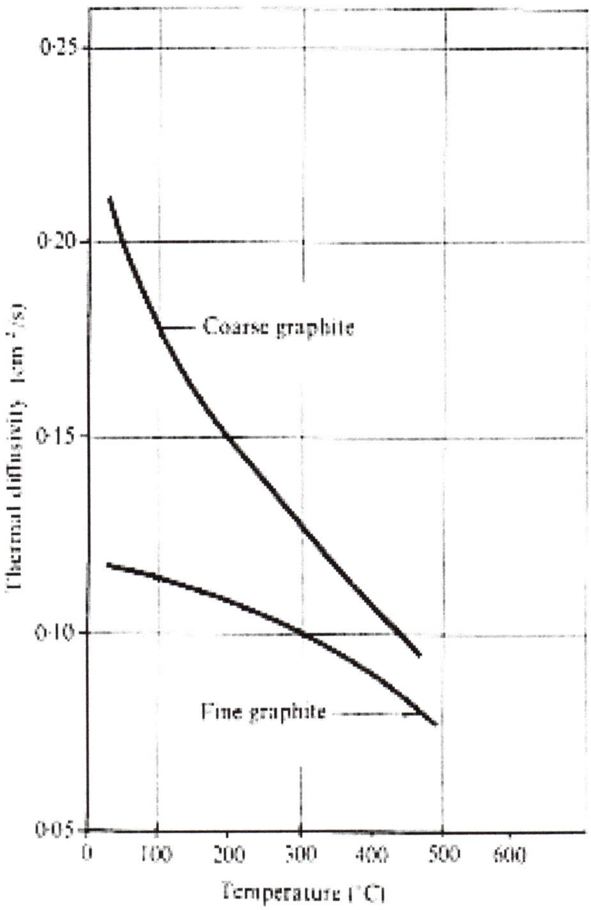

Figure A1,5. Thermal diffusivity curves for cast iron with different graphite structures

The curves show that type D graphite at 300°C has a diffusivity 25% lower than the coarse type A at the back of the mould, and hence with a fine undercooled structure throughout the section would be expected to run hotter under similar conditions than those at the coarse back.

A similar statement would apply to a mould with pearlite in

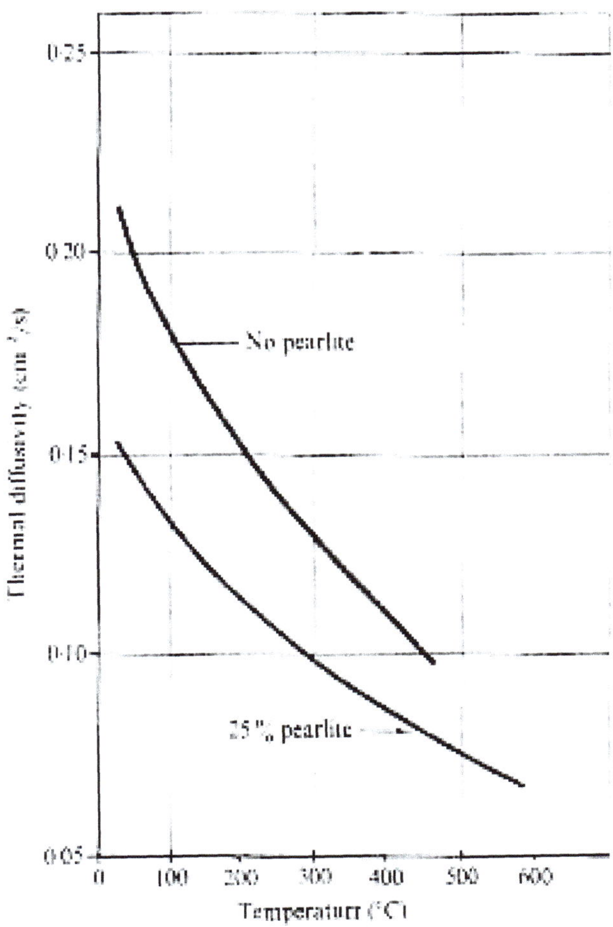

Figure A1,6. Thermal diffusivity curves of cast iron with different matrix structures and coarse graphite ferrite and ferrite + 25% pearlite

its structure, since 25% pearlite in a ferritic matrix decreases the diffusivity by approximately 25%.

Cast irons of similar structural features with and without 0.5% molybdenum show very little difference in thermal diffusivity, see Figure A1,7.

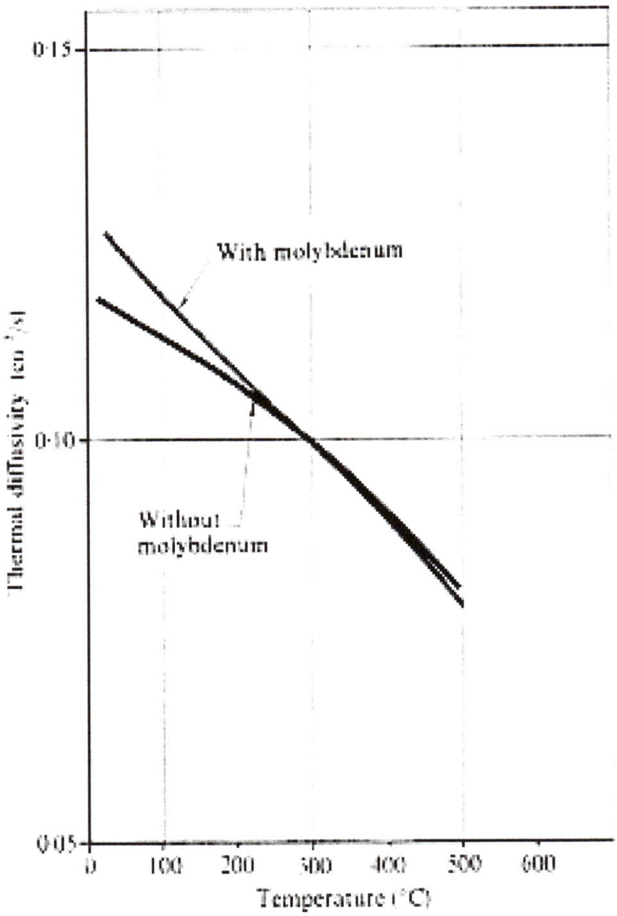

Figure A1,7. Thermal diffusivity curves of cast iron with and without 0.5% molybdenum

Materials used for mould making.

F. V. Tooley in The Handbook of Glass Manufacture listed the following materials that could be used to make glass mould equipment today they would include materials made by the following methods;

1. Unalloyed high-carbon grey iron produced by the cupola process.
2. Low-alloyed high- and low-carbon grey iron produced by the cupola, electric furnace, gas furnace, air furnace or by a duplexing method.
3. High-alloyed high- and low-carbon irons produced by the electric, gas or air furnace, or by a duplexing method.
4. Precision cast or semi-precision cast iron and steel alloys.
5. Special iron alloys, such as nodular graphite type structures.
6. Plain carbon or low alloyed steels.
7. High alloyed steels, including stainless and heat resisting types.
8. Copper base, aluminium base, or other non-ferrous alloys.
9. Pure nickel or nickel-base alloys.
10. Composite or liner style moulds.
11. Non-metallic moulds.
12. Combinations of many of the above with plating treatments or other coatings.
13. Some of the above materials with surface impregnation by alloying gases or metallic elements.
14. Several of the above with metals with specialised heat treatments.

He then says that the chemical analysis of any of the above, especially numbers 1 through to 4, is not the prime factor in the determination of their usefulness to the glass industry. Other considerations need to be taken into account, are for example;

A. Microscopic characteristics.
    i. Graphite size
    ii. Graphite formation

iii. Pearlite type
iv. Grain size
v. Kind and number of inclusions
vi. Quantity and location of free ferrite
vii. Presence of dendritic structures

B. Physical tests.
   i. Tensile strength
   ii. Transverse strength
   iii. Hardness
   iv. Impact strength

C. Heat treatment.

Although it does not fully satisfy all of the requirements, grey cast iron is by far the most common material used to make moulds. There are two basic types of cast iron, the first is cast against a chill, and the second where it is cast directly into moulding sand. The castings made with the chill have the advantage of a fine grained structure on the chilled surface, the amount of chill should be adequate in size so that the chilled surface is not machined away in the preparation of the moulds. In general casting made against a chill should all be heat treated before machining to relieve stresses set up by the drastic cooling of the chill.

All casting should undergo some form of stress relief heat treatment before leaving the foundry, the casting should be held at least two to three hours at a temperature no higher than 650°C. and be allowed to cool down with the furnace. Barton also states how graphite in cast iron can occur as flakes or in the nodular form and the flakes can be either large or small depending upon the melting history and the casting method. It is generally accepted that large course flakes of graphite are detrimental as this allows oxidation and chemical attack to take place.

The types of iron used are generally described as a soft, high carbon iron, cast against a chiller so that it has a dense, close grain structure. The percentage composition range of cast irons used for mould compoments has been published by Edgington as follows;

Element	Percentage
Carbon	3.00-3.80
Silicon	1.20-2.40
Manganese	0.60-1.00
Sulphur	0.10-1.20
Phosphorus	0.05-0.10

Table A1,1. Range of analysis for a typical cast iron by Edgington. Given at a Symposium on Moulds held at Sheffield City Hall on the 29th, 30th November and 1st December 1960

Some analyses of cast iron compositions by country

	Carbon	Silicon	Manganese	Sulphur	Phosphorus
America	3.65	1.90	0.75	0.09	0.060
England	3.51	2.06	0.73	0.21	0.180
France	3.57	1.80	0.57	0.09	0.060
Germany	3.61	2.20	0.65	0.34	0.087
Sweden	3.44	2.14	0.47	0.09	0.090

Table A1.2 Some compositions of cast irons by Renn published 1960

Heat treatment and alloying elements

There is very little difference between the basic compositions of cast iron shown above between the different countries, but there is a difference of opinion concerning the use of alloying element. The alloying elements commonly used are nickel up to 3%, chromium up to 1.5% and molybdenum up to 1%, usually in combination, in some instances small amounts of titanium and vanadium would be added.

The Overmyer Mould Company of America considered that without the addition of 0.2-0.25% of chromium and 0.75% of nickel, or 0.25% of molybdenum, or both, the iron should not be used for parts coming into contact with the glass surface. This work was supported by Cumberland, Vanick and Guillamon, who suggested that nickel, chromium and molybdenum were necessary for good performance from glass moulds.

The mould structure depends upon whether they are to be

used 'as cast' or heat treatment. The varying heat treatments that they get alter their structure, treatments that they can get are, normalised, stress relieving or annealing. Normally an 'as cast' iron mould or one which has been treated to relive stress, will contain a high proportion of pearlite, whereas an annealed mould would be practically all ferritic.

In the amounts of elements added to alloy the irons have different effects;

Chromium will promote and stabilise a matrix consisting of pearlite, hence it assists in reducing the growth which occurs due to pearlite breakdown; similarly it will increase oxidation resistance, and reduce the scaling which occurs, even when the mould is ferritic.

Nickel would not greatly affect the heat resistance, and it is usually added in combination with chromium to counteract the formation of hard eutectic carbides which can give rise to machining problems.

Molybdenum tends to form a matrix of pearlite or an acicular transformation product, and increase growth resistance by stabilising pearlite at temperatures up to 550°C.

Titanium promotes a fine, uniformly distributed graphite structure. Vanadium increases the wear and heat resistance slightly.

In bottle making cast iron is the normal material used in the making of moulds, other mould materials used are basically the various alloys of iron, where the additions of the various alloying metals improve the life of the equipment. The cast iron has to be able withstand the cyclic temperatures during its operation, for working temperatures between 500 and 700°C ordinary unalloyed cast iron is normally used, above this temperature up to about 1000°C, nickel and chromium alloys or pure nickel and chromium are used. In every instance heat resistant grey cast iron is required and it is characterised by a relatively high carbon content, mostly in the graphite form, so graphitising elements are important. The additions to the iron play different roles the main ones are;

Carbon

This primarily determines the properties of the cast iron, if the eutectoid content becomes more than 0.86% then the irons strength decreases. The hypoeutectic of the cast iron should be between 4.15 and 4.30.

Graphite

A decrease in the strength of cast iron is caused by the notch effects of the graphite inclusions. This can be improved by the development of graphite that has special or discontinuous lamellar shape. The moulds tendency to scaling decreases with the addition of graphite.

Silicon

This is the second element after graphite to exert the greatest influence on the irons properties. It has a graphitising effect up to 3.5%, it also forms ferrites and in doing so decreases the hardness.

Manganese

This forms and stabilises the carbide above 0.5%, it increases strength and hardness and bonds sulphur in to the material. It increases the effect to the heat resistance of the iron and to its resistance to scaling.

Phosphorus

This improves the wear resistance but it also increases the brittleness of the iron and therefore its tendency to crack. Under certain circumstances it can effect the sticking of the glass by improper release.

Sulphur

This is innocuous up to 0.15%, it favours carbide formation, but in so doing it increases brittleness and the occurrence of hard spots.

The effectiveness of the alloying steel additives are in the following directions;

- Al up to 4.5%, Ti, Ni up to 4.0%, and Cu up to 3.0% are graphitising.
- Cr is an outstanding carbide former.
- V, and also Ni, Mo, and Ti maintain carbide or pearlite and refine them.
- Cr, Ni, Mo, V, and Cu improves the strength of the iron, its hardness, scaling resistance and its constancy of volume.

The addition of Al improves the scaling resistance but below 0.6% and in the absence of Cr, volume constancy is impaired and the metal becomes hard to machine. In general as the heat resistance improves so do the mechanical properties.

Other materials used for mould parts

These include stainless steels, nickel based alloys and copper based alloys. A brief résumé is a follows;

Stainless steels

In the UK the use of stainless steel was first used during World War I and Hailwood patented the use of stainless steel or stainless iron for glass forming moulds - British Patent Number 193 536.

The two main steels used for special mould parts are the martensitic (AISI 416) and the austenitic (AISI 431). In bottle moulds their use has been confined principally to small inserts for either increased wear or heat resistance purposes. There main uses have been in plungers and moulds particularly in the tableware and oven ware side of the business. Similarly the types AISI 430 (12-14% Cr.) and AISI 310 (25% Cr / 20% Ni) have also been used. The alloys have very high heat resistance but very low thermal diffusivity of around 0.03 to 0.05 cm^2/s.

Stainless steels as a group offer corrosion resistance, high strength, good oxidation resistance at high temperatures and toughness, unfortunately all of these requirements cannot be found in any one steel. There are a large number of different compositions in this family of alloys but all of them have relatively low thermal conductivities. Table A1.3 gives a listing of some of them along with the different countries standard nomenclature. The straight chromium types have higher thermal conductivities and their thermal expansion is less than that of the nickel - chromium types.

Appropriate content		**Reference code**			
Chromium	Nickel	American AISI	German DIN	French AFNOR	British
13%	---	420	1.4021	Z20C13	420 S 37
16%	---	430	1.4016	Z8C17	430 S 15
16%	2%	431	1.4057	Z15CN17.03	431 S 29
25%	20%	310	1.4841	Z15CNS25.20	310 S 24
15%	35%	330	1.4864	Z12ncs37.18	EG 60

Table A1,3. Selected mould making steels from the family of stainless steels

Copper based alloys

These alloys have received considerable attention because their high thermal conductivities is claimed to cool glassware more rapidly, thus giving higher production speeds while still producing good quality ware.

There are two main types. The first contains copper, aluminium, zinc and nickel and is known by such names as Corundal alloys, Minox, and aluminium bronzes. While the second type is Incrament 800 which contains copper, aluminium, nickel and cobalt (2%) but no zinc. It is thought that the zinc may be lost at high temperatures.

The first type came out best of the samples tested for the effect on the surface finish of the glass while the second type was claimed that to cast iron it gave a 15-25% faster operation, reduced downtime and break-in time about three times the surface life, easier repairs and an product quality.

A copper alloy that has been used for a considerable number of years is an aluminium bronze containing 14-16% nickel, this has been variously referred to as Corundal GZ, Minox, ABX, and XX as well as other names. One development has been in the increasing use of continuous cast bar of this material for producing neck rings. The main advantage of this material is its high thermal diffusivity at temperatures that are typical mould working temperatures. This means that more heat can be extracted from the glass than with other mould materials.

If the structure of the material in the glass contact area is too course then premature pitting will take place. A specification governing the size of the primary phases at the surface has therefore been developed and all supplies ofthis material should conform to this specification. An attempt has been made to reduce the nickel content by using British Standard alloys

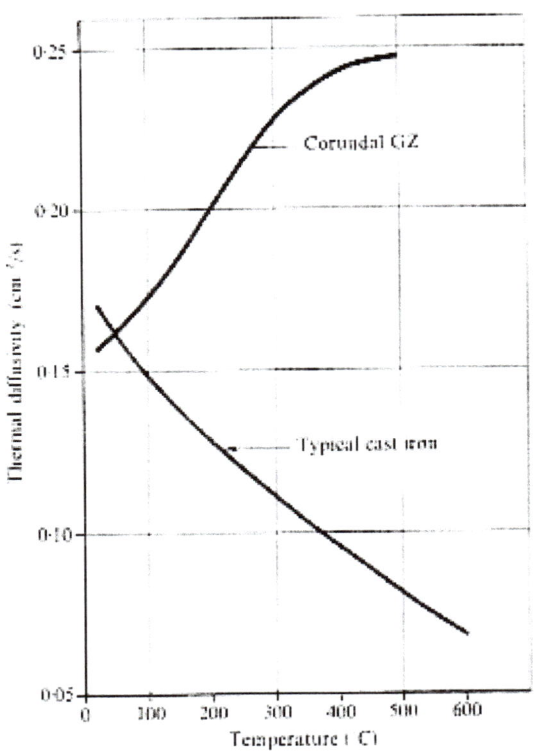

Figure A1,8. Thermal diffusivity of cast iron and Corundal GZ

AB1 and AB2, which contain 1 and 4% nickel respectively, but both have been found to be more susceptible to pitting than the normal material.

The chemical analyses of these copper based materials show that basically the materials called 700 XX, Minox, and Corundal GZ are similar and Veralloy 93 and NC 4 are almost identical. Incramet 800, NCVB and Corundal TX are slightly different, but their use at present is not so widely spread. The addition of zinc improves the castability and the final surface finish condition. Sometimes a little lead is added to improve the machinability.

Survey of Special Mould Alloys

Name	Composition Ni	Co	Al	Fe	Zn	Mn	Cu
700 XX	14.0-15.5	---	8.5-10.5	<1.0	7.5-10.0	<1.5	Bal.
Minox	14.0-16.0	---	8.5-10.5	<0.9	7.5-9.5	<0.15	Bal.
Corundal - GZ	14.0-15.0	---	9.5-10.5	<1.5	8.5-9.5	<0.4	Bal.
Veralloy 93	4.0-6.0	---	8.5-10.5	4.0-6.0	---	<2.0	Bal.
NC 4	5.0	---	10.0	4.0	---	<2.0	Bal.
Incramet 600	14.0	1.7	8.0	0.8	---	---	Bal.
NCVB	7.5	1.0	11.0	5.0	---	---	Bal.
Corundal - TX	<1.0	---	9.0-10.0	1.0-2.0	---	---	Bal.

Table A1,4. Survey of special mould alloys for mould equipment. None of them contained chromium or beryllium

The applications of these materials are;

Veralloy 93, NC 4 and Corundal - TX are used for bottom plates especially ones with a deep push-up. Because of the low nickel content the material is too soft for neck rings. Occasionally they have been used successfully as mono-block moulds (dual cavity) for IS, R7 and S10 machines.

700 XX, Minox and Corundal - GZ are frequently used for neck ring equipment.

A warning on the use of bronze alloys is that any contact with sulphur or sulphur containing glass 'dopes' especially one that contains zinc will cause pitting and in some cases equipment failure.

Nickel alloys

Numerous nickel-based alloys have been used for mould components and include nickel A inserts for pressed moulds, beryllium nickel for various mould components. Table A1.5 shows information supplied in the Dameron catalogue which states the different usages for different grades.

Dameron Developed Cast Alloys

HR 22	HR 27	HR 30	HR 34	HR 40	HR 44
Small moulds Small blanks	Small moulds Small blanks	Baffles Blow/blow plungers Guide rings Thimbles Mould repair items	Baffles Blow/blow plungers Guide rings Thimbles Cup moulds	Baffles Plungers Guide rings Solid blanks	Baffles Plungers

Table A1,5. The different uses of Dameron alloys for glass mould manufacture. HR 34 and HR 44 are sulphur resistant

Another alloy used for plungers is Monel K-500 which is a nickel-chrome alloy and is used for plungers.

Composition % and Rockwell C hardness

Alloy	Si	B	Cr	Max Fe	Max C	Hardness
HR 22	2.00-2.60	1.50-2.00	1.00 (max)	1.00	0.20	19-22
HR 27	3.10-3.35	1.10-1.35	4.75-5.25	1.00	0.20	28-30
HR 30	2.75-3.25	1.75-2.25	1.00 (max)	1.00	0.20	28-32
HR 34	2.75-3.25	1.75-2.25	3.80-4.20	1.00	0.20	30-34
HR 40	3.25-3.75	2.00-2.50	1.00 (max)	1.00	0.20	38-42
HR 44	3.25-3.75	2.00-2.50	3.80-4.20	1.00	0.20	40-44
HR 50	4.00-4.50	2.00-2.60	1.00 (max)	1.00	0.20	47-49
HR 54	4.00-4.50	2.00-2.60	3.80-4.20	1.00	0.25	49-54

Table A1,6. Shows the percentage composition and Rockwell C hardness of Dameron alloys

Mould coatings and inserts

There are two types of coatings the first where the mould equipment is 100% coated over the glass contact surface, this is a specialist function and it is not normally done at the factory, while the second is the hard facings that are put onto joints to extend the life of the equipment, this is usually spray welded technique which is done at the factory.

In his 1970 article Ensor states, "in recent years coatings have been used on cast iron parts subjected to high temperatures, corrosion, and abrasive attack to obtain longer life. In particular plungers have been coated all over and it is notable that the base material for the application is now mild steel. Hard facing alloys of nickel, chromium, boron, of silicon are spray welded on to the surface of plungers, post type baffles, bottom plates, press moulds, and the edges of split moulds to increase their abrasion and heat resistance".

A little later he goes on to say, "The welding of seam edges is now carried out by the powder welding process which make use of a metallic powder in the same way as spray welding, but the powder is fed through a metering device into the oxygen stream passing through the weld torch. This preheats the powder so that when it reaches the surface of the base metal it is readily melted to form a weld by diffusion and surface alloying".

In 1978 he said that, "The use of powder welding for the seam edges of moulds and spray/fuse welding for the complete glass contact face of plungers is still widely practised. Nickel based hard facing alloy powders with hardness's ranging from 15 to 40 Rockwell C is used for these applications. The main difficulty in using these coatings for the glass contact area of moulds and neck rings is in re-machining the contact face to the correct dimensions; consequently parts have been very expensive and their use limited".

It is also possible to use hard facing alloys in the form of castings, a particular example of this is the guide ring, or guide plate, used in a neck ring assembly. The knife edge of the guide ring can easily become damaged in use and an 'as is' cast insert made if a hard facing alloy reduces this type of damaged considerably. Plunger tips are another item of mould equipment for this type of treatment, where the tip initially pierces the glass in the blow and blow process using a hard faced alloy reduces the wear on this piece of equipment considerably.

Summary

First we will have a recap on thermal diffusivity and then list some of the various mould making materials and what they are used for. Firstly thermal diffusion which is an indicator of the amount of heat that a particular alloy will remove. This is shown graphically in Figure A1,9.

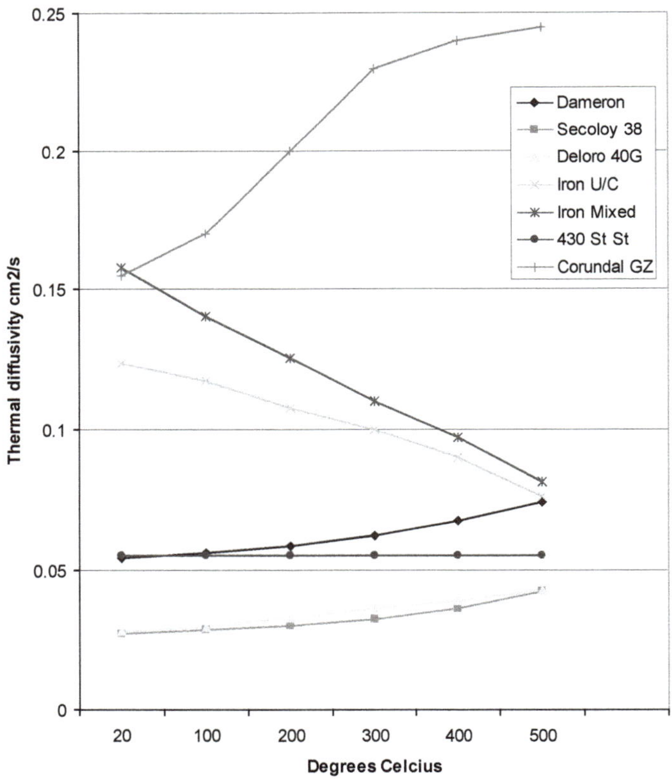

Figure A1.9. The effect of temperature on the thermal diffusivity of some of the common glass contact alloys

Table A1,7 Lists the thermal diffusivity at both 20°C and at 500°C for most of the different glass making mould materials. Some of the more typical uses of the various grades of metals used in mould making for bottle production can be seen in Table A1,8 by market names or makers is shown in Tables A1,9 to A1,16.

Thermal Diffusivity of Mould Materials

Material	Thermal diffusivity (M/s x 10^{-3}) 25°C	500°C
Copper	0.122	0.102
Corundal GZ	0.0155	0.0247
Monel Alloy K500	0.005 (20°C)	0.007 (426°C)
Stainless Steel 310	---	0.004
'Cast Iron'	0.013	0.010
No. 11 iron under cooled	0.0123	0.0077
Vantage iron undercooled	0.0118	0.0078
Dameron HR40	0.00544	0.0074
Semmet 40	0.0049	0.00685
Secoloy 38	0.00270	0.0043
Deloro Stellite 40G	0.00275	0.0043

Table A1,7. Table of Thermal Diffusivity values at 25°C and 500°C. Note that at 500°C the values of iron and Dameron are nearly identical

Material	**Applications**
Cast iron without or with small additions of Ni, Cr, Cu, Mo, Ti, Al.	Press and hollow ware moulds, bottle glass moulds.
Heat resistant steels; Ni-Cr-Fe, V5M, Ni-Cr-Wo-Cu-Fe, Cr-Co-Mn-Mo-Si-Fe.	Hollow and press ware - plungers, rings, and inserts.
Vanadium steels.	Table rollers.
Cr-Fe, Cr-Ni-Fe, Cr-Ni-Ti-Fe, Cr-Al-Ti	
Bronze with or without addition of Al	Suction moulds, plungers, neck rings.
Nickel-chromium	Moulds for flacons
Monel (Ni-Cu)	Small moulds

Table A1,8. Commonly used mould materials with some of their applications

Cast Grey mould Iron

Names	Jorad105 Vantage	Jorad 106 No 11 iron	Jorad 110 96A iron
Brinell Hardness	130-150	140-160	150-185
Thermal Expansion Co.eff. x10^{-6}/°C	10.9	11.6 (to 500°C)	
Estimated heat removing ability at 500°C 1=greatest.	3	4	6
Uses	Neck rings	Blanks & moulds Bottom plates Baffles More durable neck ring alloy	Blanks & moulds Bottom plates Baffles

Table A1,9. Commonly used cast iron for moulds

Wrought Stainless Steels

	Austenitic, 18/8 Free machining	Martensitic	Martensitic
Name	303, BS970:Part 4 1970:303S24	420, BS970:Part 4 1970:420S37	431, 1976 BS970:431S29
Hardness, Brinell	183 max.	Soft: 229 max. Hardend: 201-277	Soft: 277 max. Hardend: 248-302
Thermal Expansion Co-eff x 10^{-6}/°C	18.3 (to 500°C)		
Estimated heat removing ability (1=greatest)	11	9 (approx.)	9 (approx.)
Uses	'51' Process plugs	'41' Process blanks	Blow & blow plugs Top plates

Table A1,10. Commonly used wrought stainless steels

Copper Alloys

	Aluminium-nickel bronze	Aluminium bronze
Name	Corundal GZ, Minox, XX, Definox	AB2 BS1400:1973-AB2
Hardness, Brinell	150-200	130
Thermal Expansion Co-eff x 10^{-6}/°C	17	18.2
Estimated heat removing ability at 500°C (1=greatest)	2	1
Uses	Durable and large neck rings, bottom plates	Durable and large neck rings, bottom plates

Table A1,11 Copper alloys used for glass making mould equipment

Cast Nickel Alloys - containing Chromium

	Nickel-Silicon-Chromium-Boron Alloy	Nickel-Silicon-Chromium-Boron-Iron Alloy
Name	Deloro 40G, Colmonoy 4C*	Delstan
Hardness, Rockwell C	40G: 29-38 4C: 33-37	35-37
Thermal Expansion Co-eff x 10^{-6}/°C	13.4 to 500°C for 40G	
Estimated heat removing ability at 500°C	10	10
Uses	'62' process plungers, Blow & blow plugs Top plates	Guide plates, baffles (Cheaper alloy due to iron content)

Table A1,12. Chromium containing Nickel alloys.
* Similar but not identical alloys
Also Secoloy a modified version of 40G containing iron as a replacement for some of the nickel: used for guide rings

Cast Nickel Alloys without Chromium

	All are Nickel-Silicon-Boron Alloys			
Name	Semmet 40 Dameron HR40	Colmonoy 185	Semmet 30 Dameron HR30	Semmet 20 Dameron HR22
Hardness, Rockwell C	33-40	33-40	~30	20-25
Estimated heat-removing ability at 500°C	5	7	5	5
Uses	Plungers Baffles '41' process blanks	Plungers Baffles	Plungers Baffles	'51' process thimbles

Table A1,13. Chromium free cast nickel alloys

Cabot Stellite (UK) Ltd

Name of	PW22	SF40	Deloro Alloy 40G	Semmet
Powder Process	Power weld	Spray-fuse and power weld	Spray-fuse	Spray-fuse
Hardness of deposit, RC	18-22	35-39	34 (nom.)	35-40
Application	Cast iron	Steel or 40G castings	Steel or 40G castings	Semmet castings

Table A1,14. Cabot Stellite (UK) Ltd. powders

Wall Colmonoy Ltd

Name of powder	Colmonoy No. 4	Colmonoy 185	Colmonoy 27
Process	Spray-fuse and Powder weld	Spray-fuse	Powder weld
Hardness of deposit, RC	35-40	30-35	18-22
Application	Steel or cast alloys No. 4 or Deloro 40G	Colmonoy 185 castings	Cast iron

Table A1,15. Wall Colmonoy Ltd. powders

Eutectic Company Ltd

Name of powder	CPM 1202	PO 102 or Eutectic 10680	CPM 1207	Eutalloy 12494
Process	Powder weld	Powder weld	Spray-fuse	Spray-fuse
Hardness of deposit, RC	20-26	20-26	43	35-40
Application	Cast iron	Cast iron	Steel	Steel

Table A1,16. Eutectic Company Ltd. powders

Appendix 2

Forehearth design and control

Flow Theory - Some Studies in Gob Formation

While there may not be a direct relationship between the gob forming variables and the flow of glass through a tubular orifice, there are some correlations. Work carried out by Richardson on orifices where the diameter was less than one inch was reported as:

1. The flow (in pound per minute) varies directly with the diameter to the power of 2.5.
2. The flow varies inversely with the orifice length to the power of 0.5.
3. The flow varies inversely with the viscosity to the power of 1.6.
4. The flow varies directly with the depth to the power of 0.5.

Likewise Fanning arrived at his equation as;

$$M = [1.315 \times D^{2.5} \times h^{0.5}] / [fL^{0.5}]$$

Where M = flow in pounds per minute
D = diameter in inches
h = glass depth in inches
L = length of tubular orifice in inches
f = friction factor, with f being a function of M/DZ

where Z = viscosity in poises.

If you plot the following you get:

- Plotting the flow rate versus temperature it gives a slight concavity of the curve upwards but below 1200°C the relationship is nearly linear.
- Plotting the flow rate versus glass depth it is almost linear between 2.5 and 5.0 inches deep.
- Plotting the flow rate versus viscosity in poise gives a highly concave curve upwards with almost linear relationship with viscosities greater than 2000 poise.

- The relationship between f and M/DZ is linear if both are plotted on a logarithmic basis.

As a last statement some of the flow relations gave 'turbulent flow' conditions so these should be discarded, the formula is just a 'good guide' to flow through an orifice it is not 'fool proof'.

During a production run the relationships show that the glass weight is affected by changes to the depth of glass in the spout and changes to the temperature of the glass - temperature and glass viscosity are a related function. Kriz and Krivanek reported from studies in a commercial glass manufacture environment that a decrease in temperature of 11 to 15°F decreases the weight by 6 per cent. A similar increase in glass temperature increases the glass weight by 5 to 6 per cent.

The shape of the gob also varies with the glass temperature (see Figure A2,1 after Kriz and Krivanek) which shows the relationship between the gob shape (as the ratio of length to diameter) versus gob temperature.

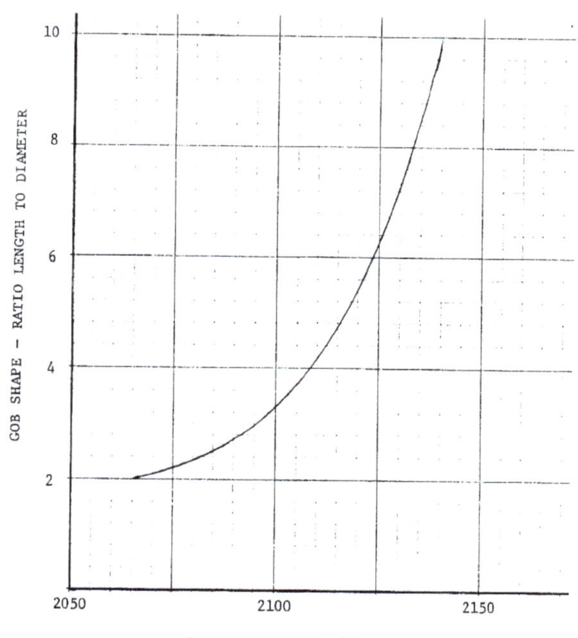

Figure A2,1. Gob shape versus temperature

For comparable conditions the curve shows that at a temperature of around 2065°F the ratio of length to diameter is 2, at 2095°F the ratio becomes 3, at around 2125°F it is 6 while at 2140°F it becomes 10.

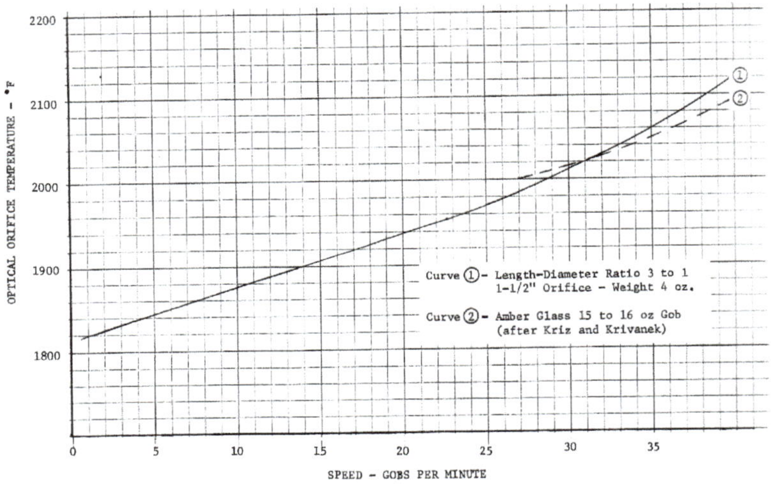

Figure A2,2. Relationship between optical orifice temperature and the machine speed in gobs per minute, (after Kris and Krivanek)

What this curve shows is that there is a possible relationship between the machine speed in gobs per minute versus the glass orifice temperature. For curve 1 the length to diameter of the gob was kept the same, orifice diameter $1\frac{1}{2}$" glass weight 4 ounce. The curve shows a slight concavity, it cannot be used as a guide for glass making because containers could not be made over the whole range. What it does exhibit is how the gob temperature has increase to keep the same length to diameter ratio as the speed increases.

For example, on the straight portion of Curve 1, for the same gob shape (length to diameter), a temperature of 1840°F. with a speed of 4.5 gobs per minute would require a temperature of 1940°F for a speed of 20 gobs per minute at to keep the same shape. Similarly on Curve 2, with amber glass a temperature of 2000°F. may be required for a speed of 28 gobs per minute but for similar conditions a temperature of 2095°F. may be required at a speed of 40 gobs per minute.

Schupbach gives changes to gob shape in a pictorial fashion see Figure A2,3 and Figure A2,4 shows what needs adjusting to adjust the gather.

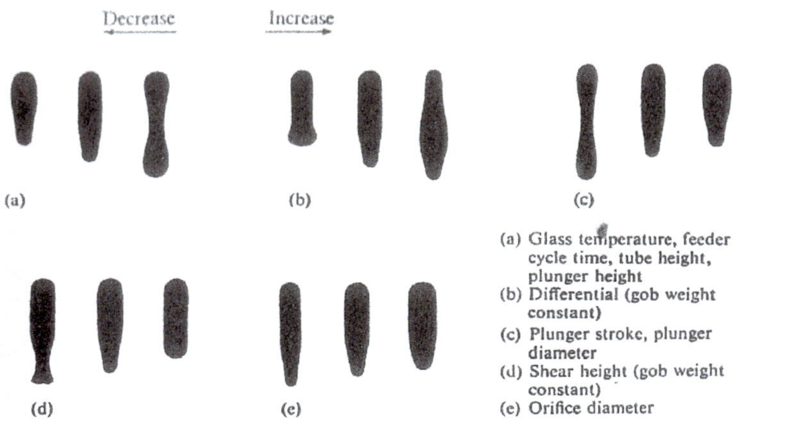

Figure A2,3. Gob shape as influenced by feeder variables. The weight changes according to the direction of change

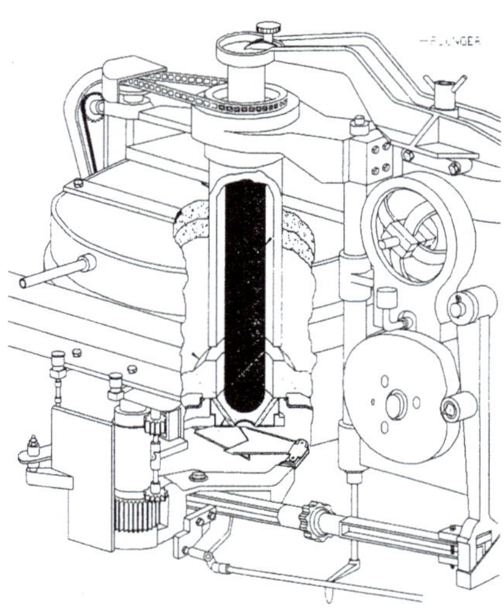

Figure A2,4. A view of a single 144 type feeder mechanism

Forehearth operation

Before we go on to the design of a forehearth here are some general observations on forehearths and their adjustment. The factors that affect glass conditioning are:- glass colour, fore-heath pull, gob temperature, glass level, refiner temperature (working end temperature) and operation, forehearth pressure, forehearth size, condition of refractories, control systems, combustion system, and cooling wind systems.

These are controlled by various departments within the factory;

Production planning

Glass colour - this is determined prior to the furnace rebuild, some furnaces may be colour flexible which is taken into consideration during their design. Glass colour as it is a 'fixed' parameter does not affect the hour to hour operation of the forehearth.

Forehearth pull - determined by the production requirements and machine type. Fixed for any individual job run so again does not affect the hour to hour operation.

Furnace pull - determined by production schedule. Fixed for that job.

Furnace Department

Refiner temperature The temperature of the glass entering the refiner is determined by the furnace pull. The higher the pull the hotter the glass is when entering the furnaces throat meaning that more heat has to be removed.

The temperature of the glass leaving the refiner depends upon whatever heating and cooling systems are available. These can be under the control of the Furnace Department or the Forming Department - depends upon the individual factory structure.

Glass level - depends upon where the measuring point is; if it is in the refiner then the measurement will not be affected during a job change where as if measured in a forehearth then variations in glass level will affect all the other machines unless the furnace man adjusts the control setting while the job change is taking place on the measuring forehearth.

Engineering Department

Forehearth size - this comes under the Design Department although these days it is usually 'farmed out' to a specialist forehearth designer who would them supply drawings, equipment and possibly refractories.

Combustion and cooling wind systems - maintained by either the Engineering Department or by the Forming Department if it is large enough (i.e. the number of machines that it controls).

Forming Department

Refractories - the forehearth has a number of replaceable refractories called expendables such as orifice ring that are usually changed at a job change. Next come plungers, tubes, stirrers and rotors, followed by feeder spouts which are usually changed about every year and finally, skimmers, mantle and back blocks. It is usually the Forming Departments duties to change all of these with help from the Engineering Department for, for example when a spout needs replacing. Skimmer blocks need monitoring for wear and lowered as required

Forehearth pressure - the back block should have been sealed after the furnace warm up and other than monitoring its condition it should not present any problems. If this block is broken or missing then any changes in the refiner pressure will be transmitted into the forehearth with the resulting effect on the forehearth temperatures.

Considering the forehearth proper it is important that it is kept at a positive pressure all the time. This is done by adjustment to either the cooling air and/or the dampers. The inspection holes between the burners should also be pulled at all times.

Temperature settings - including the gob temperature, each job has its own individual temperature settings which are set at each job change. Very occasionally they will be adjusted during the run. More about why later.

The above are the usual breakdown of duties that affect the forming operation and it is important that to get good results all are functioning efficiently. Every now and again the glass does not seem to be working as it usually does - the forming department will say that it's the glass composition but an analyses of the glass shows within the limits of the analysis that it is just the same i.e the composition is on the target composition. But the production efficiency has fallen and more than normal forming adjustments have to be made.

So what else could have changed? Is it the glasses water content, is it micro-inhomogeniety within the glass, or is it temperature. Are they acting individually or collectively. Lets look at what affects the gob weight - at the beginning of this Appendix we listed what affects the flow through an orifice.

According to both Richardson and Fanning for any orifice a number of the parameters are fixed, the only variables which will affect the weight of a gob are i.e. the throughput are;

1. Changes to the amount of glass in the spout - it needs to be varying as a constant change to the 'head loss' would not vary except that once.
   Both Fanning and Richardson say that the flow varies as the inverse of the glass head to the power of $\frac{1}{2}$.
2. Changes to the glass viscosity which can be variable if the temperature varies i.e it is cycling.
   Fanning says that it viscosity (Z) is part of the 'friction factor' in his equation calculated from M/DZ were M is the glass flow through an orifice of diameter D and Z is the glasses viscosity. Whereas Richardson for orifice diameters of one inch or less said that it varied with the inverse of the viscosity to the power 1.6.

Provided that the glass level controller is working correctly and controlling to 'fine limits' - better than ± one sixteenth of an inch, then variations due to changes in glass level will have a minimal effect on the production efficiency.

The forming process is all about removing heat in a controlled manner and a bottle making machine can be likened to a sophisticated 'heat exchanger' so any upset during the heat extraction process will affect the forming efficiency.

Changes in temperature are a different matter and some people just do not know how important is that the temperature is constant, i.e. the gob temperature needs to be controlled so that the temperature drift per hour is less than 2°C while

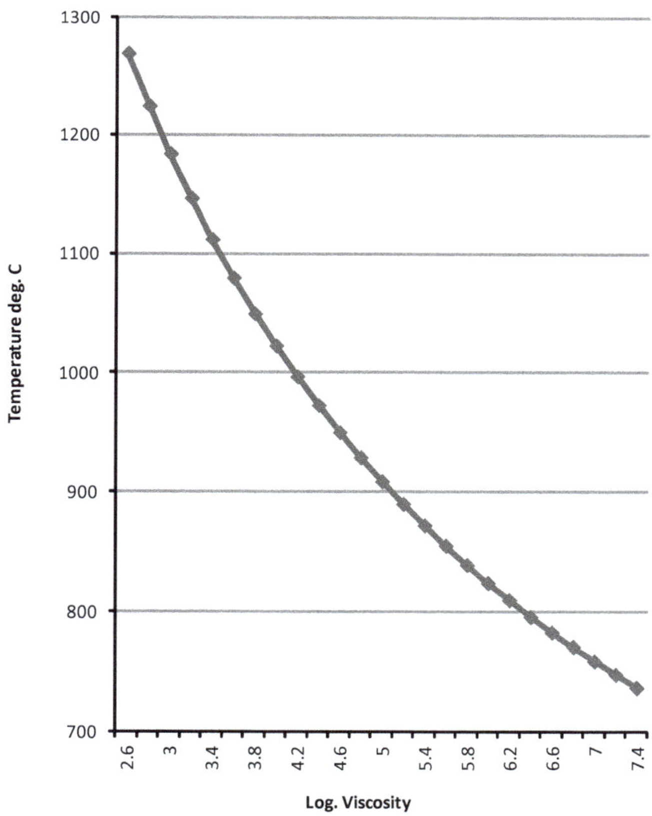

Figure A2,5. Temperature versu Log. Viscosity, i.e the working range for a container glass

the overall drift should be less than 8°C. Let us look at some instances, the viscosity of the 'working range' of a glass is that which lines between Log. 3.0 and Log. 7.0 where Log. 3.0 is the 'gob temperature' and Log. 7.0 is the temperature where the forming ceases see Figure A2.5. If we look at what a ±5°C change to the 'gob temperature we can see that if the gob temperature cycles by this amount then the maximum difference in viscosity between 1179°C and 1189°C is 115 poise which is a variation of 11.5% which is significant.

Temperature (°C)	**Log. Viscosity**	**Viscosity poise**	**Difference from target**	**Percent difference**
1189	2.975	944	-56	-5.6
1184	3.000	1000	0	0
1179	3.025	1059	+59	+ 5.9

Table A2.1 Changes in viscosity for a change of ± 5°C. on the Log. 3 temperature of a typical container glass composition

The temperature of Log. 3.0 is too high for current forming techniques and Log. 3.3 the 'Ideal Gob Temperature' is a more realistic one. If we make the same calculations for Log. 3.3±5°C (see Table A2.2) we can see that although the percentage difference is virtually the same 11.6% as against 11.5% the difference between the top and bottom of the viscosity value is much larger 229 poise as against the 115 poise at Log. 3.0 so, although the percentage differences are the same the effect of the viscosity change on the gob weight is in effect double - 115 vis 229.

Temperature (°C)	**Log. Viscosity**	**Viscosity poise**	**Difference from target**	**Percent difference**
1133	3.275	1834	-111	-5.7
1129	3.300	1995	0	0
1124.5	3.325	2113	+118	+ 5.9

Table A2.2 Changes in viscosity for a change of ±5°C on the Log. 3.3 temperature of a typical container glass composition

If we now consider the opposite end of the forming range we see in Table A2.3 the effect which is much larger, the percentage difference between maximum and minimum is 3798 poise and the percent difference is 38.0%.

Temperature (°C)	Log. Viscosity	Viscosity poise ×1000	Difference from target ×1000	Percent difference
764	6.915	8222	-1778	-17.8
759	7.000	10 000	0	0
754	7.080	12 020	+2020	+ 20.2

Table A2,3 Changes in viscosity for a change of ±5°C on the Log. 7.0 temperature of a typical container glass composition

These obviously do not affect the gob weight but they affect the glasses setting temperature and the bottles shape.

Finally for these viscosity calculation we show the viscosity variations at Log. 3.3 for changes, both ±, for each 1°C. from 0 to 5.

Temperature (°C)	Log. Viscosity	Viscosity poise	Difference from target	Percent difference
1133	3.275	1884	111	+5.7
1132	3.280	1905	90	+4.5
1131.5	3.285	1928	67	+3.4
1131	3.290	1950	45	+2.3
1130	3.295	1972	23	+1.2
1129	3.300	1995	0	0
1128	3.305	2018	23	-1.2
1127	3.310	2042	47	-2.4
1126	3.315	2065	70	-3.5
1125	3.320	2089	94	-4.7
1124.5	3.325	2113	118	-5.9

Table A2.4 Changes in viscosity for a change of one degree through to five. The glass composition is for a typical container glass

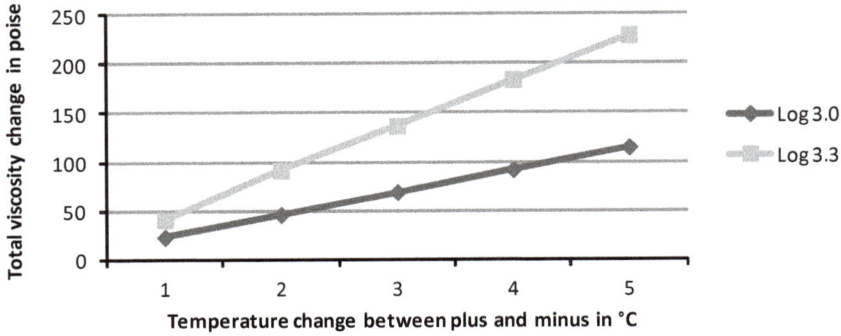

Figure A2.6 Change in viscosity between maximum and minimum temperature

Figure A2.6 shows the same information in pictorial form the line for Log. 3.0 is included. Earlier we stated that Kriz and Krivanek reported changes in weight for a corresponding temperature change. If we plot this as a crude relationship we get the following, starting with A2.7, showing changes in weight percent versus temperature.

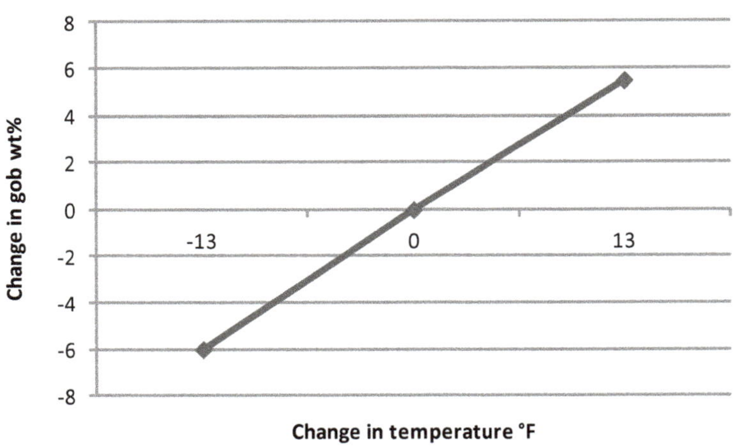

Figure A2,7 Interpretation of Kriz and Krivanek studies on variations in gob weight with changes in gob temperature

If we then superimpose this relationship to a gob that has a bottle weight of 25 ounce we get the following.

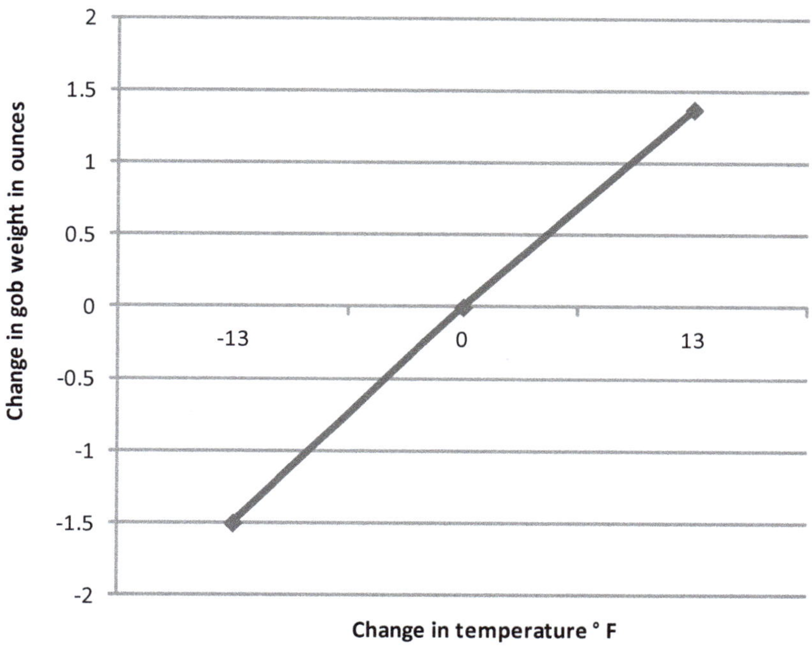

Figure A2.8 Changes in bottle weight with temperature according to Kriz and Krivanek

To make it more meaningful as the Kriz and Krivanek information was not specifically to a certain set of figures for example the 13°F quoted in the graph is given as a decrease of 11 to 15°F. If we change this to 13.5°F then it is easy to change the graph to degrees Centigrade. See Figure A2.9 overleaf.

Next comes Figure A2.10 which gives the variations in tolerance in ml for different bottle capacities., while Figure A2.11 shows the relationship between weight tolerance and bottle capacity. This is followed by Figure A2.12 showing the maximum weight allowance in grams versus bottle capacity.

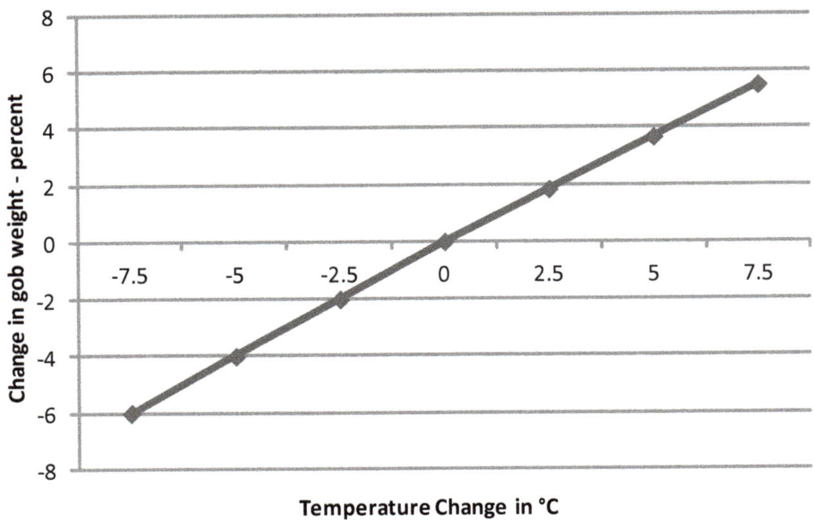

Figure A2.9 Change in gob weight versus temperature in degrees C

Emhart in their statistics divide the bottles according to their weight to capacity ratio. Up to 0.59 are classed as light-weight bottles, 0.6 to 0.79 as medium weight bottles and above

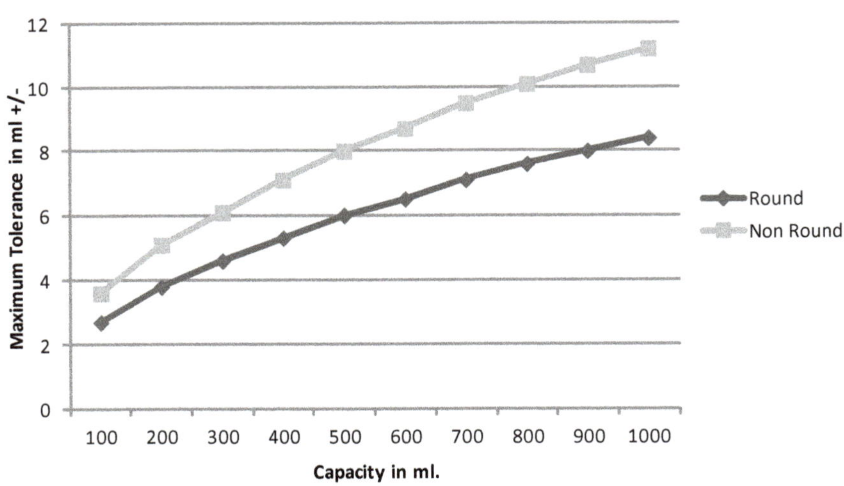

Figure A2.10 Maximum tolerance versus bottle capacity

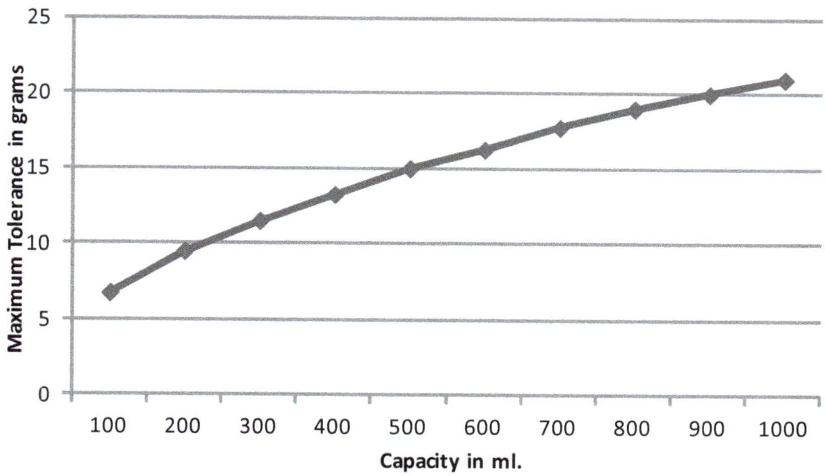

Figure A2.11 Maximum weight tolerance versus bottle capacity

0.8 as heavyweight bottles. Using these ratios we can plot the maximum allowance for each category, see Figure A2.12 overleaf which shows the relationship between the weight change for a five degree change in temperature and the bottle capacity in ml. for round bottles. There also is plotted the maximum weight tolerance for round bottles and non-round bottles.

The heavyweight curve, for round bottles the weight to capacity ratio of 1.0, shows that above approximately 270 ml. capacity a five degree change in temperature puts these bottles outside the weight tolerance. For a weight to capacity ratio of 0.7 the bottles become out of tolerance at approximately 550 ml. while the lighter weight bottles, weight to capacity ratio of 0.5, go out of tolerance at a bottle capacity just over 1000 ml. For non-round bottles the heavyweights exceed the tolerance at about 500 ml., for the mid weight ratio they reach the line at around 1000 ml. while the lightweight bottles never reach the tolerance line.

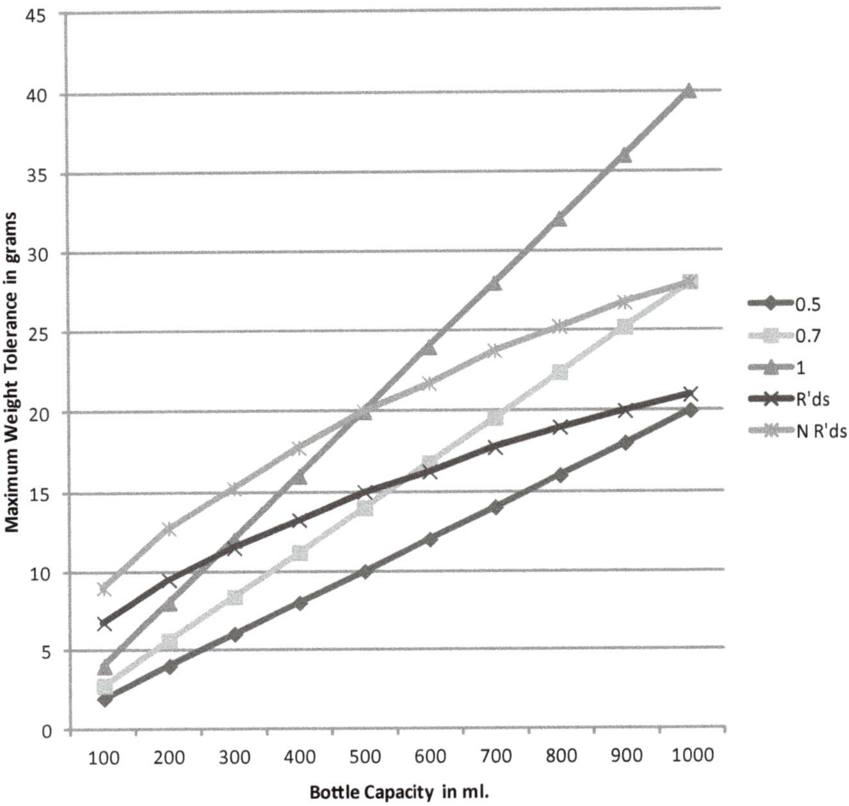

Figure A2.12 Graph showing the maximum weight tolerance for different weight to capacity ratios and the weight tolerance for round and non round containers

Earlier we said that checks on the chemical composition of the glass usually showed that it was on the target composition. To make a change of 5°C in temperature to the glass the individual oxides would have to change by the following percentages, see Table A2.5 overleaf. These are major changes in the composition even the 0.3 % change of silica for soda are way outside the control limits.

So temperature is not the main factor for the poor production figures and neither is the composition, but change must effect the machines ability to make bottles efficiently.

There are many reasons for this but here are some of the temperature related ones.

	Log 3.0	Log 3.3	Log 7.0	Log 7.65
Al_2O_3 for SiO_2	0.9%	1.2%	1.3%	1.2%
SiO_2 for CaO	0.5%	0.55%	6.75%	20%
SiO_2 for Na_2O	0.3%	0.8%	0.5%	0.5%

Table A2.5 Percentage change in oxide to increase the temperature by 5°C

Glass too cold

Corkage checks, spilt finishes, chipped finishes, broken finishes, finishes not filled up, finish checks, shoulders not blown up, light or thin bottoms, broken ware, cold appearance of bottle, wavy, uneven distribution, shear marks, brush marks,

Glass too hot

Chokes in neck, light shoulders, heavy bottoms, bottom sunk, bottles out of shape, bottles stuck to each other or have come into contact with each other.

These are all temperature related faults, for temperature think about the effects of viscosity, other faults which can be added to the above are those which are affected by changes in the temperature relationship between the glass temperature and the cooling of the moulds.

Glass Conditioning

Coloured glasses are much more difficult to condition than flint glass. With flint glass heat can penetrate through the body of the glass to the bottom layer, whereas in coloured glasses it is much more difficult.

Consider a white flint forehearth the glass enters the rear of the forehearth at around 1200°C (top glass 1195°C bottom glass 1200°C), by the end of the first cooling section the temperature of the top glass is around 1095°C whereas the bottom glass has only cooled down to 1125°C. The amount of cooling air is now reduced which allows the top glass to absorb some heat from the bottom glass and by the end of the second cooling section the temperatures are now 1095°C top and 1100°C bottom. The equalising section is then fired to increase the top temperature to over 1100°C and by the end of this section the temperatures should be the same at 1100°C with the glass flowing into the spout in a reasonable homogeneous temperature. That basically is what conditioning is about.

Temperature gradients in the forehearth.

- The temperature gradient through the depth of glass in the cooling and equalising sections affects the glass flow in the channels - the hottest glass flows faster than the coldest glass.
- If too much cooling is applied then the top glass can become so cold that the surface flow is very slow but underneath the glass is hot which leads to erratic conditions in the spout.
- There is also a cross channel gradient where the glass that is in contact with the refractories are colder than the glass in the centre of the forehearth. By firing the sides with sufficient heat to warm the sides the gradient can be reduced.

As the pulled varies from job to job the same forehearth has to cope with many different pulls, for example it may be pulling 60 tons per day at a gob temperature of 1080°C on the current job while the next job scheduled requires a gob temperature of 1120°C for a pull of 35 tons per day. The larger tonnage will require more cooling than the lower one.

The following tables, numbered A2.6/1 to 3 and Table A2.7 are a guide to the different forehearth settings that are required for different tonnages and glass colours. They are very general and you will notice that there are no referenced to either the forehearth width or to its length, neither is there any references to the width of the equalising section or to the type of spout being used.

They are from before the time when vast improvements were made in the design of the modern forehearths, for example the forehearth firing uses low pressure gas whereas most modern forehearths used high pressure gas firing or even oxy-gas. They are before the improvements to the insulation were made and other such things as centre line cooling and moveable roof blocks to allow cooling by direct radiation. But 'all in all' they give us a lot of useful information.

		Glass Colour	
Forehearth	Green	Amber	Flint
Pull range and tonnage	Low under 35 T/d	Low under 35 T/d	Low under 35 T/d
General glass temperature characteristics	Top hot Bottom cold	Top hot Bottom cold	Top cold Bottom hot
Rear cooling section	Heat bottom 1200-1230°C 4"-6" of fire	Heat bottom 1220-1245°C 4"-6" of fire	Reduce heat to cool bottom 1200-1230°C 2"-3" of fire
Front cooling section	Heat bottom 1190-1210°C 3"-5" of fire	Heat bottom 1220-1245°C 2"-3" of fire	Heat for middle 1190-1210°C 2"-4" of fire
Equalising section	Reduce top heat 1175-1200°C 2"-4" of fire	Reduce top heat 1190-1210°C 2"-4" of fire	Heat for top 1175-1190°C 3"-5" of fire
Spout	Heat spout refractories 1"-3" of fire	Heat spout refractories 1"-3" of fire	Heat spout refractories 1"-3" of fire

Table A2,6/1 Guide to what settings are required and what is happening to the glass in each section. The firing pressures are for a low pressure forhearth

Table A2,6/2 (top) Guidelines for mid range tonnages and Table A2,6/3 (bottom) Guidelines for higher tonnages are shown on the next page

	Glass Colour		
Forehearth	Green	Amber	Flint
Pull range and tonnage	Medium 35-70 T/d	Medium 35-70 T/d	Medium 35-70 T/d
General glass temperature characteristics	Top hot Bottom cold	Top hot Bottom cold	Top cold Bottom hot
Rear cooling section	Cooling for bottom 1190-1220°C 2"-4" of fire	Heat for bottom 1200-1230°C 3"-5" of fire	Cooling for bottom 1190-1220°C 1"-3" of fire
Front cooling section	Cooling for bottom Heating for middle 1160-1190°C 2"-3" of fire	Heating for bottom Cooling for middle 1175-1190°C 2"-3" of fire	Cooling for bottom Heating for middle 1160-1190°C 1"-3" of fire
Equalising section	Heat for top 1160-1190°C 1"-3" of fire	Heat for top 1160-1190°C 2"-4" of fire	Heat for top 1160-1190°C 2"-4" of fire
Spout	Heat spout refractories 1"-3" of fire	Heat spout refractories 1"-3" of fire	Heat spout refractories 1"-2" of fire

Forehearth	Green	Amber	Flint
Pull range and tonnage	High 70 T/d and over	High 70 T/d and over	High 70 T/d and over
General glass temperature characteristics	Top cold Bottom hot	Top hot Bottom cold	Top cold Bottom hot
Rear cooling section	Cooling for bottom 1175-1210°C 1"-2" of fire	Heat for bottom Cool middle and top 1175-1210°C 2"-4" of fire	Cooling for bottom 1175-1210°C 1"-2" of fire
Front cooling section	Cool for middle 1150-1175°C 1"-3" of fire	Heat for bottom Cool for middle 1150-1175°C 1"-3" of fire	Cool for middle and bottom 1150-1175°C 1"-3" of fire
Equalising section	Heat for top 1150-1175°C 1"-2" of fire	Heat for top 1150-1175°C 1"-2" of fire	Heat for top 1150-1175°C 1"-3" of fire
Spout	Heat spout refractories 1"-2" of fire	Heat spout refractories 1"-2" of fire	Heat spout refractories 1"-2" of fire

Some furnaces have alcoves, the conditioning of an alcove is the same for all colours, seeTable A2.7.

Tonnage	Alcove
Up to 35 tons/day	Heat to stabilise 1230-1260 3″ to 5″ of fire
35 to 70 tons/day	Heat to stabilise 1215 to 1230 32 to 52 of fire
Above 70 tons/day	Heat to stabilise 1200 to 1220 2″ to 4″ of fire

Table A2.7 Guidelines for alcove conditioning

Forehearth sizing is critical to the glass making process; it must always be able to cater for the required pull. If a forehearth is required to work outside of its design parameters then the production efficiency will suffer. The following information is the average type of throughputs used by forehearth manufacturers when sizing a forehearth and feeder.

When designing a forehearth and feeder system the first thing that needs to be done is to identify the job mix along with the projected tonnages. Once this information is available then the first item that needs identifying is the spout; it has to be able to meet the whole of the tonnages required and the projected job mix. Next is the forehearth's length and width - it has to be able to cool the glass down from the incoming temperature to the forehearth down to the gob temperature and do it in a manner that will give gobs that are at a constant temperature and are of a consistent thermal homogeneity. The glass conditioning has to be well inside the parameters of the forehearth so that it can cope with tonnage increases.

Possibly the first item is to decide upon is the size of the spout. Spout dimensions of those supplied by Emhart are shown in Table A2.8 along with some average spout capacities.

Next step is the forehearth sizing, the cooling section comes in widths of 16" for the 'KU' and 26", 30", 36", 43", 48", 54", 60" and 72" for the 'KW', although the 54", 60" and 72" sizes are normally used for distributors. The length averages around 20 to 24 feet. The glass depth is commonly 6 inches but 5 inches may sometimes be used for coloured glasses.

From around the 1960s due to the larger machines being installed the design of the forehearth was being studied and design changes were made. At this time the developments were being company led, many 'rules of thumb' regarding the sizing of forehearths were developed, remember that in those days that the forehearths used low pressure gas, cooling air blew from the sides into the centre and the forehearths had low levels of insulation, some of the rules are as follows:

- using plug flow - the maximum residence time should be no more than 60 minutes and no less than 30 minutes.
- a headloss of 2 inches is adequate.
- the maximum amount of heat that can be extracted is 3 tons per foot run for a K26, 4 tons per foot run for a K30 and 5 tons per foot run for a K36. These figures are now obsolete with the modern designed forehearth. The K30 is obsolete.
- another one on residence times is that for flint glass is should be 40 minutes and for green and amber glasses it

		Spout Width	Tons/day (flint)
Type 144	standard	14"	5-30
	stretched	16"	
Type 81	standard	16"	5-35
	stretched	18"	18-90
Type 194		18"	
Type 503		22"	35-90
Type 515		24"	60-150
Type 555		18"	

Table A2.8 Spout widths with some of their capacities in tons per day

should be 50 minutes. This is a bit spurious as it can only be right for one tonnage.

- make the equalising section as wide as possible by removing the narrowing portions.

There are many others, rules of thumb used by forehearth designers for white flint glass depth 6 inches are to be found in Table A2.9.

Forehearth width	Tons per day
26"	5-60
36"	35-90
43"	50-120
48"	60-150

Table A2.9. Average design parameters for forehearth sizing. Glass colour white flint, coloured forehearths would be wider. Here the higher tonnages may be using a deep spout

As you can see in Tables A2.8 and A2.9 that there is no mention of the forehearth's length or the forehearths capability of cooling the glass satisfactory which is of paramount importance.

A forehearth must always be able to supply glass to the orifice at a specific uniform temperature that is stable, and it also must be able to repeat these temperatures at any time, i.e for the same temperature setting the gob temperature must always be the same.

Let us look at some of the design parameters and what affects the choice;

- The majority of the head loss takes place in the conditioning or equalising section. The amount depends upon both the width and depth of the section. On a level forehearth it is designed to have, at its maximum pull, a maximum head loss of two inches (2") 1960 standards. The maximum of two inches of headloss in now obsolete.
- Maximum pull is affected by the type of spout being used. The pull is affected by the cross sectional area of the spout entrance, so for the same width a deep spout

is capable of higher pulls than the standard spout. For example a 81 spout would be capable of pulling 50 tons per day with 6" of glass and 65 tons per day with 9.5 inches of glass, a percentage increase of 30%.

- The maximum pull from a spout depends upon the tube diameter - the bigger the diameter the greater the pull. For example a standard 81 spout would be capable of pulling 40 tons per day with a 6" tube and 50 tons per day with a 7" tube, an increase of 25%.
- For any pull the length of the cooling section depends upon its width. So for a pull of 60 tons per day a K26 would be 24 foot long where as a K36 would only be 16 feet long.
- The amount of head loss depends upon the forehearths length, its width and the tonnage being pulled.

In the beginning of the life of the modern forehearth, the pulls were low and forehearths were shorter and narrower than is at the present time. The length of a 26" KW forehearth would have been around 12 feet maximum, the equalising section would be four feet long the width tapering from 16 to 14 inches. Initially the glass depth was either 6 inches or 5 inches with a gradual slope from 6 to 5 inches in the equalising section. There was also the KU which was 16 inches wide and had a glass depth of 10 inches, this type of forehearth is now looked upon as a special.

It was from these simple forehearth of the 1940s and 1950s that the 100 to 150 tonnes per day single forehearth and the plus 200 tonnes per day tandem forehearths have developed. Later we will discuss the tandem forehearth but here for design purposes we are concentrating on the single forehearth.

It is very difficult to make a kind of a 'check list' designer sheet for the design of a forehearth, there are so many variables along with site limitations and fuel efficiency. Some of the decisions have already made for you by the customer, which can either make your job more difficult or make it easier an example of this is that the forehearth has to fit into a confined space. To avoid any infringements with modern day practice we will be using designs from about 30 to 40 years ago and stating some of the changes made since then.

You will have given your client a check list starting with the size of orders that are to be made, the container speed and its weight and also the glass colour, the range of temperatures that it has to cope with both inlet and gob. The type of machine that will be used and what the manufacturing process is, i.e. blow and blow, press and blow or narrow neck press and blow. Then whether or not it uses Emhart's 'vertiflo' cooling or a similar process supplied by another manufacturer. Or the client would maybe just supply tonnage details - say to manufacture a range of jobs pulling between 50 and 100 tons per day. Also you would require the glass depth and the forehearth inlet temperature and whether it is controlled or not.

So first decide upon the spout size unless the customer decides upon a particular type or manufacturer. Let us use 40 to 80 tons per day as the pull, and we will divide it into three maximum tonnages, namely 35, 60 and 85 tons per day. Actually you should design for a higher maximum pull than specified to make an allowance for future machine speed increases. Emhart terminology will be used when quoting parts. Spout sizes that meet these maximum tonnage requirements are;

- Up to 35 tons per day maximum - both standard and deep 144s with a 14 inch entrance, and a standard 81 with a 16 inch entrance.
- Up to 60 tons per day maximum - a 194 with a spout entrance of 18 inches, a 16 inch wide 81 deep spout and a stretched 81, i.e. 18 inch at spout entrance.
- Up to 85 tons per day maximum - either a 115, a 503 or a 513 deep spout, entrance width 22 inches.

Having decided upon which spout you are going to use the next thing is to decide upon the foreheaths width and length. Using the 'rule of thumb' measures Figure A2.13 shows the maximum amount of glass that one of those forehearths could cool. To this I have added lines for a 43 and 48 inch forehearth for completeness.

This decision is a little bit more difficult to make maybe the first consideration should be as to what width should the

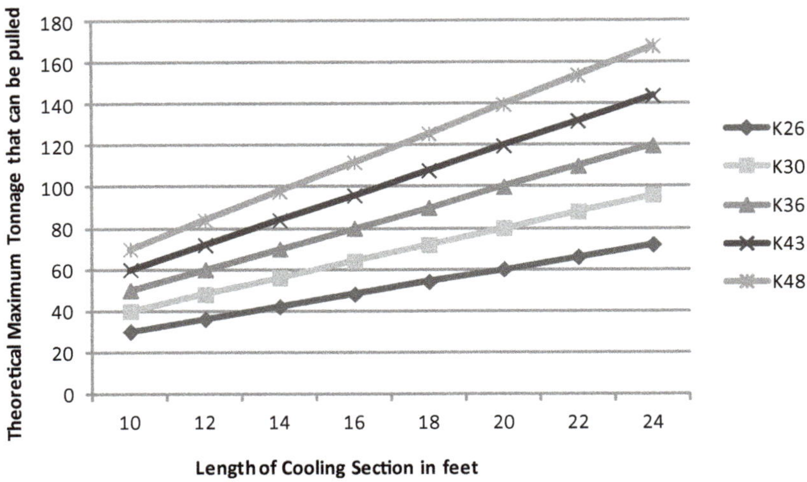

Figure A2.13 Maximum tonnage that can pulled for different lengths of forehearth and of different widths

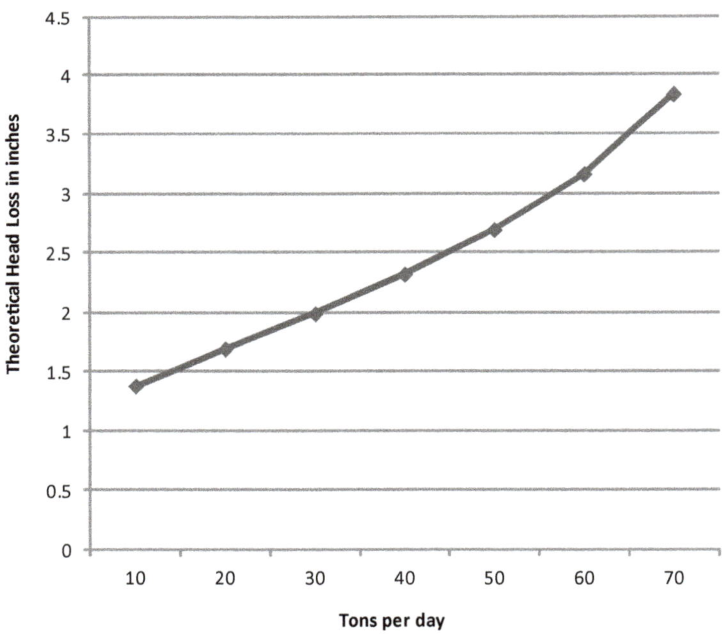

Figure A2.14 Theoretical Head loss from a 14ft 7½ in. forehearth with a 144 feeder attached. There is no consideration as to whether the glass can be cooled. The actual maximum would be 30 tons per day

equalising section. Here we are designing on six inches of glass and remember that the maximum head loss should be two inches. First to illustrate the effect of forehearth design and the glass tonnage pulled Figure A2.14 shows the head loss change with increasing pull. This is a fictitious example to illustrate the point, it has been done on a 14 foot forehearth who's maximum design limit would be 30 tons per day.

The majority of the head loss occurs in the equalising section of the forehearth. The differences between 14 and 16 inch spouts for different equalising configurations are shown in Figure A2.15 overleaf. So to minimise the head loss you have to design the equalising section to obtain what you want. First you start to remove the amount of 'narrowing' that occurs in this area but sometimes one of the things that you will come across is even when all of the 'narrowing' parts have been removed that a flat bottom equalising section will not supply the amount of glass that is required. To overcome this you will have to increase the depth of glass at the entrance to the spout which will also need to be made deeper. There are two ways to do this the first is to slope the bottom from the 'entrance to the exit' of the equalising section or to chamfer the last three inches of the channel bottom at 45 degrees thereby increasing the depth by three inches. Nowadays the latter option is the most popular.

At this time the common equalising widths were;

- 16 inches inlet to 14 inches outlet - the last channel of the cooling section reduced from 26 inches wide to 16 inches wide.
- 16 inches inlet to 16 inches outlet - the last channel of the cooling section reduced from 26 inches wide to 16 inches wide.
- 22 inches inlet to 14 inches outlet - the last channel of the cooling section reduced from 26 inches wide to 22 inches wide.
- 22 inches inlet to 16 inches outlet - the last channel of the cooling section reduced from 26 inches wide to 22 inches wide.

Not long after came the 22 inch wide spout with corresponding changes being made to facilitate its width.

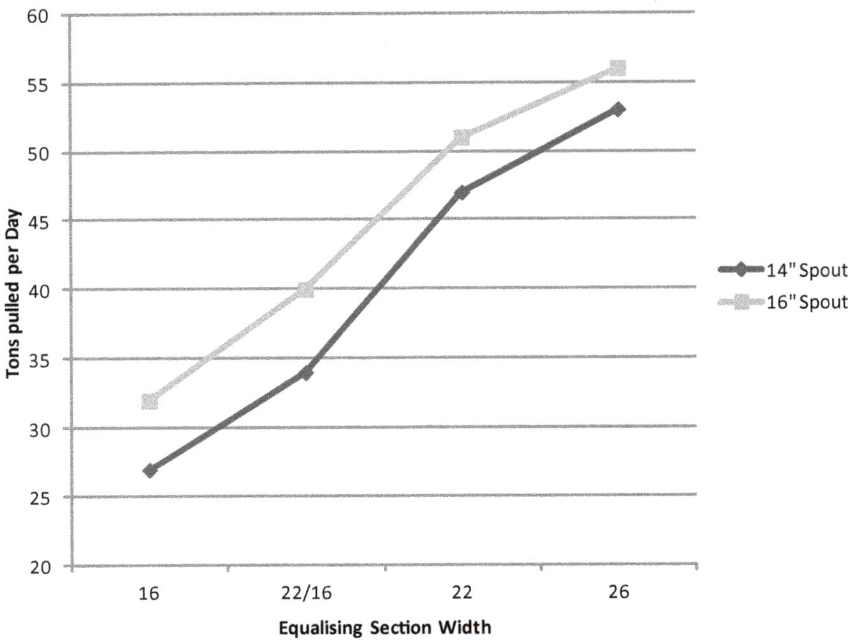

Figure A2.15 Maximum pull for maximum head loss of 2" for a 14 inch 144 spout and a 16 inch 81 spout with different equalising widths. All are for a K 26 inch cooling section

Figure A2.15 illustrates the maximum tons pulled per day for a maximum head loss of 2 inches for 144 and 81 spouts, for different equalising configurations and a K26 inch wide cooling section. Figure A2.16 shows for a headloss of 2 inches the maximum pull for different sized spout widths. All the equalising section widths are the same as the cooling sections except for the last channel where the reduction is made, and Figure A2.17 shows for a 22 inch and a 26 inch wide equalising section the maximum pull at a two inches headloss for various spout attachments.

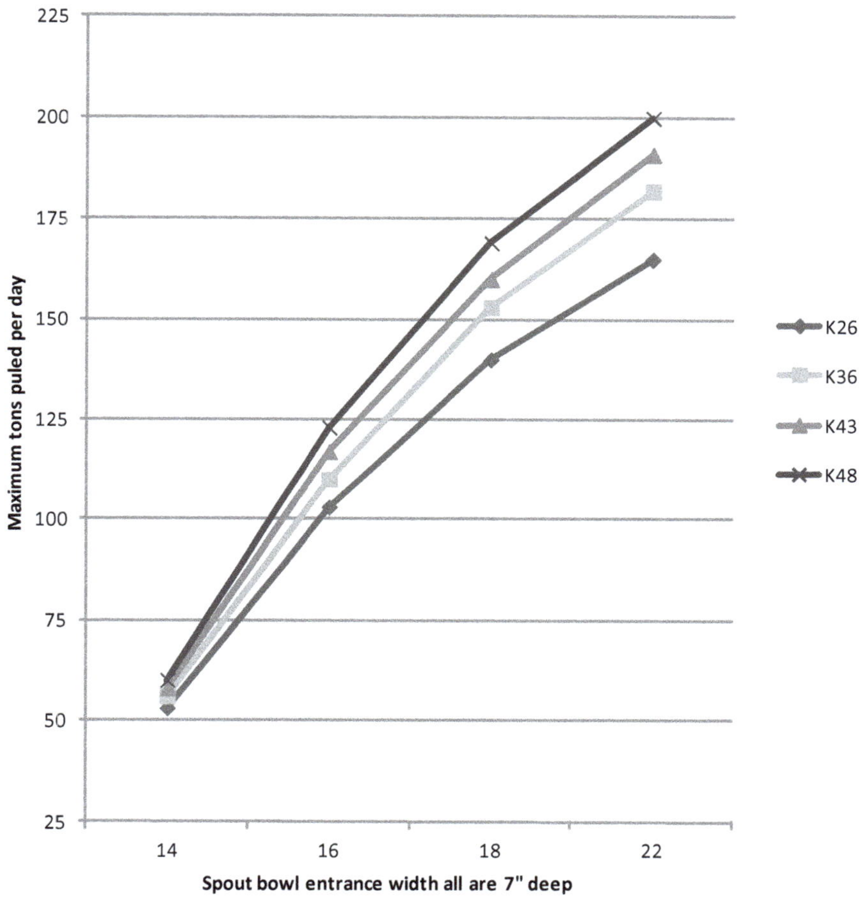

Figure A2.16 Maximum pulls for different spout widths with different sized forehearths. All of the forehearths were 36 feet long to make sure that they could cool that amount of glass. All equalising sections are the same dimensions as the forehearth cooling section

Finally after designing your forehearth calculations should be made to ensure that it meets the criteria for headloss and tons pulled at the specified temperatures. As before the maximum tons pulled should be about 10 to 15% more than the customer has asked for.

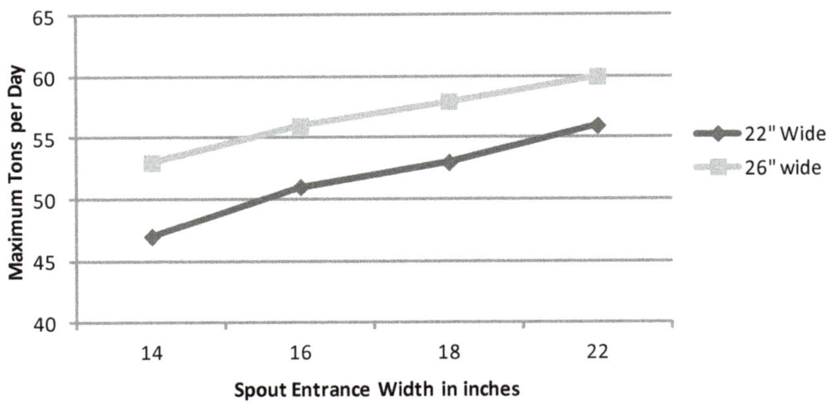

Figure A2.17 Maximum pull per day for 22″ and 26″ wide equalising sections for different spout entrance widths

The above has shown what needs to be done to met the design criteria without much of the detail - this is because you need specialised computer programmes for the calculations. That was then, nowadays forehearths are designed for a headloss of just less than one inch and the forehearth is then lowered at the spout end by whatever the headloss is. Unless specially designed spouts are being used a forehearth when lowered for a headloss of one inch will have the glass level at brimful during a job change or when the furnace is being idled for holidays, in this case a safer way would be to lower the glass level slightly.

Besides a programme for 'headloss' you now need programmes to calculate the heat loss per zone, the maximum rate of cooling per zone and the glasses overall residence time, fuel consumption, viscosity versus temperature and the effect on the transverse temperatures if the forehearth is 'cranked'.

Typical calculations that would be carried out during the design of a forehearth are the range of residence times, the maximum headloss, maximum cooling rate, all being done with a fixed equalising section length. Then by analysis of the results the design can be 'fine tuned' to give the best fit for the design criteria. This would be checked at various forehearth inlet and gob temperatures, as the range could be quite large, for example on a 60 to 100 ton per day forehearth the inlet temperatures may range from 1170 and 1270°C depending upon how hard the furnace was being pulled while the gob temperatures may vary from 1100 to 1220°C. depending upon the weight of the container.

Lets look at the reasons for these programmes in more detail and the changes in design, we'll start with those that do with energy input.

Reduction in energy usage

By the use of insulation it is possible to design a forehearth to cool only, i.e. there would not be any provisions for heating or additional cooling. From a production point of view this approach is too restrictive, the range of gob temperatures that the glass could be conditioned to would be very limited. The forehearth could not cope with changing inlet temperatures caused by a change in pull on an adjacent forehearth or a tonnage change to the furnace pull. Problems would also occur when the flow is interrupted at job changes. This limitation indicates that for the foreseeable future provisions for both the heating and cooling of the glass will have to be provided.

Some of the programmes that are required for the calculation of the fuel consumption are heat loss to the atmosphere from the glass, through the substructure, i.e. from the bottom and through the sides, from the superstructure including radiation losses and heat that has been removed by the cooling air. Normally they would be done for each section of the forehearth.

Changes to the original design are that some people cement hard brick to cover all of the channel joints to help with possible leakage problems. The insulation to the sidewalls and to a lesser extent the bottom has been upgraded although there was a move at one time into cooling the underside of the channel but this has gone out of favour due to it only cooled the glass adjacent to channel block, it did very little for the glass further away. Likewise the insulation levels of the superstructure have been upgraded.

Residence time

There are minimum and maximum residence times that are applied to forehearth design but the minimum time is the most important one due to the maximum one being governed by the forehearths length which will be kept to a minimum due to fuel constraints. Residence times vary according to the glass colour in flint a minimum of 30 minutes has been quoted, but 40 to 45 seem to be a more favourite or ideal choice. Coloured glass times would be longer say of the order of 50 minutes. These differences in residence times are due to the different coloured glasses ability to transmit heat, if we say that flint glass radiates heat at a rate of 100, then amber would be approximately 84% of that, emerald green at about 72% of that and dark green at 66% of that. This is why some coloured forehearth are designed with a shallower depth of glass.

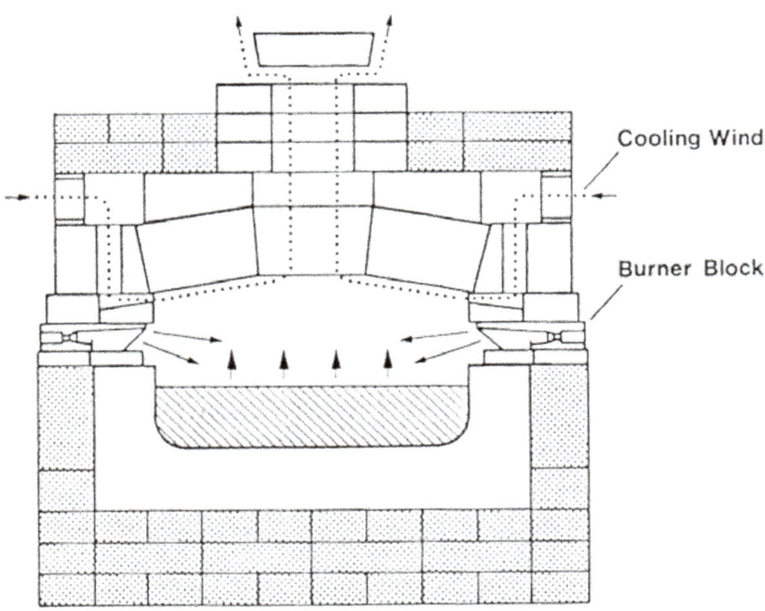

Figure A2.18 cross section of an earlier designed cooling section

Centreline cooling

In 1982 B-HF signed an agreement to market St. Gobain's SG2 forehearth which included their patented centreline cooling. The original design was similar to that shown in Figure A2.18 where the cooling air entered the cooling section under the roof blocks and above the burners. St. Gobain's new design changed all of this.

The main design change was to blow the cooling air down the centre of the arch and venting the combustion products out from the sides instead as out through the centre.

Figure A2.19, a longitudinal section, shows how this is done in more detail.

The fault with the original design was that extra fuel had to be burnt to overcome the forced cooling just above the burners. Here the cooling effect would be at a maximum due the temperature of the cooling air, it rises as it cools the glass so that on leaving the forehearth through the air and waste gas vent it will be at its hottest.

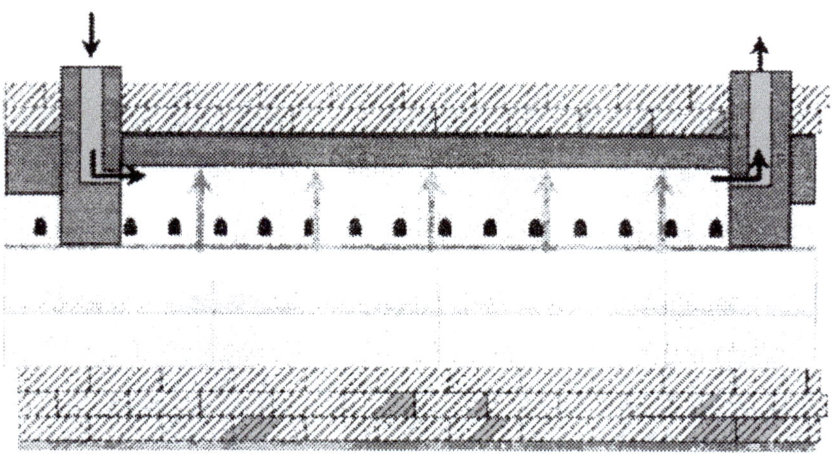

Figure A2.19. A longitudinal section of a modern style cooling section. The distance between inlet and outlet of the cooling air would be around four feet

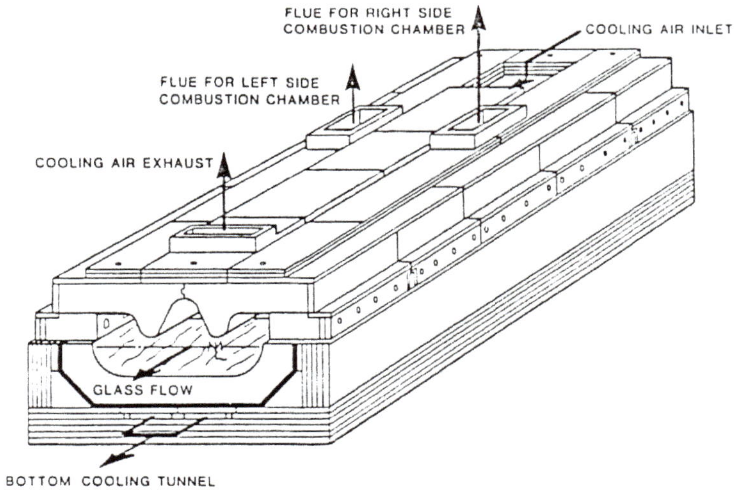

Figure A2.20 A schematic of a Emhart 540 cooling section circa mid 1980s. Note that this design incorporates bottom cooling

The changed design uses a flat roof block of either a single block or one that uses a joint in the middle, you can see this in Figure A2.20. The profile of underside of the roof block looks something like a 'W' which keeps the cooling air and the products of combustion separate, this makes both more efficient and reduces the amount of fuel required. The centre portion gets wider with forehearth width.

The mass heat extractor

This is the name given to the equipment which slides or lifts the roof blocks to open up the glass surface so that it can radiate its energy directly to the atmosphere. They can be positioned either in the distributor, in an alcove or in a forehearth, generally they would be located in the roof of a 54 inch wide or wider forehearth channel. Three photographs of a B-HF 'hinged' mass heat extractor can be found in Chapter 6, Figure A2.21 shows a Sorg Mass Heat Extractor situated in a distributor and Figure A2.22 shows one located in a forehearth. In both of the Sorg installations the roof blocks slide horizontally.

Figure A2.21 A Sorg Mass Heat Extractor situated in a distributor

Figure A2.23 shows a longtiudinal section through a mass heat extractor and its location depends upon the distributors design and whether or not there are two or four forehearths. They are not placed directly over the 'font' i.e. directly were the glass

Figure A2.22 A Sorg Mass Heat Extractor situated this time in the rear of a forehearth

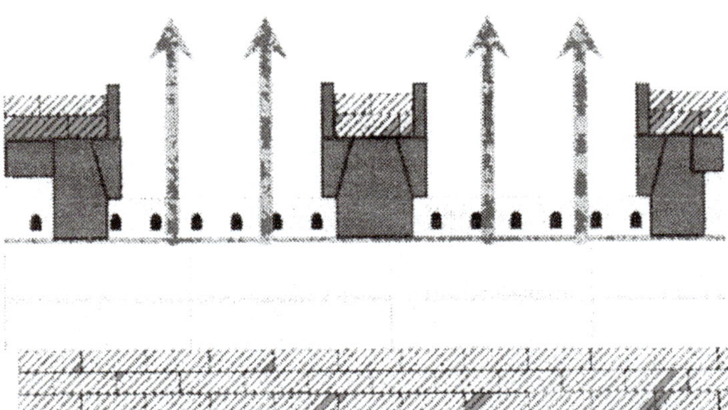

Figure A2.23 Longitudinal section through a Mass Heat Extractor

from the throat meets the surface for the first time. They are placed between the first forehearth entrance and the 'font' or they maybe located in the rear of the forehearth proper, this depends upon the forehearth temperature requirements. The glass opening needs to be at least 66% of the forehearth width, with a slot width of at least a foot and there would be at least two of them per location.

Distributors and Alcoves

The 'distributor' is what has replaced the working end, they are usually 'T' shaped and run at 90 degrees to the flow of glass as it leaves the melting chamber. The length of the arms depends upon how many machines the furnace is supplying, if there are say four machines then the two inner one forehearths would be connected directly to the distributor, while the outside two would be connected to an alcove, see Figure A2.24.

An alcove is a means of moving the glass further out from the throat before it connects, usually at right angles, to the forehearth, if a tandem machine is connected to the distributor then it would be connected to the distributor were it acts like a mini distributor. The glass flow would be split into two and then at the required distance the flow would be 'cranked' at 90 degrees to give the orifices at the correct centres, see Figure A2.25. So

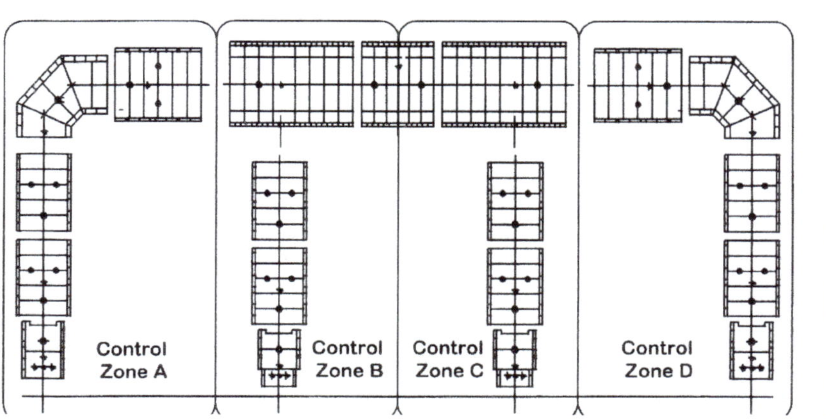

Figure A2.24 Distributor with two alcoves attached

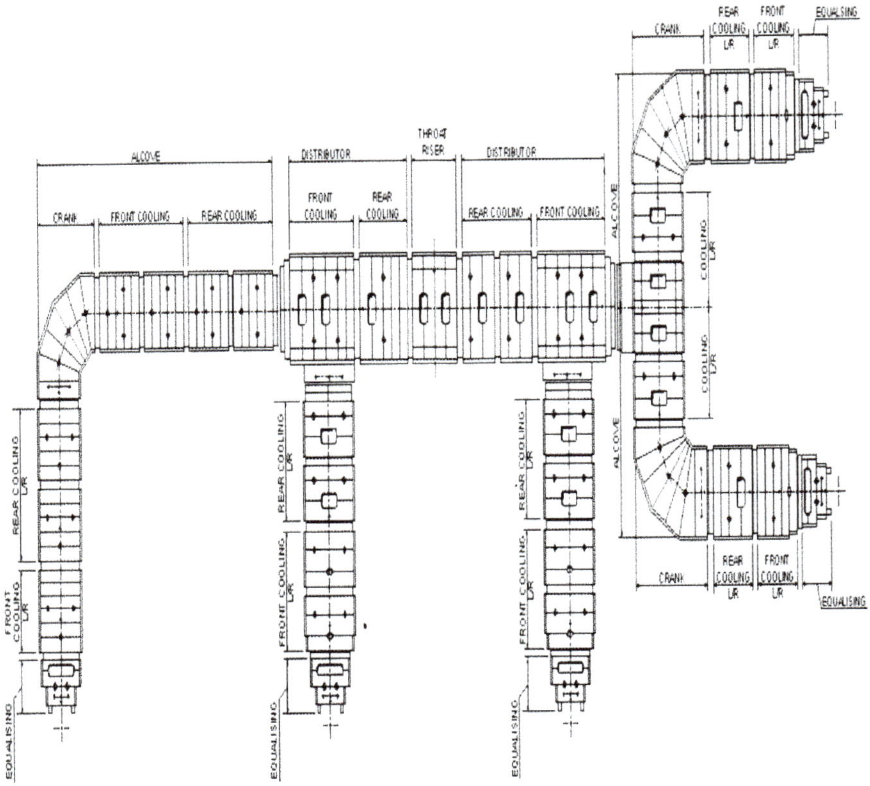

Figure A2.25 Distributor with one alcove and one tandem forehearth attached

the design changes now mean that the machines are positioned where they are planed to be, their position is not dictated by the shape of the old design working end.

The depth of glass in a distributor is much less than in an old designed working end, it would be around 20 inches (400 to 500 mm) in the distributor as against 36 inches in a working end. The distributor will then shallow to around 12 to 14 inches (300 to 350 mm) in the cooling section and get narrower after each forehearth exit. The depth of an alcove would be between 7 and 10 inches (178 to 254 mm) The depths are calculated using a headloss programme the exit depths would normally be 7 inches, i.e. 6 inches of glass cranked portions would be 7 inches deep.

Forehearths

Forehearth cooling sections are designed in 1220 mm. sections - this maximum length is the most efficient length of the centre line cooling air way before its cooling effect starts to deteriorate. They can be added together two make an 8 foot cooling section, two of these make a 16 foot cooling section add on a 4 foot equalising section and you have a 20 foot forehearth.

Forehearth widths vary according to the forehearths length, the tonnage to be pulled and whether or not the depth has been made shallower due to the glass colour. An example of the forehearth shallowing for the darker colours is say a flint K36 at 6 inches of glass forehearth was made one inch shallower due to a change in colour then the width would need increasing to 43.2 inches so the new fore hearth would be a K43.

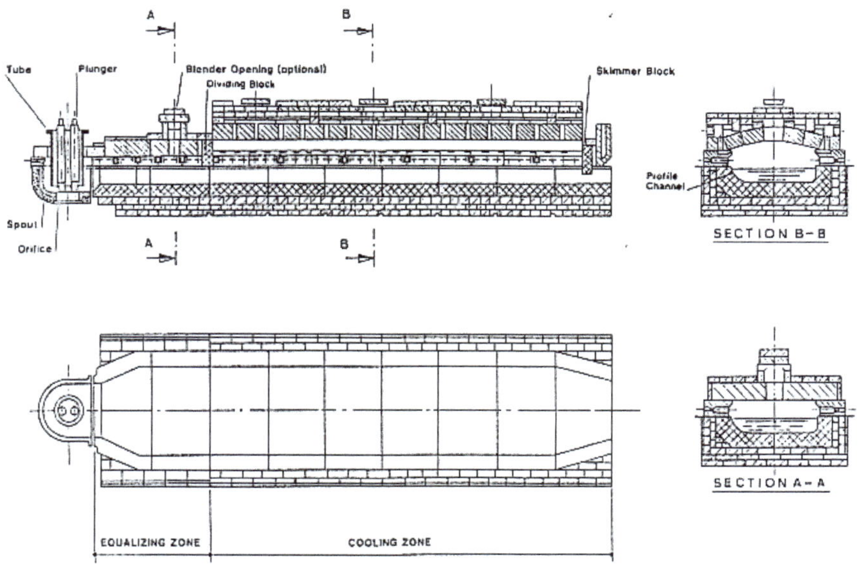

Figure A2.26 Forehearth circa 1970-1980s showing a conversion to a wider forehearth made on the run, possibly for a larger machine

Glass depths are usually changed at the forehearth entry where is would be made shallower and at the exit where it would be modified to attach a deep spout. If the forehearth included a fritting section then the portion underneath where the frit is added in some designs would be shallower.

Figure A2.26 shows a forehearth upgrade made during the campaign, this type of design, i.e. the forehearth widening after the forehearth entrance block, would never be envisaged on a new furnace. The forehearth would be designed with an entrance block the same width as the cooling section.

Some examples of forehearth throughputs are to be found in Table A2,10 while spout throughputs can be found in Table A2.11

Forehearth width in inches	Tonnage range in tons per day
K26	5-60
K36	35-90
K43	50-120
K48	60-150

Table A2.10 Average tonnages for a 24 foot flint forehearth. A coloured forehearth would be wider

Spout type	Tonnage range in tons per day
81 Standard spout	5-35
81 Deep spout	18-60
503	35-90
515	60-150

Table A2.11 Range of tonnages for different spouts

Forehearth control

The energy input to a forehearth can be either electrical, by radiation from above, or by conduction through the glass from electrodes submerged in it, or by using one of the fossil fuels either as directly obtained from a 'well' or manufactured from another fuel i.e. towns gas from coal, or it can be a mixture of the two fuels. Irrespective of the type of fuel there needs to be a good temperature measuring system and a good automatic control system.

Temperature measurement is done using either submerged thermocouples or one where they are suspended from above or by sensors that measure the temperature by radiation. The vertical thermocouples may be of a single type, where its height from the bottom is fixed, or a triple type (triplex) where the thermocouple junctions a set at one inch from the bottom, three inches from the bottom and five inches from the bottom. This only applies to when the glass depth is six inches, the separation between the thermocouples would be greater for nine inches of glass and less for shallower depths. Basically the upper and lower thermocouples would be one inch from the glasses boundary and the middle one would be equidistant between them.

Radiation pyrometers have both the least maintenance and overall cost but their drawback is that they only measure the glass surface temperature or short way into the glass at best, the depth of penetration depends upon the glass colour. The drawback with the submerged thermocouple is that they may be adequate for the spout but as they only measure the temperature at the edges of the channel it becomes meaningless for control purposes which requires the temperature in the centre of the forehearth. Other surface measuring sensors are the fibre optical types.

The aim of the current designer is to give equal temperatures across a nine point grid located nine inches back from the spout entrance. The reason that its nine inches back is that the rotating tube can cause a build up of 'cold' glass on the exit side of the tube rotation and effect the temperature readings, i.e. on the side were the flow is opposite to the main direction of flow. To keep this within acceptable limits a tube speed of 3 revolutions per minute should be considered as maximum.

A well used measure of the forehearth's efficiency is a formula which was developed by Owens-Illinois in the late 1970s, it is used by several glass companies worldwide, and it is based upon the readings from three triplex thermocouples spaced equally across the forehearth with each giving information at three depths. They are arranged like this, looking from the spout;

UL	UC	UR
ML	MC	MR
LL	LC	LR

and the formula is;

Thermal Homogeniety Efficiency (%) = $[1 - (\Delta H + \Delta V)/MC] \times 100$

The reference temperature is taken as the value of MC as variations to this temperature do not affect the final calculation. Other variations of this calculation are used the main one being the use of five measuring positions arranged like this;

	UC	
	MC	
LL	LC	LR

The triangle of glass defined by the boundaries LL - UC - LR is considered to be where the primary glass comes from to make the gobs.

While an efficiency of plus 90% would be satisfactory for both blow and blow and press and blow manufacture a forehearth efficiency of at least 95% is required for successful narrow neck press and blow manufacture, so to achieve this a good forehearth design and operation is required.

The simplest and probably the most widely used control system is a single infrared sensor sited on the glass centreline at the end of each control section, the output from this sensor is fed into a recorder controller the output from a PID controller then adjusts the fuel input to the section, adjustments to the cooling air are made manually. The next stage is to control the air input to the cooling sections as well, so a PID heat/cool controller is used.

A more sophisticated control system is to use cascade control, for this you need tri-level thermocouples and it is a quite sophisticated system to use. If there are large differences between centre and side temperatures then a bias control system can be used, this can only work if the side firing of each zone can be controlled independently and there is a side positioned tri-level thermocouple to act as the sensor.

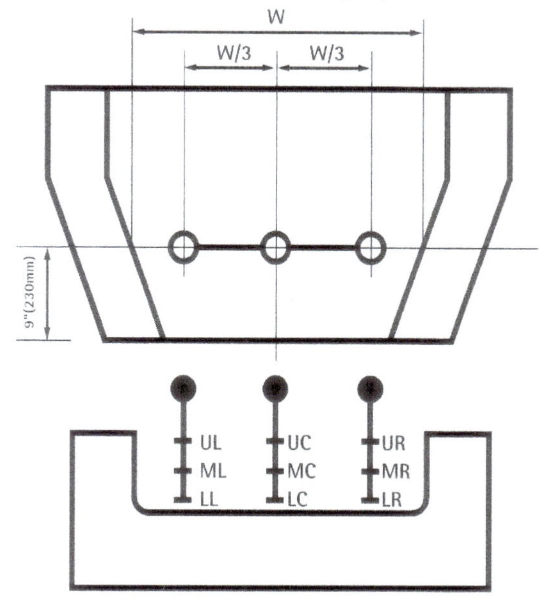

Figure A2,27. Schematic of the positions of a tri-level thermocouple installation

Finally in coloured glasses if there are large temperature differences at the exit from the equalising section then electrical energy can be added through submerged side electrodes. The following two tables illustrate the improvement to a green glass equalising section.

Table A2,13 shows a marked improvement over Table A2,12 but the average temperature would need reducing to give the same workable gob.

	Left	Middle	Right
Upper	1205	1210	1202
Middle	1198	1214	1192
Lower	1184	1210	1175

Table A2,12 Before boost in equalising section. Thermal Eff. 86.5%

	Left	Middle	Right
Upper	1213	1212	1211
Middle	1217	1217	1213
Lower	1212	1212	1214

Table A2.13 After boost in equalising section. Thermal Eff. 98.3%

Appendix 3

The Identification of Glass Bottles

1.0 History

Glass has been made for at least 3500 years, originally as glazes for pottery and beads. The earliest glass finds are from the Tigris-Euphrates region circa 2500 B.C. For around the next 1000 years there is a "knowledge" gap which could be classified as the "dark ages", very little is known until the records of Pliny, Herodotus and Threphastus showed that in 1500 B.C. there was a thriving glass industry around the Eastern Mediterranean. They were all unanimous in their praise of the glass making at Alexandria, Sidon, Tyre and Thebes.

The earliest "hollow ware" was made by the "sand core" method. A simple description of this multi-stage process is as follows;

1. On the end of a rod a former following the internal profile of the vessel was made using a sand and clay mixture.
2. A thread or string of molten glass was then carefully wrapped or coiled around it.
3. The glass was then re-heated until soft and then rolled on a smooth stone slab until the surface was smooth and even.
4. After some form of annealing the rod and the core was carefully removed. As the neck of the bottle was often waisted the core could only be removed by picking at it.

This method of manufacture was limited to the making of small pots and vases, the walls of the vessels were very thick which limited their usefulness. This method of manufacture was used from the 5th to the 1st Century BC when "free blown" glass started to be made.

2.0 British Made Bottles

Up to the beginning of the 17th Century bottles in this country were mainly made of leather or earthenware. The earliest glass wine bottles were made during the early 1600s, they were "free blown" i.e. made without a mould, and are called "shaft and globe" because of their shape. Their shape is a globular body and a long neck, they had no standing base so they were often put into small wicker baskets which kept the necks vertical, which was a necessity once the seal had been removed. They then gradually evolved in to the, so called, "onion" shape of the 1700s where the globe became squatter with a flat bottom which eventually became pushed up in the centre and the neck became shorter.

From the early part of the 18th Century wooden moulds were used which gave the bottles a more regular and consistent shape. The next stage in the evolution of the wine bottle was to make the neck longer and tapered; this shape is called "the mallet" which gradually became a more straight sided cylindrical shaped bottle by 1750.

Some important dates to help with the identification of bottles.

1755 Printed or impressed name marks start to appear on bottles from this date.

1792 The first soft drinks were bottled in earthenware bottles by Jacob Schweppes of London.

1800 Printed marks generally appear after this date.

1802 Charles Chubsee of Stourbridge produced a bottle making mould which opened mechanically in to three parts. Embossing containing the Royal Coat of Arms appears after this date.

1814 William Hamilton patented an "egg" shaped bottle which became known as the "Hamilton." As the bottle had no standing base it had to be stored on its side, this prevented the cork from drying out and maintained an air proof seal.

1821 Rickett patented his method of making bottles using a three piece mould. Which allowed large quantities to be made for the first time.

1842 The Diamond shaped registration marking system started at this time. This system of registration was used from 1842 to 1883, from 1842 to 1867 the year letter appears at the top section of the diamond. From 1868 to 1883 the letter appears in the right hand corner. The system was used for both glass and pottery; III is for glass and IV for pottery.

Lettering incorporating "Limited" appears after this date.

1857 The snap was invented so the use of the pontil iron gradually ceased.
Lettering incorporating "Trade Mark" appears after this date.

1870 Hiram Codd of Camberwell patented the first version of the famous "Codd Bottle".
The internal screw finish is from now.

1875 The invention of the swing stopper.

1884 "Rd. No." - Registered Number is from now. Rd. No. 1 was registered 1 Jan 1884.

1891 The word "England" is from this date.
The crown finish does not appear before this date.

1920 Hand applied lip made to circa this date, these became fewer and fewer as more and more bottles were made on automatic machines.

3. American made bottles

This is from an American internet site with some additions in brackets; there are three keys to help in dating most bottles:

1. Side seams

- **None**: The bottle may be free blown, in which case it has a very uneven shape and dates before 1860. Or the bottle may have a nice even shape, but was spun in the mould to smooth out the seams; a practice common around 1900-1920.
- **BIM:** The side seams run from the base and end below the top of the lip, which is the result of "blowing in the mould" (BIM). Can usually tell that the lip is crude and was applied by hand.
- **3PM:** 3 piece mould. The bottom half (from base to shoulder) has no seams, then there is a seam near the shoulder which runs completely around the circumference of the

bottle. From this shoulder seam are two side seams that run up the neck and end below the top of the lip. 3PM was primarily in use from 1840 - 1870. Can usually tell that the lip is crude and was applied by hand.
- **ABM:** If the side seams run through the top of the lip, then the bottle is ABM (made on an automatic bottle machine). The first ABM bottles started appearing in 1905, and by 1920 most bottles were being made this way.

2. **Base type**
    - **Opened pontilled:** usually date before 1860.
    - **Iron pontilled:** usually date 1845-1870.
    - **Smooth:** usually date 1870 or later.
3. **Top type**
    - **Matthews gravitating stopper:** date after 1864.
    - **Codd stopper:** date after 1873 when Hiram Codd invented this bottle.
    - **Hutchinson blob top:** date after 1879.
    - **Lightning stopper:** date after 1880.
    - **Crown top:** date after 1892, (when crown top invented by William Painter).
    - **Screw top:** usually date after 1920 [external screw?].
    - **Applied lip:** date before 1900.
    - **Tooled lip:** date after about 1890.
    - **Other:** There were many varieties of tops, so if the top neither crown or screw then rely on the side seams and the bottles base type.

British Patent Office, Design Registration Marks

These diamond shaped marks are to be found on a variety of Victorian bottles that were made between 1842 and 1883. The mark showed that the design had been registered with the London Patent Office and gave copyright protection for a period of three years. The marks give the earliest date that the bottle could have been manufactured. Not all bottle designs were registered most of these designs would be bottles that could be classed as "general trade", common design used by more than one filler.

Figure A3.1 shows the two variations of the diamond registration that were used, the first from 1842 to 1867 where the year letter is to the top and the second from 1868 to 1883 where the year letter is to the right. The class of article is shown at the very top, III for Glass and IV for ceramics, i.e. Pottery and Porcelain.

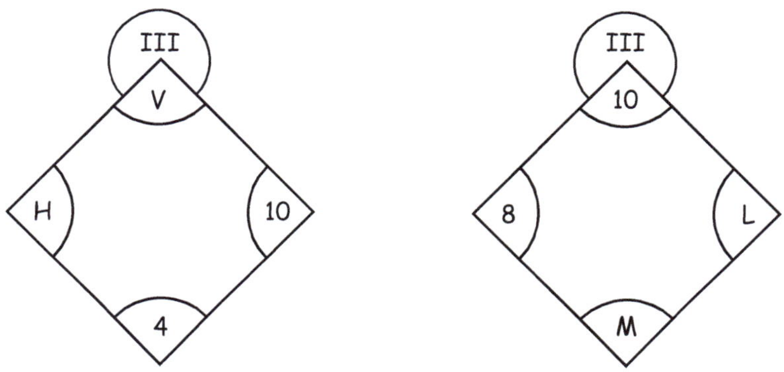

Used until 1868 Used from 1868

Year letters

Year	Letter	Year	Letter	Year	Letter
1842	X	1856	L	1870	C
1843	H	1857	K	1871	A
1844	C	1858	B	1872	I
1845	A	1859	M	1873	F
1846	I	1860	Z	1874	U
1847	F	1861	R	1875	S
1848	U	1862	O	1876	V
1849	S	1863	G	1877	P
1850	V	1864	N	1878	D
1851	P	1865	W	1879	Y
1852	D	1866	Q	1880	J
1853	Y	1867	T	1881	E
1854	J	1868	X	1882	L
1855	E	1869	H	1883	K

Month letters

Jan	C	May	E	Sep	D
Feb	G	Jun	M	Oct	N
Mar	W	Jul	I	Nov	K
Apl	H	Aug	R	Dec	A

Figure A3.1 Shows variations of the diamond registration

Registration numbers

From 1st January 1884 the design of any article, not just glass, could be registered, each design was allotted a unique registration number on the day that the design was accepted. The numbers were sequential and ran from 1884 through to 1909. These numbers can be used to identify the year that the design was registered therefore determining that the article could not have been made before that particular year.

After 1st January 1884 some bottles manufactured in the United Kingdom displayed the designs registered number, which can be seen either on the body (usually at the rear) or on the base. The following listings are the registered numbers used for each year up to 1909. Each year starts January 1st.

Year	From	To	Year	From	To
1884	1	19753	1897	291241	311658
1885	19754	40479	1898	311658	331706
1886	40480	64519	1899	331707	351201
1887	64520	90482	1900	351202	368153
1888	90483	116647	1901	368154	385087
1889	116648	141272	1902	385088	402912
1890	141273	163766	1903	402913	425016
1891	163767	185712	1904	425017	447547
1892	185713	205239	1905	447548	471485
1893	205240	224219	1906	471486	492486
1894	224720	246974	1907	492487	518414
1895	246975	268391	1908	518415	534962
1896	268392	291240	1909	534963	and so on

Changes to bottle designs with time.

The following sketches show some of the time lines of the various types or families of bottles. There is also two depicting the evolution of the finishes and of the mould seams which are definitely a guide when attempting to date bottles.

Mould seams can only give an approximate guide to the age of a bottle because there was much overlapping of techniques and mould improvements. In many cases it is safer to say that the bottle was made after some date, i.e. a Codd bottle could not have been made before 1875 when it was 'invented' so it could only be made after that date.

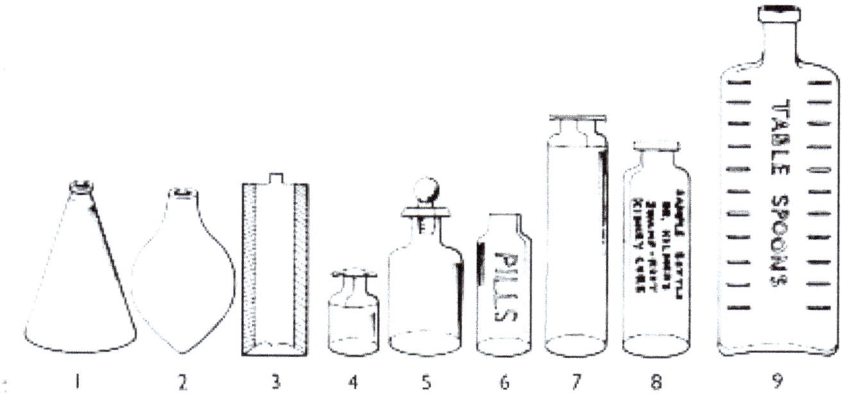

Figure A3.2. Evolution of Medicine bottles. Legend;
1. 16th century earthenware bottle. 2. Early free blown flask.
3. Case bottle of the 18th century. 4. Flared lip.
5. Ground glass stopper. 6. Early 19th century sheared lip.
7. 3 piece mould made after 1820. 8. Early 19th c. embossed medicine bottle.
9. 20th century medicine bottle

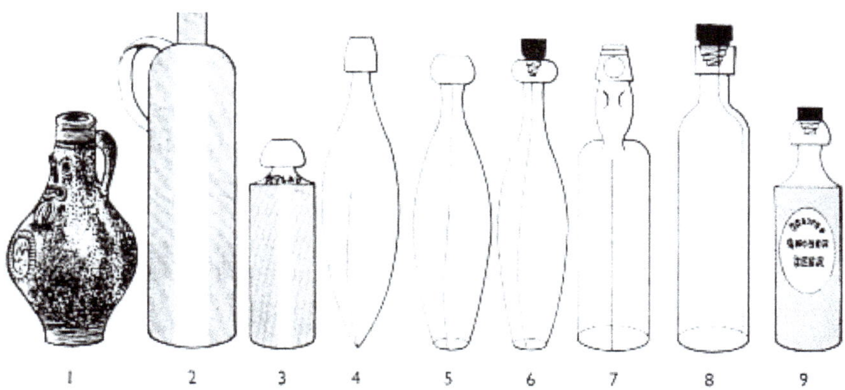

Figure A3.3. Evolution of Mineral Bottles. Legend;
1. 17th century Bellarmine Flemish jug. 2. 18th century English stoneware.
3. 19th century stone ginger beer. 4. Hamilton bottle of 1814
5. Flat bottom Hamilton circa 1870s. 6. No 5 with internal screw after 1872.
7. Codd's glass marble bottle of 1875. 8. 1890s vertical sided internal screw.
9. Early 20th century stone ginger beer

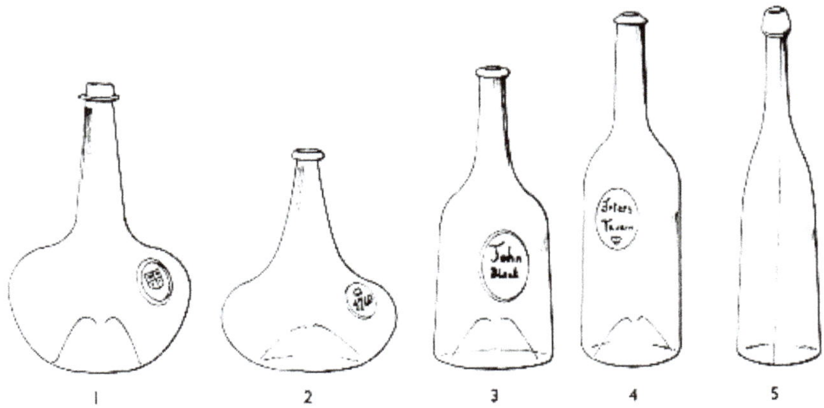

Figure A3.4. Evolution of wine bottles. Legend;
1. - 1650. 2. - 1700. 3. - 1750. 4. - 1800. 5 - 1850

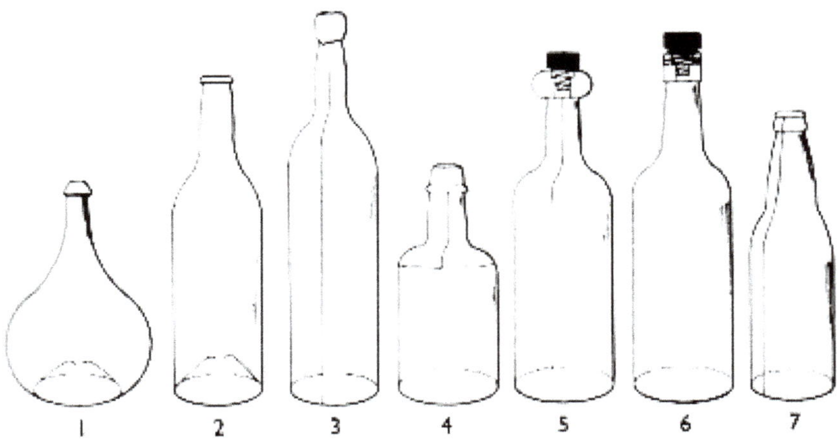

Figure A3.5. Evolution of beer bottles. Legend;
1. 1700 early type beer. 2. 1800 straighter sided beer.
3. 1850 moulded bottle. 4. 1820s 3 piece moulded bottle.
5. 1872 internal screw – 'blob top'. 6. 1910 improved internal screw.
7. modern beer bottle

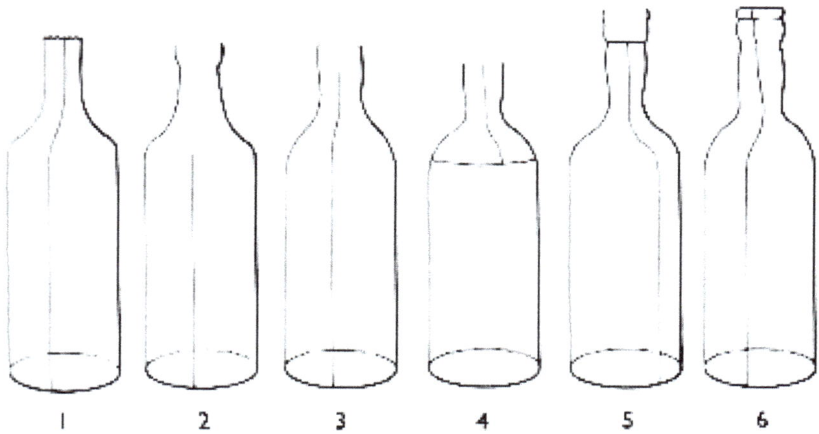

Figure A3.6. Mould seams. Legend;
1 sheared lip. 2. made before 1820. 3. made after 1820.
4. 3 piece mould made after 1820. 5. applied lip bottle made after 1872.
6. modern machine made bottle with seam passing through lip

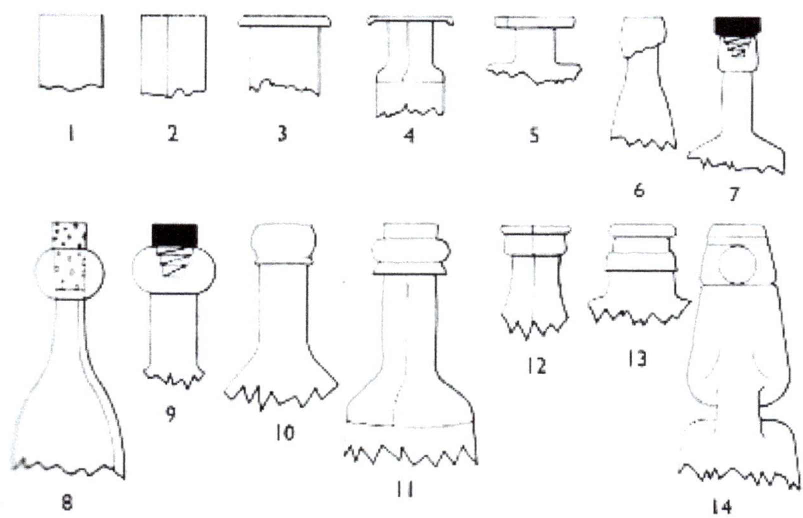

Figure A3.7. Evolution of Lips. Legend;
1. Free blown. 2. Sheared. 3. Rolled. 4. Flared. 5. Laid on ring.
6. Applied. 7. Internal screw. 8. 'Blob top' to take cork stopper. 9. 'Blob top' to take internal screw. 10. Improved applied lip on 2 piece mould. 11. Improved lip on 3 piece mould.
12. 20th century lip. 13. Collar below lip. 14. Codd's bottle

A Bottle Makers and Collectors Dictionary

The following information which is about glass containers not stoneware is to help in their identification; it is more than a glossary it is a mixture of bottle collectors terms, definitions of old bottles, the types that can be found with in cases their rarity. These old bottle collecting definitions and terminology have been obtained from a number of sources, when there is more than one definition all have been included. All are original none have been changed except for grammatical errors. Some are technically incorrect, sufficiently so as to make the Glass Technologist wince.

3PM

3 piece mould. Dated 1820-1880.

A.B.M.

(1) Automatic bottle machine.

(2) Automatic bottle machines started 1903. Bottle made by ABM can be identified by seam that goes all the way to the top of the mouth.

In the US by 1913 most bottles were made by ABMs, here a little later.

(3) Automatic bottle machine, developed by Michael Owens in 1903, and gradually led to the demise of hand made glass bottles.

A.C.L.

Applied colour label, also called painted label or decorated.

Acid cleaned

A glass bottle which has been cleaned with a very strong dangerous acid to remove "sickness" caused by burial for a long period in the ground which has reacted with the original make up of the glass.

Acid etching

This is the process of applying a name etc. to a glass bottle using acid rather than having the bottle embossed during manufacture. This is a secondary process carried out after the bottle has been made and annealed.

Anneal/Annealing

(1) Allowing a bottle to slowly cool room temperature enabling the glass blower to continue working on the piece.
(2) The slow cooling of glass in an oven after re-heating to reduce brittleness.
(3) An essential glass manufacturing process gradually cooling items to room temperature in an oven or lehr.

Applied lip

(1) A type of lip that was applied to the main body of the bottle after it had been mould made.
(2) This denotes partly hand made bottles where the lip of the bottle was "applied" by hand after the body and neck had been formed.
(3) Lip formed by adding a ring of metal around the top of the neck of the bottle. Once added, the glass is shaped to form the bottle lip.

Applied double collar or double tapered lip circa 1815–1885 (note wider time span) made its first appearance on 'black'

1840–1870
Applied round band

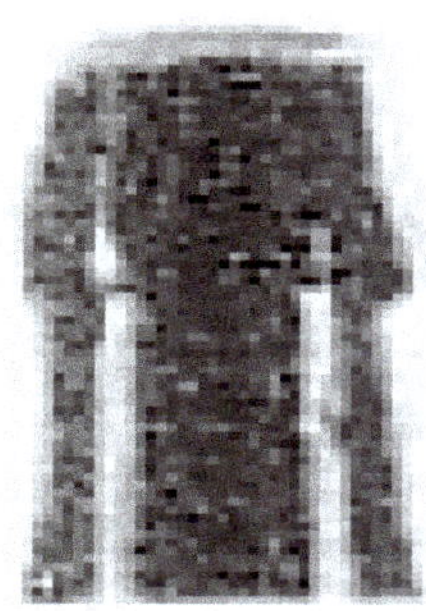

1840-1870
Applied taper

1840-1870
Applied double collar

1870-1910
Tooled with applied ring

wines and beer bottles. The second taper was used to hold wire that was wrapped round the bottle and over the cork to hold it in place.

Applied tapered lip circa 1844-1855 lip used to support string or wire that was used to keep the cork in position.

Applied top

Blobish style; also called Blob top. Dated 1840-1880.

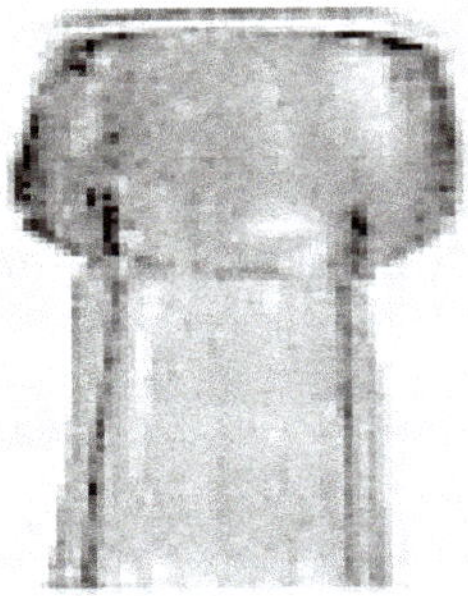

Rounded taper lip - circa 1847-1920. First used on sodas and later beers.

Aqua

Cheaply made Victorian or Edwardian glass tinged with light green / light blue appearance named after its resemblance to water.

Baby feeder

A glass bottle use to administer milk to a child, also known as murder bottles due to high mortality rate caused by diseased ridden bottles which in most cases were near impossible to sterilise after use. Baby feeders are found in a number of shapes. American term - Nursers

Batch

(1) Raw glass making material.
(2) The raw materials from which the glass is made.
(3) Mix of raw materials prepared for heating in the furnace to produce the metal.
(4) Mixture of raw materials ready for melting.

Battle Dore

A wooden paddle used for flattening parts of a hand made bottle.

B.I.M.

Blown in mould, hand made. Can tell if a bottle is BIM or BIMAL by side seams which stop before they reach the lip.

B.I.M.A.C.

Blown in a mould with an applied crown lip.

B.I.M.A.L.

(1) Blown in a mould with an applied lip.
(2) Blown in mould, applied lip. Also called two - piece mould. Dated 1850-1900.

Bird swing

Strand of glass that goes between two sides - a manufacturing defect (not considered damage).
In modern manufacture this is considered to be a "critical defect" especially in beverages or foods. In baby food manufacture if the strand is broken a baby could swallow it.

Bitters

An early patent medicine containing alcohol.

Black glass

Very dark glass, usually green or brown when held up to the light. This is the cheapest type of container glass.

Blob top/BT

(1) A large bulbous lip to take a cork, frequently used on aerated water bottles.

(2) Lip formed by applying a quantity of metal around the neck of the bottle and forming a rounded lip. See applied top above.

Blob lip/blob top

Heavy knob-like finish to pottery or glass, developed for regular use.

Blob top on a mineral bottle

Blob top on a 'coke' bottle

Bloom

An all-over film of opaque dullness occasionally found on the surface of flint glass. English and Irish alike especially Irish - this applies to lead glass but similar faults apply to 'bottle glass'. The effect is caused by the use of high sulphur fuels in the lehr. If it is cleaned off it will return.

Blow over

A seam mark caused by glass expanding over the shoulder of the mould when blown in a dip mould.

Blow pipe/blowing iron

(1) The hollow rod use to blow bottles.
(2) A hollow iron pipe with flared end used for glass blowing.
(3) Hollow iron rod used for blowing glass.

Blown mouldings

(1) One-piece mould were first used in which inexpensive hollow ware was blown and shaped. Pre pressed ware. The glass was blown using a mould of shape smaller than the finished article, then it was blown to the finished size, the base being slightly smaller than the top.
(2) The two -piece open and shut mould invented in 1802 by Charles Chubsee of Stourbridge.
(3) The three-piece mould, with one fixed and two hinged sections, patented by H. Ricketts in 1821.

Bottle glass

See green glass.

Bottles

In the United States, with window glass, the commercial product of eighteenth and early nineteenth centuries; of dark olive-green or olive-amber metal, blown in full sized two piece moulds. In England the use of bottles became widespread by the late seventeenth and early eighteenth centuries. Many were manufactured in Bristol and Nailsea. Shapes and sizes of English and American bottles are listed below.

Calabash [USA]

Ovid body tapering into a cylindrical neck with collored lip; blown in full sized two piece moulds with intaglio designs falling into the pictorial, decorative, and historical categories; c. 1850–70.

Carboy [USA]

Large demijohn; usually set into a wooden tub; used mainly for corrosive liquids, such as aqua fortis.

Carboy [English]

A large vessel made of green bottle glass to carry corrosive acids for industrial purposes, They date from the late eighteenth century and were packed originally in wicker containers and more recently in steel frames stuffed with straw.

Chestnut [USA]

Somewhat carelessly free-blown bottles with long neck and fat chestnut shaped body, frequently full of bubbles and asymmetrical; without uniformity of size, ranging from a few ounces to a gallon or more in capacity; ranging, in natural bottle colours, from ambers through olive ambers to greens; called also Ludlow because of a tradition of having been made in a Ludlow, Massachusetts glasshouse, and junk; made generally, eighteenth -early nineteenth centuries.

Cylinder [English]

Date from the late 1730s shaped in earthenware moulds. By 1750 one-piece gunmetal moulds had been introduced, These bottles measured 5 inches [12.7 cm] in diameter, reduced to about 4 inches [10.2 cm.] in the 1770s. In cylindrical bottles until the 1790s a distinct bulge encircled the base of the body. By the early 1790s bottle glass could be blown to a thinner section in the mould without loss of strength and diameter and was reduced to 3 inches [7.6 cm], with the shoulder high and less pronounced than formerly. From 1820 shoulders were still further accentuated. Mechanically moulded bottles dated from the 1850s.

Demijohn [USA]

Mainly a storage and shipping bottle, often with a wicker jacket; freeblown or moulded for symmetrical, globular, or oval form, with long neck and lip usually collared; quart to twenty gallons sizes. In Britain capacity not less than 3 gallons and not more than 10 gallons.

Nursing [suckling] [USA]

Flattened ovoid flask rounded at the end, short neck with sheared, sometimes flaring, lip; plain and pattern moulded in ribs, flutes and diamonds; eighteenth - early nineteenth centuries.

Onion [England]

A free blown bottle shaped like an onion, c 1690-1720.

Pocket bottle [USA]

Flask of about half pint capacity.

Shaft and Globe [England]

Sealed and dated examples show that they developed in four main chronological groups from about 1620 to 1730:

1620-60, bulbus body, low kick and long neck encircled half an inch below the smooth, flat surfaced lip with a thin, sharp edged string ring.

1650, more pronounced, the sides of the body sloping more steeply inward towards the base, making the kick narrower. The knife edge was retained on the string ring.

1680-1715, the body was progressively widened and became more squat, with a high kick. The neck, shorter and tapering, joined the body in a smooth curve harmonising with the more rounded shape of the body.

1710-40, the sides of the body became perpendicular or with a slight outward slant, curving into the neck with a shoulder more square and pronounced than formerly. The neck was wide at the shoulder junction and the string rim raised to a position immediately below the mouth. These forms were abandoned from the late 1730s in favour of cylindrical bottles.

Swirl [USA]

A Midwestern type, pattern moulded in vertical ribs or flutes, swirled usually to the right, occasionally to the left; wide colour range, from aquamarines to deep greens, olive greens, citron, ambers,and blues; two distinct shapes:

globular, nearly spherical body, slender neck, slightly tapering or straight, collared lip;
short cylindrical body, sides often tapering slightly, sloping shoulder, short neck, collared lip; proportions, distinguished features; some forms also with vertical and broken-swirl ribbing or fluting.

Bottoligist
Collective name given to a collector of bottles.

Bristol blue
Possibly originating in Bristol but made in most glass making centres, may be grouped into basic qualities;
1760-1790, intense dark blue with a faintly purplish hue.
1790-1805, harsh dark blue, less intense than (a).
1804-1820, as (a) but without the purplish tint.
1821-1840s, a costly near royal blue known as King's blue, as (a) but in decorative forms and not flawed.
These may be subdivided into different qualities which varied with the technical factilities available at the individual glass house. Bottle glasses would be of a lesser quality than fine tableware.

Bruise
Same as "fish eye", except some bruises may be more clear than whitish. A faint bruise is more clear, and a more pronounced bruise is whitish (i.e. a fish eye).

Bubble
A bubble is an air bubble that became trapped in the glass during the manufacturing process. Bubbles are not considered damage. In fact bubbles usually add to the appeal of old glass. Glass made after about 1920 does not usually have bubbles. Crude glass almost always has bubbles. An "open bubble" may be considered damage. A lot of small elongated bubbles, closely grouped together, are called seed bubbles.

Bullet stopper

(1) A term applied to bullet - shaped internal stoppers used in mineral water bottles.
(2) A bottle manufactured for the aerated water trade, has an internal stopper usually wood which can traverse the entire length of the bottle freely.
(3) A bullet - shaped plug used in pressurised bottles. Manufactured using either wood, glass, porcelain, or a compound, often with a rubber washer around the centre.

c.c. (crown cap)

Invented in 1892. A metal closure with a cork liner, which clasped round the thin glass or pottery rim, opened with a bottle opener.

C.T.

Crown top.

(cu)

Curved embossing.

Calabash

A gourd shaped flask.

Carboy

A large bottle containing 5-10 gallons of liquid, usually onion shaped.

Case gin

(1) A square shaped bottle designed to pack easily into a case.
(2) Square bottle which tapers from the shoulder down to the base to aid space when packing. Many examples came from The Netherlands (Holland).

Case wear

Scuffs, small scratches, and wearing away of the embossed lettering caused by the process of recycling (re-use) the bottles. A case wear ring may be seen around soda bottles that have been rolled frequently across the rollers of a recycling machine.

Caster

Bottle form with perforated cap, usually metal, also called a 'shaker'.
Frame for condiment bottles or containers such as casters, cruets, and mustard pots.

Caster bottles

Bottles to set in a caster or cruet frame.

Chair

(1) The bench used by the maker.
(2) The team of men making flint bottles.

Chip

A spot on a bottle where a piece or chunk of glass has been lost due to damage.

Coated

A shellac or other type of coating that covers the bottle, usually in an attempt to conceal heavy case wear. Sometime mineral oil is used to coat the bottle, which leaves it with a distinctive oily feel but it does hide the case wear.

C.L.

Crown lip.

Cocks

Faulty products, rejects.

Collar rib

Moulded rib simulating round plain collar; mainly on blown three-mould decanters and toilet bottles.

Collars

U.S. term synonymous with neck rings. Heavy applied thread or ribbon, plain or tooled, laid on around:

(1) lips of bottles and flasks

(2) necks of decanters

a. plain round, single or double,
b. rigaree, single or double,
c. triangular, usually singular, rarely double,
d. triple ring, wide single medial rib,
e. chain, rare,

(3) around stem or shaft, also part of composite stem or shaft; vermicular [wavy], rare in American glass.

Cords

(1) Slight striae discernible to the fingers on the surface of the glass.

(2) Apparent undulating markings within the metal, perfectly vitrified and transparent, show the metal to be of uneven composition because insufficiently molten before working.

N.B. In England 'striae' or 'stria' is a term used in the making of optical glass; defined as a cord of low intensity generally of interest in optical glass manufacture. While cord itself is defined as - an attenuated glassy inclusion possessing optical and other properties differing from those of the surrounding glass, and 'ream' is defined as - an imperfection, non-homogeneous layers in flat glass. In the container industry there is a fault called a 'cat scratch' that is similar to the original definition.

In glass making occasionally the glass is not chemically homogeneous so the streak of cord has a different chemical composition and hence properties to those of the main glass. In the flat glass industry this bends the light passing through the sheet of glass - giving rise to distortions and hence an un-saleable product. While in the container industry once again it distorts the light but more importantly if strong enough it can affect the glass's strength again giving rise to un-saleable glass.

Codd

Term used to describe any aerated water bottle with a glass ball stopper retained in its neck.

Codd bottle

(1) A glass bottle with an internal glass ball stopper trapped in the neck chamber, which was forced into the mouth of the bottle due to gas pressure during filling. The ball pressing against a washer held in a lip recess forming the seal. Invented/patented by Hiram Codd in 1872, has numerous other patents other than the original. Bottles used in Britain over a 50 year period in the aerated water trade. Available in aqua, amber, olive, green, cobalt blue and black.

(2) A mineral water bottle first patented by Hiram Codd of Camberwell in 1870. It used a glass marble stopper held in place against a rubber washer in the neck of the bottle by the pressure of gas to fill it. The stopper was trapped in the neck of the bottle by indents on one side whilst the bottle was emptied. The term is generally used for other patent bottles which used a stopper trapped in the neck.

Codd stopper

A bottle closure using a marble in the neck.
The closure was called a long tapered lip circa 1875-1930.
The Americn patent April 29th 1873.

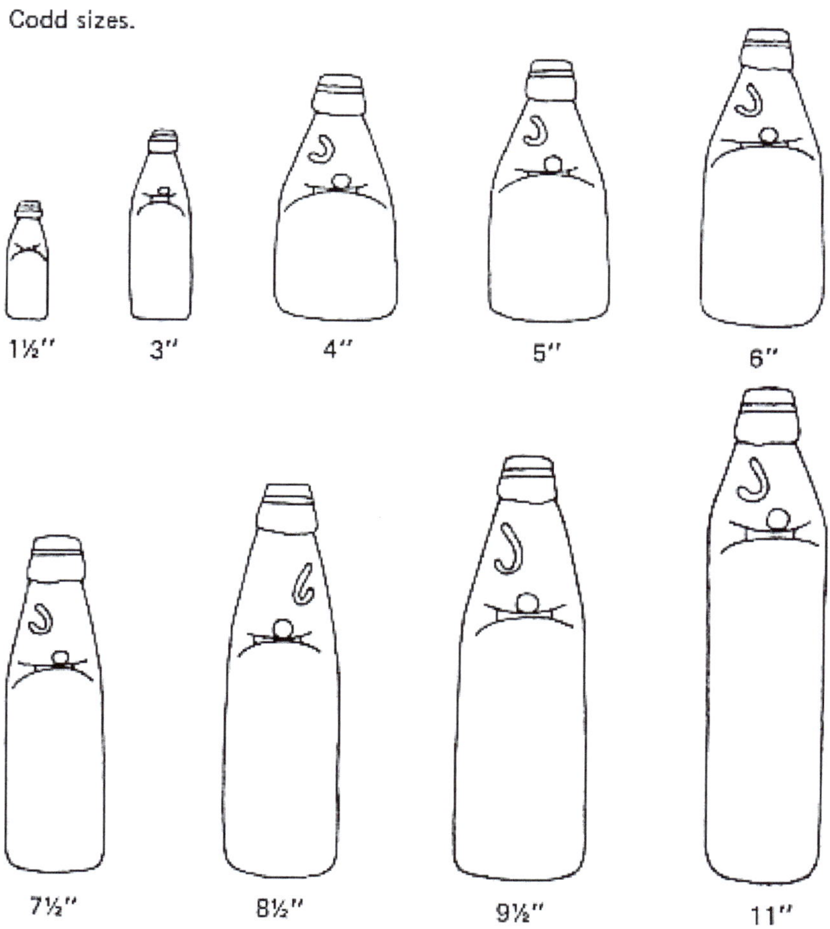

Figure A3.8. Different sizes of Codd bottles

Cork

A type of tree bark that is made into stoppers.

Crack

Usually an uneven line inside the glass that is the result of damage.

Crazing

Vertical lines in the neck of the bottle that almost look like faint cracks but are the result of twisting and cooling process during manufacture.

Crown cap

A metal cap with a corrugated edge which fits on a crown lip, it was patented by William Painter in America in 1892. A cork with a serrated metal affixed to the top. In common use in the UK post-1915-1920. (Stoneware Ginger Beer bottles with this form of closure were prone to damage on opening).

Crown lip

Term use to describe a bottle designed to take a crown cap, an indicator of a bottles age.

Cruet

Lipped bottle with or without handle.
Caster bottle

Cucumber

1. A round bottom cylinder with ridges running down the length of the bottle forming panels.
2. A round base cylindrical bottle with flattened side panels, from the 1880s period used in the aerated water trade, commonly found in aqua although green, amber and cobalt blue examples are recorded in limited numbers.

Cullet

1. Broken glass mixed with raw materials to make a batch of glass
2. Broken glass added to the batch to assist melting.
3. Broken glass or glass waste recycled and used in a batch.
4. Cleaned, broken glass used in all new mixtures to promote fusion and improve quality of the metal.

Cup bottomed mould

A mould with a cup indentation on the bottom.

Cure

A patent or proprietary medicine sold as a remedy. "Quack" cures were sold without prescription and usually did not fulfill any of their makers claims.

Decolourised glass

Glass made with a manganese, seleniun or arsenic additive.

Demijohn

A large bladder - shaped bottle designed to hold from 1 to 10 gallons.

Ding

Small impact mark, larger than a ping but less than $\frac{1}{4}$ inch (5 to 6 mm).

Note that the glass is crunched at the impact mark. A ding larger than a $\frac{1}{4}$ inch is called a "bruise" or "fish eye" if the glass remains intact, or it is called a "chip" if the glass is removed from the impact spot.

Dip mould

Early type of one piece mould with the top slightly larger than the bottom so that the bottle could be removed.

Dumpy

(1) A bottle which is short with an over wide body for its height, i.e. Dumpy Codd, Dumpy Beer, etc.

(2) This term usually refers to the shape of certain types of bottles, i.e. Seltzers, mineral waters and ginger beers, usually shorter and wider than the more common bottles of the same type.

Dug

A bottle is denoted as "dug" to indicate that it will have a film or water stain on the glass. Such stains can be removed professionally.

Embossed/Embossing

(1) A glass bottle having raised lettering.
(2) Any raised letter or design on the bottle.
(3) Raised lettering.

Empontilling

Attaching a metal rod to the bottom of the bottle so that it may be held during finishing.

Engraving

Decorating glass by cutting with diamond or copper tools.

Etching

Decorating glass by using hydrofloric acid.

F.B.

Free blown means that glass was blown by a glassmaker, using a blowpipe; it was not poured into a mould.

F.B.H.

Flat bottom Hamilton. An aerated water bottle ovoid in shape with a very narrow base allowing it to stand upright. A development of the Hamilton bottle which had a pointed end (base).

Finish

The process of forming the neck of the bottle.

Fire grenade

An early type of fire extinguisher which was thrown at the fire, contained dangerous chemicals by today's standards.

Fire polishing

A method of processing blown moulded ware to obliterate tool marks and produce a smooth even finish.

Fish eye

Whitish circular mark on the surface of the glass caused by impact.

Flake

Very small and shallow chip, usually less than ¼ inch (5 to 6 mm.) and less than 1/16 inch deep (1 to 2 mm.).

Flared lip

(1) A bottle lip pre-1850 which was "splayed out".

(2) Lip formed by stretching out the neck of a bottle.

Flash

A very faint "crack" - bottle sometimes has to be turned a certain way to see it.

Flasks

American types

(1.) Chestnut. Free blown and patterned moulded, rarely blown three mould, of the timeless and universal shape resembling a slender chestnut; without uniformity or size, ranging from a few ounces to over a quart capacity, majority of pocket sized bottles mainly from the late 18th and early 19th C.

(2.) Decorative. Blown in two piece moulds bearing intaglio devices such as the sunburst, urn of fruit and cornucopia; mainly half pint and pint sizes, c 1815-1840.

(3.) Grandfather. Midwestern chestnut flask of a quart or more capacity; pattern moulded in ribbed designs.

(4.) Historical. Moulded like (2); designs of historical importance; commemorative of events and public figures, emblems, slogans, candidate in presidential campaigns; emblems and designs related to economic life; also national and other symbolic emblems; c. 1815–late 19th century, majority before 1870.

(5.) Masonic. Moulded like (2); designs with (a) Masonic emblems each side, (b) Masonic emblems one side; reverse, different design. American Eagle most common; c 1815-30.
(6.) Pictorial. Moulded like (2); designs depicting people and /or flora and fauna; c. 1815-70.
(7.) Pitkin. Generic term for pocket bottles or flasks, blown by the German half-post method and pattern moulded in ribbed designs - vertical, swirled, and broken swirl; eighteen early nineteenth century; first identified by tradition with the Pitkin's glassworks, East Hartford, Connecticut, c. 1790-1830.

Flint glass

(1) A high quality glass, normally clear.
(2) In the context of the bottle industry, any colourless glass.
(3) Now termed lead crystal, developed by George Ravenscroft [1618–1681], in England, who was granted a seven year patent [No. 176] in May 1674 to make glass in which the silica was derived from calcined flints. In 1675 he first used lead oxide as a flux in place of vegetable potash. This produced a glass which was denser, heavier, softer, and with greater refractive brilliance than had been produced before. After improvements to the process in the 1680s the world glass trade became an English monopoly for a century and a half.

Free blown

(1) An early bottle blown "by hand" without the use of a mould.
(2) A bottle which was made without the use of any mould.
(3) A vessel blowing without the use of moulds.
Wine bottles that fall in this category are Onion, Mallet and Cylinders, 16th, 17th, and 18th century all are irregular in shape, size and capacity. Bottles were very expensive in the early days of glass bottle manufacture hence the more desirable examples for todays collectors have applied body or shoulder seals. A small blob of glass with the original owners mark, initials, monogrammed, crest full name or date impressed on to it. All free blown bottles have a rough pontil scar in the

bottom were the pontil rod was affixed to aid the blower when forming the lip.

Wine bottles - changes in shape

The shape of wine bottles have gradually changed with time, from the original shaft and globe (basically a circular bulb on the end of a neck) to the modern straight sided bottle.

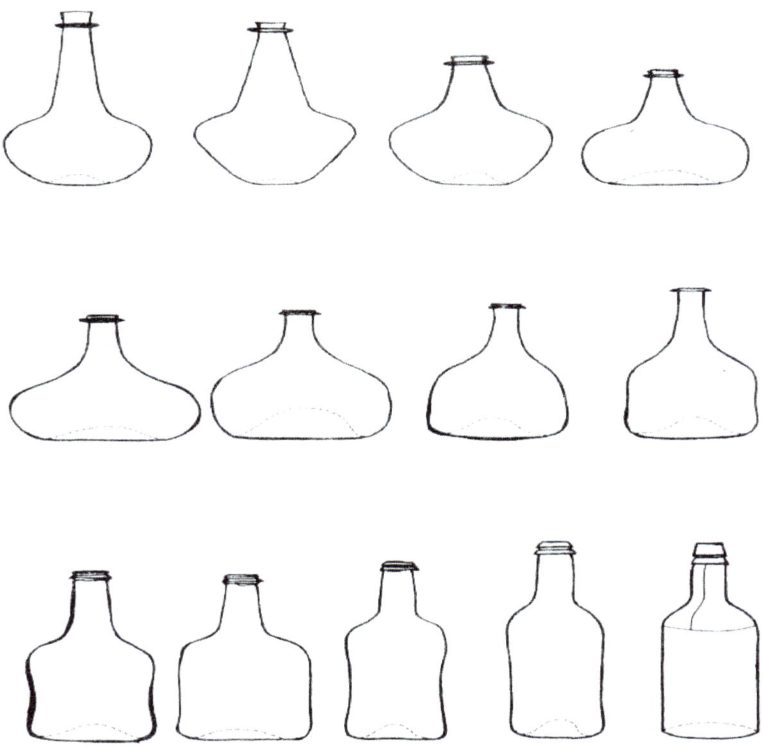

Figure A3.9. Changes in shape to wine bottles with time

Top Row, l to r.	Middle Row, l to r.	Bottom Row, l to r
c. 1630-1660 Shaft & globe	c. 1700-1718 Onion	c. 1750-1760
c. 1660-1670	c. 1715-1720 Onion	c. 1770-1780
c. 1670-1680 Onion	c. 1720-1730	c. 1780-1800
c. 1680-1700 Onion	c. 1730-1750	c. 1800-1820
		c. 1820-

Frit

A flux of low temperature melting silicates used in the manufacture of glass before the invention of the high temperature furnace in 1734.

Full sized mould

Mould composed of two or more hinged pieces (leaves), having inner surface with or without decorative design, and the size and form of the article; glass thus moulded characterised by concavo-convex surfaces.

G.B.

Ginger beer one of the most popular forms of non-alcoholic beverages available in the United Kingdom during the 19th and early 20th centuries. Originally packaged in stoneware bottles, and later in glass. Very popular and highly collected.

Gather

Uninflated and uniform blob of metal taken from the pot or furnace working chamber on the end of a blowpipe.

Glory hole

A small furnace for re-heating the neck of the bottle.

Gob or gather

(1) The amount of metal gathered to make one bottle.
(2) Molten glass taken from the furnace on a blowing iron.

Gouge

An area of glass that is missing from the bottle, typically when a "fish eye" is removed.

Graphite pontil scar

The open rough area of glass to be found on the base of free blown bottles caused by the glass blower holding the bottle with an iron rod while forming the lip. For collectors the larger and rougher the pontil scar the better

Green glass

Glass in its natural colour, neither rendered colourless nor artificially coloured; generally made from coarser and less pure materials than those used for fine wares; soda, potash or lime the principal alkaline base; many bottles and window glass made from this glass.

Ground wear

A bottle that has been cleaned may still retain "pitting" and other marks that are the result of the bottle being buried underground.

H.S.

Hutchinson stopper. Circa 1879 - 1915. American patent granted April 4th 1879. No. 2130992

Hand Finished.

Bottle and lip blown in a mould, then hand finished. Dated 1890-1911.

Hairline crack

A very thin and short crack.

Hamilton

(1) A term used to describe a point ended aerated water bottle which will not stand, sometimes referred to as ovate.
(2) A round point ended bottle for the aerated water trade, attributed to Paul Hamilton of Belfast in the late 17th century. The contents were always in contact with the cork, preventing

it from drying out hence a perfect seal was guaranteed at all times.

Common in aqua, scarce in amber, green, brown, cobalt blue and black. Common amber Hamiltons are Leigh & Co. and Groves Whitnal Salford. As the bottle developed into the early 20th century the pointed end was gradually replaced by a small flat base. Known as Flat Bottomed Hamiltons these were in common use well into the 1930s.

(3) According to McKearin's "American Bottles & Flasks" The torpedo or "egg" bottle was patented in 1809 by William Hamilton of Dublin.

The bottles curved bottom shaped had several advantages;

(a) forcing the bottle to lie on its side kept the liquid in constant contact with the cork and thereby prevented the leakage of air.

(b) shape allowed it to be much stronger than conventional flat bottom bottles of equal weights so it was less likely to crack/break under the pressure of the carbonation.

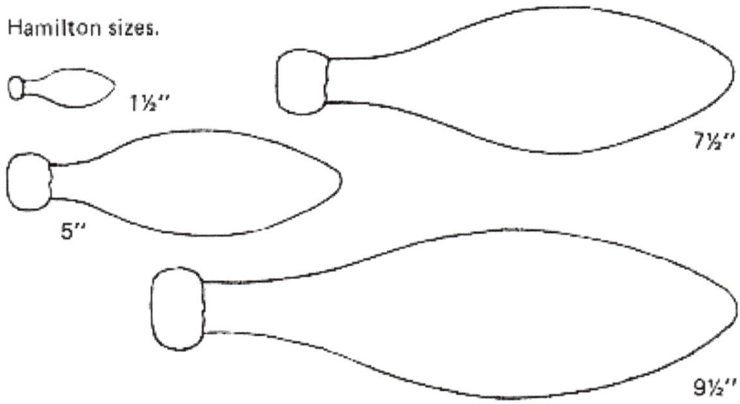

Figure A3.10. Different sizes of Hamilton's

(c) easier to pack in to crates.

(d) hard to set down an opened bottle until you had finished the contents, so customers drank faster.

Handbown

Freeblown; used in contra-distinction to machine blown.

Hock

A tall cylindrical bottle from Germany which contained Hock Wine, usually plain (no embossing or decoration), found in red or turquoise. Hock is not a recognised German brand name it is one that the British gave to the wine growing district around Hockenheim.

Hole

(1) The aperture in the furnace wall where gathering took place.
(2) The team of men who made the bottles at the hole.

Hybrid

(1) The cross between a Codd and a Hamilton bottle, Codd neck and shoulder and a Hamilton pointed base.
(2) Aerated water bottle combining a Codd closure (neck, ball and lip) with a Hamilton pointed end. Usually found in aqua, scarce are coloured green and amber bottles along with some coloured lips.

Internal screw lip/top

Bottle with an internal threaded lip, common on beer, mineral water and some ginger beer bottles, etc.

Irridescent/Irridescence

(1) Decay occuring where a buried piece of glass reacts with the soil conditions and flakes form and come away.
(2) A type of stain that has an opaline or rainbow colour. Sometimes this stain is very hard to clean.

Journey

A shifts work. One days work.

Kick up

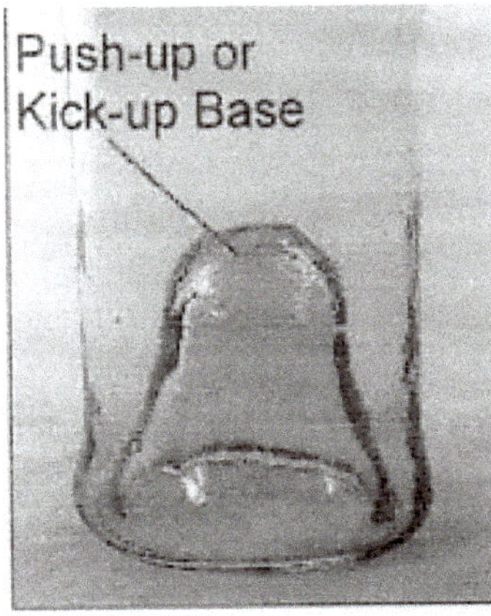

(1) An inverted sage base found on most free blown bottles caused during manufacture when the pontil rod was applied, also the area to locate the pontil scar.

(2) Indentation at the base of a hand blown bottle created by pushing the base upwards with the pontil. The kick-up prevents the pontil mark making the bottle unstable and scratching a table surface.

(3) The conical indentation found in the base of a of an early bottle or decanter. This was essential for proper annealing when glass workers had only limited means of toughening their glass.

L.S.

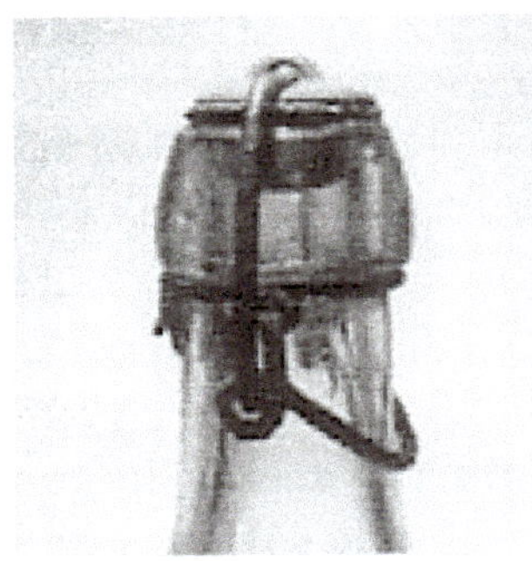

Lightning stopper. Invented by Charles De Quillfeldt. American patent January 5th, 1875 number 158 406

Laid on ring

Lip formed by applying a washer-like disc of metal to the top of the bottle neck.

Lead glass

Glass containing lead oxide as a flux; also called flint glass.

Lear, leer or lehr

An oven for annealing glassware.

Lime glass

Glass containing lime; first produced by William Leighton, Birmingham, in 1864; the metal is as clear as lead glass and cheaper, but not as resonant or heavy.

Lug

Depressed indent in the neck of a Codd bottle to catch and retain the glass marble stopper when the contents are being poured.

Machine made

A bottle which is entirely mould made, with seams running through the lip and the side of the item flush, product of an automatic bottle machine enabling vast quantities of near perfect bottles to be produced on a hourly basis, a very good indicator of the bottles age.

M.B.

Mould blown means that it was blown into a mould as part of the forming process.

Maker

The head of the bottle making team, himself responsible for shaping the neck.

Mallet

Free blown wine bottle shape, c1725-60, shaped like a woodworker's mallet.

Marver

1. A smooth stone or iron slab against which the gob was shaped.
2. Slab of stone or iron used for rolling the gather on.
3. Polished marble slab supported by a frame on which a gather of metal is rolled.

Metal

1. Molten glass.
2. Molten glass ready for working.
3. Glass either in the molten or hardened state.

Moils

Fragments of glass left on blowing pipes after wetting of. They were removed and used as cullet.

Moulded glass

Glass ornamented and/or given partial or final body shape by use of a mould; applicable to pressed glass, but reserved generally for blown-moulded.

Nick

Very tiny chip, less than 1/16 inch (1 to 2 mm).

Nursing bottles

See Bottles.

Number

The minimum quantity of bottles to be made in a journey.

Onion

Free blown glass bottle shape, c. 1690–1720, like an onion in shape.

Opalescence

A thin film that develops on the glass surface due to the conditions of burial, some collectors prefer bottles in this condition.

Open bubble

A bubble that has become open to the outside air, either inside or outside of the bottle. Outside open bubbles are sometimes considered damage since they may result from impact to the bubble area, but open bubbles are also seen as the result of the crude manufacturing process.

Overwork

Bottles made in excess of the number.

Parison

1. The midway stage in the formation of a bottle.
2. Inflated, unformed gather of metal.

Ping

Small impact mark, typically smaller than 1/16 inch (1 to 2 mm). Smaller than a "ding".

Pitting

Small little pin-size holes in the outside of the glass that are the result of glass being underground.

Poison

Covers a huge subject grouping. The commonest being rectangular, octagonal, triangular flat back, and the Lysol format. They carry embossed Warning Ribs, Stars or Cross Hatching along with Not to be taken / Poison. Found in cobalt blue, green, amber and brown, aqua and clear glass. Hundreds of patent designs recorded.

Pontil knock-off

Waste glass removed from the end of a pontil after the removal of the worked glass.

Pontil mark

1. A rough mark on the base of a piece of worked glass where it was fixed to a pontil.
2. Scar on the bottom of the bottle, usually a rough circular glass mark. It was made by the punty rod that held the glass for the glassblower. If the scar was rough it is called

an open pontil; if it is smoothed out, it is called a ground pontil or refired pontil depending upon how the smoothing was done. Date pre 1850s.

3. A scar left on blown glass when the punty is broken off. Generally found on the base of a glass. Ground and polished to a smooth depression, usually from about 1750, and invariably so on fine glass from about 1780. Seemingly less frequently ground of and polished on early US ordinary wares than on the British.

Popped-out fish eye

A bruise (or "fish eye") that has popped out, leaving a curved chip in side the bottle.

Pot

A large refractory clay vessel in which the glass was made in a furnace.

Potstone

A piece of sand/rock inside the glass that is the result of the original manufacturing process.

Punty or pontill

(1) A rod on which the bottle was held for its finish.

(2) Solid iron rod used to hold the glass after being transferred from the blowing iron.

(3) Glass rod pressed into the bottom of a hand made bottle to support it while the lip area was finished. When snapped off the pontil left a rough and uneven finish, sometimes jagged and sharp.

(4) A long iron rod attached to one end of blown glass during the finishing process after removal from the blowpipe.

Repair

Epoxy is sometimes used to repair a bottle by filling in the missing piece with epoxy resin that closely matches the original

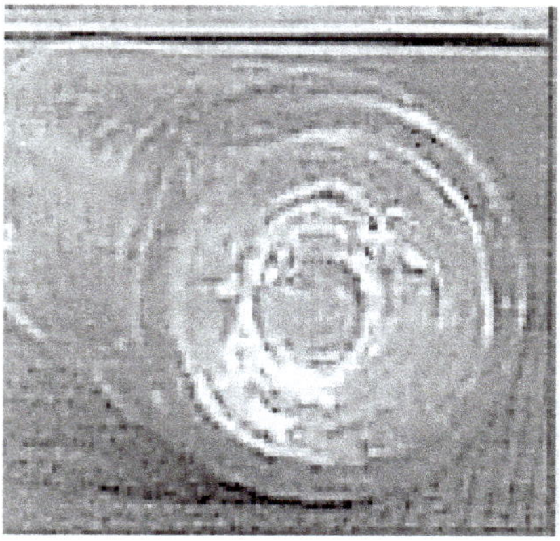

After the 1850s, a new type of punty rod was developed that left a mark known as an "iron pontil". This mark is round and smooth and left an iron residue which turned red if it rusted.

Figure A3.11. Open pontil 1825-1845

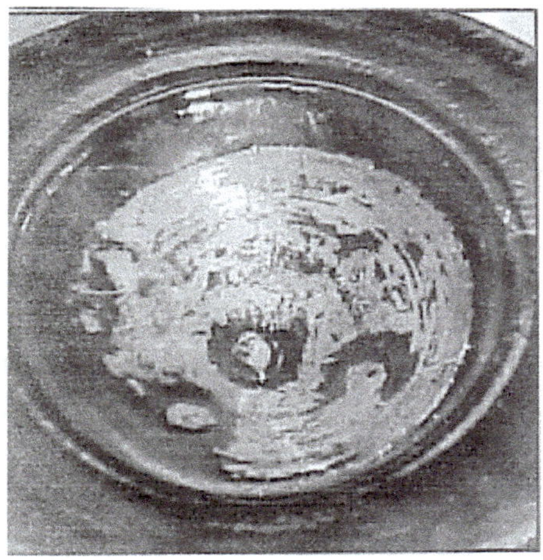

Figures A3,12 and A3.13 Iron pontil with graphite still intact, and iron pontil with non-rusted graphite residue

colour of the glass. These repairs are sometimes done so well that they cannot be detected by collectors. To be sure, expose the bottle in the dark to a strong black light – any epoxy repairs will glow a bright white while the rest of the bottle looks dark. Portable hand held black lights are not strong enough.

Rolled lip

Lip formed by shaping and rolling the neck of the bottle back on itself.

Rolling pins

Were originally made as salt containers during the Napoleonic wars when the salt tax was thirty times greater than the salt itself, which was sold by the bottle. The first of these rolling pins were produced in about 1800, were of thick bottle glass, the open end being tightly stoppered. Each end was provided with a solid knob so that they could be safely hung in a dry place. By 1820 salt bottles were being made of coloured glass purple, blue, mottled, and striped. The cold glass tube weighted with salt proved an excellent rolling pin for pastry. Hanging conspicuously on the kitchen wall, the rolling pin became a field for homely decoration, such as text painted on a background of opaque white glass, and during the 1820s it tended to lose its purely utilitarian purpose and became to be regarded as a lucky mascot, its filling chosen to the occasion. Gilded, painted, and enamelled, they were inscribed with mottos and good wishes. Biblical quotations, the name of the recipient, and a date; others were decorated with seafaring subjects. Some were ornamented with the sharpened end of a especially hardened steel tool, the design being portrayed by means of small, closely spaced dots. Decorations included hounds chasing a hare, ploughing, and paddle steamers. They were made at the glass making centres of Bristol, Birmingham, Sunderland and Stourbridge; later 1840s onwards the bottle makers made them in ordinary green bottle glass, as common kitchenware. Reproductions have been made in tens of thousands since 1910.

Scent bottles

In England scent bottles of coloured and of opaque white glass are not uncommon. Some of the latter can be attributed to Bristol manufactories because of their creamy colour and their complete opacity, and because the painting on them is related to that on more important specimens of white glass. Of coloured glass scent bottles, it is not possible to say more than they were probably made in Bristol. Many of the surviving examples resemble closely in their decoration the work to be seen on watch cases and other articles known to have been made in London. As it is known that glass 'smelling bottles' were being made there as early as 1752, perhaps a London origin is common to them all. Scent bottles continued to be made in the prevailing type and shapes until the 1830s when for the next 50 years or more they were made in huge numbers and in every conceivable shape and colour.

Scrape

A mark on the outside of the glass, sometimes caused by a shovel scraping the bottle.

Scratch

Small lines on the outside of the bottle caused by the glass being scratched by a sharp object.

Screw stopper

A composite stopper that was screwed into the lip of a bottle., which had screw threads on the inside.

Scratch

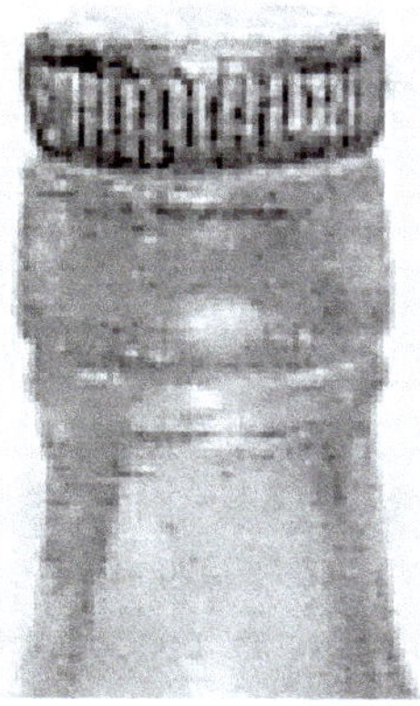

Screw stopper

Sealed glasses

Early English flint glass tableware to which small glass discs impressed with the makers mark were impressed: raven's head, George Ravenscroft, September 1675-1681; the King's arms, Henry Holden, glass maker to the king from 1683; Lion and Coronet, Duke of Buckingham.

Seals

The circular glass seals were almost certainly impressed with an original intaglio made of brass. This was either made by a professional engraver working on his own behalf or, in the case of a glassworks specialising in sealed bottles, by one of the staff.

Glass seals fall into three categories, whether they bear dates or not, and may be conveniently classified as: armorial, name or initial, and 'others', the last including merchants marks, Masonic signs, and the names of houses.

Seeds

Minute air bubbles in the metal, indicating that the glasshouse could not raise the furnace temperature high enough or wait long enough to eliminate all of the air bubbles trapped among the raw materials and all of the gases given off by the fusing together of the raw materials.

Sheared lip

(1) The top of the bottle was cut the blowpipe with shears.

(2) Jagged lip finish created in manufacture to enable corks to be pressed into the top to seal, before being dipped into a wax covering.

Shear lip ink

Small glass ink bottle, were cheap to produce, found in a variety of shapes, square, round, boat, cotton reel, tent etc., and in a number of colours the commonest being aqua.

Shearing scar

Shearing scars originate when the knife severs the glass on a suction machine. In America they are often referred to as an 'Owens ring' and all are after 1903.

Figure A3.14. Base of an Owens bottle showing the uneven scar from where the knife has severed the glass during manufacture

Sick/Sickness

(1) The specter of bottle collectors everywhere. Most bottles were manufactured from very poor quality base materials often soda, lime and silica, and after years of burial often in very acidic and/or damp conditions the bottle surface becomes cloudy and in extreme cases it can be a thick layer with flaky scaling. This cloudiness detracts from the original beauty of the object such as making a once magnificent cobalt blue glass bottle take on a scruffy white appearance. Sickness can effect all types of bottles and every colour of glass.

(2) Bottles recovered from a dump where ground acids have altered the surface to an unpleasant, white, milky finish. The normal bottle cleaning process will not remove this sickness, it can be removed by acid cleaning but this can be a dangerous process so should only be carried out by an expert. The acids used for this type of cleaning are very powerful and work by dissolving (eating away) the surface of the glass. If spilt they can eat away the human body even quicker.

Slug plate

Metal plate inserted into a bottle mould to provide product specific information to an otherwise generic mould. Also refers to the mark left on a bottle by such a plate.

Smooth lip

Smooth rounded finish made by hand.

Spirit flask

A glass bottle usually in aqua coloured glass for the sale in small quantities of whisky, gin, brandy and rum etc. Found in a number of shapes pumkin seed, coffin, and square being the commonest. Bottles will carry retailers name, address and town on them either embossed or acid etched. Sometimes referred to as hip flasks.

Soda syphon

Large heavy glass bottle for dispensing soda water, retailers name usually acid etched on to the body of the bottle. Bottles were often found in public bars. Recorded in clear, amber, green, blue and pink coloured glass.

S.S. Swing stopper

In 1875 Charles de Quillfeldt patented a metal lever device, with a pottery lid which swung open and closed easily. Used by Corona into the 1960s, and still found on some trendy European lagers and beers.

Stain

A whitish film on the bottle that typically results from mineral deposits while the bottle was underground. This stain can be cleaned.

Bottles with labels may also have original content stain which is the remains of the original product (e.g. medicine, ink, etc.) that has long since evaporated.

Stones

1. Red and black specks within the fabric of early flint glass, the result of imperfect fusion between the oxide of lead and silica.
2. An inclusion that is infused in the body of the glass, they are normally of refractory origin rarely they can be metal.

String course or string rim

1. Glass ring around the neck of a bottle used for tying down corks.
2. The raised band near the top of the neck of a bottle, which provided a grip under which the string for securing the cork or other cover may be fastened.

In the seventeenth century bottles this took the form of a single band of glass about a quarter of an inch below the orifice, and it remains a feature of subsequent bottles.

Taker in

A youth employed to carry finished bottles to the lehr.

Tale glass

A second-quality metal taken from the top of the pot, and sold more cheaply than the lower, finer metal.

Tear

Air bubble found in glass, especially in earlier glass.

Teazer

The furnace man, stoker.

Tint

A residual colour tinge inherent on the ingredients from which the metal is composed.

Torpedo

Bottle used in the aerated trade, long cylindrical bottle with a rounded end (not pointed as a Hamilton is). Common in aqua glass, scarce in all other colours.

Three piece mould

A bottle mould in three parts patented by Henry Ricketts Glassworks of Bristol. England in 1829, the first time that bottles could guarantee a correct capacity was held, a very important device in the evolution of bottle manufacturing techniques.

Trade mark

A personalised design, monogram, picture, etc. which appeared as company logo's after 1862.

Virol pot/jar

Trade name for the Virol Company, a manufacturer of bone marrow extract, very popular towards the end of the 19th century through into the 1940s. Originally the pots followed a similar format to cream pots, stoneware, cream glaze with a black under glaze print. Available in several sizes. In the late 1920s the company disregarded the stone pot for brown glass jars.

W.C.F.

Wire cork fastener.

Wetter off

A youth employed to free the bottle from the blowing iron.

Whiskey flask

A glass spirit flask usually in aqua glass with retailers details embossed upon them, recorded in a number of shapes pumpkinseed, coffin and square being the commonest.

Whittled

Curved line (or stretch) marks that run vertically along the neck of the bottle, caused by hand finishing of the glass when lip is twisted on to the neck.

Wiped top

The mould lines end before the top since the neck was wiped smooth after the top was tooled on. Dated pre-1915.

Zara

A liquor bottle with an applied seal to the shoulder its name from Zara, Dalmatia. Seal depicts the double headed eagle, found in square and cylindrical format and often in turquoise coloured glass.

Rating the condition of antique bottles

The condition of a bottle has a dramatic impact upon its value, once a common bottle gets a crack or a chip it usually loses half or more of its value. Damage can be tolerated on the more rarer bottles but it still has an impact on the price.

The following are some terms to identify the condition of a bottle, they are from the internet and are the combined opinions of many experience collectors.

- **Mint** - Bottle is in brand new condition with absolutely no damage. Also called "attic mint" because most of the bottles in this condition must have been stored in an attic to have survived in such perfect condition.
- **Near mint** - Bottle would be mint except for some very minor damage like a small stain, tiny nick, light scratch, small open bubble, or tiny patches of case wear.
- **Excellent** - Between "Near Mint" and "Very Good". Bottle may have some light stain and/or light case wear and /or light scratches but no chips or cracks.
- **Very good** - Some scratches and medium to heavy case wear, some inside and/or outside staining. May have some minor nicks and no more than one or two small chips.
- **Good** - Heavy scratches, substantial case wear, heavy staining, some chips.
- **Poor** - Cracked and/or chipped.
- **Damaged** - Large cracks and/or large pieces chipped away.

Fakes and forgeries

Some dealers are unscrupulous so buyers must be aware that all that is out there is not always what is seams. Some of the more obvious frauds are;

Black glass sealed wines

Be aware of old seals (from broken bottles) re-glued to plain unsealed shapes.

Codds

Still made in the Far East in a variety of forms, some even with plastic lips. The heavier examples emanate from India, still sold by street vendors, and come with blue marbles in the neck or coloured body swirls. There are even black glass examples. Bases on these tend to be thick and uneven.

Coloured lacquer

By far the most commonly encountered alteration to low cost aqua items is to make the bottles like higher-valued rarities by applying an outer lacquered coloured coating, which makes the surface less cold and when held up to natural light reveals inconsistent colour density. A gentle scrape, underneath, with a metal object (coin) should scratch the fake surface. Inks, syphons, and Codds, are the most popular choices.

Eye baths

Modern coloured examples in glass (some with swirls) and plastic are easily avoided with a little knowledge.

Hamiltons

Pink, and other colours, have come from the Far East, with incorrect blob lips. Schweppes embossed aqua hamiltons tend to feature plenty of air bubbles, and are slightly longer than the originals.

Night lights

Coloured figurals are recorded, but not easily detected. Non-figurals, made for modern day use, tend to be much heavier and minus the rough / sheared lip.

Warner's safe cure

Three different copies are known:

- Green pint, with label. Base embossed with gift company who commissioned them.
- Frank's Kidney & Liver Cure is simply a fatter-looking bottle, intended to deceive nobody.
- A range of oddly coloured Warners from the Far East all featuring the wrong top; the inside echoes the outside curve shape, instead of the heavy chunky blob lip.

Appendix 4

Selected Abstracts from The Journal of the Society of Glass Technology

1919. Volume III.

#124. Automatic Glass Bottle Machines. (Brit. Pat No. 111 868). A. Wilzin (Oct.11th , 1916, No. 14 445).

#125. Automatic Glass Gathering. (Brit. Pat. No. 111 701. A. Wilzin (Dec. 4th, 1916, No. 17 382).

#126. Glass Making Machinery. (Brit. Pat. No. 112 007). E. Roirant (Sept. 14th, 1917, No. 13 190). Convention date Dec. 4th, 1916. Not yet accepted. Abridged as open to inspection under Sect. 91 of the Act.

#127. Automatic Gathering of Glass by Suction. (Brit. Pat. No. 111 890) C. .F. Cox (Dec 11th, 1916 No. 17 791)

#128. Semi-automatic Bottle-making Machine. (Brit. Pat. No. 111 231). A. E. Clegg (April 10th, 1917, No. 4 983).

#201. Glass-shaping Machines. Hartford-Fairmont Co., USA, (assignees of K. E. Peiler and E. H. Lorenz) (Brit. Pat. No. 120 907, November 11th, 1918, No. 18 460. Convention date, November 20th, 1917. Not yet accepted. Abridged as open to inspection under Sect. 91 of the Act.

#202. Semi-automatic Bottle Making Machine. E. H. Langwell and E. Durand (Brit. Pat. No. 114 193, March 21st, 1917, No. 4 114).

#203. Bottle Making Machine. T. B. Kitson and G. O. Togue (Brit. Pat. No. 114 477, April 5th, 1917, No. 4 927).

#204. Machinery for Glass Bottle Manufacture. SHAW-BRUNER, Ltd., G. C. Pyle and G. C. Dager (Brit. Pat. No. 114 709, June 1st, 1917, No. 7 879).

#205. Glass Bottle Machines. W. Forster (Brit. Pat. No. 114 678, April 21st, 1917, No. 5 604).

#211. Feeding Device in Glass Manufacture. R. E. McCauley, USA (Brit. Pat. No. 1140583, October 20th, 1917, No. 15 247).

#212. Automatic Flowing Devive for Feeding Molten Glass. G. E. Howard, USA (Brit. Pat. No. 120 744, August 10th, 1917, No. 11 513).

#216. Feeding Device for Bottle Machines. J. Forster (Brit. Pat. No. 113 797, December 14th, 1916, No. 17 970).

#281. Making Glass Bottles, etc. Sykes & Sugden and T. Jackson (Brit. Pat. No. 124 118, June 8th, 1918 No. 9 462).

#282. Glass-making Machinery. A. E. Clegg (Brit. Pat. No. 124 356, July 12th, 1918, No. 11 420).

#350. Method and Machine for Gathering Glass. K. E. Peiler, Hartford Fairmont Co. (USA Pat., December 9th, 1919, No. 1 324 464. Filed December 4th, 1916, No. 134 827).

#351. Glass Working Apparatus and Process. F. L. O. Wadsworth, Ball Bros. Glass Manufacturing Co. (USA Pat., July 29th, 1919, No. 1 311 474. Filed April 24th, 1914, No. 834 230).

#354. Process and Apparatus for feeding Molten Glass. B. D. Chamberlin, Empire Machine Co. (USA Pat., December 2nd, 1919 No. 1 323 450. Filed May 1st, 1916).

#355. Apparatus for Feeding Molten Glass. W. A. Lorenz, Hartford Fairmont Co. (USA Pat., April 8th, 1919, No. 1 300 180. Filed August 7th, 1917 No. 184 874).

1920 Volume IV.

#54. Bottle Blowing Machinery. S. D. Olsen (Brit. Pat., June 17th, 1918, No. 9 981).

#55. An Automatic, Self-feeding, Bottle-making Machine. T. Stenhouse (USA Pat. March 18th, 1919, No. 1 297 412. Filed November 12th, 1914. No. 871 659).

#56. Glass Pressing Machines. W. J. Miller USA, (Brit. Pat., August 8th, No. 19 289/19). Also (Brit. Pat., August 8th, 1918, No. 19 290/19). (Brit. Pat., August 8th, 1918, No. 19 291/19), and (Brit. Pat., August 8th, 1918, No. 19 292/19).

#57. Glass Pressing Machines. W. A. Lorenz (USA Pat., January 1919, No. 1 291 952. Filed December 26th, 1913. No. 808 803).

#58. Machinery for Moulding Glass. W. J. Miller, Swissvale, Pennsylvania, USA (Brit. Pat. No. 118 282.) (August 8th, 1918, No. 12 883).

#59. Timing Devices for Glass Pressing Machinery. Hartford-Fairmont Co., USA (Brit. Pat., July 28th, 1919, No. 18 714).

#60. Feeding Glass Machines. R. E. McCauley, London. (Brit. Pat. No. 118 585.) (May 31st, 1918, No. 9 030).

#62. Device for Promoting Uniformity of Flow in Glass Feeder. R. E. McCauley, London. (Brit. Pat. No. 118 581.) (May 21st, 1918, No. 8 472).

#77. Automatic Take-in. Richard La France (USA Pat. No. 1 296 818, March 11th, 1919, Filed September 15th, 1913. No. 789 779).

#134. Glass Blowing Machine. O. Imray, London (Trehand Verrinigung Akt.-Ges., Berlin) (Brit. Pat. No. 133 071, June 14th, 1917, No. 8 559).

#135. Glass Blowing Machines. W. McNeilage, Spotswood, Victoria, Australia (Brit. Pat. No. 132 589, September 9th, 1918, No. 14 643).

#140. Semi-automatic Bottle-blowing Machine. James W. Lynch (USA Pat., March 18th, 1919, No. 1 297 706 Filed September 28th, 1916, No. 122 658).

#141. E. Roirant St. Ouen, Seine, France (Brit. Pat. No. 134 152, March 10th, 1919, No. 5 868).

#158. Glass Feeding Device. O. M. Tucker and W. A. Reeves, Columbus, Ohio, USA (Brit. Pat. No. 131 586, August 5th, 1919, No. 19 288).

#159. Glass Working. F. L. T. Wadsworth (Ball Bros.' Glass Manufacturing Co.) (USA Pat. No. 1 323 507, December 2nd, 1919. Filed February 23rd, 1915, No. 9 878).

#160. Glass Feeding Device. British Thomson-Houston Co., London, and J. Gray (Brit. Pat. No. 131 756, September 12th, 1918, No. 14 862).

#161. Apparatus for Feeding Molten Glass. W. A. Lorenz, Hartford

Fairmont Co. (USA Pat. No, 1 300 181, April 8th, 1919. Filed August 7th, 1917, No. 184 875).

#162. Gathering by Suction. Emile Roirant (USA Pat., July 29th, 1919, 1 311 249. Filed November 14th, 1918, No. 262 591).

#163. Automatic Glass Feeder. S. D. Olsen, Leeds (Brit. Pat. No. 132 898, October 3rd, 1918, No. 16 089).

#165. Glass Delivering Apparatus. Emile Roirant (USA Pat., June 17th, 1919, No. 1 307 150. Filed August 13th, 1918. No. 249 706).

#166. Glass delivery Apparatus. F. L. O. Wadsworth (Ball Bros.' Glass Manufacturing Co.) (USA Pat. No. 1 307 527, June 24th, 1919. Filed July 30th, 1914, No. 854 024).

#167. Glass Discharging Mechanism. J. Rau (USA Pat. February 3rd, 1920, No. 1 329 851. Filed September 10th, 1918, No. 25 363).

#168. Mould for Glassware. K. E. Peiler and W. A. Lorenz (Hartford Fairmont Co.) (USA Pat. No. 1 331 471, February 25th, 1920. Filed March 26th, 1915, No. 17 181).

#246. Bottle-blowing Machine. Michael J. Owens (USA Pat., November 25th, 1919, No. 1 322 726. Filed October 26th, 1914, No. 868 730).

#247. Semi-automatic Bottle-blowing Machines. John Rau (USA Pat., December 16th, 1919, No. 1 325 331. Filed January 12th , 1917, No. 142 023).

#254. Apparatus for Transferring Glass to Moulds. R. La France (Assignor to the Owens Bottle Machine Co.) (USA Pat., February 24th, 1920, No. 1 331 511. Filed June 17th, 1918).

#255. Apparatus for Transferring Charges of Glass. J. M. Lents (Assignor to the Owens Bottle Machine Co.) (USA Pat., February 24th, 1920, No. 1 331 512. Filed May 20th, 1918).

#257. Means for Transferring Molten Glass to Moulds. L. D. Soubier (Assignor to the Owens bottle Machine Co.) (USA Pat., February 24th, 1920.No. 1 331 536. Filed July 8th. 1918.

#258. Mechanism for Delivering Glass. R. Netter USA Pat., September 2nd, 1919 No. 1 314 691. Filed November 12th, 1915, No. 61 015.

#260. Apparatus for Delivering Mould Charges from a Glass

Furnace. S. Davidson and A. L. Schram (Assignors to Schram Glass Manufacturung Co.). (USA Pat., April 27th, 1920, No. 1 338 214. Filed July 21st 1917, No. 182 046).

#263. Shearing Hot Metal. W. A. Lorenz (Assignor to The Hartford-Fairmont Co.) (USA Pat., December 30th, 1919 No. 1 326 460. Filed March 27th, 1917, No. 157 668).

#331. Making Narrow-necked Bottles. J. Pullon and A. E. Clegg (Brit. Pat., April 7th, 1919, No. 139 654).

#332. Machine for Making Narrow-necked Bottles. Naamlooze Vennootschop Glasfabriek Leerdam Vooreen Jeekel, Mijnssen, & Co., Leerdam, Holland (Brit. Pat. No. 136 518. March 31st, 1919, No. 8051).

#334. Automatic Glass-blowing Machine. W. J. Miller, Pa., USA (Brit. Pat., October 30th, 1918, No. 142 156).

#335. Automatic Gathering and Pressing Machine. W. E. Evans and M. J. Owens (USA) (Brit. Pat., August 27th, 1919, No. 142 007).

#343. Pressing Plunger Mechanism. W. J. Miller (USA Pat., December 30th, 1919, No. 1 326 365. Originlly filed August 11th, 1917, No. 185 647. Divided, and this application filed June 22nd, 1918, No. 241 431).

#344. Means for Conveying Ware to Lehrs. R. W. Niver. (Assignor to Owens Bottle Machine Co.) (USA Pat., October 19th, 1920, No. 1 356 157. Filed August 19th, 1918, No. 250 424).

#347. Glass-working Machine. (F. L. O. Wadsworth, assignor to Ball Bros. Glass Co.) (USA) Pat. December 30th, 1919, No. 1 326 235. Filed April 24th, 1914, No. 834 233).

#363. Apparatus for Feeding Molten Glass. (B. D. Chamberlain, assignor to the Empire Machine Co.) (USA Pat., February 19th, 1918, No 1 256 979. Originally filed June 2nd, 1911, No. 630 944).

#364. Automatic Glass Feeding Device. R. E. McCauley, Stockton, California, USA (Brit. Pat. No. 135 454. January 6th, 1919, No. 3600).

#365. Feeding Device for Glass-making Machinery. Hartford-Fairmont Co., New York, USA (Brit. Pat., November 12th, 1919, No. 141 320).

#366. Glass Feeding Device. Empire Machine Co., New York, USA (Brit. Pat., February 9th, 1920, No. 138 895).

1921 Volume V

#64. Improvements in Glass-blowing Machines. W. J. Miller, USA (Brit. Pat., October 30th, 1918, No. 143 827. No. 9235/20).

#65. Semi-automatic Glass-blowing Machine. F. O'Neill, USA (Brit. Pat., Feb. 6th, 1919, No. 144 344, No. 2932).

#68. Glass Delivery Device. S. D. Olsen (Brit. Pat., February 22nd, 1919, No. 143 606. Provisional No. 4476).

#69. Glass Delivery from a Tank Furnace. W. L. Pratt, London (C.H. Rankin, USA) (Brit. Pat., December 30th, 1919, No. 143 159. Provisional No. 32 634).

#70. Glass Delivery from a Tank Furnace. W. L. Pratt (Brit. Pat., November 6th, 1919 No. 144 552. Provisional No. 27 378).

#71. Glass Feeding Device. Hartford-Fairmont Co., New York, USA (Assignees of K. E. Peiler, Connecticut, USA) (Brit. Pat., January 15th, 1920. No. 142 785. Provisional No. 1364. Convention date, May 5th, 1919. Not yet accepted).

#72. Means for Transferring Glass to Moulds. L. D. Soubier (Assignor to the Owens Bottle Machine Co.) (USA Pat., October 19th, 1920, No. 1 356 174. Filed August 267th, 1918, No. 251 530).

#73. Glass Automatic Feeding Device. Hartford-Fairmont Co., USA (Assignees of K. E. Peiler, Connecticit, USA) (Brit. Pat., January 15th, 1920, No. 142 786. Provisional No. 1371. Convention date, May 5th, 1919. Not yet accepted).

#146. Bottle-blowing machine. W. J. Miller, Swissvale, Pa. USA (Brit. Pat., September 24th, 1920, No. 152 597. Provisional No. 27 263. Convention date, October 16th, 1919. Not yet accepted).

#147. Bottle-blowing Machine. S. E. Winder, USA, and H. C. Daubenspeck (Brit. Pat. No. 152 414, July 12th, 1919 No. 17 493).

#148. Improved Owens Machine, W. E. Evans, London (Europaischer Verband der Flashenfabriken Ges., Germany) (Brit. Pat. October 16th, 1919. No. 151 831. Provisional No. 25 423).

#149. Automatic glass-blowing Machine. Wilzen Automatic Bottle-making Machine Co., London (Assignees of E. Roirant, St. Ouen, France) (Brit. Pat., No. 150 679, August 17th, 1920. No. 23950. Convention date, September 3rd, 1919. Not yet accepted).

#150. Modified Boucher Bottle Machine. Boucher et Fils, France (Brit. Pat., No. 144 691, June 9th, 1920. No. 15 647. Convention date, December 28th, 1918. Not yet accepted).

#151. Press and Blow Machine. Wilcox Machine Co., USA (Assignees of A.C. Wilcox) (Brit. Pat., July 8th, 1920. No. 147 635. Provisional No. 19 594. Convention date, March 6th, 1917, Not yet accepted).

#155. Glass Pressing Machine. H. C. Daubenspeck, London, and S. E. Winder, Chicago, USA (Brit. Pat. No. 146 602, April 8th, 1919, No. 8841).

#156. Paste Mould Machine. W. J. Miller (Brit. Pat. No. 151 606, September 13th, 1920, Provisional No. 26 248).

#159. Glass Blowing Device. O. M. Tucker, W. A. Reeves and J. M. Beatty, USA (Brit. Pat. No. 148848, July 10th, 1920, No 20 556. Convention date, April 10th, 1916. Not yet accepted).

#161. Glass Feeding Device. J. Wolf, Germany (Brit. Pat. No. 148 556, July 10th, 1920. Provisional No. 20 551. Convention date, May 8th, 1919. Not yet accepted).

#162. Glass Feeding Device. W. J. Miller, Swissvale, Pa., USA (Brit. Pat. No. 151 605, September 13th, 1920, Provisional date, September 23rd, 1919. Not yet accepted).

#164. Glass Feeding Device. O. M. Tucker, W. A. Reeves and J. M. Beatty, Ohio, USA (Brit. Pat. No. 147 657, July 8th, 1920. Provision No. 19 500. Convention date, October 24th, 1913. Not yet accepted).

#165. Transfer Device. H. C. Daubenspeck and S. E. Winder, USA (Brit. Pat. No. 149 402, May 8th, 1919. provisional No. 11 470).

#260. An Automatic Bottle-blowing Machine. F. L. O. Wadsworth, Pittsburgh, Pa., Assignor to Ball Bros. Glass Manufacturing Co., Muncie, Indiana (USA Pat. No. 1 333 143, March 9th, 1920. Filed April 24th, 1914, No. 834 233. Divided and this application filed June 27th, 1918, No. 242 263).

#262. The Wilzin Automatic Bottle-making Machine. Wilzin Automatic Bottle-making Machine Co., Westminster, London, Assignees of E. Roirant, St. Ouen, Seine, France (Brit. Pat. No. 154 915, august 17th, 1920. Convention date September 3rd, 1919. Not yet accepted).

#264. Pressing Glass Coffins. W. J. Miller, Swissvale, Pa. USA (USA Pat., No.1 342 821, June 8th, 1920. Filed June 17th, 1918, No. 240 481).

#356. Glass-blowing Machine. H. Fairbrother, London (Brit. Pat., No. 160 366 (Lynch Glass Machinery Co.; Jackson Street, Anderson, Indiana, USA), April 6th, 1920).

#358. Press and Blow Machines. W. J. Miller, Swissvale, Pa., USA (Brit. Pat., No. 156 569, September 24th, 1920. Convention date, January 5th, Not yet accepted).

#360. Glass-pressing machine. H. C. Daubenspeck, Blackheath, London, and S. E. Winder, Chicago, USA (Brit. Pat., No. 158 711, November 12th, 1919).

#361. Glass Delivery Devices. K. E. Peiler, Hartford, Connecticut, Assignor to Hartford-Fairmont Co. Canajoharie, New York, USA (Brit. Pat., No. 157 159, January 8th, 1921. Convention date, December 4th, 1916. Not yet accepted).

#362. Feeding Charges of Glass to Machines. K. E. Peiler, Hartford, Connecticut, Assignor to Hartford-Fairmont Co., Canajoharie, New York, USA (Brit. Pat., No. 157 160, January 8th, 1921. Convention date, March 17th, 1914. Not yet accepted).

#363. Device for Varying the Weight of Charges Delivered by Glass-feeding Machines. British Hartford-Fairmont Syndicate,Ltd., and J. H. O. Bunge, 148, Audrey House, Ely Place, Holborn, London (Brit. Pat. No. 161 001, January 23rd, 1920).

1922 Volume VI

#63. Bottle-making Machine. E. S. Hutton, Lapel, Indiana (U.A.S. Pat., No. 1 394 092, October 18th, 1921. Filed August 2nd, 1919, No. 314 853).

#64. Bottle-blowing Machine. M. Meunier, Reims, France (Brit. Pat.,

No. 162 193, May 8th, 1920).

#65. Automatic Bottle-blowing Machine. R. F. Hall, Moseley, Birmingham (Brit. Pat., No. 164 675, January 1st, 1918.
Also includes;
Brit. Pat., No. 164 763, July 10th, 1917.
Brit. Pat., No. 164 764, September 19th, 1917.
Brit. Pat., No. 165 112, July 10th, 1917.
Brit. Pat., No. 165 453, July 10th, 1917.
Brit. Pat., No. 165 799, January 21st, 1918.
Brit. Pat., No. 165 800, July 10th, 1917.

#66. Improvements in Automatic Glass Bottle Machinery. T. C. Moorshead, London, (Brit. Pat., No. 165 875. October 8th, 1919).

#68. Improved Semi-automatic Bottle Machine. E. Hughes, London, J. Middleton, Leeds, and B. Wilson, London and Glassworks Ltd., Edmonton, London (Brit. Pat., No. 162 029, December 19th, 1919).

#70. Conveyor for Bottles. A. J. H. Haddan, London, and the Owens Bottle Co., Toledo, O., USA (Brit. Pat., No. 165 350, August 5th, 1920).

#71. A New Type of Press and Blow Machine. R. F. Hall, Birmingham, England (Brit. Pat., No. 164 365, January 23rd, 1918: USA Pat., No. 1 393 118, October 11th, 1921. Filed January 15th, 1919, No. 271 223).

#72. Machine for Making Pressed Glassware. T. C. Moorshead, London (Brit. Pat., No. 166 272, October 6th, 1919)

#75. Glass Gathering and Shaping Machine. M. J. Owens, Toledo, Ohio, Assignor to Toledo Glass Co. (USA Pat. No. 1 364 514, January 4th, 1921. Filed April 18th, 1912, No. 691 589).

#76. Glass Feeding Device. F. O'Neill, Toledo, Ohio. USA (Brit. Pat., No. 166 605, January 28th, 1920).

#163. An Automatic Bottle-blowing Machine. C. F. Cox (USA Pat. No. 1 385 428, July 1921. Filed June 4th, 1912, No. 701 506).

#164. Improvements in Glass Bottle-blowing Machines. C. J. Dowdall, Dublin (Brit. Pat., No. 168 167, June 3rd, 1920).

#166. Press and Blow Machines. W. J. Miller, London (Brit. Pat., No. 171 420, May 10th, 1920).

#167. Transfer Device for Press and Blow Machines. W. J. Miller, London (Brit. Pat., No. 168 111, April 26th, 1920).

#168. Glass-making Machinery for Use with Feeding Devices. W. J. Miller, (Brit. Pat., No. 169 944, April 1st, 1920).

#169. Conveying Device for Transferring Glass Bottles to Lehrs. F. W. Danberg, Toledo, Ohio, USA (Brit. Pat., No. 168 121, May 17th, 1920).

#266. Bottle-blowing Machine. E. Hughes, London, J. Middleton, Leeds, B. Wilson, London, and Glassworks, Ltd., London (Brit. Pat., No. 174 880, April 21st, 1921).

#267. Automatic machine for Blowing Bottles. H. Hillman, Gross-Räschen, Niederlausitz, Germany (Brit Pat., No. 172 947, December 2nd, 1921. Convention date, December 18th, 1920).

#268. Semi-automatic Bottle-blowing Machine. W. J. Mellersh-Jackson, London, and C. J. Root, Terre Haute, Indiana, USA (Brit.Pat., No. 176 017, August 24th, 1920).

#269. Glass Delivery Device. Howard Automatic Glass Feeder Co., and G. E. Howard, both in Butler, Pa., USA (Brit. Pat., No. 174 311, July 13th, 1920).

#270. A New Glass Feeding Apparatus. K. E. Peiler, Hartford, Connecticut, Assignor to Hartford-Fairmont Co., Canajoharie, New York (USA Pat., No. 1 401 921, December 27th, 1921. Filed December 30th, 1916, No. 139 767).

#271. Apparatus for Gathering Molten Glass. J. B. Graham, Evansville, Indiana, Assignor to the Owens Bottle Machine Co., Toledo, Ohio (USA Pat., No. 1 353 953, September 28th, 1920. Filed January 18th, 1918, No. 143 002).

#272. Glass Delivery Device. A. N. Cramer, Toledo, Ohio, USA Assignor to the Owens Bottle Co., Toledo (Brit. Pat., No. 175 949, May 30th, 1921. Convention date, February 23rd. Not yet accepted).

#273. Glass Delivery Device. W. J. Miller, Swissvale, Pa., USA (Brit. Pat., No. 172 634, December 6th, 1921).

#274. Process and Apparatus for Feeding Glass. W. J. Miller, Swissvale, Pa., USA (USA Pat., No. 1 413 753, April 25th, 1922. Filed December 6th, 1920, No. 428 753).

#275. A Glass Feeding Device. F. L. O. Wadsworth, Pittsburg, Pa., Assignor to Ball Bros. Glass Manufacturing Co., Muncie, Indiana, (USA Pat., No. 1 375 336, April 19th, 1921. Filed June 27th, 1918, No. 242 266).

#276. A Glass Feeding Device. K. E. Peiler, Hartford, Connecticut, Assignor to Hartford-Fairmont Co., Canajoharie, New York (USA Pat., No. 1 349 551, August 10th, 1920. Filed December 30th, 1916, No. 139 768).

#277. Apparatus for Feeding Charges of Hot Glass to Machines. R. La France, Toledo, Ohio, (USA Pat., No. 1 390 448, September 13th, 1921. Filed March 20th, 1918, No. 223 548).

#278. Method for Automatically Controlling the Temperature and Shape of Charges of Glass. J. B. Graham, Evansville, Indiana, Assignor to Owens Bottle Machine Co., Toledo, Ohio (USA Pat., No. 1 353 907, September 28th, 1920. Filed June 14th, 1917, No. 174 697).

#279. A Method for Forming and Transferring Gobs of Molten Glass. J. F. Rule, Toledo, Ohio, Assignor to the Owens Bottle Co., Toledo, Ohio (USA Pat. No. 1 413 788, April 25th, 1922. Filed October 1st, 1919 No. 327 648).

#280. Means of Delivering Molten Glass to Machines in Continuous Motion. R. La France, Toledo, Ohio, Assignor to the Owens Bottle Co., Toledo, Ohio (USA Pat. No. 1 416 725, May 23rd, 1922. Filed July 10th, 1918 No. 244 135).

#283. Means of Transferring Charges of Hot Glass to Machines. (M. J. Owens, Assignor to the Owens Bottle Machine Co., Toledo, Ohio)(USA Pat. No. 1 350 464, August 24th, 1920. Filed April 15th. 1918, No. 228 583).

#285. A Glass-discharging Mechanism. E. S. Hutton, Lapel, Indiana (USA Pat. No. 1 391 957, September 27th, 1921. filed March 26th, 1920, No. 368 867). Also reference to (USA Pat. No. 1 394 677, October 25th, 1921. Filed July 6th, 1920, No. 394 241).

#286. Glass Flow Controller. F. O'Neill, Toledo, Ohio (USA Pat. No. 1 143 183, April 18th, 1922. Filed October 17th, 1921, No. 508 $321\frac{1}{2}$).

#290. A Mould Clamp for Glass-forming Machines. E. G. Bridges,

Anderson, Indiana (USA Pat. No. 1 398 670, November 29th, 1921. Filed March 24th, 1920, No. 368 208).

#291. Air Cushioning Device for the Plunger of a Pressing Machine. W. J. Miller, Swissvale, Pa., (Brit. Pat. No. 174 644, September 24th, 1920. Convention date, October 16th, 1919. Not yet accepted).

#398. Hand-operated Bottle Machine with suction Gathering. J. Wolf, Cologne, Germany (Brit. Pat. No. 179 422, April 27th, 1921).

#400. Apparatus for Conveying Molten Glass. J. B. Graham, Evansville, Indiana, USA, and R La France, Toledo, Ohio, USA, assignors to the Owens Bottle Machine Co. (USA Pat., No. 1 350 448, August 24th, 1920. filed April 17th, 1918, No. 229 010).

#403. Apparatus for Feeding Molten Glass. K. E. Peiler, West Hartford, Connecticut, USA, Assignor to Hartford-Fairmont Co., Canajoharie, New York, USA (USA Pat., No. 1 405 936, February 7th, 1922. Filed July 17th, 1920, No. 396 935).

#404. Glass Distributing Device. A. R. Hunter Hartford, Connecticut, USA, Assignor to Hartford-Fairmont Co., Canajoharie, New York, USA (USA Pat., No. 1 353 115, September 14th, 1920. Filed November 12th, 1917, No. 201 469).

#405. A Feeder for Molten Glass. K. E. Peiler, Hartford, Connecticut, USA, Assignor to Hartford-Fairmont, Canajoharie, New York, USA (USA Pat., No. 1 277 255, August 27th, 1918. Filed August 7th, 1912, No. 713 738).

#406. Mechanism for Flow Feeding Device. T. Stenhouse, Washington, Pa. USA, Assignor to Hazel-Atlas Glass Co., Wheeling, W. Va., USA (USA Pat., No. 1 265 827, May 14th, 1918. Filed October 23rd, 1914, No. 868 169. Renewed September 23rd, 1916, No. 121 907).

#407. Drop Feeding Device for Glass Machines. A. L. Schram, Hillsboro, Illinois, USA, Assignor to Drey Automatic Glass Machine Co., St. Louis, Missouri, USA (USA Pat., No. 1 353 276, September 21st, 1920. Filed August 28th, 1915, No. 47 719).

#409. Shears for Glass Feeding Devices. M. White, York and A.E. Clegg, Leeds (Brit. Pat., No. 181 233, May 28th, 1921).

#411. Glass Feeding Device. F. O'Neill, Toledo, Ohio, USA (Brit. Pat., No. 179977, January 12th, 1921).

1923 Volume VII

#90. Improvements in Hand-operated Bottle Machines. J. Middleton, 202 High Street, Borough, B. Wilson, 22, Dysons Road, Edmonton, and GLASSWORKS, Ltd., Kimberley Road, Edmonton, all in London (Brt. Pat., June 14th, 1921, No. 182 664).

#91. An Automatic Glass-forming Machine. H. A. Genest, Hartford, Connecticut, Assignor to the Hartford-Fairmont Co., Canajoharie, New York, USA (Brit. Pat. July 25th, 1922, No. 184 172. Convention date August 8th, 1921).

#92. An Automatic Glass Bottle-making Machine. E. H. Lorenz, West Hartford, Connecticut, Assignor to the Hartford-Fairmont Co., Canajoharie, New York, USA (Brit. Pat., July 25th, 1922, No. 184 171. Convention date August 8th, 1922).

#99. Improvements in the Mounting of Moulds on Glass Bottle Machines. S. C. Halse, Albion Street, and T. Jones, 27, Nicholson Street, both in Castleford, Yorkshire (Brit. Pat., August 23rd, 1921, No. 184 699).

#190. Automatic Bottle-blowing Machine. Soc. Anon. D'Etudes et de Constructions d'Appareils Mecaniques pur la Verrerie, Paris, assignors to the Wilzin Automatic Bottle Making Machine Company, Ltd., London (Brit. Pat., No. 191 012, December 7th, 1922. Not yet accepted).

#191. A Take-out Mechanism for Glass-blowing Machines. O. Y. Imray, London, and the Obear-Nester lass Co., St. Louis, USA (Brit. Pat., No. 190 224, September 12th, 1921).

#192. A Steadying Device for Glass-blowing Machines. O. Y. Imray, London, and the Obear-Nester Glass Co., St. Louis, USA (Brit. Pat., No. 190 225, September 12th, 1921).

#194. Device for Removing Glass Articles from Moulds. R. F. Hall, Wythall, near Birmingham (Brit. Pat., No. 189 918, June 7th, 1921).

#199. An Improved Paste Mould Machine. W. J. Miller, Swissvale,

Pa., USA (USA Pat., No. 1 428 855, September 12th, 1922. Filed September 23rd, 1919, No. 325 617).

#283. A Bottle Blowing Machine. R. La France, Toledo, Ohio, USA (Brit. Pat., No. 192 347, March 16th, 1922. Convention date, January 30th. Not yet accepted).

#284. Improvements in Glass Bottle Blowing Machines. O. Y. Imray, London, and the Obear-Nester Glass Co., St. Louis, USA (Brit. Pat., No. 190 080, September 12th, 1921).

#286. An Automatic Feeding Device. E. O. Hiller, Hartford, Connecticut, USA, Assignor to the British Hartford-Fairmont Syndicate, Ltd., (Brit. Pat., No. 195 927, October 25th, 1922. Convention date, April 6th. Not yet accepted).

#287. Improvements in Feeding Devices. K. E. Peiler, Hartford, Connecticut, USA, Assignor to the Hartford-Empire Co., Hartford, Connecticut, USA (Brit. Pat., No. 195 376, March 19th, 1923. Convention date, March 25th, 1922. Not yet accepted).

#288. A Device for Delivering Charges of Molten Glass. H. H. Pitt, Lesbury, Northumberland (Brit. Pat., No. 195 448, December 29th, 1921).

#392 Machine for the Manufacture of Bottles and other Blown Glassware. Emille Roirant, St. Ouen, France, Assignor to Arthur Wilzen, St. Ouen, France (USA Pat. December 12th, 1922, No. 1 438 65. Filed April 10th, 1919, No. 289 012).

#393 A New Press-and-blow Machine. E. H. Lorenz, West Hartford, Connecticut, USA, Assignor to The Hartford-Empire Co., Hartford, Connecticut, USA (Brit. Pat., No. 199 377, May 29th, 1923. Convention date, June 14th, 1922. Not yet accepted)

#394 An Automatic Feeding Device. C. H. Rankin, Wellsburg, West Virginia, USA, (Brit. Pat., No. 198 952, September 28th, 1922).

#399 Means for Feeding Charges of Molten Glass. R. La France, Toledo, Ohio, Assignor to The Owens Bottle Company, Toledo, Ohio (USA Pat., No. 1 438 369, December 19th, 1922. Filed August 11th, 1920, No. 402 816).

#401 Apparatus for and Method of Producing Charges of Molten Glass. L. D. Soubier, Toledo, Ohio, Assignor to the Owens Bottle

Company, Toledo, Ohio (USA Pat., No. 1 428 994, September 12th, 1922. Filed December 20th, 1920, No. 413 848).

#402 **A Plunger, for Use in an Automatic Feeding Device.** F. O'Neill, Toledo, Ohio, USA (Brit. Pat., No. 199 805, March 28th, 1922).

#403 **Guides for Transferring Glass to Moulds.** L. D. Soubier, Toledo, Ohio, Assignor to The Owens Bottle Company, Toledo, Ohio (USA Pat., No. 1 439 384, December 19th, 1922. Filed April 30th, 1920, No. 377 770).

#404 **A Device for Ejecting Finished Ware from Moulds.** E. H. Lorenz and K. E. Peiler, and The Hartford-Fairmont Company, Canajoharie, New York (USA Pat., No. 1 406 045, February 7th, 1922. Filed March 29th, 1917, No. 158 242).

#407 **A Shearing Mechanism for Glass-making Machines.** R. F. Hall, Wythal, near Birmingham (Brit. Pat., No. 199 414, December 23rd, 1921).

www.ingramcontent.com/pod-product-compliance
Ingram Content Group UK Ltd.
Pitfield, Milton Keynes, MK11 3LW, UK
UKHW020129290926
15471UKWH00003B/376

9 780900 68276